Introduction to

Time Series Analysis and Forecasting

with Applications of SAS and SPSS

Introduction to
Time Series Analysis and Forecasting

with Applications of SAS and SPSS

Robert A. Yaffee
Statistics and Social Science Group
Academic Computing Service of the Information Technology Services
New York University
New York, New York
and
Division of Geriatric Psychiatry
State University of New York Health Science Center at Brooklyn
Brooklyn, NY

with

Monnie McGee
Hunter College
City University of New York
New York, New York

ACADEMIC PRESS, INC.
San Diego   London   Boston   New York   Sydney   Tokyo   Toronto

Academic Press
A Harcourt Science and Technology Company
525 B Street, Suite 1900, San Diego, California 92101-4495, USA
http://www.academicpress.com

Academic Press
Harcourt Place, 32 Jamestown Road, London NW1 7BY, UK
http://www.hbuk.co.uk/ap/

Library of Congress Catalog Card Number: 99-62662

International Standard Book Number: 0-12-767870-0

PRINTED IN THE UNITED STATES OF AMERICA
00 01 02 03 04 05 MM 9 8 7 6 5 4 3 2 1

For
Liz and Mike

Contents

Chapter 3

Introduction to Box–Jenkins Time Series Analysis

Chapter 4

The Basic ARIMA Model

Chapter 5

Seasonal ARIMA Models

Chapter 6

Estimation and Diagnosis

Chapter 7
Metadiagnosis and Forecasting

Chapter 8
Intervention Analysis

Chapter 9

Transfer Function Models

Chapter 10

Autoregressive Error Models

Chapter 11
A Review of Model and Forecast Evaluation

Chapter 12
Power Analysis and Sample Size Determination for Well-Known Time Series Models

Monnie McGee

Preface

This book is the product of an intellectual odyssey in search of an understanding of historical truth in culture, society, and politics, and the scenarios likely to unfold from them. The quest for this understanding of reality and its potential is not always easy. Those who fail to understand history will not fully understand the current situation. If they do not understand their current situation, they will be unable to take advantage of its latent opportunities or to sidestep the emergent snares hidden within it. George Santayana appreciated the dangers inherent in this ignorance when he said, "Those who fail to learn from history are doomed to repeat it." Kierkegaard lemented that history is replete with examples of men condemned to live life forward while only understanding it backward. Even if, as Nobel laureate Neils Bohr once remarked, "Prediction is difficult, especially of the future," many great pundits and leaders emphasized the real need to understand the past and how to forecast from it. Winston Churchill, with an intuitive understanding of extrapolation, remarked that "the farther back you can look, the farther forward you can see."

Tragic tales abound where vital policies failed because decision makers did not fathom the historical background—with its flow of cultural forces, demographic resources, social forces, economic processes, political processes—of a problem for which they had to make policy. Too often lives were lost or runied for lack of adequate diplomatic, military, political, or economic intelligence and understanding. Obversely, policies succeeded in accomplishing vital objectives because policy makers have understood the likely scenarios of events. After we learned from the past, we needed to study and understand the current situation to appreciate its future possibilities and probabilities. Indeed, the journalistic and scientific quest for "what is" may reveal the outlines of "what can be." The qualitative investigation of "what has been" and "what is" may be the mere beginning of this quest.

The principal objective of this textbook is to introduce the reader to the fundamental approaches to time series analysis and forecasting. Although the book explores the basic nature of a time series, it presumes that the reader has an understanding of the methodology of measurement and scale construction. In case there are missing data, the book briefly addresses the imputation of missing data. For the most part, the book assumes that there are not significant amounts of missing data in the series and that any missing data have been properly replaced or imputed. Designed for the advanced undergraduate or the beginning graduate student, this text examines the principal approaches to the analysis of time series processes and their forecasting. In simple and clear language, it explains moving average, exponential smoothing, decomposition (Census X-11 plus comments on Census X-12), ARIMA, intervention, transfer function, regression, error correction, and autoregressive error models. These models are generally used for analysis of historical, recent, current, or simulated data with a view toward forecasting. The book also examines evaluation of models, forecasts, and their combinations. Thus, the text attempts to discuss the basic approaches to time series analysis and forecasting.

Another objective of this text is to explain and demonstrate novel theoretical features and their applications. Some of the relatively new features include treatment of Y2K problem circumventions, Census X-12, different transfer function modeling strategies, a scenario analysis, an application of different forecast combination methods, and an analysis of sample size requirements for different models. Although Census X-12 is not yet part of either statistical package, its principal features are discussed because it is being used by governments as the standard method of deseasonalization. In fact, SAS is planning on implementing PROC X12 in a forthcoming version. When dealing with transfer function models, both the conventional Box–Jenkins–Tiao and the linear transfer function approaches are presented. The newer approach, which does not use prewhitening, is more amenable to more complex, multiple input models. In the chapter on event impact or intervention analysis, an approach is taken that compared the impact of an intervention with what would have happened if all things remained the same. A "what if" baseline is posited against which the impact is measured and modeled. The book also briefly addresses cointegration and error correction models, which embed both long-run and short-run changes in the same model. In the penultimate chapter, the evaluation and comparison of models and forecasts are discussed. Attention is paid to the relative advantages and disadvantages of the application of one approach over another under different conditions. This section is especially important in view of the discovery in some of the forecast competitions that the more complex models do not always provide the best forecasts. The methods as well as the relative advantages and disadvantages of combining forecasts to improve forecast accuracy are also analyzed. Finally, to dispel erroneous conventional wisdom concerning sample size, the final chapter empirically examines the selection of the proper sample size for different types of analysis. In so doing, Monnie McGee makes a scholarly methodological contribution to the study of sample size required for time series tests to attain a power of 0.80, an approach to the subject of power of time series tests that has not received sufficient discussion in the literature until now.

As theory and modeling are explained, the text shows how popular statistical programs, using recent and historical data are prepared to perform the time series analysis and forecasting. The statistical packages used in this book—namely, the Statistical Analysis System (SAS) and the Statistical Package for the Social Sciences (SPSS)—are arguably the most popular general purpose statistical packages among university students in the social or natural sciences today. An understanding of theory is necessary for their proper application under varying circumstances. Therefore, after explaining the statistical theory, along with basic preprocessing commands, I present computer program examples that apply either or both of the SAS Econometric Time Series (SAS/ETS) module or the SPSS Trends module. The programming syntax, instead of the graphic interfaces, of the packages is presented because the use of this syntax tends to remain constant over time while the graphical interfaces of the statistical packages change frequently. In the presentation of data, the real data are first graphed. Because graphical display can be critical to understanding the nature of the series, graphs of the data (especially the SAS Graphs) are elaborately programmed to produce high-resolution graphical output. The data are culled from areas of public opinion research, policy analysis, political science, economics, sociology, and even astronomy and occasionally come from areas of great historical, social, economic, or political importance during the period of time analyzed. The graphs include not only the historical data; after Chapter 7 explains forecasting, they also include forecasts and their profiles. SAS and SPSS computer programs, along with their data, are posted on the Academic Press Web site (http://www.academicpress.com/sbe/authors) to assist instructors in teaching and students in learning this subject matter. Students may run these programs and examine the output for themselves. Through their application of these time series programming techniques, they can enhance their capabilities in the quest for understanding the past, the present, and to a limited extent, *que sera.*

This text is the product of an abiding interest in understanding longitudinal data analysis in general and time series processes in particular. I could not have accomplished this work without the help of many people. Working on three other projects at the time I was writing this book, I asked Professor

Monnie McGee to help me expedite a time consuming analysis of the sample size and statistical power of common time series models. Although Monnie used S plus and I use SAS and SPSS in the rest of the book, researchers and students will find that many of her findings apply to other statistical packages as well. Therefore, a key contributing scholar is Professor Monnie McGee, who contributed an important chapter on a subject that must be a concern to practitioners in the field, sample size and power of times series tests.

There are a number of scholars to whom I owe a great intellectual debt for both inspiration and time series and forecasting education. Although I have quoted them freely, I would like to give special thanks to George E. P. Box, Gwilym Jenkins, Clive W. J. Granger, Robert F. Engle, Paul Newbold, Spyros Makridakis, Steven Wheelwright, Steven McGee, Rob Hyndman, Michelle Hibon, Robert Fildes, Richard McCleary, Richard Hay, Jr., Wayne Fuller, David A. Dickey, T. C. Mills, David F. Hendry, Charles Ostrom, Dan Wood, G. S. Maddala, Jan Kamenta, Robert Pindyck, Daniel Rubenfeld, Walter Labys, George G. Judge, R. Carter Hill, Henri Theil, J. J. Johnston, Frank Diebold, Bill Greene, David Greenberg, Richard Maisel, J. Scott Armstrong, David F. Hendry, Damodir Gujarati, Jeff Siminoff, Cliff Hurvitch, Gary Simon, Donald Rock, and Mark Nicolich.

The research and writing of many other scholars substantially influenced this work. They are numerous and I list the principal ones in alphabetical order. Bovas Abraham, Sam Adams, Isaiah Berlin, Bruce L. Bowerman, Lynne Bresler, Peter J. Brockwell, Courtney Brown, Brent L. Cohen, Jeff B. Cromwell, Russell Davidson, Richard A. Davis, Gul Ege, Dan Ege, Donald Erdman, Robert Fildes, Phillip H. Francis, W. Gilchrist, Jennifer M. Ginn, A. S. Goldberger, Jim Granato, William E. Griffiths, Damodar N. Gujarati, J. D. Hamilton, D. M. Hanssens, Eric A. Hanusheck, Averill Harriman, Andrew C. Harvey, K. Holden, C. C. Holt, R. Robert Huckfeldt, G. B. Hudak, Rob J. Hyndman John Jackson, J. J. Johnston, M. G. Kendall, Paul Kennedy, Minbo Kim, Lyman Kirkpatrick Jr., P. A. Klein, C. W. Kohfeld, Stanley I. Kutler, Walter Labys, Johannes Ledolter, Mike Leonard, R. Lewandowski, Thomas W. Likens, Charles C. Lin, Mark Little, L.-M. Liu, Jeffrey Lopes, Hans Lütkepohol, James MacKinnon, David McDowell, V. E. McGee, G. R. Meek, Errol E. Meidinger, G. C. Montgomery, Meltem A. Narter, C. R. Nelson, M. P. Neimira, Richard T. O'Connell, Keith Ord, Sudhakar Pandit, Alan Pankratz, H. Jin Park, D. A. Peel, C. I. Plosser, James Ramsey, David P. Reilly, T. Terasvirta, Michel Terraza, J. L. Thompson, George C. Tiao, R. S. Tsay, Walter Vandaele, Helen Weeks, William W. S. Wei, John Williams, Terry Woodfield, Donna Woodward, and Shein-Ming Wu.

There are other scholars, writers, statesmen, and consultants whose data, activities research, and teachings directly or indirectly contributed to the writing of this text. They include D. F. Andrews, Professor Courtney Brown, Dr. Carl Cohen, Professor Jim Granato, Dr. Stanley Greenberg, the Honorable Averill Harriman, Steven A. M. Herzog, R. Robert Huckfeldt, Professor Guillermina Jasso, President Lyndon Baines Johnson, Professor Lyman Kirkpatrick, Jr., Professor Stanley Kutler, Mike Leonard, Dr. Carol Magai, Robert S. McNamara, McGeorge Bundy, Professor Mark Nicolich, David P. Reilly, Professor Donald Rock, Gus Russo, John Stockwell, Professor Pamela Stone, Professor Peter Tuckel, and Donna Woodward.

I also owe a debt of deep gratitude to key people at Academic Press. To Senior Editor Dr. Scott Bentley and his assistants Beth Bloom, Karen Frost, and Nick Panissidi; to production editors and staff members Brenda Johnson, Mark Sherry, and Mike Early; and to Jerry Altman for posting such accompanying teaching materials as the computer program syntax and data sets at *http://www.apnet.com/sbe/authors*, I remain truly grateful. For her invaluable editorial help, I must give thanks to Kristin Landon. Nor can I forget Lana Traver, Kelly Ricci, and the rest of the staff of the PRD Group for their cooperative graphics preparation and composition; I must express my appreciation to them for their professional contribution to this work.

To the very knowledgeable and helpful people at SAS and SPSS, I must acknowledge a debt for their gracious and substantial assistance. SAS consultants, Donna Woodward, Kevin Meyer, and SAS developer Michael Leonard were always gracious, knowledgeable, and very helpful. Other consultants have helped more obliquely. For their very knowledgeable and personal professional assistance, I remain deeply indebted.

I would also like to thank the people at SPSS, to whom I owe a real debt of gratitude for their knowledgeable and professional assistance. Tony Babinec, Director of Advanced Marketing Products; Mary Nelson and David Cody, Managers in charge of Decision Time development; Dave Nichols, Senior Support Statistician; Dongping Fang, Statistician; and David Mathesson, from the Technical Support staff, provided friendly and knowledgeable assistance with various aspects of the SPSS Trends algorithms. Nor can I forget Andy Kodner or Dave Mattingly, who were also very helpful. To David Mattingly and Mary Nelson, I want to express my thanks for the opportunity to beta-test the Trends module. To Mary Nelson and David Cody, I would like to express my gratitude for the opportunity to beta-test the Decision Time Software in the summer of 1999.

The roots of my interest in forecast go back to the mid to late 1960s and 1970s, when my friends from those years saw my concern with forecasting emerge. To Lehigh University professors Joseph A. Dowling, Jerry Fishman, John Cary, and George Kyte, I remain grateful for support and guidance. To Roman Yuszczuk, George Russ, and other dear friends, to whom I confided those ominous forecasts, there is no need to say that I told you so. In my lectures, I explained what I observed, analyzed, and foresaw, daring to be forthright in hopes of preventing something worse, and risking the consequences. Many people were wont to say that if you remember the 1960s you were not there. However, we sought to understand and do remember. To Professors Stanley Tennenbaum, John Myhill, and Akiko Kino, at the State University of New York at Buffalo Mathematics Department during the late 1960s, a word of thanks should be offered. For other friends from Buffalo, such as Jesse Nash and Laurie McNeil, I also have to be thankful.

To my friends at the University of Waterloo, in Ontario, Canada, where I immersed myself in statistics and its applications, I remain grateful. To the Snyder family, Richard Ernst, Bill Thomas, Professor June Lowe, Professor Ashok Kapur, Professor Carlo Sempi, and Drs. Steve and Betty Gregory, I express my thanks for all their help. Moreover, I confess an indirect obligation to Admiral Hyman Rickover, whose legendary advice inspired me not to waste time.

To Lt. Col. Robert Avon, Ret., Executive Director of the Lake George Opera Festival, who permitted me to forecast student audience development for the Lake George Opera Festival, I still owe a debt of gratitude, as well as to those friends from Skidmore College Professors Daniel Egy, Bill Fox, Bob Smith, and Bob Jones. To librarians Barbara Smith and Mary O'Donnell, I remain grateful. Nor are Jane Marshall and Marsha Levell forgotten.

From the University of Michigan, where I spent several summers, with financial assistance from the Inter-University Consortium for Political and Social Research (ICPSR) and New York University, I am indebted to Hank Heitowit and Gwen Fellenberger for their guidance, assistance, and support. With the inspiration and teachings of Professor Daniel Wood, John Williams, Walter C. Labys, Courtney Brown, and Jim Granato, along with the assistance of Genie Baker, Dieter Burrell, Professor Xavier Martin, and Dr. Maryke Dressing, I developed my knowledge of dynamic regression and time series to include autoregression and ARIMA analysis. As for all of my good friends at and from Ann Arbor, both identified and unnamed, I remain grateful for their contribution to a wonderful intellectual milieu in which our genuine pursuit of knowledge, understanding, and wisdom was really appreciated and supported.

I am deeply grateful for the support of New York University as well. From the Academic Computing Facility, I gleaned much support. To Frank Lopresti, head of the Statistics and Social Science Group, I owe a debt of gratitude for his cooperation and flexibility, as well as to Dr. Yakov Smotritsky, without whose help in data set acquisition much of this would not have been possible. To Burt Holland for his early support of my involvement with ICPSR and to Edi Franceschini and Dr. George Sadowsky, I remain indebted for the opportunity to teach the time series sequence. To Judy Clifford, I also remain grateful for her administrative assistance. For collegial support, instruction, and inspiration, I must specifically thank the New York University Stern School of Business Statistics and Operations Research (including some already acknowledged) faculty, including Professors Jeff Siminoff, Gary Simon, Bill Greene, James Ramsey, Cliff Hurvich, Rohit Deo, Halina Frydman, Frank Diebold, Edward Melnick, Aaron Tenenbein, and Andreas Weigand. Moreover, the support and inspiration of the Sociology Department Faculty including Professors Richard Maisel, David Greenberg, Jo Dixon, Wolf Hydebrand, and Guillermina Jasso, was instrumental in developing my knowledge of longitudinal analysis in areas of event history and time series analysis. These people are fine scholars and good people, who help constitute and maintain a very good intellectual milieu for which I am grateful.

A number of intellectual friends from other fields were the source of inspiration and other kinds of assistance. In research in the field of addictions research. Valerie C. Lorenz, Ph.D., C.A.S., and William Holmes, of the Compulsive Gambling Center, Inc. In Baltimore, Maryland; Robert M. Politzer, Sc.D. C.A.S., Director of Research for the Washington Center for Addictions; and Clark J. Hudak, Jr. Ph.D., Director of the Washington Center for Addictions proved to be wonderful research colleagues studying pathological gambling. Professor John Kindt of the University of Illinois at Champaign/Urbana; Professor Henry Lesieur, formerly of the Department of Sociology at St. Johns University; and Howard Shaffer, Ph.D., C.A.S., Editor-in-chief of the Journal of Gambling Studies, and research assistants Mathew Hall, Walter Bethune, and Emily McNamara at Harvard Medical School, Division of Addictions, have been good, efficient, and supportive colleagues. Nor can I neglect the supportive assistance of Dr. Veronica J. Brodsky at New York University in this area.

In the area of drug addiction research, I thank Steve Titus of the New York University Medical Center Department of Environmental Medicine for the opportunity to assist in structural equation modeling of drug abuse predispositions research. Among former colleagues in sociomedical research, I thank Dr. Karolyn Siegel and Dr. Shelly Kern, formerly of the Department of Social work Research at Memorial Sloan Kettering Cancer Center; Dr. Ann Brunswick, Dr. Peter Messeri, and Dr. Carla Lewis at Columbia University School of Public Health; Dr. Stephanie Auer, Dr. Steven G. Sclan, and Dr. Bary Reisberg of the Aging and Dementia Research Center at New York University Medical School; and more recently Dr. Carl Cohen and Dr. Carol Magai, State University Health Science

Center at Brooklyn. It was a pleasure to work with John Stockwell in researching developing patterns of the AIDS crisis. His findings proved invaluable in analysis of the gathering of data and its evaluation.

JFK assassination study contributed to my analysis of the Watergate scandal. Among those who did prodigious amounts of investigative work in this area were Gus Russo, John Davis, Robert Blakey, Gaeton Fonzi, John Stockwell, Jim Marrs, Dick Russell, and Mary Nichols. Jim Gray and Gilbert Offenhartz should also be mentioned. Special thanks must also go to Michael Bechloss for transcribing the LBJ White House tapes that explained the official basis of the Warren Commission position.

In the fields of political history, political science, and public opinion research, I also thank Professor Marina Mercada, whose courses in international relations permitted me to present some of my former research, out of which my interest in longitudinal analysis and forecast grew. In the field of political science, Professors Adamantia Pollis, Aristide Zolberg, Richard Bensel, and Jacob Landynski of the Graduate Faculty of the New School for Social Research are wonderful people. In the area of International economics, I also thank Professor Giuseppe Ammendola for his recent assistance. As persons who helped in the more quantitative dimension, Professors Donald Rock and Mark Nicolich provided great inspiration and statistical advice, and to both of them I will always remain indebted. To Professors Dan Cohen and Pam Stone, former chairpersons of the Computer Science and Sociology Departments of Hunter College, respectfully, and to Talbot Katz, Ph.D., I am thankful for much assistance during my years of teaching at Hunter College. For inspiration in political history, political polling, and public opinion research, I have to thank Professor Richard Maisel, Professor Kurt Schlicting, Professor Peter Tuckel, and Dr. Stanley Greenberg for their inspiration and cooperation in these areas of research.

Others whose cooperation and support were critical at one time or another included Frank LaFond, Dr. Winthrop Munro, Les Giermanski, Maria Ycasiano, Nancy Frankel, Ralph Duane, Ed DeMoto, Peggy McEvoy, and Professor Yuko Kinoshita. Special thanks must also be given to publishers John Wiley and Sons, Inc., and Prentice Hall, Inc., for permission to publish and post authors' data.

Throughout this wonderful odyssey, I have been blessed with meeting and working with many wonderfully capable, talented, and ethically fine people. This journey has been exhilarating and fascinating. For their understanding of and supportive forbearance with the demands imposed by this project, I must thank the members of my family—Dana, Glenn (and Barb and Steve)—and my parents Elizabeth and Michael Yaffee, to whom I dedicate this book. If this text helps students, researchers, and consultants learn time series analysis and forecasting, I will have succeeded in substantially contributing to their education, understanding, and wisdom. If the contents of this book inspires them to develop their knowledge of this field, I will have truly succeeded in no small part due to the assistance of those just mentioned. For the contents of the book, I must nonetheless take full responsibility.

Robert A. Yaffee, Ph.D.
Brooklyn, New York
September 1999

About the Authors

Robert A. Yaffee, Ph.D., is a senior research consultant/statistician in the Statistics and Social Science Group of New York University's Academic Computing Facility as well as a Research Scientist/Statistician at the State University of New York Health Science Center in Brooklyn's Division of Geriatric Psychiatry. He received his Ph.D. in political science from Graduate Faculty of Political and Social Research of The New School for Social Research. He serves as a member of the editorial board of the *Journal of Gambling Studies* and was on the Research Faculty of Columbia University's School of Public Health before coming to NYU. He also taught in the Statistical packages in the Computer Science department and the Empirical Research and Advanced Statistics in the Sociology Department of Hunter College. He has published in the fields of statistics, medical research, and psychology.

Monnie McGee, Ph.D., is an assistant professor in the Department of Mathematics and Statistics at Hunter College, City University of New York, New York. She received her Ph.D. in statistics from Rice University, Houston, Texas. She has worked as a bio-statistical consultant for The Rockefeller University and as a computational statistician for Electricité de France. She has published in the areas of time series and medical statistics. Her hobbies include ice-skating and foreign languages.

Chapter 1

Introduction and Overview

1.1. PURPOSE

The purpose of this textbook is to introduce basic time series analysis and forecasting. As an applied approach to time series analysis, this text is designed to link theory with programming practice. Utilizing two of the most contemporaneously popular computer statistical packages—namely, SAS and SPSS—this book may serve as a basic text for those who wish to learn or do time series analysis. It may be used as a reference for persons using these techniques for forecasting.

The level of presentation is kept as simple as possible to make it useful for undergraduates as well as graduate students. Although the presentation primarily concentrates on the theory of time series analysis and forecasting, it contains samples of SAS and SPSS computer programs and analysis of their output. The discussion of these programs and their output demonstrate their application to problems and their interpretation to persons unfamiliar with them.

Over time, the computer program interfaces and menu options change as new versions of the statistical packages appear on the market. Therefore,

1

I have decided not to depend on the graphical user interface or menus selections peculiar to one version; instead, I concentrate on the use of program command syntax. Both SAS and SPSS allow the user to apply the command syntax to run a program. As a matter of fact, SPSS provides more options to those who employ that syntax than those who use the menus. In short, knowledge of the command syntax should have a longer useful life than knowledge of the menu selections from the graphical user interface of a particular version of the software.

At this time, SAS has an automatic time series and forecasting system and SPSS has a module, DecisionTime®, under development. After these systems allow the researcher to submit the series or event variables to be analyzed, they purport to automatically test different models, select the best model according to specified criteria, and generate a forecast profile from it. They allow custom-design of the intervention or transfer function model. To date, these automatic systems have neither been entered into international competition, nor have they been comparatively evaluated.

Because of their pedagogical utility for teaching all of the aspects of time series analysis and forecasting, this book focuses on the program syntax that can be used in SAS and SPSS. As it is explained in this text, the student can learn the theory, its decision points, the options available, the criteria for making those decisions, and how to make the proper decisions at each step. In this way, he can learn the model and forecast evaluation, as well as the proper protocol for applying time series analysis to forecasting. If he needs to modify the model, he will better know how to alter or apply it. For these reasons, the SAS and SPSS syntax for programming time series analysis and forecasting is the focus of this book.

This book does not require a very sophisticated mathematical background. A background knowledge of basic algebra, statistics, and matrix algebra is needed. A knowledge of basic statistics is also presumed. Although the use of calculus is generally avoided, basic calculus is used in Chapter Six to explain the statistical estimation of Box–Jenkins time series analysis. Therefore, advanced undergraduate students, graduate students, postgraduate students, and researchers in the social sciences, business, management, operations research, engineering, or applied mathematics fields should find this text easy to read and understand.

1.2. TIME SERIES

Granger and Newbold (1986) describe a time series as "... a sequence of observations ordered by a time parameter." Time series may be measured continuously or discretely. Continuous time series are recorded instantane-ously and steadily, as an oscillograph records harmonic oscillations of an audio amplifier. Most measurements in the social sciences are made at regular intervals, and these time series data are discrete. Accumulations of rainfall measured discretely at regular intervals would be an example. Others may be pooled from individual observations to make up a summary statistic, measured at regular intervals over time. Some linear series that are not chronologically ordered may be amenable to time series analysis. Ideally, the series used consist of observations that are equidistant from one another in time and contain no missing observations.

1.3. MISSING DATA

If some data values are missing, they should be replaced by a theoretically defensible algorithm. If some social or economic indicators have too much missing data, then the series may not be amenable to time series analysis. Much World Bank and United Nations data come from countries that for one reason or another did not collect data on particular problems or issues regularly for a long enough period of time for it to be useful.

When a series does not have too many missing observations, it may be possible to perform some missing data analysis, estimation, and replace-ment. A crude missing data replacement method is to plug in the mean for the overall series. A less crude algorithm is to use the mean of the period within the series in which the observation is missing. Another algorithm is to take the mean of the adjacent observations. Missing value replacement in exponential smoothing often applies one-step-ahead forecasting from the previous observation. Other forms of interpolation employ linear splines, cubic splines, or step function estimation of the missing data. There are other methods as well. Both SAS and SPSS provide options for missing data replacement. Both warn the user that the series being analyzed contains missing values and then estimate values for substitution (Ege *et al.,* 1993; SPSS, 1996). Nonetheless, if there are too many observations missing, the series may simply be unusable.

1.4. SAMPLE SIZE

As a rule, the series should contain enough observations for proper parameter estimation. There seems to be no hard and fast rule about the minimum size. Some authors say at least 30 observations are needed. Others say 50, and others indicate that there should be at least 60 observations.

If the series includes cycles, then it should span enough cycles to precisely model them. If the series possesses seasonality, it should span enough seasons to model them accurately; thus, seasonal processes need more observations than nonseasonal ones. If the parameters of the process are estimated with large-sample maximum likelihood estimators, these series will require more observations than those whose parameters are estimated with unconditional or conditional least squares. For pedagogical reasons as well as reasons of scholarly interest, I may occasionally use series with fewer than 50 observations. Because the resolution of this issue may be crucial to proper modeling of a series, Monnie McGee in the last chapter gives a power and sample size analysis suggesting that these figures may not always be large enough. Not all series of interest meet these minimal sample size criteria, and therefore they should be modeled with reservation. Clearly, the more observations, the better. For details of determining the approximate minimal length of the series, see the final chapter.

1.5. REPRESENTATIVENESS

If the series comes from a sample of a population, then the sampling should be done so that the sample is representative of the population. The sampling should be a probability sample repeated at equal intervals over time. If a single sample is being used to infer an underlying probability distribution and the sample moments for limited lengths of the series approach their population moments as the series gets infinitely large, the process is said to be ergodic (Mills, 1990). Without representativeness the sample would not have external validity.

1.6. SCOPE OF APPLICATION

Time series data abound in many different fields. There are clearly time series in political science (measures of presidential approval, proportion of the vote that is Democratic or Republican). Many series can be found in economics (GPI, GNP, GDP, CPI, national unemployment, and exchange rate fluctuations, to name a few). There are multiple series in sociology (for example, immigration rates, crime rates of particular offenses, population size, percentage of the population employed). There are many series in psychology (relapse rates for drug, alcohol, or gambling addictions are cases in point). There are many time series in biomedical statistics (pulse, EEG waves, and blood pressure, for example). In meteorology, one may monitor temperatures, barometric pressures, or percent of cloud cover. In astronomy one may monitor the sunspot activity, brightness of stars, or other phenomena. Depending on the nature of the time series and the objective of the analysis, different approaches are used to study these data.

1.7. STOCHASTIC AND DETERMINISTIC PROCESSES

A series may be an observed realization of an underlying stochastic process. The underlying data-generating process is not observed; it is only more or less imperfectly represented in the observed series. Time series are realizations of underlying data-generating processes over a time span, occurring at regular points in time. As such, time series have identifiable stochastic or deterministic components. If the process is stochastic, each data value of the series may be viewed as a sample mean of a probability distribution of an underlying population at each point in time. Each distribution has a mean and variance. Each pair of distributions has a covariance between observed values. One makes a working assumption of ergodicity— that, as the length of the realization approaches infinity, the sample moments of the realized series approximate the population moments of the data-generating process—in order to estimate the unknown parameters of the population from single realizations.

Those series that are not driven by stochastic processes may be driven by deterministic processes. Some deterministic processes may be functional relationships prescribed by the laws of physics or accounting. They may indicate the presence or absence of an event. There may be any number of processes that do not involve probability distributions and estimation. Phenomena that can be calculated exactly are deterministic and not stochastic.

1.8. STATIONARITY

Time series may be stationary or nonstationary. Stationary series are characterized by a kind of statistical equilibrium around a constant mean level as well as a constant dispersion around that mean level (Box and Jenkins, 1976). There are several kinds of stationarity. A series is said to be stationary in the wide sense, weak sense, or second order if it has a fixed mean and a constant variance. A series is said to be strictly stationary if it has, in addition to a fixed mean and constant variance, a constant autocovariance structure. When a series possesses this covariance stationarity, the covariance structure is stable over time (Diebold, 1998). That is to say, the autocovariance remains the same regardless of the point of temporal reference. Under these circumstances, the autocovariance depends only on the number of time periods between the two points of temporal reference (Mills, 1990, 1993). If a series is stationary, the magnitude of the autocorrelation attenuates fairly rapidly, whereas if the series is nonstationary or integrated, the autocorrelation diminishes gradually over time. If, however, these equally spaced observations are deemed realizations of multivariate normal distributions, the series is considered to be strictly stationary.

Many macroeconomic series are integrated or nonstationary. Nonstationary series that lack mean stationarity have no mean attractor toward which the level tends over time. Nonstationary series without homogeneous stationarity do not have a constant or bounded variance. If the series has a stochastic trend, then the level with an element of randomness, is a function of time. In regressions of one series on another, each of which is riven with stochastic trend, a spurious regression with an inflated coefficient of determination may result. Null hypotheses with T and F tests will tend to be overrejected, suggesting false positive relationships (Granger and Newbold, 1986; Greene, 1997). Unstable and indefinitely growing variances inherent in nonstationary series not only complicate significance tests, they render forecasting problematic as well.

Nonstationary series are characterized by random walk, drift, trend, or changing variance. If each realization of the stochastic process appears to be a random fluctuation, as in the haphazard step of a drunken sailor, bereft of his bearings, zapped with random shocks, the series of movements is a random walk. If the series exhibits such sporadic movement around a level before the end of the time horizon under consideration, it exhibits random walk plus drift. Drift, in other words, is random variation around a nonzero mean. This behavior, not entirely predictable from its past, is sometimes inappropriately called a stochastic trend, because a series with trend manifests an average change in mean level over time (Harvey, 1993). When a disequilibrium of forces impinges on the series and stochastically brings about a change in level of the series, we say that the series is characterized by stochastic trend (Wei, 1990). Deterministic trends are systematic changes of the mean level of a series as a function of time. Whether or not these trends are deterministic or stochastic, they may be linear or curvilinear. If they are curvilinear, trends may be polynomial, exponential, or dampened. A trend may be short-run or long-run. The level of the series may erratically move about. There may be many local turning points. If the data have a stochastic trend, then there is a change of level in the series that is not entirely predictable from its history. Seasonal effects are annual fluctuations that coincide with period(s) of the year. For example, power usage may rise with heating costs during the winter and with air conditioning costs during the summer. Swimsuit sales may peak during the early or middle summer. The seasonal effects may be additive or multiplicative. Cyclical or long-wave effects are fluctuations that have much longer periods, such as the 11-year sunspot cycle or a particular business cycle. They may interact with trends to produce a trend cycle. Other nonstationary series have growing or shrinking variance. Changes in variance may come from trading-day effects or the influence of other variables on the series under consideration. One says that series afflicted with significantly changing variance have homogeneous nonstationarity. To prepare them for statistical modeling, series are transformed to stationarity either by taking the natural log, by taking a difference, or by taking residuals from a regression. If the series can be transformed to stationarity by differencing, one calls the series difference-stationary. If one can transform the series to stationarity by detrending it in a regression and using the residuals, then we say that the series is trend-stationary.

Time series can be presented in graphs or plots. SPSS may be used to produce a time sequence plot; SAS may be used to produce a timeplot or graphical plot of the series. The ordinate usually refers to the level of the series, whereas the abscissa is the time horizon or window under consideration. Other software may be used to produce the appropriate time sequence charts.

1.9. METHODOLOGICAL APPROACHES

This book presents four basic approaches to analyzing time series data. It examines smoothing methods, decomposition models, Box–Jenkins time series models, and autoregression models for time series analysis and forecasting. Although all of the methods may use extrapolation, the exponential smoothing and calendar-based decomposition methods are sometimes called extrapolative methods. Univariate Box–Jenkins models are sometimes called noncausal models, where the objective is to describe the series and to base prediction on the formulated model rather than to explain what influences them. Multivariate time series methods can include the use of an intervention indicator in Box–Jenkins–Tiao intervention models to examine the impact of an event on the time series. They can also include transfer function models, where an input series and a response series are cross-correlated through some transfer function. Both the intervention and transfer function models are sometimes referred to as causal models, where change in the exogenous variable or series is used to explain the change in the endogenous series. The exogenous component can consist of either a dummy indicator of the presence of an impact or a stochastic series that drives the response series. Such models are used to test hypothesized explanatory interrelationships between the time-dependent processes. Causal models may include autoregression models, where the endogenous variable is a function of lags of itself, lags of other series, time, and/or autocorrelated errors. Throughout the book, there are examples of forecasting with these models. Also, in Chapters 7 and 10, regression and autoregression models are used to combine forecasts to improve accuracy. After a discussion of model and forecast evaluation, the book concludes with a sample size and power analysis of common time series models by Monnie McGee. With these approaches, this book opens the door to time series analysis and forecasting.

The introduction considers the nature of time and a time series. The first section of the book addresses means of measuring those series. It discusses the extrapolation methods. These methods begin with the single moving average, the double moving average, and the moving average with trend. They extend to single exponential smoothing, double exponential smoothing, and then more advanced kinds of smoothing techniques. The time series decomposition methods include additive decomposition, multiplicative decomposition, and the Census X-12 decomposition.

The next section addresses the more sophisticated univariate Box–Jenkins models. This section begins with a consideration of the assumptions of Box–Jenkins methodology. The assumption of stationarity is discussed in detail. Various transformations to attain stationarity—including, logarithms, differencing, and others— are also addressed. Simple autoregressive process and moving average processes, along with the bounds of stationarity and invertibility, are explained. The section continues with an explication of the principles of autoregressive moving average (ARMA), autoregressive integrated moving average (ARIMA), seasonal ARIMA, and mixed multiplicative models coupled with examples of programming in both SAS and SPSS. The computer syntax and data sets will be found on the Academic Press Web Site (World Wide Web URL: http://www.academicpress.com/ sbe/authors/). After a consideration of the identification of models, a discussion of estimation and diagnosis follows. The section concludes with a treatment of metadiagnosis and forecasting of the univariate noncausal models.

The third section focuses on multivariate causal time series models, including intervention and transfer function models. This treatment begins with the multivariate Box–Jenkins–Tiao approach to impact analysis. The presence or absence of deterministic events is coded as a step or pulse, and the impacts as response functions. The responses are formulated as functions of those step or pulse input variables over time. The treatment of multiple time series continues with a consideration of transfer function (sometimes called TFARIMA or ARMAX) models, which model the transfer function between the input and the output time series. Both conventional prewhitening and the linear transfer function modeling approaches are presented. Other causal models include regression time series models. The problems encountered using multiple regression and correctives for those problems are also reviewed. Autoregressive models, including distributed lag and ARCH models, are also considered. Following a chapter on model and forecast evaluation, Monnie McGee provides an assessment of minimal sample requirements.

1.10. IMPORTANCE

What this book is not about is important in delimiting the scope of the subject matter. It avoids discussion of subjective methods, such as the Delphi technique of forecasting. It focuses on discrete time series and it concentrates on the time, not the frequency, domain. It does not attempt to deal with all kinds of multiple time series, nor does it address vector autoregression, vector autoregressive moving average, or state space models. Although it briefly discusses ARCH and GARCH models with regard to forecasting, it does not examine all kinds of nonlinear models. It does not attempt to deal with Bayesian models, engineering control systems, or dynamic simultaneous equation models. For reasons of space and economy, these models remain beyond the scope of this book and are left for a more advanced text.

To understand the nature of time series data, one needs to describe the series and then formulate it in terms of a statistical model. The time-ordering and temporal dependence pose unique problems for statistical analysis, and these problems must be taken into consideration. In order to forecast these processes, policies, and behaviors, corrections have to be developed and implemented for these problems. Forecasting is often necessary to understand the current situation when there is a time lag between data collection and assessment. Forecasting is also necessary for tactical planning and/or strategic planning. Moreover, forecasting may be essential to process engineering and control as well. These methods are essential for operations research in many areas. Whether the objective is description, explanation, prediction, monitoring, adaptation, or control, the study of time-ordered and -dependent phenomena is important.

1.11. NOTATION

1.11.1. GENDER

A few words about the basic notation used in this work are now in order. Although reference is made to the researcher in the masculine sense, no gender bias is implied. Researchers may indeed be female and often are. The masculine attribution rests purely on convention and convenience: No invidious bias is intended.

1.11.2. SUMMATION

The data are presumed to be discrete. The text makes use of subscript, summation, expectation, lag, and difference notation. To present, develop, and explain the processes discussed, a review of the elements of this notation is in order. The summation operator symbolizes adding the elements in a series and is signified by the capital Greek letter sigma (Σ). When it is necessary to indicate temporal position in the series, a subscript is used. If the variable in a time series is indicated by y_t, then the subscript t indicates the temporal position of the element in the series. If t proceeds from 1, 2, ... , T, this series may be represented by y_1 through y_T. The summation operator usually possesses a subscript and a superscript. The subscript identifies the type and lower limit of the series to be summed, whereas the superscript indicates the upper limit of the series to be summed. For example,

$$\sum_{t=1}^{T} y_i = y_1 + y_2 + \ldots + y_T \tag{1.1}$$

has a subscript of t = 1 and a superscript of T. The meaning of this symbol is that the sum of the y values for period 1 to T inclusive is calculated. T is often used as the total number of time periods. It is often used instead of n to indicate the total sample size of a series. Single summation is thereby indicated.

Double summation has a slightly more complicated meaning. If a table of rows and columns is being considered, one may indicate that the sum of the rows and the columns is computed by two summation signs in tandem. The inside (rightmost) sum cycles (sums) first.

$$\sum_{c=1}^{C} \sum_{r=1}^{R} x_{rc}$$

$$= x_{11} + x_{12} + \ldots + x_{1C}$$

$$+ x_{21} + x_{22} + \ldots + x_{2C} \tag{1.2}$$

$$\cdot \quad \cdot \quad \ldots \quad \cdot$$

$$+ x_{R1} + x_{R2} + \ldots + x_{RC}.$$

The double sum means that one takes row 1 and sums the elements in the columns of that row, then takes row 2 and sums the elements in the columns of that row, and iterates the process until all of the elements in the table have been summed. A triple sum would involve summing by cycling through rows, columns, and layers for the summing. The sums would be taken by iterating through the layers, columns, and rows in that order. When the elements in all of the combinations of rows, columns, and layers would be summed, the process would be completed.

If the data were continuous rather than discrete, then the integration sign from calculus would be used. A single integration sign would represent the area under the function that follows the integration sign. With the discrete time series used here, the summation sign is generally appropriate.

1.11.3. EXPECTATION

Expectation is an operation often performed on discrete variables used in the explanations of this text. Therefore, it is helpful to understand the meaning of the expected value of a variable. The expected value of a discrete random variable is obtained by multiplying its value at a particular time period times its probability:

$$E(Y) = \sum_{i=1}^{T} Y_i p(Y_i), \tag{1.3}$$

where

$E(Y)$ – expected value of discrete variable Y,
Y_i = value of Y at time period i, and
$p(Y_i)$ = probability of Y at periods 1 through T.

The expected value of a continuous random variable is often called its mean.

$$E(Y) = \int_{-\infty}^{\infty} Yf(Y) \, dy, \tag{1.4}$$

where $E(Y)$ is the expected value of random variable Y.

There are a few simple rules for expectation. One of them is that if there is a constant k, then $E(ky) = kE(y)$. Another is that if there is a random variable x, then $E(kx) = \sum kx \, p(kx)$. Also, $E(k + x) = k + E(x)$. If there are two random variables, x and y, then $E(x + y) = E(x) + E(y)$ and $E(xy) = E(x)E(y)$. The variance of a variable is often defined in terms of its expectation. $\text{Var}(x) = E[x - E(x)]^2$. The covariance of two variables is defined by $\text{Cov}(x, y) = E[x - E(x)]E[(y - E(y)]$. As these basic equations may be invoked from time to time, it is useful to be somewhat familiar with this notation (Hays, 1973).

1.11.4. LAG OPERATOR

The lag operator, symbolized by L, is also used for much of this analysis. Originally, Box and Jenkins used a B to designate the same operator, which they called the backshift operator. The lag operator used on a variable at time t refers to the value of the same variable at time $t - 1$; therefore, $Ly_t = y_{t-1}$. Similarly, $2LY_T = 2Y_{t-1}$. The lag operator backshifts the focus one lag or time period. The algebra of the lag is similar to that of the exponential operator. More generally, $L_n L_m(Y_t) = L_{n+m}(Y_t) = Y_{t-n-m}$.

Powers of the lag operator translate into periods of lag: $L^6 = y_{t-6}$; $L(Ly_t) = L^2 y_t = y_{t-2}$. Inverses of lags exist as well: $LL^{-1} = 1$. Inverses of expressions involving lags invert the expression: $z_t(1 - L)^{-1} = z_t/(1-L)$. It is also interesting that inverse of the first lag may result in series of infinite differences. We refer to the inverse of differencing as summing or integration because $1/(1 - L) = (1 - L)^{-1} = (1 + L + L^2 + L^3 + \ldots + L^{n-1} + L^n + \ldots)$. Using the lag operator facilitates explanations of differencing.

1.11.5. THE DIFFERENCE OPERATOR

The difference operator, del, is symbolized by the ∇. The first difference of y_t is given by the following expression: $w_t = \nabla y_t = y_t - y_{t-1}$. Another way of expressing this first difference is $w_t = \nabla y_t = (1 - L)y_t$. The second difference is the first difference of the first difference: $\nabla^2 y_t = \nabla(\nabla y_t) = (1 - L)(1 - L)y_t = (1 - 2L + L^2)y_t = (y_t - 2 y_{t-1} + y_{t-2})$.

These brief introductory explanations should enable the reader previously unfamiliar with this notation to more easily understand the following chapters. Previewing these matters of mathematical expression now will facilitate later analysis.

1.11.6. MEAN-CENTERING THE SERIES

There are some circumstances in which the centering of a series is advisable. Series are often mean-centered when complicated models, intervention models, or multiple input transfer function models are developed in order to save degrees of freedom in estimating the model as well as to simplify the model. To distinguish series that have been centered from those that have not, a difference in case is used. In this text, a capital Y_t will be used to denote a series that has been centered, by the subtraction of the value of the series mean from the original series value at each time period, whereas a small y_t will be used to denote a series that has not been mean-centered.

REFERENCES

Box, G.E.P. and Jenkins, G.M. (1976). *Time Series Analysis Forecasting and Control.* San Francisco: Holden Day, p. 21.

Diebold, F.X. (1998). *Elements of Forecasting.* Cincinnati: Southwestern College Publishing, p. 130.

Ege, G., Erdman, D.J., Killam, R.B., Kim, M., Lin, C.C., Little, M.R., Sawyer, D.M., Stokes, M.E., Narter, M.A., & Park, II.J. (1993). *SAS ETS/User's Guide.* Version 6, 2nd ed. Cary, NC: SAS Institute, Inc., pp. 139, 216.

Goodrich, R. (1992) *Applied Statistical Forecasting.* Belmont, MA: Business Forecast Systems, pp. 10–11.

Granger, C.W.J. and Newbold, P. (1986) *Forecasting Econometric Time Series,* New York: Academic Press, p. 1.

Greene, W. H. (1997). *Econometric Analysis* 3rd ed. Englewood Cliffs, NJ: Prentice Hall, p. 844.

Harvey, A.C. (1993) *Time Series Models,* Cambridge, MA: Cambridge University Press, pp. 10–11.

Hays, W. (1973). *Statistics for the Social Sciences.* 2nd ed. New York: Holt, Rhinehart, and Winston, pp. 861–877 presents the algebra of summation and expectation. Also, Kirk, R. (1982). Experimental Design. 2nd ed. Belmont, CA: Brooks Cole, p. 768.

Mills, T.C. (1990). *Time Series Techniques for Economists.* New York: Cambridge University Press, pp. 63–66.

Mills, T.C. (1993). *The Econometric Modeling of Financial Time Series.* New York: Cambridge University Press, p. 8.

SPSS, Inc. (1996). *SPSS 7.0 Statistical Algorithms.* Chicago, Ill: SPSS, Inc., p. 45.

Wei, W.S. (1990). *Time Series Analysis Univariate and Multivariate Methods* Redwood City, CA.: Addison-Wesley, p. 70.

Chapter 2

Extrapolative and Decomposition Models

2.1. INTRODUCTION

This chapter examines exponential smoothing and decomposition models. It begins with an introduction of statistics useful in assessment time series analysis and forecasting. From an examination of moving average methods, it develops an explanation of exponential smoothing models, which are then used as a basis for expounding on decomposition methods. The decomposition methods used by the U.S. Bureau of the Census and Statistics Canada to decompose series into their trend, cycle, seasonal, and irregular components are now used around the world to remove the seasonal component from these series preparatory to making them available for public use. Even though these methods are early ones in the development of time series and forecasting, their current applications give them pedagogical and contemporary practical value (Holden *et. al.,* 1990).

2.2. GOODNESS-OF-FIT INDICATORS

Many researchers seek to analyze time series data by detecting, extracting, and then extrapolating the patterns inherent in time series. They

15

may try to decompose the time series into additive or multiplicative component patterns. The preliminary toolkit used for inspection of a series includes a number of univariate assessment-of-fit indicators. The construction and formulation of these indicators are examined so the reader will see how they can be applied to the comparative analysis of fit, explanation, and accuracy in the methods of analysis.

After fitting a time series model, one can evaluate it with forecast fit measures. The researcher may subtract the forecast value from the observed value of the data at that time point and obtain a measure of error or bias. The statistics used to describe this error are similar to the univariate statistics just mentioned, except that the forecast is often substituted for the average value of the series. To evaluate the amount of this forecast error, the researcher may employ the mean error or the mean absolute error. The mean error (ME) is merely the average error. The mean absolute error (MAE) is calculated by taking the absolute value of the difference between the estimated forecast and the actual value at the same time so that the negative values do not cancel the positive values. The average of these absolute values is taken to obtain the mean absolute error:

$$\text{Mean absolute error} = \sum_{t=1}^{T} \frac{|e_t|}{T} \tag{2.1}$$

where t = time period, T = total number of observations, and e_t = (observed value − forecasted value)$_{\text{at time } t}$. To attain a sense of the dispersion of error, the researcher can examine the sum of squared errors, the mean square error, or the standard deviation of the errors. Another statistic commonly used to assess the forecast accuracy is the sum of squared errors. Instead of taking the absolute value of the error to avoid the cancellation of error caused by adding negative to positive errors, one squares the error for each observation, and adds up these squares for the whole forecast to give the sum of squared errors (SSE):

$$\text{Sum of squared errors} = \sum_{t=1}^{T} e_t^2 \tag{2.2}$$

When the sum of squared errors is divided by its degrees of freedom, the result is the error variance or mean square error (MSE):

$$\text{Mean square error} = \sum_{t=1}^{T} \frac{e_t^2}{T - k} \tag{2.3}$$

where T = total number of observations, and k = number of parameters in model. When the square root of the mean square error is computed, the result is the standard deviation of error, sometimes referred to as the root mean square error (RMSE):

Standard deviation of errors

$$(\text{root mean square error}) = \sqrt{\sum_{t=1}^{T} \frac{e^2}{T-k}} \qquad (2.4)$$

Any of these standard statistics can be used to assess the extent of forecast error in a forecast.

There are a number of proportional measures that can also be used for description of the relative error of the series. The percentage error, the mean percentage error, and the mean absolute percentage error measure the relative amount of error or bias in the forecast. The percentage error (PE) is the proportion of error at a particular point of time in the series:

$$\text{Percentage error} = \frac{(x_t - f_t)}{x_t} \times 100 \qquad (2.5)$$

where x_t = observed value of data at time t, and f_t = forecasted value at time t.

Although the percentage error is a good measure of accuracy for a particular forecast, the analyst may choose to analyze relative error in the entire series. The average percentage error in the entire series is a general measure of fit useful in comparing the fits of different models. This measure adds up all of the percentage errors at each time point and divides them by the number of time points. This measure is sometimes abbreviated MPE:

$$\text{Mean percentage error} = \sum_{t=1}^{T} \frac{PE_t}{T} \qquad (2.6)$$

where PE_t = percentage error of data at time t. Because the positive and negative errors may tend to cancel themselves, this statistic is often replaced by the mean absolute percentage error (MAPE):

$$\text{Mean absolute percentage error} = \sum_{t=1}^{T} \frac{|PE_t|}{T} \qquad (2.7)$$

where PE_t = percentage error, and T = total number of observations.

With any or all of these statistics, a time series forecast may be described and comparatively evaluated (Makridakis et al., 1983).

2.3. AVERAGING TECHNIQUES

2.3.1. THE SIMPLE AVERAGE

For preliminary description and analysis, summary measures may be used to describe a series spanning a number of time periods. Some of these summary statistics—for example, a simple average, a single moving average, a centered moving average, or possibly a double moving average—can be used to smooth a series. To smooth a time series, the analyst may wish to express the general level of the series as a simple average or the changing level over time as a moving average. The general level may serve as a baseline against which to describe fluctuations. The simple average can be used to describe a series that does not exhibit a trend; it gives each observation an equal weight in the computation. The simple average is helpful in designating and comparing the general levels of different series, each of which may have a constant mean:

$$\text{Simple Average} = \bar{y} = \sum_{t=1}^{T} \frac{y_t}{T} \tag{2.8}$$

2.3.2. THE SINGLE MOVING AVERAGE

When a researcher analyzes a time series, he may be more interested in a sliding assessment of the level of the series. He may use one of several linear moving average methods, including a single or double moving average for forecasting. The single moving average is a mean of a constant number of observations. This mean is based on the same number of observations in a sliding time span that moves its point of origin one time period at a time, from the beginning to the most recent observations of the series. The number of observations used for the computation of the mean is called the order of the series. The mean is computed and recorded for this number of observations from the beginning until the end of the series, at which point the calculations cease. Each of the observations in the calculation of the moving average is given an equal weight when the simple average is calculated. In the formula for the moving average, shown in Eq. (2.9), the subscript i is replaced by t, and the n from the simple average becomes a t as well. The span from t_1 to t_3 embraces three time periods.

A single moving average of order three

$$MA(3) = \sum_{t=t_1}^{t_3} x_t \tag{2.9}$$

The cumulative effect of the moving average, however, gives more weight to the central observations than to those at the ends of the series.

The effect of the single moving average is to smooth out irregular fluctuations, sometimes referred to as the hash, of the time series. This moving average may also smooth out the seasonality (characteristic annual variation, often associated with the seasons of the year) inherent in the series. The extent of smoothing depends on the order of the series: The more time periods included in this order (average), the more smoothing takes place. A moving average of order 1, sometimes referred to as a naive forecast, is used as a forecast by taking the last observation as a forecast for the subsequent value of the series.

As an illustration, a moving average of order 3—that is, MA(3)—is used for forecasting one-step-ahead; this kind of moving average is often used for quarterly data. This moving average takes the average of the three quarterly observations of that year, thereby effectively smoothing out additive seasonal variation of that year. This average is set aside in another column. At the next calculation of this moving average, the starting point for calculation begins with the value of the observation at the second time period in the observed series. The sum of the three observations, beginning with that second time period, is taken and then divided by 3. The mean that is calculated and recorded as the second observation in the column for the single moving average series. The third observation of the new single moving average series is calculated using the third observation of the original series as the first of three consecutive observations added together before dividing that sum by 3. Again this mean is set aside in the column reserved for the new series consisting of single moving averages. The sequence of means computed from an average based on consecutive observations moving over time constitutes the new series that is called the single moving average. The computation of this kind of moving average lends more weight to the middle observations in the series than does the simple average. Table 2.1 shows the computations for a moving average of order 3 of household heating units sold.

Note that the moving average does not extend for the whole time span of the series. The moving average of order T begins after t periods have elapsed. In this example, $T = 3$ and the moving average is a mean of the three preceding periods. Nonetheless, some persons opt for some smoothing of irregular variation and prefer this kind of moving average for naive forecasting of the next observation of the series from last single moving average value.

Table 2.1

Forecasting with Single Moving Average Smoothing

Monthly Time Periods	Sales in Heating Units	Single Moving Average ($T = 3$)	Error
1. January	10 units		
2. February	9		
3. March	8		
4. April	7	$(10 + 9 + 8)/3 = 9.00$	−2.00
5. May	3	$(9 + 8 + 7)/3 = 8.00$	−5.00
6. June	2	6.00	−4.00
7. July	1	4.00	−3.00
8. August	0	2.00	−2.00
9. September	1	1.00	−0.00
10. October	5	0.67	4.33
11. November	12	2.00	10.0
12. December	14	6.00	8.0
Forecast		10.33	

2.3.3. CENTERED MOVING AVERAGES

Calculation of the moving average differs depending on whether the series contains an even or odd number of observations. Many series have even numbers of observations, such as those spanning a year of 12 months. A T-period moving average should really be centered at time period $(T + 1)/2$, and this centering is sometimes called the correction for lag (Makridakis *et al.*, 1983). Rarely, however, is the naive forecast using a single moving average tendered except as an impromptu approximation for general purposes. In cases such as this, centering the moving average solves the problem. Centering involves taking an additional moving average of order 2 of the first moving averages. The resulting moving average is a mid-value for the first moving average of an even order. The researcher would take the moving average for the period before the midpoint and the moving average for the period after that midpoint, and then take the average of those two scores to obtain the centered moving average. This is a common means by which moving averages of even-numbered series are handled.

2.3.4. DOUBLE MOVING AVERAGES

A double moving average may be used for additional smoothing of a single moving average. Computing the double moving average is simple: First a single moving average series is computed. Then a second moving average series is calculated from the first moving average. The double moving average is distinguished from the single moving average by beginning T periods after the starting point of the series. The first moving average is of order T. A second moving average, made up of the components of the first moving average, is of order N. In other words, the double moving average takes the average of N of the first moving averages. The double moving average begins $T + N$ time points after the beginning of the first point in the series. It results in more smoothing of the first smoothing moving average. The extent of this smoothing depends on the lengths of the first and second moving average. For long series, with much irregularity or incremental error, this kind of smoothing facilitates elimination of short-run fluctuations. This double moving average is called a moving average of order T by N, denoted by $(T \times N)$. Let $T = 3$ and $N = 3$ in the example in Table 2.2.

The use of double moving averages permits calculation of intercept and trend for the basic formula by which exponential smoothing forecasts are generated. In Table 2.2, note that the forecast is constructed with the aid of the double moving average. The double moving average can be used to compute an intercept and a trend coefficient, the average change over h periods, which are added together to obtain the forecast, F_{t+h}, for h steps ahead:

$$F_{t+h} = a_t + b_t h \tag{2.10}$$

Table 2.2

Double Moving Average Forecasting with Linear Trend

A Time Periods T	B Data Series	C Single Moving Average MA(3)	D Error, B − C	E Double Moving Average MA(3 × 3)	F Error, C − E	G Trend	H Prediction, E + F + G
1	34						
2	36						
3	38	36	2				
4	40	38	2				
5	42	40	2	38	2	2	42
6	44	42	2	40	2	2	44
7	46	44	2	42	2	2	46
8	48	46	2	44	2	2	48
9	50	48	2	46	2	2	50
10	52	50	2	48	2	2	52
							54

The intercept, a_t, is simply two times the single moving average minus the double moving average. The trend coefficient, b_t, is the average difference between single moving average and the double moving average from one time point to the next, and this is computed by subtracting the double from the single moving average, multiplying the difference by 2, and dividing by $T - 1$. To obtain the forecast, the error between the single and double moving average is added to the sum of the single moving average and the trend. This process has been called a moving average with linear trend. It is helpful to consider the calculations of the moving average with linear trend for the forecast. For longer series, this process may reduce the minimum mean square error of the series and hence render a more accurate forecast than the earlier naive one (Makridakis *et al.*, 1983).

2.3.5. WEIGHTED MOVING AVERAGES

Although the simple averages use equally weighted observations at first, they end up with equally weighted observations except for the endpoints. However, double moving averages have substantially unequally weighted observations, with potentially problematic consequences for prediction. A double moving average of 3 by 3 provides a good illustration of this problem. The weighting of the observations in a double moving average gives the middle observations more influence than more recent observations because the middle values in the series are used in the calculation of the final mean more than the observations at either the original or the recent tail of the time series. The more recent observations have more effect on future observations than those in the more distant past, so a linearly weighted series might be of greater utility than a conventional moving average.

Double moving average:

$$
\begin{aligned}
MA(3)_1 &= X_1 + X_2 + X_3 \\
MA(3)_2 &= X_2 + X_3 + X_4 \\
MA(3)_3 &= X_3 + X_4 + X_5 \\
\hline
X_1 + 2X_2 &+ 3X_3 + 2X_4 + X_5
\end{aligned}
$$

$$X_1 + 2X_2 + 3X_3 + 2X_4 + X_5$$

Forecast (double moving average)$_{t+1} = x_t + \frac{2}{9}X_{t-1} + \frac{1}{3}X_{t-2} + \frac{2}{9}x_{t-3} + \frac{1}{9}x_{t-4}$

Forecast (linearly weighted moving average)$_{t+1} = x_t + \frac{3}{4}x_{t-1} + \frac{2}{4}x_{t-2} + \frac{1}{4}x_{t-3}$

$$(2.11)$$

This forecast is characterized by a linear decrement of the weights as the time period is extended into the past. This weighting scheme gives $1/T$ less importance to each of the T values as one proceeds back along the time path. The effect of the past observations on the future ones may actually decline nonlinearly as one proceeds into the past. To compensate for the irregular weighting of observations, exponential smoothing is introduced.

2.4. EXPONENTIAL SMOOTHING

2.4.1. SIMPLE EXPONENTIAL SMOOTHING

Exponential smoothing is a method, conceived of by Robert Macaulay in 1931 and developed by Robert G. Brown during World War II, for extrapolative forecasting from series data. The more sophisticated exponential smoothing methods seek to isolate trends or seasonality from irregular variation. Where such patterns are found, the more advanced methods identify and model these patterns. The models can then incorporate those patterns into the forecast. When used for forecasting, exponential smoothing uses weighted averages of the past data. The effect of recent observations is expected to decline exponentially over time. The further back along the historical time path one travels, the less influence each observation has on the forecasts. To represent this geometric decline in influence, an exponential weighting scheme is applied in a procedure referred to as simple (single) exponential smoothing (Gardiner, 1987).

Suppose that a prediction is going to be based on a moving average. The moving average prediction will be called MA_{t+1} and the previous moving average will be called MA_t. If the moving average under consideration is made up of 10 observations, then the easiest way to update the moving average is to slide it along the time path, one time period at a time. At each time period, the average of the 10 observations will be taken. Another way to conceptualize this same process is to take 1/10 of the value of the observation at time t and to subtract 1/10 of the moving average formed from the ten most recent observations before combining them to produce a new moving average prediction (Brown, 1963):

$$MA_{t+1} = (1 - 1/10)MA_t + (1/10)x_t \qquad (2.12)$$

In this example, the moving average consists of 10 observations, though the moving average may be made up of any number of observations. The proportion of the latest observation taken is called a smoothing constant, α. The formula representing this simple smoothing is

$$\begin{aligned} MA_{t+1} &= (1 - \alpha)MA_t + (\alpha)x_t \\ &= (\alpha)x_t + (1 - \alpha)MA_t \end{aligned} \qquad (2.13)$$

In view of the fact that this moving average is a smoothing function that may be applied for the purpose of forecasting, a forecast, F_6, may be substituted for the moving average, MA_t, in this formula to obtain a formula for forecasting:

$$\hat{F}_{t+1} = \alpha x_t + (1 - \alpha)F_t. \tag{2.14}$$

Extending this expression one step along the time line into the past, one obtains the expansion:

$$\begin{aligned}
\hat{F}_{t+1} &= \alpha x_t + (1 - \alpha)[\alpha X_{t-1} + (1 - \alpha)] F_{t-1} \\
&= \alpha x_t + \alpha(1 - \alpha)X_{t-1} + (1 - \alpha)^2 F_{t-1}.
\end{aligned} \tag{2.15}$$

If this expression is extended two and then n steps into the past, it becomes

$$\begin{aligned}
\hat{F}_{t+1} &= \alpha x_t + (1 - \alpha)[\alpha X_{t-1} + (1 - \alpha)] F_{t-1} \\
&= \alpha x_t + \alpha(1 - \alpha)X_{t-1} + (1 - \alpha)^2 F_{t-1} + \alpha(1 - \alpha)^3 X_{t-3} \\
&\quad + \cdots + \alpha(1 - \alpha)^{n-1} X_{t-n-1} + (1 - \alpha)^n F_{t-n-1}.
\end{aligned} \tag{2.16}$$

At this point, the meaning of the smoothing weight, α, is modified slightly from the meaning it had in the first example. The magnitude of the smoothing constant ranges between 0 and 1.0. A smaller smoothing constant gives more relative weight to the observations in the more distant past. A larger smoothing constant, within these bounds, gives more weight to the most recent observation and less weight to the most distant observations. The smaller the smoothing weight, the more weight is given to observations in the past and the greater the smoothing of the data. In this way, the smoothing constant, α, controls the memory of the process.

Two choices must be made before simple exponential smoothing is possible: the initial value of the smoothing weight and the final value of the smoothing constant.

First consider the choice of the optimal smoothing constant. This constant may be found by graphical or statistical comparison. Any of the goodness-of-fit indicators discussed earlier can be applied to objectively compare one forecast error with another. Often, the better smoothing weight is less than 0.5 and greater than 0.10, although this need not be the case. For graphical presentation and evaluation, a spreadsheet may be used to generate the predictions and chart them. The smoothing constant of a simple exponential smoothing of these data can be chosen by visual inspection.

A manager planning his inventory might decide that he should use a particular smoothing constant to estimate his needs. Three smoothings, along with their smoothing constants, are shown in Figure 2.1. The data are represented by the heavy line in the background, while the different

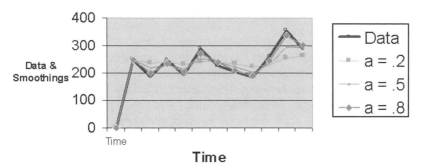

Figure 2.1 Single exponential smoothing with various smoothing parameters.

single exponential smoothings constructed with the three different smoothing constants shown in the legend are shown as lines broken by symbols. Based on the level of smoothing, the manager can decide which suits his needs optimally.

Makridakis *et al.* (1983) suggest four ways to choose an initial value of the series. The forecaster should acquire enough data to divide his data set into at least two segments. The first (historical or training) segment is for estimating initial values and model parameters. The second (hold-out) data set is used for validation of the explanatory or forecasting model. It is common to select the average of the estimation sample as the starting value of the smoothing constant to be used for forecasting over the span of the validation sample. If the series is an extension of a previous series, then the average of that previous series may be used. Alternatively, the mean of the whole series may be used as a starting value. Different starting values may be employed while some measure of forecasting accuracy may be compared to see which will be ultimately chosen as the best. Finally, backcasting, using an ARIMA model to be discussed in Chapters Four through Seven, may be employed. Based on the existing data in the series, the analyst may forecast backward to obtain the initial value.

When these matters are resolved, the equation that emerges from a simple (single) exponential smoothing is a linear constant model. The model has merely a mean and an error term. Note that this model accounts for neither trend nor seasonality.

$$Y_t = \mu + e_t \tag{2.17}$$

Models that accommodate trend and seasonality will be discussed shortly.

2.4.1.1. Single Exponential Smoothing: Programming Syntax and Output Interpretation

Both SAS and SPS have programs that perform single exponential smoothing. Both programs allow the user to select the initial value for the smoothing, but by default select the mean of the series for an initial value. Neither program tolerates missing values within the series for exponential smoothing. The SPSS program permits the user to insert his own smoothing weight, α, or permits a grid search of the sum of squared errors to find the optimal smoothing weight value for smoothing or forecasting.

If a retailer examined the proportion of available space in his warehouse over the past 262 days to estimate how freely he could procure stock over the next 24 days, he might employ this simple extrapolative method. If he were using SAS, his prepared program might contain explanatory annotations in the program bracketed by /* and */ and title statements that begin the documentation of the exponential smoothing procedure invoked by a PROC FORECAST statement.

Each SAS program has a DATA step, a PROC step, and a data set. The data step defines the variables and their locations. It performs any transformations of them prior to the inclusion of the data under a CARDS or DATALINES statement. The statistical PROCedure usually follows the preliminary data preparation. Title statements are often inserted underneath the procedure referred to, and such titles will appear at the top of the pages containing the statistical output of that procedure.

```
Title 'SAS program file: C2pgm1.sas';
Title2 'Simple Exponential Smoothing';
title3 'Free Warehouse Space';
title4 'for Stock Procurement';
data one; retain time(1);   /* time is initialized at 1 */
      input invspace;        /* variable is defined */
    time + 1;                /* time counter constructed */
cards;                       /* the data follow */
data go here
proc print data=one;        /* check of program construction */
title 'data one';           /* gets data from data set one */
run;

proc forecast data=one method=expo trend=1 lead=24
  outall outfitstats out=fore outest=foretest;
  var invspace; /* ************************** */
     id time; /* Explanation of Proc Forecast   */
run;           /* proc forecast does expo smothg  */
               /* method=exp uses simple exponential smoothing */
               /* trend = 1 uses a constant model   */
               /* lead=24 forecasts 24 days ahead   */
```

```
                /* outall produces actual forecast 195 u95 std vars */
                /* outfitstats produces forecast eval stats */
                /* out=fore produces output data set    */
                /* outest=foretest produces forecast eval data set */
                /* invspace is the variable under examination */
                /* id time uses the time counter as the day var */
                /* ********************************** */
proc print data=foretest; * prints the evaluation stats */

title 'Forecast Evaluation Statistics';
run;

data all;      /* Merges original data with generated data/*
 merge fore one; by time;
run;

symbol1 i=join c=green; /* sets up the lines for the plot */
symbol2 i=join c=red;
symbol3 i=join c=blue;
symbol4 i=join v=star c=purple;
axis1 order=(.10 to .50 by .02) label=('Inv Space'); /*creates axis */
proc gplot data=all;    /* gplot gets merged data */
plot invspace*time=_type_/overlay vaxis=axis1;/* plots space v time */
where _type_ ^='RESIDUAL' & _type_ ^= 'STD'; /* drops nuisance vars */
title 'Exponential Smoothing Forecast of';
title2 'Free Inventory Space';
run;
```

The SAS forecast procedure produces two output data sets. The forecast values are produced in an output data set called FORE and the fit statistics in a data set called FORETEST. The fit statistics, based on the fit of the model to the historical data, are first listed below and defined for the reader.

```
          Evaluate Series
OBS _TYPE_ TIME INVSPACE
```

OBS	_TYPE_	TIME	INVSPACE	
1	N	262	262	sample size
2	NRESID	262	262	number of residuals
3	DF	262	261	degrees of freedom
4	WEIGHT	262	0.2	smoothing weight
5	S1	262	0.2383997	smoothed value
6	SIGMA	262	0.0557227	standard deviation of error
7	CONSTANT	262	0.2383997	constant
8	SST	262	1.2794963	total sum of squares
9	SSE	262	0.8104098	sum of squared errors
10	MSE	262	0.003105	mean square error

```
11 RMSE       262  0.0557227 root mean square error
12 MAPE       262 13.317347  mean absolute percent error
13 MPE        262 -3.795525  mean percent error
14 MAE        262  0.0398765 mean absolute error
15 ME         262 -0.003094  mean error
16 MAXE       262  0.363617  maximum error
17 MINE       262 -0.163524  minimum error
18 MAXPE      262 61.318205  maximum percentage error
19 MINPE      262 -95.07207  minimum percentage error
20 RSQUARE    262  0.3666181 r square
21 ADJRSQ     262  0.3666181 adjusted r square
22 RW_RSQ     262  0.3720846 random walk r square
23 ARSQ       262  0.3617646 Amemiya's adjusted r square
24 APC        262  0.0031169 Amemiya's Prediction Criterion
25 AIC        262 -1511.983  Akaike Information Criterion
26 SBC        262 -1508.414  Schwartz Bayesian Criterion
```

When the data in the FORE data set are graphed with the plotting commands, the plot in Fig. 2.2 is produced.

In Figure 2.2 the actual data, the forecast, and the upper and lower 95% confidence limits of the forecast are plotted in a time sequence plot. From an inspection of this chart, the manager can easily decide what proportion of space will be available for inventory storage in the next 24 days.

The SPSS command syntax for simple exponential smoothing of these inventory data and a time sequence plot of its predictions follows. In both

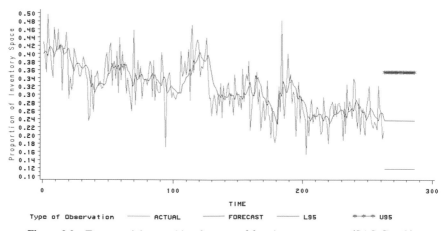

Figure 2.2 Exponential smoothing forecast of free inventory space (SAS Graph).

SPSS and SAS, comments may be indicated by a statement beginning with a single asterisk at the left-hand side of the line. SPSS commands begin in the left-most column of the line. The usual command terminator (in the case of SPSS, a period; in the case of SAS, a semicolon) ends the comment. Continuations of SPSS commands are indicated by the / at the beginning of the next subcommand.

```
* SPSS Program file: c2pgm2.sps.
get file='c2fig3.sav'.
TSET PRINT=DEFAULT NEWVAR=ALL.
PREDICT THRU 300.
EXSMOOTH /VARIABLES=invspace
  /MODEL=NN
  /ALPHA=GRID(0 1 .1)
  /INITTAL=CALCULATE.
Execute.
* iterative replacement of missing time values for predicted periods.
if (missing(time)=1) time=$casenum.
*Sequence Charts.
TEMPORARY.
COMPUTE #OBSN = #OBSN + 1.
COMPUTE MK_V_# = ( #OBSN < 261 ).
TSPLOT VARIABLES= invspace fit_1
  /ID= time
  /NOLOG
  /MARK MK_V_#.
* the following command tests the model for fit.
Fit err_1 /dfe=261.
```

These SPSS commands invoke simple exponential smoothing of the variable invspace. Based on the 262 cases (days) of the invspace variable describing the proportion of available inventory space, these commands request predicted values through 300 observations. The MODEL subcommand specifies the type of trend and seasonal component. Because this is simple exponential smoothing, the NN designation stands for neither trend nor seasonal component. In other words, the first of these letters is the trend parameter. Trend specification options are N, for none, L for linear, E for exponential, and D for dampened. The types of seasonal component options available are N for none, A for additive, and M for multiplicative. The smoothing weight, ALPHA, is found by a grid search over the sum of squared errors produced by each iteration of alpha from 0 to 1 by a step value of 0.1. The smoothing weight yielding the smallest sum of squared errors is chosen by the program as the alpha for the final model. The /INITIAL =CALCULATE option invokes the mean of the series as a starting value. Alternatively, the user may enter his own choice of starting value. Two variables are constructed as a result of this analysis, the fit, called fit_1, and the error, called err_1. These variables are placed at the end of the system file, which is given a suffix of .sav. These newly constructed variables contain the predicted and residual scores of the smoothing process. Because of the forecast into the future horizon, the output specifies how many cases have been added to the dataset. From this output, it may be seen that the grid search arrives at an optimal alpha of 0.2 on the basis of an objective criterion of a minimum sum of squared errors.

```
MODEL: MOD_2.c2pgm2.sps.
Results of EXSMOOTH procedure for Variable INVSPACE
MODEL= NN (No trend, no seasonality)

   Initial values:    Series        Trend
               .31557     Not used
DFE = 261.
The 10 smallest SSE's are:    Alpha   SSE
                  .2000000     .83483
                  .3000000     .84597
                  .1000000     .86395
                  .4000000     .87678
                  .5000000     .91916
                  .6000000     .97101
                  .7000000    1.03279
                  .8000000    1.10613
                  .9000000    1.19352
                  .0000000    1.27950
The following new variables are being created:
  NAME     LABEL
  FIT_1    Fit for INVSPACE from EXSMOOTH, MOD_2 NN A .20
  ERR_1    Error for INVSPACE from EXSMOOTH, MOD_2 NN A .20
24 new cases have been added.
```

The goodness of fit of the model is tested with the next command, FIT err_1 /DFE=261. For a correct test of fit, the user must find or calculate the degrees of freedom (DFE = number of observations minus the number of degrees of freedom for the hypothesis) and enter them (*SPSS 7.0 Algorithms,* 1997). They are given in the output of the smoothing and can be entered in this command thereafter. The output contains a few measures of fit, such as mean error, mean absolute error, mean square error, and root mean square error. These statistics are useful in comparing and contrasting competing models.

FIT Error Statistics

Error Variable		ERR_1
Observed Variable		N/A
N of Cases	Use	262
Deg Freedom	Use	261
Mean Error	Use	-.0015
Mean Abs Error	Use	.0404
Mean Pct Error	Use	N/A
Mean Abs Pct Err	Use	N/A
SSE	Use	.8348
MSE	Use	.0032
RMS	Use	.0566
Durbin-Watson	Use	1.8789

At the time of this writing, SPSS can produce a time sequence plot. Figure 2.3 graphically presents the actual proportion of inventory space available along with the values predicted by this computational scheme. A line of demarcation separates the actual from the predicted values.

Single exponential smoothing is advantageous for forecasting under particular circumstances. It is a simple form of moving average model. These

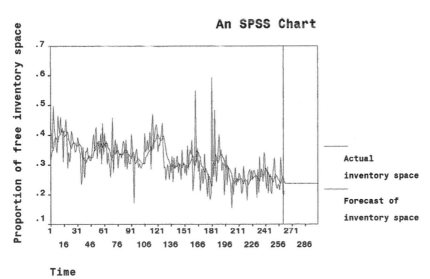

Figure 2.3 A single exponential smoothing of free inventory space (SPSS Chart).

models lack trend or seasonality, and they do not require a long series for prediction. They are easy to program and can be run with a spreadsheet by those who understand the process. However, they lack flexibility without more development, and are not that useful for modeling a process unless they themselves yield a mathematical model of the process. The output from this procedure, a mean plus some random error, is merely a smoothed set of values plus a sequence of predictions, which leaves something to be desired if one needs to explain a more complicated data-generating process.

2.4.2. HOLT'S LINEAR EXPONENTIAL SMOOTHING

In failing to account for trends in the data, simple exponential smoothing remains unable to handle interesting and important nonstationary processes. E. S. Gardiner expounds on how C. C. Holt, whose early work was sponsored by the Office of Naval Research, developed a model that accommodates a trend in the series (Gardiner, 1987). The final model for a prediction contains a mean and a slope coefficient along with the error term:

$$Y_t = \mu_t + \beta_t t + e_t \tag{2.18}$$

This final model consists of two component equations for updating (smoothing) the two parameters of the equation system—namely, the mean, μ, and the trend coefficient, β. The updating equation for the mean level of the model is a version of the simple exponential smoothing, except that the trend coefficient is added to the previous intercept to form the component that receives the exponential decline in influence on the current observation as the process is expanded back into the past. The alpha coefficient is the smoothing weight for this equation:

$$\mu_t = \alpha Y_t + (1 - \alpha)(\mu_{t-1} + b_{t-1}) \tag{2.19}$$

The trend coefficient is also updated by a similar exponential smoothing. To distinguish the trend updating smoothing weight from that for the intercept, γ is used instead. The values for both smoothing weights can range from 0 to 1.0.

$$b_t = \gamma(u_t - u_{t-1}) + (1 - \gamma)b_{t-1} \tag{2.20}$$

In the algorithm by which this process works, first the level is updated. The level is a function of the current value of the dependent variable plus a portion of the previous level and trend at that point in time. Once the new level is found, the trend parameter is updated. Based on the difference between the current and previous intercept and a complement of the previous trend, the current trend parameter is found. This process updates the coefficients of the final prediction equation, taking into account the mean and the trend as well. If one were to compute the forecast equation h steps ahead, it would be $\hat{Y}_t(h) = u_t + ht_t$.

Holt's method can be applied to prediction of trust in government. Either SAS or SPSS may be used to program this forecast. In the American National Election Study, political scientists at the Institute of Social Research at the University of Michigan have studied attitudes of the voting public, including trust in government. The aggregate response to this indicator functions as a feeling thermometer for the political system. The public are asked "How much do you trust the government to do the right thing?" The possible answers are Don't know, Never, Some of the time, Most of the time, and Almost always. The percentage of people having a positive attitude—that is, trusting government to do the right thing most of the time or almost always—is examined over time. If the percentage of responses shows a serious decline in public trust, then the political climate may become too hostile for the political process to function and it may break down. When the series of biennial surveys was examined, a short series was constructed that lends itself to some exponential smoothing.

A preliminary review of the series revealed the presence of a negative trend in public trust in government. A plot of percentage of positive trust in government vs time was constructed with SAS Graph and is shown in Fig. 2.4.

From the series depicted in the graph, it appears that decline in public trust set in during domestic racial strife, urban riots, and escalation of the

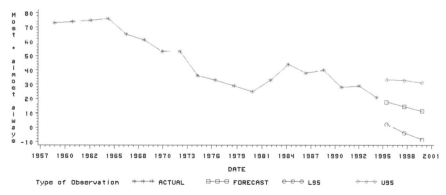

Figure 2.4 Holt Smoothing Forecast of Public Trust in Government National Election Study Table 5A.1 (1958–1994). Source: Inter-University Consortium for Political and Social Research, Institute of Social Research, University of Michigan. Data (SAS Graph).

Vietnam War. A credibility gap developed over inflated body counts and over-optimistic official assessments of allied military operations in Vietnam. Public trust in government declined after the Tet Offensive in 1968 and continued to slide until 1970, when the slippage seemed to let up. In 1972, during the Watergate scandal, trust in government plummeted. When President Richard Nixon left office in 1974, the steepness of the decline abated, but President Gerald Ford's pardon of Nixon and the intelligence agency scandals kept trust in government slipping. It was not till Reagan came into office that trust in government began to grow, and grow it did for 5 years. The Iran–Contra scandal probably produced another crisis that made people politically cynical again. The first 2 years of the Bush tenure experienced an improvement in public trust, but it began to decline until the Gulf War. During the Gulf War, there was a growth in trust in government. But during the economic doldrums of the last year of the Bush term, trust declined again. This slippage was turned around by the first 2 years of the Clinton administration. In 1994, the Republicans gained control of Congress, and by the end of 1995 had shut down the government while demanding huge reductions in taxes and Medicare spending. Trust in government began to fall again.

What is the prediction, ceteris paribus, of how much the public would trust government in the last years of the Clinton presidency? A glance at the output statistics for this model discovers that the trend is a -2.94 and that the constant for the model is 76.38. The R^2 for this and the no-trend model were compared, and the linear trend model was found to have the better fit. It is interesting to note that the farther into the future the prediction is extended, the wider the confidence interval. Contrary to this statistical prediction and those of political pundits, by 1998, President William J. Clinton, in spite of a campaign to tarnish his reputation with allegations of one scandal or another, enjoyed the highest public job approval rating since he came into office: more than 63% of the public approved of his job performance according to both the Gallup and *Washington Post* polls (spring 1998).

It is easy to program a Holt linear trend exponential smoothing model in SAS. The basic difference between this program and the previous SAS program is that in the earlier program, the TREND option was assigned a value of 1 for a constant only. In this program the TREND option has a value of 2, for two parameters. The first parameter is a constant, and the second is that of a linear trend. A regression on a constant, a time and a time-squared parameter reveals whether the constant, linear, and/or quadratic terms are significant. The value of the TREND option represents the highest number of parameters used for testing the time trend. Therefore, if there were a quadratic trend, then a value of 3 would be used for the

TREND option. In this case, the optimal coefficient value for TREND is 2. In this case, the linear component was the significant one, which is why the Holt model was chosen for smoothing and forecasting 6 years ahead with such a short series.

```
SAS PROGRAM SYNTAX:
/* c2pgm3.sas */
title'National Election Study Table 5A.1 v604';
title2 'Percentage who Trust in Federal Government 1958-1994';
title3 'How much of the time do you trust the gvt to do whats right';
title4 '1960 and 1962 values are interpolations';

data trust;                      /* name of data set */
  input date: monyy5. none some most almalwys dk; /* variable definitn */
label none='None of the time' /* var labels */
 some='Some of the time'
  most='Most of the time'
  almalwys='Almost always';
Positive = most + almalwys; /*construction of test variables */

Negative = none + some;

year = year(date);              /* construction of time trend vars */
yearsq = year**2; output;

label Positive = 'Most + almost always'
  Negative = 'None + sometimes';
cards;
  data go here
proc print label;               /* check of data */
run;


/* The Graphical Plot */
axis1 label=none order=(0 to 80 by 10); /* Sets up axis for plot */
symbol1 v=star c=brown i=join;          /* defines the lines in plot */
symbol2 v=square c=black i=join;
symbol3 v=circle c=cyan i=join;          /* Gplot to examine raw data */
symbol4 v=diamond c=green i=join;
symbol5 v=triange c=blue i=join;
footnote justify=L 'Star=none Square=some Circle=most Diamond=almost
always';
proc gplot; /* Examine the answer battery re trust in govt */
  plot (none some most almalwys) * date /overlay
  vaxis=axis1 ;
format date monyy5. ;
run;


 /* collapsing the plot into positive & negative */
```

```
axis1 order=(20 to 80 by 10) label=none
  offset=(2,2) width=3 ;
axis2 order=(1958 to 2000 by 10) label = ('Year');
symbol1 v=star c=blue i=join;
symbol2 v=square c=red i=join;
footnote justify=L 'Star=% Most or Almost always Square= % None or some
Trust';
proc gplot;
  plot (positive negative) * year/ vaxis=axis1
  haxis=axis2 overlay;
run;

proc reg;
  model positive = year yearsq;
title5 'Test of Type of Trend in Trust in Government';
run;

proc forecast data=trust trend=2 interval=year2 lead=3 out=resid
 outactual outlimit outest=fits outfitstats;
 var positive;  /* trend = 2 for Holt Linear model */
 id date;
format date monyy5.;
title5 'Holt Forecast with Linear Trend';
run;
proc print data=fits;  /* checking fit for this model */
title5 'Goodness of fit';
run;
proc print data=resid;  /* printing out the forecast values
 */
 run;


proc gplot data=resid;  /* generating forecast interval plot */
 plot positive * date = _type_ /
 haxis = '1958 to 2000 by 2';
format date year4.;
footnote justify=L ' ';
title5 'Percentage Trusting Government Most of time or Almost always';
run;

SPSS PROGRAM SYNTAX:

* C2PGM4.SPS .
* Example of Holt Linear Exponential Smoothing Applied to short.
* series from American National Election Study Data.
title'National Election Study Table 5A.1 v604'.
subtitle 'Percentage who Trust in Federal Government 1958-1994'.
* 'How much of the time do you trust the gvt to do whats right'.
* '1960 and 1962 values are interpolations'.
DATA LIST/DATE 1-5(A) NONE 8 SOME 9-14 MOST 15-18 ALMALWYS 19-23 DK 25-2
VAR LABEL NONE 'NONE OF THE TIME'/
```

```
  SOME 'SOME OF THE TIME'/
  MOST 'MOST OF THE TIME'/
  ALMALWYS 'ALMOST ALWAYS'.
COMPUTE POSITIVE=SUM(MOST,ALMALWYS).
COMPUTE NEGATIVE=SUM(NONE,SOME).
STRING YEARA(A2).
COMPUTE YEARA=SUBST(DATE,4,2).
RECODE YEARA (CONVERT) INTO ELECYEAR.
COMPUTE ELECYEAR = 1900 + ELECYEAR.
FORMATS ELECYEAR(F4.0).
VAR LABELS POSITIVE 'MOST + ALMOST ALWAYS'
  NEGATIVE = 'NONE + SOMETIMES'.
FORMATS NONE,SOME,MOST,ALMALWYS,DK(F4.1).
BEGIN DATA.
  Data go here
END DATA.
LIST VARIABLES=ALL.
EXECUTE.
DATE YEAR 1952 BY 2.
EXECUTE.
* EXPONENTIAL SMOOTHING.
TSET PRINT=DEFAULT NEWVAR=ALL .
PREDICT THRU 22 .
EXSMOOTH /VARIABLES=POSITIVE
 /MODEL=HOLT
 /ALPHA=GRID(0 1 .1)
 /GAMMA=GRID(0 1 .2)
 /INITIAL=CALCULATE.
Execute.
Fit ert 1 /DfE=17.
Execute.
*Sequence Charts .
TEMPORARY.
COMPUTE #OBSN = #OBSN + 1.
COMPUTE MK_V_# = ( #OBSN < 19 ).
TSPLOT VARIABLES= POSITIVE FIT_1
 /ID= YEAR
 /NOLOG
 /FORMAT NOFILL NOREFERENCE
 /MARK MK_V_#.
EXECUTE.
```

The SPSS output follows. The output shows the initial values of the constant and the linear trend parameter, and it provides, in decreasing order of quality, the 10 best alphas, gammas, and their associated sums of squared errors. The first line contains the best parameter estimates, selected for the smoothing predicted scores. The fit and error variables are then constructed. The output, shown below, specifies the model as a HOLT model with a linear trend, but no seasonality. Values for the mean and trend parameters are given. The fit statistics, not shown here, would follow the listed output.

The command syntax for a time sequence plot of the actual and predicted values, with a vertical reference line at the point of prediction, concludes SPSS program C2pgm4.sps.

```
Results of EXSMOOTH procedure for Variable POSITIVE
MODEL= HOLT (Linear trend, no seasonality)

 Initial values: Series   Trend
                74.44444 -2.88889


DFE = 17.

The 10 smallest SSE's are:  Alpha      Gamma         SSE
                           1.000000  .0000000   819.92420
                            .9000000  .0000000   820.80126
                            .8000000  .0000000   844.68187
                            .7000000  .0000000   891.48161
                            .9000000  .2000000   948.54177
                           1.000000  .2000000   953.90135
                            .6000000  .0000000   961.79947
                            .8000000  .2000000   983.15832
                            .9000000  .4000000  1006.43601
                            .8000000  .4000000  1024.63425
The following new variables are being created:

NAME LABEL

FIT_1  Fit for POSITIVE from EXSMOOTH, MOD_8 HO A1.00 G .00
ERR_1  Error for POSITIVE from EXSMOOTH, MOD_8 HO A1.00 G .00

3 new cases have been added.
```

For models with a constant, linear trend and no seasonality, the Holt method is fairly simple and may be applied to stationary or nonstationary series. It is applicable to short series, but it cannot handle seasonality. If the series has significant seasonal variation, the accuracy of the forecast degrades and the analyst will have to resort to a more sophisticated model.

2.4.3. THE DAMPENED TREND LINEAR EXPONENTIAL SMOOTHING MODEL

Although taking a linear trend into account represents an improvement on simple exponential smoothing, it does not deal with more complex types of trends. Neither dampened nor exponential trends are linear. A DAMPENED trend refers to a regression component for the trend in the updating equation. The updating (smoothing) equations are the same as in the linear Holt exponential smoothing model (Equations 2.19 and 2.20), except that the lagged trend coefficients, b_{t-1}, are multiplied by a dampening factor, ϕ^i. When these modifications are made, the final prediction model for a dampened trend linear exponential smoothing equation with no seasonal component (SAS Institute, 1995) follows:

$$Y_{t+h} = \mu_t + \sum_{i=0}^{h} \phi^i b_t \qquad (2.21)$$

with ϕ^i = dampening factor.
Otherwise, the model is the same.

Alternatively, the model could have an EXPONENTIAL trend, where time is an exponent of the trend parameter in the final equation : $Y_{t+h} = \mu_t b_t^t$. Many series have other variations in type of trend. It is common for series to have regular annual variation that also needs to be taken into account. For exponential smoothing to be widely applicable, it would have to be able to model this variation as well.

2.4.4. EXPONENTIAL SMOOTHING FOR SERIES WITH TREND AND SEASONALITY: WINTER'S METHODS

To accommodate both tend and seasonality, the Winters model adds a seasonal parameter to the Holt model. This is a useful addition, insofar as seasonality is commonplace with many kinds of series data. Many goods and services are more frequently produced, sold, distributed, or consumed during specific times of the year. Clearly, management, planning or budgeting that involves these goods might require forecasting that can accommodate seasonal variation in the series. This accommodation can be additive or multiplicative. In the additive model, the seasonal parameter, S_t, is merely added to the overall Holt equation to produce the additive Winters model:

$$Y_{t+h} = \mu_t + b_t t + S_{t-p+h} + e_t \qquad (2.22)$$

The subscript p is the periodicity of the seasonality, and $t = h$, is the number of periods into the forecast horizon the prediction is being made. Each of the three parameters in this model requires an updating (smoothing) equation: The updating equation for the mean is

$$\mu_t = \alpha(Y_t - S_{t-p}) + (1 - \alpha)(\mu_{t-1} + b_{t-1}) \qquad (2.23)$$

Meanwhile, the trend updating equation is given by

$$b_t = \gamma(\mu_t - \mu_{t-1}) + (1 - \gamma)b_{t-1} \tag{2.24}$$

and the seasonal updating is done by

$$S_t = \delta(Y_t - \mu_t) + (1 - \delta)S_{t-p} \tag{2.25}$$

The seasonal smoothing weight is called delta, δ. The seasonal factors, represented by S_{t-p}, are normalized so that they sum to zero in the additive Winters model. All together, these smoothing equations adjust and combine the component parts of the prediction equation from the values of the previous components (SAS Institute, 1995). By adding one more parameter to the Holt model, the Winters model additively accommodates the major components of a time series.

The multiplicative Winters model consists of a linear trend and a multiplicative seasonal parameter, δ. The general formula for this Winters model is

$$\hat{Y}_t = (\mu_t + b_t t)S_{t-p+h} + e_t \tag{2.26}$$

As with the additive version, each of the three parameters is updated with its own exponential smoothing equation. Because this is a multiplicative model, smoothing is performed by division of the seasonal component into the series. The mean is updated by the smoothed ratio of the series divided by its seasonal component at its periodic lag plus smoothed lagged linear and trend components:

$$\mu_t = \alpha \left(\frac{Y_t}{S_{t-p}} \right) + (1 - \alpha)(\mu_{t-1} + b_{t-1}) \tag{2.27}$$

The trend is smoothed the same way as in the Holt model and the additive Winters version:

$$b_t = \gamma(\mu_t - \mu_{t-1}) + (1 - \gamma)b_{t-1} \tag{2.28}$$

The seasonal smoothing follows from a portion of the ratio of the series value over the average plus a smoothed portion of the seasonality at its periodic lag. The seasonal component is normalized in the Winters models so that the seasonal factors, represented by S_t, average to 1.

$$S_t = \delta \left(\frac{Y_t}{u_t} \right) + (1 - \delta)S_{t-p} \tag{2.29}$$

2.4.4.1. PROGRAM SYNTAX AND OUTPUT INTERPRETATION

If seasonality resides or appears to reside within a series, regardless of whether a series exhibits a trend, the use of a Winters model may be appropriate. The model may be additive or multiplicative. From data on U.S. young male unemployment (ages 16–19) from 1951 through 1954, it can be seen that there are significant seasonal variations in the unadjusted data. The variable for this male unemployment is called "maleun." There is a month variable in the data set and a summer dummy variable is constructed out of that. The date variable is constructed in SAS with the intnx function. There was significantly more annual unemployment among these youths during the summer when they entered the workplace than when they were out of the workplace and in school. To handle this seasonal variation, a 6-month forecast of young male unemployment is generated with a Winters multiplicative seasonal model.

```
SAS program syntax:
/* c2pgm5.sas */
options ls=80;
title 'Young US Male Unemployment(in thousands)';
title2 '16-19 years of age, data not pre-seasonally adjusted';
title3 'Andrews & Herzberg Data p 392';
title4 'Springer-Verlag 1985 Table 65.1';

Data munemp;
 input maleun 6-8 year 43-46 month 51-52 summer 57-60;
 date = intnx('month','01jan1948'd,_n_-1); /* creation of date var*/
 Summer = 0;     /* creation of summer */
 if month > 5 and month < 9 then summer=1; /* dummy variable */
 format date monyy5.;
if year > 1951;
cards;
the data go here
proc print;
run;


symbol1 i=join v=star c=red;
proc gplot;
 plot maleun * date;
run;


proc forecast data=munemp interval=month lead=6 outactual outlimits
 out=pred outest=est outfitstats
 method=winters seasons=12 trend=2 ;
 id date;
 var maleun;

proc print data=est;
title 'Fit Statistics';
run;
```

```
symbol1 i=join c=green;
symbol2 i=join c=blue;
symbol3 i=join c=red;
symbol4 i=join c=red;
proc gplot data=pred;
 plot maleun * date = _type_ ;
title 'Forecast Plot of Young US Male Unemployment';
title2 '16-19 years of age, data not pre-seasonally adjusted';
title3 'Andrews & Herzberg Data p 392';
title4 'Springer-Verlag 1985 Table 65.1';
run;
```

The principal difference between the Winters model and the Holt model in SAS is the METHOD=WINTERS option in the forecast procedure and the SEASONS=12. The user can set the number of seasons or use terms such as QTR, MONTH, DAY, or HOUR. This model is set to accommodate a constant and a linear trend in addition to these seasons. Its R^2 value is 0.43. This forecast procedure produces the fit statistics in the output file called EST and the forecast values are called PRED in the output file. When PRED is plotted, it appears as shown in Fig. 2.5.

SPSS program syntax for the summer U.S. male unemployment series is given next.

```
SPSS program syntax:
*C2pgm6.sps.
* Same data source: Andrews and Herzberg Data Springer 1985.
```

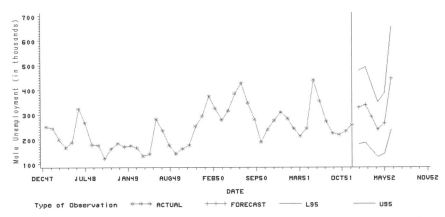

Figure 2.5 Winters forecast of young (ages 16 through 19) U.S. male unemployment. Data from Andrews and Herberg (1984), *DATA*, New York: Springer-Verlag, Table 65.1, p. 392. (SAS Graph).

```
Title 'Young US Male Unemployment'.
Subtitle 'Ages 16-19 from Data '.
Data list free /maleun year month.
Begin data.
 Data go here
End data.
List variables=all.
execute.
* Exponential Smoothing.
TSET PRINT=DEFAULT NEWVAR=ALL .
PREDICT THRU YEAR 1956 MONTH 8 .
EXSMOOTH /VARIABLES=maleun
 /SEASFACT=month
 /MODEL=WINTERS
 /ALPHA=GRID(0 1 .1)
 /DELTA=GRID(0 1 .2)
 /GAMMA=GRID(0 1 .2)
 /INITIAL=CALCULATE.
FIT ERR_1/DFE=35.
execute.
*Sequence Charts .
TSPLOT VARIABLES= maleun fit_1
 /ID= date_
 /NOLOG
 /MARK YEAR 1955 MONTH 12 .
```

This SPSS program performs a multiplicative Winters exponential smoothing on the young male unemployment data used in the SAS program, after listing out the data. Like the previous SAS program, it uses monthly seasonal components to model the seasonal variation in the series. From the optimal sum of squared errors, the model settles on an alpha updating parameter value of 0.9, a gamma trend updating value of 0.0, and a seasonal delta parameter value of 1.0, after employing a series mean of 193.2 and a trend of 192.3 as starting values.

The fit of this model is very good, as can be seen from the forecast plot. Without the ability to model the seasonality, the forecast could have a worse fit and less accuracy. The deviations of the fit from the actual data can be seen in Fig. 2.6, in which the seasonal fluctuation manifests itself in the fit of this model.

2.4.5. BASIC EVALUATION OF EXPONENTIAL SMOOTHING

Exponential smoothing has specific applications of which the analyst should be aware. These methods are easy to learn, understand, set up, and use. Because they are based on short-term moving averages, they are good

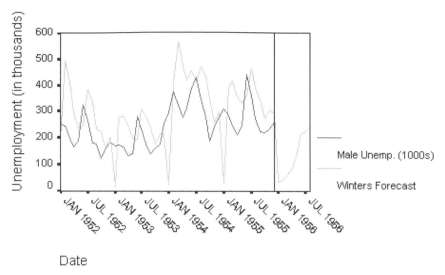

Figure 2.6 Winters forecast of young U.S. male unemployment. Data from Andrews and Herberg (1984), *DATA*, New York: Springer-Verlag, Table 65.1, p. 392. (SPSS Chart).

for short-term series. They are easy to program, especially with SAS, SPSS, and other statistical packages designed to analyze time series. They are economical in consumption of computer time and resources and they are easily run on personal computers. They are easy to monitor, evaluate, and regulate by adaptive procedures—for example, parameter selection with grid searches of sums of squared errors. They do well in competition, but they are not good at predicting all kinds of turning points (Fildes, 1987).

How do the various exponential smoothing methods fare in competition with one another? Single exponential smoothing is often better than the naive forecast based on the last observation. Single exponential smoothing generally forecasts well with deseasonalized monthly data. When yearly data are analyzed, single exponential smoothing often does not do as well as the Holt or Winters methods, where trends or seasonality may be involved. The Winters method does the best when there is trend and seasonality in the data. For short, deseasonalized, monthly series, single exponential smoothing has done better than Holt or Winters methods, or even the Box–Jenkins approach discussed in the next chapter (Makridakis *et al.*, 1984). If the series data are more plentiful and the inherent patterns more complex, then other methods may be more useful.

2.5. DECOMPOSITION METHODS

Before exploring exponential smoothing, it is helpful to examine a time series and come to an appreciation of its component parts. An examination of what these components are, how they are formulated, how they may differ from one to another, how one tests for their presence, and how one estimates their parameters is in order. When early economists sought to understand the nature of a business cycle, they began studying series in search of calendar effects (prior monthly and trading day adjustments), trends, cycles, seasonal, and irregular components. These components could be added or multiplied together to constitute the time series. The decomposition could be represented by

$$\hat{Y}a_t = \hat{P}_t \times \hat{D}_t \times \hat{T}_t \times \hat{S}_t \times \hat{C}_t \times \hat{I}_t$$
$$\text{or}$$
$$\hat{Y}m_t = \hat{P}_t + \hat{D}_t + \hat{T}_t + \hat{S}_t + \hat{C}_t + \hat{I}_t \tag{2.30}$$

where $\hat{Y}a_t$ = additively composed time series, $\hat{Y}m_t$ = multiplicatively composed time series, $\hat{P}_t$ = prior monthly adjustments, $\hat{D}_t$ = trading day adjustments, $\hat{T}_t$ = trend, $\hat{S}_t$ = seasonality, and $\hat{I}_t$ = irregularity. The multiplicative decomposition multiplies these components by one another. In the additive decomposition, component predictor variables are added to one another. This multiplicative process is the one usually used by the U.S. Bureau of the Census and is generally assumed unless the additive relationship is specifically postulated.

2.5.1. COMPONENTS OF A SERIES

Whether the process undergirding an observed series is additive or multiplicative, one needs to ascertain whether it contains a trend. Calendar effects consist of prior monthly and trading day adjustments. In the X-12 version being developed now, there will also be a leap year and moving holiday adjustment (Niemira and Klein, 1994).

The trend is a long run tendency characterizing the time series. It may be a linear increase or decrease in level over time. It may be stochastic, a result of a random process, or deterministic, a result of a prescribed mathematical function of time. If nonlinear, the trend could be fitted or modeled by a polynomial or fractional power. It might even be of a compound, logistic, or S-shaped nature. Seasonal components or signals, by contrast, are distinguishable patterns of regular annual variations in a series. These may be due to changes in the precipitation or temperature, or to legal or academic requirements such as paying taxes or taking exams. Cycles, however, are more or less regular long-range swings above or below some equilibrium level or trend line. They have upswings, peaks, downswings, and troughs. They are studied for their turning points, durations, frequencies, depths, phases, and effects on related phenomena. Fluctuation of sunspot activity take place in an 11-year cycle, for example. Business cycles are postulated recurrent patterns of prosperity, warning, recession, depression, and recovery that can extend for periods much longer than a single year, for another example (Makridakis *et al.*, 1983). What is left over after these components are extracted from the series is the irregular or error component. For the most part, these four types of change make up the basic components of a series.

2.5.2. Trends

Trends, whether deterministic or stochastic, have to be considered for extracting, fitting, and forecasting. A deterministic trend may derive from a definition that prescribes a well-defined formula for increment or decrement as a function of time, such as contractual interest. The cost of a 3-year loan may increase by agreement at a simple 2% per year. The interest on the loan by agreement is 0.02% per year. The amount of interest in effect is determined by agreement on a formula and hence deterministic.

A stochastic trend is due to random shift of level, perhaps the cumulative effect of some force that endows the series with a long-run change in level. Trends may stem from changes in society, social movements, technology, social custom, economic conditions, market conditions, or environment (Farnum and Stanton, 1989). An example of a stochastic, nonlinear historical trend is the growth after 1977 in the number of international terrorist incidents until a peak was reached in 1987, after which this number declined.

The trend is quadratic. It rises to an apex at 1987 and then declines, perhaps owing to the increased international cooperation against such persons and their cells or organizations. These tallies exclude incidents of intra-Palestinian violence. (Wilcox, 1997). Insofar as the trend represents a shift of the mean, it needs to be detected, identified, and modeled or the series may become unamenable to modeling, fitting, and forecasting. Usually the series can be detrended by decomposing it into its components of variation and extracting these signals.

Regression may be used to test and model a trend. First, one plots the series against time. If the trend appears linear, one can regress it against a measure of time. If one finds a significant and/or substantial relationship with time, the magnitude of the coefficient of time is evidence of a linear trend. Alternatively, some trends may appear to be nonlinear. For example,

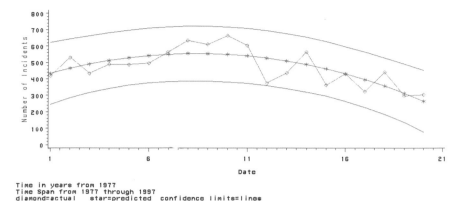

Time in years from 1977
Time Span from 1977 through 1997
diamond=actual   star=predicted   confidence limits=lines

Figure 2.7 Number of international terrorist incidents: Source: U.S. Department of State (http://www.state.gov/www/global/terrorism/1997Report/incidents.html) (SAS Graph).

one can construe a plot of the number of international terrorist incidents between 1977 and 1996 against time as a linear or a quadratic trend, depending on how time is parameterized. If time is measured by the year, there is a negative linear trend, but if trend is measured by the number of years since 1977, when the data began to be collected, then there appears to be a nonlinear trend, rising until 1987 when an apex is reached and then declining. To test the existence of a statistically significant quadratic trend, a regression model was specified with both a linear and a quadratic time component—for example, time and time squared. The dependent variable was the number of international terrorist incidents; the independent variables were a count of the number of years and a squared count of the number of years since the inception of the series. Both the linear and the squared term were found to be statistically significant predictor variables. Assuming that both linear and quadratic coefficients are significant, the higher coefficient will determine whether the trend is more quadratic than linear, or vice versa. The signs of the linear coefficients determine whether the curve is sloping downward or upward; the signs of the quadratic coefficients determine whether the function is curved upward or downward. A statistically significant quadratic trend curving downward has been found to characterize this number of international terrorist incidents over time (Wilcox, 1997). In this case the quadratic model was plotted against the year along with its upper and lower confidence intervals to see if the actual series remained bracketed by them.

When nonlinear relationships exist, one can transform them into linear ones prior to modeling by either a natural log transformation of the dependent variable or a Box–Cox transformation.The series may be detrended by regression or transformation.

   If the functional form of the trend is more complicated, the researcher designates the real data as the series of interest $C(t)$ and the functional form $Y(t)$. He may compute the sum of squared errors (SSE) as follows: $SSE = [C(t) - Y(T)]^2$. This is the unexplained sum of squares. The proportion of variance explained for the model, R^2, may be computed as follows: $R^2 = 1 - (SSE/SS\ Total)$. R^2 is the objective criterion by which the fit is tested. This is the sort of curve fitting that SPSS performs with its CURVEFIT procedure.

   A researcher may opt for all of these tests. The R^2 and significance tests for each parameter indicate which are significant. The functional form with the highest R^2 indicates the best fit and the one that should be chosen.

   If the researcher is using SAS, he may use linear regression to test each of these functional forms. In so doing, he should regress the series of interest as a dependent variable upon the time variable. The R^2 may be output as values of the output data set and printed in a list linking each variable for each model. He may compare these R^2 values to determine which fits best. The following SAS code tests the R^2 for several of these models and finally prints R^2 for each model.

```
Options ls=80 ps=55;
title 'c2pgm7.sas Functional Forms of Various Trends';
data trend;
  do time = 1 to 200;
  a = (1/1000)*time;
  b = .001*time;
  linear = a + b*time;
  square = a + b*time + time**2;
  power = a*time**b;
  compound = a*(b**time);
  inverse = a + b/time;
  Ln1 = a + b*log(time);
  growth = Exp(a + b*time);
  exponen = a*exp(b*time);
  Sshape = exp(a + b/time);
  output;
  end;
symbol1 i=join c=green;
symbol2 i=join c=red;
symbol3 i=join c=blue;
proc gplot;
  plot (square power) * time;
run;
```

```
proc gplot;
 plot (compound inverse growth exponen sshape)*time;
run;
```

Once one identifies the nature of the trend, one usually needs to detrend the series for subsequent analysis. An appropriate transformation may render the relationship linear, and one can perform a regression with the transformed dependent variable.

If the trend is stochastic, one way to detrend the series is by a difference transformation. By subtracting the value of the series one time lag before the current time period, one obtains the first difference of the series. This procedure removes a linear trend. Take the linear trend model: $Y_t = a + bt$. If one subtracts its first lag from it, the following equation is obtained: $Z_t = \nabla Y_t = Y_t - Y_{t-1} = (a - a) + [bt - b(t-1)] = b$. From this result, one can conclude that the linear (first order) trend component, t, was removed.

If the series has a higher order trend component, the first difference will not remove all of the trend. Consider a series with a quadratic trend: $Y_t = a + bt + ct^2$. By subtracting its first lag, Granger (1989) obtains its first difference, ∇Y_t, which is also designated, Z_t:

$$\begin{aligned} Z_t = \nabla Y_t = Y_t - Y_{t-1} &= (a - a) + [bt - b(t - 1)] + [(ct^2 - c(t - 1)^2] \\ &= (bt - bt + b) + [ct^2 - (ct^2 - 2ct + c)] \\ &= b + 2ct - c \end{aligned} \tag{2.31}$$

What remains is still a function of time and therefore trend. Although one has removed the quadratic trend, the linear trend remains. By taking the first difference again, one obtains the second difference and the trend disappears altogether:

$$\begin{aligned} \nabla Z_t &= b + 2ct - c - [(b + 2c(t - 1) - c)] \\ &= (b - b) + (2ct - 2ct) - (c + 2c - c) \\ &= 0 \end{aligned} \tag{2.32}$$

In other words, the second difference of the quadratic trend removes the time factor altogether:

$$\nabla^2 Y_t = 0 \tag{2.33}$$

By mathematical induction, one may infer that the power of the differencing is equal to the power of the trend. Series that can be detrended by differencing are called difference stationary series.

Other series must be detrended by regression and subtraction of the trend component from the series. These series are called trend stationary. If a series cannot be detrended by differencing, one should try regression detrending. Later, if necessary, one can difference the residuals from the regression. Once it is detrended, the series may be further analyzed.

2.5.3. SEASONALITY

When the series is characterized by a substantial regular annual variation, one must control for the seasonality as well as trend in order to forecast. Seasonality, the periodic annual changes in the series, may follow from yearly changes in weather such as temperature, humidity, or precipitation. Seasonal changes provide optimal times in the crop cycle for turning the soil, fertilizing, planting, and harvesting. Summer vacations from primary and secondary school traditionally allow children time for summer recreation. Sports equipment and clothing sales in temperate zones follow the seasons, whether for water or snow sports. Forecasting with such series requires seasonal adjustment (deseasonalization), which is discussed in more detail shortly, or seasonal variation may augment the forecast error unnecessarily.

2.5.4. CYCLES

For years economists have searched for clear cut-cycles, like those found in nature—for example, the sun-spot cycle. Economists have searched for inventory, investment, growth, building, and monetary cycles. Eventually, researchers began to look for indicators of the business cycle. They searched for leading indicators that would portend a turning point in the business cycle. Although they found a number of coincident and lagging indicators, the search for reliable leading indicators has generally been unsuccessful (Niemira and Klein, 1994). Where trend and cycle are not separated from one another, the series component is called the trend-cycle. Later in the chapter, there will be a discussion of the classical decomposition and X-11 methods of analyzing economic processes to show how the indicator can be decomposed into trend, seasonal, cyclical, and irregular components.

2.5.5. BACKGROUND

Decomposition of time series began in the 1920s with the work of Frederick R. Macaulay of the National Bureau of Economic Research on the ratio-to-moving average approach to time series decomposition. Work on decomposition was pursued by the U.S. Bureau of the Census. As each method was developed, it was given the name, "X"- hyphenated with the version number. By 1965, the U.S. Bureau of the Census proclaimed a computer-intensive calendar-based method for seasonal adjustment of original time series developed by Julius Shishkin, Allan Young, and John C. Musgrave. This method, known as X-11, became the official method for seasonal decomposition and adjustment.

The X-11 method decomposes a series into prior monthly, trading day, trend, cyclical, seasonal, and irregular variations. Prior monthly factors are calendar adjustments made for the months in which the series is under consideration. The trading day adjustment is obtained by a regression on the days of the week for the months under consideration. The seasonal component consists of the regular patterns of intrayear variations. The trend-cycle is the component of variation consisting of the long-range trend and business cycle. The irregular variations are residual effects of unusual phenomena such as strikes, political events, weather conditions, reporting errors, and sampling errors (Shishkin, et al., 1967). An estimate of the trend and cyclical factors is obtained with a moving average that extends over the seasonal period. Dividing this moving average into the original series yields seasonal irregular ratios. Each month of these ratios is smoothed over the years in the series to provide estimates of the seasonal adjustment factors. The irregular factor is filtered by the smoothing process. Dividing each month's data by these adjustment factors gives the seasonally adjusted data (Brocklebank and Dickey, 1994). The series decomposition and seasonal adjustment facilitates comparisons between sequential months or quarters as well as comparison of trends and cycles evident in these series (Ege et al., 1993). Because many of the series made publicly available by governments are now seasonally adjusted by this method, it is important to understand this method and its programming.

The decomposition may be either an additive or a multiplicative model, although the multiplicative model is most commonly applied. Although we expound the multiplicative procedure, the additive procedure merely replaces multiplication with addition and division with subtraction. In the 1980s, Estela B. Dagum (1988) and colleagues at the Time Series Research and Analysis Division of Statistics Canada began to use autoregressive integrated moving average (ARIMA) methods to extend the X-11 method. Dagum et al. (1996) applied the ARIMA procedure to X-11 to backcast starting values as well as to seasonally adjust their data. Their innovations included automatically removing trading day and Easter holiday effects before ARIMA modeling, selecting and replacing extreme values, generating improved seasonal and trend cycle weights, and forecasting more accurately over variable horizons. This chapter focuses on classical decomposition and the X-11 procedure, apart from the enhancements incorporated by Statistics Canada. The additive theory presumes that the trend, cycle, seasonality, and error components can be summed together to yield the series under consideration. The formulation of this decomposition was previously shown in Eq. (2.30). The multiplicative method of decomposition presumes that these components may be multiplied together to yield the series, previously shown in Eq. (2.30). Statistical packages usually provide the ability to model the series with either the additive or the multiplicative model. To perform decomposition, SPSS has the SEASON procedure and SAS has the X11 procedure. SEASON is a procedure for classical decomposition. Both SAS X11 and the SPSS X11ARIMA can perform the Census X-11 seasonal adjustment with ARIMA endpoint adjustment.

In connection with year 2000 (Y2K) programming problems, researchers using the latest versions can proceed with confidence. Individuals using older versions of these statistical packages have to be more careful. SAS Institute, Inc. notes that a small number of Year 2000-related problems have been reported for PROC X11 in releases 6.12 and 6.09E of SAS econometric time series software. SAS Institute, Inc. has provided maintenance for these releases which corrects these problems. Users running releases 6.12 and 6.09E should install the most recent maintenance to receive all available fixes. Users running earlier versions of SAS should upgrade to the most current release. These problems have been corrected for Version 7 and later releases of SAS software. For information on Y2K issues and fixes related to SAS software products, please see SAS Institute's web site at http://www.sas.com. Although the SPSS X11ARIMA procedure works well for series and forecasts within the 20th Century, it exhibits end-of-the-century problems and therefore the procedure has been removed altogether from version 10 of SPSS. In these respects, both SAS and SPSS have sought to free their software from end-of-the-century seasonal adjustment and forecasting complications.

2.5.6. OVERVIEW OF X-11

Shishkin *et al.* (1967) explain the decomposition process in some detail. There are basically five stages in this process: (1) trading day adjustment, (2) estimation of the variable trend cycle with a moving average procedure, (3) preparation of seasonal adjustment factors and their application to effect the seasonal adjustment, (4) a graduated treatment of extremes, and (5) the generation of component tables and summary statistics. Within stages 2 through 4, there are iterations of smoothing and adjustment.

2.5.6.1. Stage 1: Prior Factors and Trading Day Adjustment

Because different countries and months have different numbers of working or trading days, the first stage in the X-11 process adjusts the series for the number of trading days for the locale of the series under process. The monthly irregular values are regressed on a monthly data set that contains the number of days in the month in which each day occurs. The regression yields the seven daily weights. From these weights come the monthly factors, which are divided into the data to filter out trading day variation (Shiskin *et al.,* 1967). This division adjusts the data for the number of trading days. These prior factors are optionally used to preadjust the data for subsequent analysis and processing.

2.5.6.2. Stage 2: Estimation of the Trend Cycle

The trend-cycle is estimated by a moving average routine. The choice of the moving average to be used is based on a ratio of the estimate of the irregular to the estimate of the cyclical components of variation. First, this analyst obtains the ratio of a preliminary estimate of the seasonally adjusted series to the 13-term moving average of the preliminary seasonally adjusted series. The ratio is divided into high, medium, and low levels of irregularity to cyclical variation. For the higher levels of irregularity, the longer 23-term moving average is used for smoothing. For medium levels, the 13-term moving average is used, and for the smoother series, a 9-term moving average is applied. The smoother series receive the shorter moving average smoothing, although quarterly series are smoothed with a 5-term moving average. The precise weights of these moving averages are given in Shiskin *et al.* (1967).

2.5.6.3. Stage 3: Seasonal Factor Procedure

The third stage of the X-11 process entails a preliminary preparation of seasonal correction factors and the use of those factors to seasonally adjust the data. In stages 3 through 5, this method iterates through estimation of the trend cycle, seasonal, and irregular (error) components of the data. Within each iteration, the procedure smoothes the data to estimate the trend cycle component, divides the data by the trend cycle to estimate seasonal and irregular components, and uses a moving average to eliminate randomness while calculating standard deviations with which to form control limits for use in outlier identification and replacement. Each iteration also includes an estimation of the trend cycle, seasonal, and irregular components.

$$\hat{S}_t\hat{I}_t = \frac{O_t}{\hat{C}_t} \tag{2.34}$$

The hats over the terms in Eq. (2.34) designate preliminary estimates.

2.5.6.4. Stage 4: Smoothing of Outlier Distortion

Extreme values, which may arise from unusual events, are eliminated by outlier trimming. For smoothing, one takes a 3×3 moving average and use its standard deviation over a period of 5 years as a vehicle with which to determine control limits for each month. Data points residing outside these limits, usually 2.5 standard deviations away from the moving average, are deemed outliers. Outliers are replaced by a value weighted according to its closeness to the average. The farther away the observation, the less weight it is given. This has the effect of smoothing or trimming the distortion from outliers.

2.5.6.5. Stage 5 : Output of Tables, Summary Measures, and Test Statistics

The last basic stage in the process is to compute summary test statistics to confirm the seasonal adjustment. Among the test statistics computed are those from the adjoining month test, the January test, the equality test, and the percentage change test. Once the program computes these summary statistics that test for the presence or removal of seasonality, the assessment begins. For the span of interest, each year consists of a row of 12 columns. Each column contains the monthly values of the series. The bottom row contains the average monthly value of each column. In addition to the battery of test statistics mentioned, the month for cyclical dominance is also indicated. We now turn to applications of these tests and summary measures.

The first test statistic computed is the adjoining month test. One way is to look at a year by month data matrix is to examine the individual cells. For each cell in the data matrix, a ratio can be computed of the value for that month to the average of the values for the preceding and the proceding adjacent months, except when the first month is January and there is no preceding month, in which case the value of the ratio is 0. This ratio is called the adjoining month test. The adjoining month test may be used to assess residual seasonality. Nonseasonal data do not manifest much variation in the adjoining month tests, whereas seasonal data do. The adjacent month test statistics indicate whether the seasonality has been successfully removed from the series.

The January test statistic helps in evaluating the adjusted series as a standard for comparison. If the seasonally adjusted series data values are divided by the value for January of that year, and that fraction is multiplied by 100, then the series are all standardized on the same annual starting value of 100. This percentage of the January value provides a criterion for evaluating month-to-month variation in the series for residual seasonality. If there is not much fluctuation, then seasonality is no longer inherent in the series.

The equality test helps determine whether the data were properly adjusted. Dividing the 12-month moving average of the seasonally adjusted data by the 12-month moving average of the raw data yields a fraction. When one multiplies this fraction by 100, one obtains a set of ratios standardized on a base of 100. When these ratios are very close to 100, then there is negligible overadjustment. If the equality test ratios are below 90 or above 110, then there may have been seasonal overadjustment.

Among the tests that are useful in comparing the seasonally adjusted with the raw data is the percentage change test. Percentage change tests show the percentage change from one month to the next. When applied to the values in the original data, the percentage change tests provide a basis against which to compare percentage change tests of seasonally adjusted data, random components, or trend cycle components. The differences will reveal the amount of seasonality that was adjusted out of the original data.

Another measure of relative variation is the month of cyclical dominance (MCD). The ratio of the average percentage change (for all of the months) of the error (irregular) component to that of the trend cycle helps determine the month of cyclical dominance. The span of this average can extend from one to multiple months. As the number of months included in this average increases, the percentage change of error gets reduced and the percentage change in trend cycle increases. Eventually, they converge to the same level. If the span of the average is extended farther, the ratio declines to less than 1. The span in which this ratio dips below 1 is called the month of cyclical dominance. If this month of cyclical dominance is 3, then a moving average of 3 months of the seasonally adjusted data should capture the trend cycle of the series. Such an average can then be used as a basis for estimating the trend cycle of the series and forecasting future observations (Makridakis *et al.*, 1983). In these ways, the summary statistical tests can be used to evaluate the extent of deseasonalization.

2.5.6.6. X11-ARIMA

One of the problems with X-11 is that the weighting for the series is symmetric. The endpoints are asymmetrically produced, resulting in poor predictions under some circumstances. To remedy this problem, Dagum *et al.* (1996) explained how Statistics Canada used the ARIMA procedure to identify the X-11 extreme values, replace them, and develop the weights for the trend cycle and seasonal components. These innovations substantially improved the forecasting capability (Dagum, 1988). The application of the Box–Jenkins ARIMA option to X-11 decomposition is known as X11AR-IMA/88. Both SAS and SPSS X-11 programs have options for applying the ARIMA technique for such purposes, although SPSS is removing this procedure from Version 10 to assure users of Y2K compliance. The theoretical details of the ARIMA modeling procedure are discussed in the chapters immediately following. For now, it is enough to note that this is an improved method of forecasting endpoints and backcasting initial values of the model.

2.5.6.7. Statistical Packages: SAS and SPSS versions of X-11

Census X-11 can be programmed with many popular statistical packages, including SAS and SPSS. Often, it is useful to try to decompose a series prior to further analysis. Both SAS and SPSS can perform simple additive or multiplicative decomposition of the time series. SAS Proc X11 can be used to either additively or multiplicatively decompose a series into the trading day, trend cycle, seasonal, and irregular components. It can generate tables at almost every intermediate step and can seasonally adjust monthly or quarterly series that have no missing data. The program outputs tables beginning with the letters A through C containing the results of intermediate calculations in the X-11 process. Tables beginning with the letter D contain the final estimates of the series components. Table D10 contains the final seasonal factors. Table D11 contains the seasonal adjustment of the series. Table D12 contains the estimate of the trend cycle, and table D13 contains the final irregular component (Ege *et al.,* 1993). SAS can also produce these tables and/or their data in an output data set for use in subsequent analysis.

SAS Syntax for X-11

The SAS syntax for the X11 procedure can be exemplified by the price of gas per 100 therms in U.S. cities.

```
options ls=80 ps=55;
title ' C2pgm8.sas Average Price of US Utilities';
title2 'Bureau of Labor Statistics Data';
title3 'Data extracted on: August 04, 1996 (02:13 AM)';
data gas;
title4 'Gas price per 100 therms';
 input year month $ cost1;
label cost='$ Price per 100 therms';
date=intnx('month','01nov1978'd,_n_-1);
Seriesid='apu000072611';
format date monyy5.;
```

```
cost = cost1;
if month='M09' and year= 1985 then cost = 61.57;
time + 1;
cards;
 1978 M11 27.667
 ....................
 1996 M06 65.261
proc print;
run;
symbol1 i=join v=star c=green;
proc gplot ;
 plot cost1 * date;
run;
proc x11 ;
 monthly date=date;
 var cost;
 tables b1 d10 d11 d12 d13 f2 ;
 output out=out2 b1=cost d10=d10 d11=adjusted d12=d12 d13=d13 f1=f1;
run;
symbol1 i=join v=star c=red;
symbol2 i=join c=green;
legend label=none value=('original' 'adjusted');
proc gplot data=out2;
 plot cost * date=1 adjusted*date=2/overlay legend;
 plot cost * date;
run;
proc gplot data=out2;
 plot d10 * date;
title4 'Final Seasonal Factors';
run;
proc gplot data=out2;
 plot d12 * date;
title4 'Final Trend Cycle';
run;
proc gplot data=out2;
 plot d13 * date;
title4 'Final Irregular Series';
run;
proc gplot data=out2;
 plot adjusted * date;
title4 'Final Seasonally adjusted Series';
run;
```

In this example, PROC X11 gathers its data from the file, called "gas-dat." These data are the average U.S. city price in dollars per 100 therms of natural gas, an indicator of utility costs in the United States. These data are monthly and the variable containing the date is called, "date." The series with missing data, "cost1," has its missing data replaced and the series with the estimated and replaced missing data is called "cost." It is "cost" that is subjected to X-11 seasonal adjustment in this example. For simplification, only specific tables are requested: Tables B1 and D10 through D13, produced along with the month of cyclical domination in F1. Then the data are plotted against the original series and the adjusted series overlaid. The final seasonal factors and other tables are then graphed in the accompanying figures.

The first graph, in Fig. 2.8, shows the original series from Table B1 and prior factors plotted along with the final seasonally adjusted series from Table D11. The cost of gas increases until September 1983, and then begins to decline. The gradual decline continues until September 1992.

The difference between the original and the seasonally adjusted series is that a lot of the sharp peaks and valleys in the variation are attenuated. The series may be decomposed to reveal the trend cycle, from Table D12, shown in Fig. 2.9.

Decomposition of the original series also yields the final seasonal factors, from Table D10, shown in Fig, 2.10.

Also, the final irregular component of the original series may be found in Table D13, shown in Fig. 2.11.

In general, the series may be decomposed into its component parts and seasonally adjusted to smooth out the jagged edges. Not only are the seasonally adjusted data available for plotting, as shown in Figure 2.12, they are available in tabulations as shown in Tables 2.3 and 2.4 as well.

SPSS Syntax for Decomposition

SPSS also performs decomposition and seasonal adjustment. SEASON and X11ARIMA estimate seasonal factors prior to additive, multiplicative,

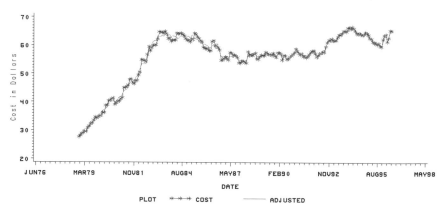

Figure 2.8 Decomposition of Average U.S. City Price of Natural Gas series into Original and Seasonally Adjusted Series (SAS Graph of PROC X11 Output). (Source: U.S. Bureau of Labor Statistics Data).

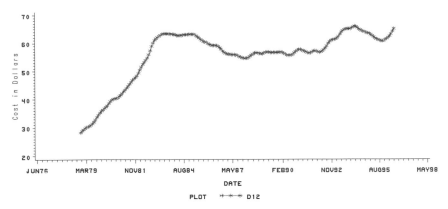

Figure 2.9 D12: Final trend-cycle of U.S. City Average Prices of Natural Gas (dollars/100 therms). A SAS Graph of PROC X11 Output. (Source: U.S. Bureau of Labor Statistics Data).

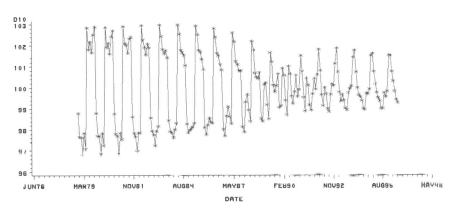

Figure 2.10 D10: Final Seasonal Factors: U.S. Average City Price of natural gas (SAS Graph of PROC X11 Output). (Source: U.S. Bureau of Labor Statistics Data).

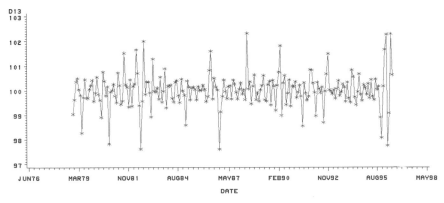

Figure 2.11 D13: Irregular Component: U.S. Average City Price of natural gas (SAS Graph of PROC X11 Output). (Source: U.S. Bureau of Labor Statistics Data).

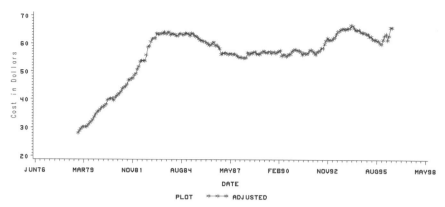

Figure 2.12   Seasonally Adjusted Series of U.S. Average City Price of natural gas (SAS Graph of PROC X11 Output). (Source: U.S. Bureau of Labor Statistics Data).

Table 2.3

Seasonally Adjusted Average U.S. City Cost of $ Price per Therm

X11 Procedure
Seasonal Adjustment of—COST $ Price per 100 Therms
D13 Final Irregular Series

Year	Jan	Feb	Mar	Apr	May	Jun
1978	.	.	.	.	.	.
1979	100.366	100.502	100.052	99.802	98.270	99.715
1980	99.574	99.905	100.556	99.858	99.601	98.913
1981	100.016	100.272	99.781	99.512	100.735	100.220
1982	100.442	99.375	100.183	100.295	101.650	100.707
1983	100.357	99.918	98.931	101.278	99.991	99.946
1984	99.305	100.187	100.190	100.228	99.696	99.536
1985	99.814	98.590	100.388	100.141	99.625	100.107
1986	100.220	100.042	99.555	99.761	100.814	101.585
1987	99.774	100.432	99.957	99.676	100.163	100.200
1988	100.312	99.866	100.031	99.636	102.293	100.075
1989	99.578	100.015	100.219	100.608	99.620	99.578
1990	100.217	100.753	101.800	98.992	100.304	100.599
1991	100.357	99.716	99.950	100.212	99.778	98.556
1992	100.298	99.913	98.969	100.319	100.075	99.902
1993	99.959	99.856	100.026	99.685	100.069	100.390
1994	99.695	99.513	100.817	100.528	99.726	99.405
1995	99.840	100.264	99.690	100.420	99.764	99.620
1996	101.634	102.238	97.759	99.066	102.267	100.616
S.D.	0.505	0.705	0.819	0.545	1.005	0.677

Total: 21201   Mean: 100.01   S.D.:   0.69879

Table 2.4

Seasonally Adjusted Average U.S. City Cost of $ Price per Therm

Year	Jul	Aug	Sep	Oct	Nov	Dec	Total
			X11 Procedure				
		Seasonal Adjustment of—COST $ Price per 100 Therms					
		D11 Final Seasonally Adjusted Series					
1978	.	.	.	.	28.008	28.758	56.767
1979	31.635	32.018	32.774	33.748	34.630	35.450	381.507
1980	39.580	40.006	40.064	40.310	39.475	40.590	462.222
1981	44.410	45.177	46.772	46.845	47.299	47.395	532.832
1982	53.565	53.440	55.653	58.479	58.807	60.403	646.967
1983	63.437	63.205	63.822	63.163	63.379	63.871	754.831
1984	63.066	63.236	62.933	62.785	63.410	63.194	755.805
1985	61.903	61.401	60.755	60.775	60.469	60.248	742.567
1986	58.955	59.183	58.560	57.907	55.886	56.244	704.570
1987	55.797	56.101	55.769	55.293	54.875	54.878	669.546
1988	56.390	56.221	56.762	56.940	56.161	56.137	669.761
1989	56.847	56.975	56.449	56.954	56.814	56.354	680.313
1990	55.363	55.706	56.255	56.161	57.190	57.618	677.342
1991	56.775	56.423	56.354	56.709	57.639	57.749	685.159
1992	58.380	58.318	59.925	61.059	61.990	61.285	703.012
1993	64.495	64.964	65.344	65.320	64.919	65.467	765.126
1994	64.751	64.993	64.266	63.863	64.050	63.785	779.843
1995	61.603	61.110	61.004	60.178	59.825	61.449	740.125
1996	.	.	.	.	.	.	381.833
Avg	55.703	55.793	56.086	56.264	54.713	55.049	
Total:	11790	Mean: 55.614	S.D.: 9.3479				

or natural log additive series decomposition into seasonal, trend, cycle, and irregular components.

For simple decomposition, users may prefer the less complicated SEASON. The SPSS program C2pgm9.sps, up to the point of the "list variables" command, is inserted into a syntax window and run. At this juncture, a date variable is created with the "define dates" option in the data window of the menu. Years and months are selected and the starting year and month are inserted into the data set. This defines the periodicity of the data set as monthly, a definition that is necessary for the remainder of the program to be run. Then the remainder of the program is run from the syntax window.

```
* SEASON program c2pgm9.sps.
*Seasonal Decomposition.
title ' Average Price of US Utilities'.
subtitle 'Bureau of Labor Statistics Data'.
```

```
* Data extracted on: August 04, 1996 (02:13 AM)'.
* Gas price per 100 therms'.
data list / year 2-5 month 8-9 cost1 15-20.
compute cost = cost1.
if (month=9 & year= 1985) cost = 61.57.
var labels cost='$ Price per 100 therms'.
* Seriesid='apu000072611'.
begin data.
 1978 M11 27.667
 Data go here
 1996 M06 65.261
end data.
list variables=all.
*at this point monthly date variables are constructed in sav file.
* Seasonal Decomposition.
TSET PRINT=BRIEF NEWVAR=ALL .
SEASON
 /VARIABLES=cost
 /MODEL=MULTIPLICATIVE
 /MA=CENTERED.
*Sequence Charts .
title 'Trend-cycle'.
TSPLOT VARIABLES= stc_1
 /ID= date_
 /NOLO
 /FORMAT NOFILL NOREFERENCE.
*Sequence Charts .
title 'Seasonal Factors'.
TSPLOT VARIABLES= saf_1
 /ID= date_
 /NOLO
 /FORMAT NOFILL NOREFERENCE.
*Sequence Charts .
title 'Irregular Series.
TSPLOT VARIABLES= err_1
 /ID= date_
 /NOLO
 /FORMAT NOFILL NOREFERENCE.
title 'Seasonally Adjusted Series.
TSPLOT VARIABLES= sas_1
 /ID= date_
 /NOLO
 /FORMAT NOFILL NOREFERENCE.
```

In this classical seasonal decomposition program, the same cost of gas series featured in the SAS program above is decomposed into a multiplicative model. The program decomposes the series into components for each of which it creates a new variable in the data set. It automatically decomposes the series into constructed a trend cycle component, stc_1, a seasonally adjusted factors series, saf_1, an irregular component, err_1, and the final seasonally adjusted series, sas_1. These generated series can be plotted against the date variable with commands at the end of the SPSS program.

2.5.7.3. SPSS X11ARIMA

Although SPSS has provided the interface, SPSS based this program on an early version of the computer code of the Statistics Canada version of the United States Bureau of the Census X-11 (SPSS Trends 6.1, 1994; Mathesson, 1994). With X11ARIMA, the user may opt for seasonal adjustment of the series by these factors after requesting the summary statistics on the seasonally adjusted series. It is possible to forecast or backcast, with a custom design of the ARIMA model, the endpoints of the series prior to seasonal adjustment to improve forecasting with the series. Other adjustments can be made to improve the model as well. The researcher has alternatives of trading day regression adjustments, moving average adjustments, replacement of extreme values, and establishment of control limits. In short, the SPSS X11ARIMA program is a good program for performing Census X-11 seasonal adjustment on 20th Century series (Monsell, 1999).

The output of the SPSS X11ARIMA program consists of 3 to 76 tables. The user has some control over which tables he would like. He may specify the number of digits to the right of the decimal place in a format statement. Four new variables are generated by the program. The seasonally adjusted series is saved with the variable name, sas_1. The effect of seasonal adjustment is to smooth frequent seasonal peaks and troughs of the seasonality while preserving the longer cyclical and trend effects. Seasonal adjustment factors are produced with the variable name of saf_1. The trend cycle smoothing of the original series is called stc_1. The error or irregular component of the series is saved as err_1. These four variables are saved at the end of the SPSS system file being used.

SPSS X11ARIMA, however, has some limitations. The periodicity of the X11ARIMA model is found from the SPSS Date_ variable. This trend cycle is based on a periodicity of 12 if the series is monthly and a periodicity of 4 if the series is quarterly. Either monthly or quarterly periodicity is required for this X11ARIMA procedure to be used. Extreme values are included in estimation of the trend cycle component. Estimation of the seasonal factors begins with a 3 × 3 moving average and finishes with a 3 ;ti; 5 moving average. Trend cycle smoothing is performed with a multiple higher order Henderson moving average. Although the program is unable to estimate series with missing data, SPSS, with the new missing values replacement module, possesses a variety of ways to estimate and replace missing values. Moreover, SPSS cannot perform X11ARIMA with dates prior to 1900, and each series must contain at least 3 years of data. For ARIMA extrapolation, 5 years of data is a minimum series length (SPSS Trends 6.1, 1994). When the series has more than 15 years of observations, backcasts are not generated by the ARIMA subcommand. To allow the backcasts to be generated, the series length must be limited. Finally, seasonal adjustment and 1 year ahead forecasts cannot extend into the 21st Century.

Although a full explanation of the ARIMA estimation backcasting and forecasting is at this point beyond the scope of the book, an example of the SPSS syntax for this procedure appears as follows:

```
* SPSS X11ARIMA c2pgm10.sps.


X11ARIMA
  /VARIABLES cost
  /MODEL=MULTIPLICATIVE
  /NOPRVARS
  /ARIMA=EXTREMES(REPLACE) BACKCAST
  /YEARTOTAL
  /NOEXTREMETC
  /MONTHLEN=TRADEDAYS
  /PERIOD=MONTHLY
  /TRADEDAYREG TDSIGMA(2.5) ADJUST
    COMPUTE(1978)
    BEGIN(1978)
  /DAYWGTS=MON(1.4) TUES(1.4) WED(1.4) THUR(1.4)
FRI(1.4)
  /NOUSERMODEL
  /LOWSIGMA=1.5 /HISIGMA=2.5
  /SAVE=PRED SAF STC RESID
  /MACURVES SEASFACT(Hybrid) TRENDCYC(12)
HENDERSON(SELECT)
  /PRINT STANDARD
  /PLOT ALL .
```

The explanation of the X11ARIMA syntax is involved. The model uses the cost variable. It can generate multiplicative, additive, or natural log additive models, although this example is a multiplicative model. The NOPRVARS subcommand specifies that no variable is used for prior adjustment of the series. The ARIMA subcommand means that the Box–Jenkins method, including extreme values in this case, will be used for backcasting starting values. It may also be used for forecasting. A discussion of backcasting and forecasting techniques follows in later chapters. The NOYEARTOTAL subcommand indicates that the calendar year totals are not preserved. NOEXTREMETC means that the extremes are not modified in the trend cycle. The MONTHLEN=TRADEDAYS option means that the month length variation is included in the trading day factors rather than in the seasonal factors. Trading-day regression estimates of the series beginning in 1978 are computed and form the control limits beyond which extreme values are replaced. The trading days are equally weighted in the estimation process. There is no custom-designed ARIMA model applied for this purpose. The Henderson weighted moving average is used for smoothing the trend cycle and the standard tables and plots are generated. The variables created may be used for decomposition or forecasting.

2.5.8. COMPARISON OF EXPONENTIAL SMOOTHING AND X-11 DECOMPOSITION

Both exponential smoothing and decomposition methods described here are forms of univariate modeling and decomposition. They deal with single series, their decomposition, and component extraction. The moving average and exponential smoothing methods are simple methods utilized for inventory control and short-term forecasting. Where trends are inherent in the data, regression trend fitting, Holt–Winters exponential smoothing, and decomposition methods prove useful. If the series are short, the moving average and exponential smoothing methods are often useful. When the forecast horizon is one step ahead or just a few steps ahead, they prove very efficient. For longer series and longer run forecasts, regression trend fitting or decomposition of the series into a trend cycle may be more useful. Where seasonality inheres within the series, Winters and decomposition methods may be useful with short forecasting horizons. Kenny and Durbin (1982) evaluated various forecasting techniques and recommend application of X-11 to a series augmented with a 12-month forecast in order to obtain satisfactory results. In Chapters 7 and 10 of this book, methods of combining techniques to improve forecasts will be discussed.

2.6. NEW FEATURES OF CENSUS X-12

The U.S. Census has developed its X-12 program, which contains some innovations over the earlier X-11 and the 1988 update, X11ARIMA, developed by E. B. Dagum *et al.* Dagum had introduced X11ARIMA to use back- and forecasting to reduce bias at the ends of the series. The new X-12 program contains more "systematic and focused diagnostics for assessing the quality of seasonal adjustments." X-12 has a wide variety of filters from which to choose in order to extract trend and seasonal patterns, plus a set of asymmetric filters to be used for the ends of the series. Some of the diagnostics assess the stability of the extracted components of the series. Optional power transformations permit optimal modeling of the series. X-12 contains a linear regression with ARIMA errors (REGARIMA) that forecasts, backcasts, and preadjusts for sundry effects. Such a procedure is discussed briefly in Chapter 11. The corrected Akaike Information Criterion is used to detect the existence of trading-day effects.

Corrected Akaike Information Criterion

$$= -2LnL + 2m \left(\frac{N}{(N - m - 1)} \right), \quad (2.35)$$

where

L = estimated likelihood function,
N = sample size, and
m = number of estimated paramters.

This REGARIMA can partial out the effects of explanatory variables prior to decomposition, as well as better test for seasonal patterns and sundry calendar effects, including trading-day, moving-holiday, and leap-year effects. In this way, it can partial out user-defined effects and thereby eliminate corruption from such sources of bias (Findley *et al.*, 1998; Makridakis *et al.*, 1997). REGARIMA provides for enhanced detection of and protection from additive outliers and level shifts (including transient ramps). Moreover, the X-12 program incorporates an option for automatic model selection based on the best corrected AIC (Findley *et al.*, 1998). X-12 may soon become the institutional standard deseasonalization for series data and find its way into the SAS and SPSS statistical packages.

REFERENCES

Andrews, D. F., and Herzberg, A. M. (1985). *Data*. New York: Springer-Verlag, p. 392. Data are used and printed with permission from Springer-Verlag.

Bressler, L. E., Cohen, B. L., Ginn, J. M., Lopes, J., Meek, G. R., and Weeks, H. (1991). *SAS/ETS Software Applications Guide 1: Time Series Modeling and Forecasting, Financial Reporting, and Loan Analysis.* Cary, NC: SAS Institute, Inc.

Brocklebank, J., and Dickey, D. (1994). *Forecasting Techniques Using SAS/ETS Software: Course Notes.* Cary, NC: SAS Institute, Inc., p. 561.

Brown, R.G. (1963). *Smoothing, Forecasting, and Prediction.* Englewood Clifs, NJ: Prentice Hall, pp. 98–102.

Dagum, E. B. (1988). *The X11ARIMA/88 Seasonal Adjustment Method: Foundations and User's Manual.* Ottawa, Ca.: Time Series Research and Analysis Division, Statistics Canada, pp. 1–3.

Dagum, E. B., Chab, N., & Chiu, K. (1996). Derivation and Properties of the X11AIMA and Census II Linear Filters. *Journal of Official Statistics.* **12**, (4), Statistics Sweden, pp. 329–348.

Ege, G. *et al.* (1993). *SAS ETS/User's Guide*, 2nd ed. Cary, NC: SAS Institute, Inc., pp. 918–939.

Farnum, N. R., and Stanton, L. W. (1989). *Quantitative Forecasting Methods.* Boston: PWS-Kent Publishing, Co., p. 151.

Fildes, R. (1987). Forecasting: The Issues. In *The Handbook of Forecasting: A Manager's Guide*, 2nd ed. (Makridakis, S., and Wheelwright, S. C. Eds.), pp. 150–172. New York: Wiley.

Findley, D. F., Monsell, B.C., Bell, W. R., Otto, M. C. , and Chen, B-C. (1998). "New Capabilities and Methods of the X-12-ARIMA Seasonal Adjustment Program." In *Journal of Business and Economic Statistics*, **16**(2), pp. 127–152.

Gardiner, E. S., Jr. (1987). Smoothing Methods for Short Term Planning and Control. In *The Handbook of Forecasting: A Manager's Guide*, 2nd ed. (Makridakis, S., and Wheelwright, S. C., Eds.), pp. 173–195. New York: Wiley.

Granger, C. W. J. (1989). *Forecasting in Business and Economics,* 2nd ed. San Diego: Academic Press, pp. 39–40.

Holden, K., Peel, D. A., and Thompson, J. L. (1990). *Economic Forecasting: An introduction.* New York. Cambridge University Press, pp.1–16.

Kenny, P. B., and Durbin, J. (1982). Local Trend Estimation and Seasonal Adjustment of Economic and Social Time Series. *Journal of the Royal Statistical Society* **145**(1), pp.1–41.

Makridakis, S., Wheelwright, s., and McGee, V. (1983). *Forecasting: Methods and Applications* New York: Wiley, pp. 44–47, 77–84, 147–178, 626–629.

Makridakis, S. *et al.* (1984). The Accuracy of Extrapolation (Time Series) Methods: Results of a Forecasting Competition. *In The Forecasting Accuracy of Major Time Series Methods* (Makridakis, S., Andersen, A., Carbone, R., Fildes, R., Hibon, M., Lewandowski, R., Newton, H., Parzen, E., and Winkler, R. (Eds), p. 127. New York: Wiley.

Makridakis, S., Wheelright, S., and Hyndman, R. J. (1997). *Forecasting: Methods and Applications,* 3rd ed. New York: Wiley, pp. 373, 520–542, 553 574.

Mathesson, D. (1997). SPSS, Inc. Technical Support, personal communication, Aug. 1, 12997.

Monsell, B. (1999). Statistical Research Div., U.S. Bureau of the Census, U.S. Department of Commerce through notes that early versions of X11ARIMA are not Y2K compliant. Later versions are Y2K compliant and therefore users are advised to upgrade to versions X-11.2, X-11Q2, or X-12-ARIMA. Personal communication, Sept. 13, 1999.

Niemira, M. P., and Klein, P. A. (1994). *Forecasting Financial and Economic Cycles.* New York: Wiley, pp. 115–139.

SAS Institute, Inc. (1995). *SAS/ETS Software: Time Series Forecasting System.* Cary, NC: SAS Institute, Inc., pp. 227–235.

Shiskin, J., Young, A. H., and Musgrave, J. C. (1967). The X-11 Variant of the Census Method II Seasonal Adjustment Program. Technical Paper No. 15, Washington, DC : Economic Research and Analysis Division, U.S. Bureau of the Census, U.S. Department of Commerce. pp. 1–17,62–63.

SPSS, Inc. (1994). *SPSS Trends 6.1*. Chicago: SPSS, Inc.

SPSS, Inc. (1997). *SPSS 7.0 Algorithms*. Chicago: SPSS, Inc.

Wilcox, P. C., Jr. (1997). International Terrorist Incidents over Time (1977–1996).?Office of the Coordinator for Counter-terrorism, U.S. Department of State. Retrieved July 8, 1999 from the World Wide Web: http://www.state.gov/www/images/chart70.gif.

Chapter 3

Introduction to Box–Jenkins Time Series Analysis

3.1. INTRODUCTION

In 1972 George E. P. Box and Gwilym M. Jenkins developed a method for analyzing stationary univariate time series data. In this chapter, the importance and general nature of the ARIMA approach to time series analysis are discussed. The novel contributions of this method and limita tions are explained. Prerequisites of Box–Jenkins models are defined and explored. Different types of nonstationarity are elaborated. We also discuss tests for detecting these forms of nonstationarity and expound on transformations to stationarity. We then review problems following from the failure to fulfill these prerequisites, as well as common means by which these problems may be resolved. Programming examples with both SAS and SPSS are included. This new approach to modeling time series is introduced.

3.2. THE IMPORTANCE OF TIME SERIES ANALYSIS MODELING

The smoothing methods were methods of extrapolation based on moving average, and weighted moving averages, with adjustments for trend and seasonality. Decomposition methods utilized these techniques to break down series into trend, cycle, season, and irregular components and to deseasonalize series in preparation for forecasting. The Box–Jenkins (AR-IMA) method differences the series to stationarity and then combines the moving average with autoregressive parameters to yield a comprehensive model amenable to forecasting. By synthesizing previously known methods, Box and Jenkins have endowed modeling capability with greater flexibility and power. The model developed serves not only to explain the underlying process generating the series, but as a basis for forecasting. Introducing exogenous inputs of a deterministic or stochastic nature allows analysis of the impulse responses of discrete endogenous response series. In fact, these processes may be used to study engineering feedback and control systems (Box *et al.*, 1994).

3.3. LIMITATIONS

There are a few limitations to the Box–Jenkins models. If there are not enough data, they may be no better at forecasting than the decomposition or exponential smoothing techniques. Box–Jenkins models usually are based on stochastic rather than deterministic or axiomatic processes. Much depends on the proper temporal focus. These models are better at formulating incremental rather than structural change (McCleary *et al.*, 1980). They presume weak stationarity, equal-spaced intervals of observations, and at least 30 to 50 observations. Most authors recommend at least 50 observations, but Monnie McGee examines this matter more closely in the last chapter of this text and shows that the recommended number of observations will be found to depend on other factors not yet fully addressed. If these assumptions are fulfilled, the Box–Jenkins methodology may provide good forecasting from univariate series.

3.4. ASSUMPTIONS

The Box–Jenkins method requires that the discrete time series data be equally spaced over time and that there be no missing values in the series. It has been noted that "... the series may be modeled as a probabilistic function of past inputs, random shocks, and outputs" (McCleary *et al.*, 1980). The choice of temporal interval is important. If the data vary every month but are gathered only once a year, then the monthly or seasonal changes will be lost in the data collection. Conversely, if the data are gathered every month, but the changes take place every 11 years, then the analyst may not see the long-run changes unless enough data are gathered.

Therefore, temporal resolution should be designed to focus on the subject or object of analysis as it changes over time: The rate of observation must be synchronized with the rate of change for proper study of the subject matter. The data are usually serially dependent; adjacent values of the series usually exhibit dependence. When the data are deterministic, the past completely determines the current or future (Gottman, 1981). Unless the series is deterministic, it is assumed to be a stochastic realization of an underlying data-generating process. The series must be long enough to provide power for testing the parameters for significance, thereby permitting accurate parameter estimation. Although conventional wisdom maintains that the series should be about 50 observations in length, series length remains a subject of controversy (Box and Jenkins, 1976; McCleary et al., 1980). If the series contains seasonal components, its length must span a sufficient number of seasons for modeling purposes. If maximum likelihood estimation is used, then the series may have to be as long as 100 observations. Sometimes series are so long that they may experience a change of definition. For example, AIDS data from the Centers for Disease Control (CDC) underwent several changes of the definition of AIDS. The characteristics of the series under one definition may well be different from those under another definition. The counts may experience regime shifts, and reference lines identifying the changes in definition should be entered in the time sequence graphs for careful delineation and examination of the regimes. Technically speaking, each segment should have enough observations for correct modeling. There is no substitute for understanding the theory and controversies surrounding the inclusion/exclusion criteria and means of data collection for the series under consideration.

The series also needs to be stationary in the second or weak sense. As was noted in Chapter 1, the series must be stationary in mean, variance, and autocovariance. The mean, variance, and autocovariance structure between the same number of time lags should be constant. The reason for this requirement is to render the general mechanism of the time series process more or less time-invariant (Granger and Newbold, 1986). Nonstationary series have permanent memories, whereas stationary series have temporary memories. Nonstationary series have gradually diminishing autocorrelations that can be a function of time, whereas stationary series have stable but rapidly diminishing autocorrelations. Nonstationary series have unbounded variances that grow as a function of time, whereas stationary series have finite variances that are bounded. Stationary processes possess important properties of convergence as the sample size increases. The sample mean converges to the true mean. The variance estimates converges to the true variance of the process. These limiting properties often do not exist in nonstationary processes (Banerjee et al., 1993). The lack of finite bounded variances can inflate forecast errors. Nonstationary series that are time dependent may have spurious correlation with one another, confounding and compounding problems with multivariate time series modeling. All of these problems plague proper modeling of the data-generating process, for which reason weak or covariance stationarity is required for this kind of time series modeling. If the series does not fulfill these requirements, then the data require preprocessing prior to analysis.

Missing values may be replaced by several algorithms. If some are missing, then they should be replaced by a theoretically defensible algorithm for missing data replacement. A crude missing data replacement method is to plug in the mean for the overall series. A less crude algorithm is to use the mean of the period of the series in which the observation is missing. Another algorithm is to take the mean of the adjacent observations. Another technique may be to take the median of nearby points. Linear interpolation may be employed to impute a missing value, as can linear trend at a point. In the Windows version of SPSS, a "syntax window" may be opened and any one of the following missing value replacement commands may be inserted, preparatory to execution of the command.

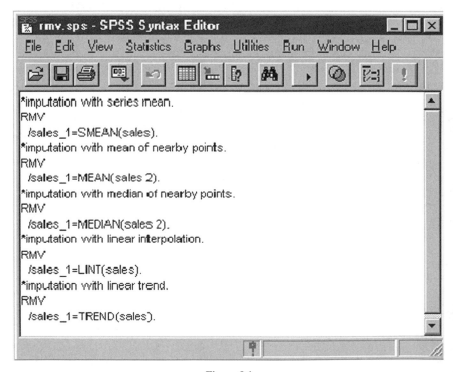

Figure 3.1

Selecting the command and running it will construct a new series, called
sales_1, which has the missing value replaced. This series may then be used
in the analysis.

With SAS, PROC EXPAND may be used to interpolate the missing
values. Interpolation may be performed by fitting a continuous curve joining
line segments. The SAS syntax for this procedure begins with an options
statement.

```
options ls=80 ps=60; /* sets column width and page length            */
data expnd;
 input Y;               /* inputs variable Y                          */
date = intnx('year','01jan1900'd, _n_-1);

                       /* ********************************************* */
                       /* INTNX function creates new variable         */
                       /* which is named DATE in form of year,        */
                       /* starting date, and increments of one        */
                       /* year for each observation in data set       */
                       /* ********************************************* */

format date year4.;    /* format for date is 19??                     */
cards;                 /* the data follow the cards statement         */
24
25
.
.
.
29
30
proc expand data=expnd out=new from = year method=join ;
 convert Y = Ynew/observed = middle;
 id date;
title 'Interpolated data observed=middle';
title2 'Method = Join';
proc print data=new; var date Y Ynew;
run;
proc expand data=expnd out=new2 from = year method=join;
 convert Y = Ynew/observed = average;
id date;
title 'Interpolated data observed = average';
title2 'Method=Join';
run;
proc print data=new2; var date Y Ynew;
run;
```

In this program, the PROC EXPAND utilizes the join algorithm to interpolate a middle and an average value for missing value in two different output
data sets. A date variable is constructed from 1900 through 1906 with the
INTNX function. This data variable is formatted to produce a yearly value
with the format date year4. command. The variable name of the series
under consideration is called Y. The data set is called expnd, and the proc
expand output data sets constructed by the middle and average values are

called new and new2, respectively. The raw data for the variable name is called Y while the interpolated series is called Ynew. The results of this interpolation are displayed in the SAS output:

```
Interpolated data observed=middle
   Method=join

OBS DATE Y YNEW

1 1900 24 24
2 1901 25 25
3 1902 26 26
4 1903 27 27
5 1904 . 28
6 1905 29 29
7 1906 30 30

 Interpolated data observed = average
Method=Join

OBS DATE Y YNEW

1 1900 24 24.0000
2 1901 25 25.0000
3 1902 26 26.0000
4 1903 27 27.0000
5 1904 . 28.0000
6 1905 29 29.0000
7 1906 30 30.0000
```

Box–Jenkins time series analysis requires complete time series. If the series has outliers, these outliers may follow from aberrations in the series. The researcher may consider them missing values and use the missing-value replacement process just described to replace them. In this way, he can prepare a complete time series, with equally spaced temporal intervals, prior to Box–Jenkins analysis.

3.5. TIME SERIES

3.5.1. MOVING AVERAGE PROCESSES

In the social sciences, time series are discrete, stochastic realizations of underlying data-generating processes. There is a constant time ordering to the data. The values of the realization are equally spaced over time. Adjacent values usually are related to one another. This process may take place in several ways. One way involves a shock, innovation, or error driving the time-ordered stochastic series. That is, a random error, innovation, or shock, e_{t-1}, at a previous time period, $t-1$, plus a shock at current time, t, drives the series to yield an output value of Y_t at time t (McCleary *et al.*, 1980). An example of this process may be epidemiological tracking of the prevalence rate of a disease. The prevalence rate is the proportion of the total population reported to have a disease at a point or interval of time. Reports of AIDS cases to the Centers for Disease Control (CDC) in Atlanta, for example, lead to a reported CDC prevalence rate in the United States. Researchers may deem the number of cases reported to the CDC as input shocks, and the CDC National Case Count as the output. The cases reported to the CDC and the number of deaths can be tallied each quarter and then graphed. When these data are modeled as a first-order moving average process, they can be used to explain the diffusion of the disease and to forecast the growth of a social problem for health care policy planning.

The growth of this series, once it has been mean centered, may follow an effect at time t, which is represented by e_t, plus a portion of the shock carried over effect from the previous time period, $t-1$. The lag of time between t and $t-1$ may not just be that of one time period. It may be that of several or q time periods. In this case e_{t-q} would be the shock that drives this series. The more cases reported to the CDC, the higher the national prevalence rate. The fewer cases reported, the less the national incidence level reported. This process may be expressed by the following moving average formula:

$$Y_t = e_t - \theta_1 e_{t-1}$$
$$= e_t(1 - \theta_1 L),$$

(3.1)

where y_t is the original series, μ is the mean of series, Y_t is the mean centered series or $Y_t = y_t - \mu$, e_t is the shock at time t, e_{t-1} is the previous shock, and θ_1 is the moving average coefficient. In this instance we observe that the current national prevalence is equal to a shock during the same time period as well as θ_1 times a shock at the previous time period.

The value of θ_1 will depend on which of these signs will be used. The computer program calculates the mean of the series, with which the series can be centered. This process, which involves a finite memory of one time lag, is called a first-order moving average and is designated as MA(1).

Higher order moving average models are also possible. A second-order moving average process, MA(2), would entail a memory for two time lags. If, hypothetically, the contemporary U.S. AIDS prevalence series had a memory that lasted for two time periods, then shocks from two time periods in the past would have an effect on the series before that effect wore off.

The model of the series would be that of a second-order moving average, formulated as

$$Y_t = e_t - \theta_1 e_{t-1} - \theta_2 e_{t-2}$$

$$= e_t(1 - \theta_1 L - \theta_2 L^2). \tag{3.2}$$

In this case, the prevalence rate is a function of the current and previous two shocks. The extent to which the effect of those shocks would be carried over to the present would be represented by the magnitudes, signs, and significances of parameters θ_1 and θ_2.

3.5.2. AUTOREGRESSIVE PROCESSES

Another type of process may be at work as well. When the value of a series at a current time period is a function of its immediately previous value plus some error, the underlying generating mechanism is called an autoregressive process. For example, the percentage of Gallup Poll respondents among the U.S. public who approve of a president's job performance is a function of its value at a lag of one time period. Therefore, the Y_t is a function of some portion of Y_{t-1} plus some error term. The nature of this relationship may be expressed as follows:

$$Y_t = \phi_1 Y_{t-1} + e_t$$

$$= \phi_1 L Y_t + e_t \tag{3.3}$$

or

$$(1 - \phi_1 L)Y_t = e_t.$$

When the output is a regression on the immediately previous output plus some error term, the portion of the previous rating carried over to the rating at time t is designated as ϕ_1. This kind of relationship is called a first-order autoregressive process and is designated as AR(1). The presidential approval series is one where approval is regressed upon a previous value of itself plus some random error. But if the effect of the presidential approval carried over for two time periods, then the autoregressive relationship would be represented by

$$Y_t = \phi_1 Y_{t-1} + \phi_2 Y_{t-2} + e_t$$

$$= (\phi_1 L + \phi_2 L^2)Y_t + e_t. \tag{3.4}$$

In this formula the current approval rating would be a function of its two previous ratings, a so-called second-order autoregressive relationship, AR(2).

3.5.3. ARMA PROCESSES

Another data-generating mechanism that may be at work is a combination of the autoregressive and moving average processes. Series that have both autoregressive and moving average characteristics are known as ARMA processes. A formulation of an ARMA process is given in Eq. (3.5):

$$Y_t = \phi_1 Y_{t-1} + \phi_2 Y_{t-2} + e_t - \theta_1 e_{t-1} - \theta_2 e_{t-2}. \tag{3.5}$$

In this case, both the autoregressive and the moving average processes are of order 2. Sometimes this process is designated as ARMA(2,2). To be sure, ARMA(1,1) processes may occur as well. Most processes in the social sciences are first- or second-order.

3.5.4. NONSTATIONARY SERIES AND TRANSFORMATIONS TO STATIONARITY

Because the Box–Jenkins method is an analysis in the time domain applied to stationary series data, it is necessary to consider the basis of nonstationarity, with a view toward transforming series into stationarity. Stationary series are found in stable environments (Farnum and Stanton, 1989). Such series may have local or global stationarity (Harvey, 1991). Global stationarity pertains to the time span of the whole series. Local stationarity pertains to the time span of a portion of the series. There is weak and strong (strict) stationarity. When a process is weakly stationary, there is stationarity in the mean, the homogeneity, and the autocovariance structure. In other words, both the mean and variance remain constant over time, and the autocovariance depends only on the number of time lags between temporal reference points in the series. Weak stationarity is also called covariance stationarity or stationarity in the second sense. For strict stationarity to obtain, another condition must be fulfilled. If the distributions of the observations are normally distributed, the series is said to possess strict stationarity (Harvey, 1991; Mills, 1990).

Perhaps the simplest of all series is a white noise process, a series of random shocks, normally and independently distributed around a mean of zero with a constant variance but without serial correlation. An example of a white noise model, with merely a mean (constant) of zero and an error term unrelated to previous errors is (Enders, 1995; Harvey, 1993)

$$E(e_t) = E(e_{t-k}) = 0$$

$$E(e_t^2) = E(e_{t-1}^2) = E(e_{t-k}^2) \tag{3.6}$$

$$E(e_t, e_{t-k}) = \begin{bmatrix} \sigma^2, & if\, k = 0 \\ 0, & if\, k \neq 0 \end{bmatrix},$$

where k is the number of lags. This process may be construed as a series of random shocks around some mean, which might be zero. Although the distinction between weak and strong stationarity may be important, references in this text to stationarity denote weak (covariance) stationarity unless otherwise specified.

Nonstationarity may follow from the presence of one or several of five conditions: outliers, random walk, drift, trend, or changing variance. The series must be examined in order to ascertain whether any of these nonstationary phenomena inhere within the series. A plot or graph of the data against time (sometimes referred to as a timeplot or time sequence plot) is constructed first. Outliers, which distort the mean of the series and render it nonconstant, often stand out in a time sequence plot of the series. If the value of the outlier indicates a typographical error, one of the missing value replacement algorithms may be invoked in order to substitute a more plausible observation. Trimming the series, by weighting the outliers, may also be used to induce mean stationarity.

If a nonstationary series is riven with trend, the series possesses an average change in level over time. The trend may be stochastic or deterministic. Consider the stochastic trend first. When a nonstationary series is characterized by a random walk, each subsequent observation of the series randomly wanders from the previous one. That is, the current observation of the series, Y_t, equals the previous observation of the series, Y_{t-1}, plus a random shock or innovation, e_t. A series with random walk follows the movement of a drunken sailor navigating his next step on dry land (McCleary et al., 1980). Without reversion to the mean, the value of this series meanders aimlessly. The formulation of this random walk process is

$$y_t = y_{t-1} + e_t,$$

so that

$$y_t - y_{t-1} = e_t \tag{3.7}$$

or

$$(1 - L)y_t = e_t.$$

The accumulation of these random variations generates meanderings of series level:

$$y_t = y_{t-1} + e_t$$

$$= y_{t-2} + e_{t-1} + e_t$$

$$= .\qquad.\qquad. \tag{3.8}$$

$$= y_{t-j} + \sum_{i=0}^{j-1} e_{t-i}.$$

Other examples of such nonconstant mean levels are birth rates and death rates. In this nonstationary process, the level meanders in accordance with time, while the variance, to^2, grows in accordance with time. As the time t approaches infinity, the variance approaches infinity. This kind of stochastic trend may be rendered stationary by differencing, however:

$$y_t = y_{t-1} + e_t$$

$$y_t - y_{t-1} = y_{t-1} - y_{t-1} + e_t \tag{3.9}$$

$$\nabla y_t = e_t,$$

where $e_t \sim N(0, \sigma_t^2)$.

To render a random walk stationary, it is necessary to transform the series. First differencing—that is, subtracting the lagged value of the series from its current value causes the resulting series to randomly fluctuate around the point of origin, which turns out to be not significantly different from the zero level. We call processes that can be transformed into stationarity by differencing, "difference stationary" (Nelson and Plosser, 1982). An example of first differencing being used to endow a difference stationary series with stability in the mean is shown with annual U.S. male (16+ years old) civilian employment in Fig. 3.2. After differencing removes the stochastic trend, the transformed series exhibits a constant mean.

If the nonstationary series is random walk plus drift, then the series will appear to fluctuate randomly from the previous temporal position, but this process will start out at some level significantly different from zero. That nonzero level around which the random walk originates can be represented by a constant term, α. That white noise process drifts around the level, α, which is significantly different from zero. Drift emanates from the accumulation or integration of successive shocks over time. An example of random walk plus stochastic drift might be the gambler's toss of a fair coin. Each toss may be a heads or a tails. Because each toss is, under fair and ideal conditions, independent of every other toss, the outcome of the flip of the fair coin is going to be a head or a tail. That is, there will be one outcome

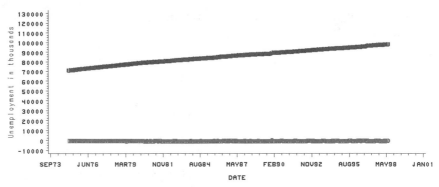

Jpper series=undifferenced series   Lower series=differenced series

Figure 3.2 Unemployment of U.S. males 16+ years old. Monthly data, not seasonally adjusted; labor force data from current population survey. Source: Bureau of Labor Statistics.

out of two possibilities. In the long-run or global perspective—that is, after many tosses—the probability of a heads will be $\frac{1}{2}$ and the probability of a tails will be $\frac{1}{2}$. In the short-run perspective, there may be a run of several random heads in a row. The number of heads or tails may drift upward or downward. A gambler commits a fallacy when believing that because there have been so many heads in a row, the next flip has to turn up a tails, because each flip is completely independent of any other. This drift is an integrative process. The formulation of this process (Hendry, 1995) is:

$$
\begin{aligned}
y_t &= y_{t-1} + \alpha + e_t \\
&= y_{t-2} + \alpha + \alpha + e_{t-1} + e_t \\
&= \ldots \\
&= y_0 + \alpha t + \sum_{k=0}^{k} e_{t-k}.
\end{aligned}
\tag{3.10}
$$

The drift is integrated into the stochastic trend described earlier, while its variance, $\mathrm{var}(y) = t\sigma^2$, approaches infinity as the time, t, approaches infinity. Hence, the variance of any forecast error, though the process is still difference stationary, increases without bound as well (Nelson and Plosser, 1982). The significance test is biased downward when drift is added to the random walk.

We call a series with a deterministic trend "trend stationary," and we can detrend by regression on a linear or polynomial time parameter (Nelson and Plosser, 1982). Regression coefficients of the time variable are trend coefficients; the significant time parameters will control for the trend, and the residuals will comprise the detrended stochastic component of the series. For example, such a trend may be formulated as

$$y_t = \alpha + b_1 t + e_t$$

whose expected value is

$$E(y_t) = \alpha + b_1 t + E(e_t)$$

$$= \alpha + b_1 t.$$

(3.11)

Examples of trend are population growth, learning, inflation/deflation, technological change, developments in social norms or customs, developing environmental conditions, and growth of public awareness of a phenomenon (Farnum and Stanton, 1989). The level, variance, and covariance of these processes are functions of time also. As the time increases, the level and variance increase. Such integrated series have no asymptotic variance limit, and the forecast error may expand indefinitely. The addition of a deterministic trend biases the significance test further downward than when there is random walk plus drift. In general, transformations to stationary are performed by differencing for stochastic trends and by regression for deterministic trends (Enders, 1995). For the most part, differencing will handle the transformations to stationarity (Nelson and Plosser, 1982).

3.6. TESTS FOR NONSTATIONARITY

3.6.1. THE DICKEY–FULLER TEST

There are objective tests that may be conducted to determine whether a series is nonstationary. The series could be nonstationary because of random walk, drift, or trend. One way to test this is to evaluate a regression that nests a mean, a lagged term (to test for difference stationarity), and a deterministic trend term (to test for trend stationarity) in one model:

$$y_t - \alpha + y_{t-1} + \beta t + e_t$$

and by taking the first difference of the y_t one finds that

(3.12)

$$\nabla y_t = \alpha + (\rho - 1)y_{t-1} + \beta t + e_t.$$

This model forms the basis of the Dickey–Fuller test. The test parameter distributions depend on the sample size and which terms are in the model. Therefore, the application of the Dickey–Fuller test depends on the regression context in which the lagged dependent variable is tested. The three model contexts are those of (1) a pure random walk, (2) random walk plus drift, or (3) the combination of deterministic trend, random walk, and drift.

To detect the stochastic form of nonstationarity, the Dickey–Fuller test entails a regression of a series on a first lag of itself to determine whether the regression coefficient of the lagged term is essentially equal to unity and is significant, under conditions (cases) of no constant, some nonzero constant, or some nonzero constant plus a deterministic trend coefficient. Consider the case called the autoregressive no constant test. The model to be tested presumes that the regression equation contains no significant constant term. The regression equation to be tested in the first case is

$$y_t = \rho y_{t-1} + e_t \tag{3.13}$$

A regression equation without a constant means that this model tests for pure random walk without drift.

$$y_t = \rho_1 y_{t-1} + e_t$$

and if $\rho_1 = 1$, then

$$y_t - y_{t-1} = e_t \tag{3.14}$$

$$(1 - L)y_t = e_t$$

$$t = \frac{\rho_1 - 1}{se_{\alpha_1}}$$

$$|t| \geq \tau_t, \tag{3.15}$$

where τ_t is the critical value of this first case. The null hypothesis is that $\rho_1 = 1$. If the null hypothesis cannot be rejected, the data generating process is inferred to have a unit root and to be nonstationary. Therefore, the two-sided significance test performed is that for the statistical significance of $\rho_1 - 1$. The test resembles a t-test. The null hypothesis that the series is a nonstationary random walk is rejected if $|t| > |\tau_1|$, where the value of τ_1 depends on the sample size and which other parameters are in the equation. Monte Carlo studies have shown that the critical values do not follow those of a t-test, however. Only when the sample is reasonably small and other parameters are not contained in the model does this distribution resemble a t distribution. In general, the smaller the sample size, the larger the critical values, and for all three models the parameter is biased slightly downward (Banerjee et al., 1993). Because of this bias, the Dickey–Fuller table of critical values for $\rho = 1$ stems is reproduced with permission from John Wiley and Sons, Inc., and Wayne Fuller in Appendix A (Fuller, 1996). Notwithstanding that, SAS has performed its own Monte Carlo studies with 10^8 replications, which is more than the 60,000 used by Dickey, for which reason the accuracy of the SAS critical values for $\rho = 1$ is expected to be much greater than those reported in earlier papers (SAS Institute, 1995).

The second Dickey–Fuller case involves a context of random walk plus drift. A regression tests the hypothesis that the series is nonstationary in the context of random walk plus drift. The regression in this second case, sometimes called the AR(1) with constant model (Greene, 1997), is

$$y_t = \alpha_0 + \rho_1 y_{t-1} + e_t. \tag{3.16}$$

The null hypothesis is that the series under consideration is a integrated at the first order, that is, $I(1)$. In other words, the null hypothesis is a test of whether $\rho = 1$. The alternative hypothesis is that the series is stationary. In the context of random walk plus drift around a nonzero mean, when the series is asymptotically stationary it has a constant mean of $\alpha_0/(1 - \rho)$ (Banerjee *et al.*, 1993). These different circumstances require that when this regression model is tested, the significance test specified in Eq. (3.16) be based not on the critical values for τ_1 but on those for τ_2. The distribution of critical values is biased downward more than those of the t distribution, and even more than those of the first case. The t_2 critical values, for the model of AR(1) with a constant, may also be found in the Dickey–Fuller Table in Appendix A. The Dickey–Fuller tests involve an individual and a joint test. There is not only the test for $\rho = 1$; there is a joint F test for the null hypothesis that $\alpha=0$ and $\rho = 1$ as well. These two tests comprise the essence of the Dickey–Fuller tests in the context of random walk plus drift.

The third Dickey–Fuller case is one with a context of random walk plus drift in addition to a deterministic linear trend. In this context, a regression, shown in Eq. (3.17), also tests the null hypothesis that the series is nonstationary. As in the earlier cases, the null hypothesis is that $\rho_1=1$, and the alternative hypothesis is that the series is stationary. If the null hypothesis for the test in Eq. (3.15) is rejected, there would be no simple differencing required. In this context, the distribution of the test statistic becomes even more nonstandard than in the first and second contexts; that is to say, the limiting distribution of critical values for τ_3 is more strongly biased downward than before. The reader may find the critical values for the τ_3 parameter in a third section of the Dickey–Fuller Table in Appendix A or from the SAS program.

$$y_t = \alpha_0 + \rho_1 y_{t-1} + bt + e_t \tag{3.17}$$

Because this version of the Dickey–Fuller test includes the lagged endogenous variable and the time trend parameter, difference stationary as well as trend stationary aspects of the series are tested at the same time with the joint F test. The joint F test for this model simultaneously tests the null hypothesis that $\rho = 1$ and $\beta = 0$. The alternative hypothesis of the joint F test is that at least one of these is not as hypothesized. Yet the Dickey–Fuller tests presume that the residuals are white noise.

3.6.2. AUGMENTED DICKEY–FULLER TEST

Not all Dickey–Fuller regression models have white noise residuals. If there is autocorrelation in the series, it has to be removed from the residuals of the regressions before the Dickey–Fuller tests are run. Under conditions of residual serial correlation, the augmented Dickey–Fuller test,

$$y_t = \alpha_0 + \rho_1 y_{t-1} + \sum_{j=2}^{p-1} \beta_j \nabla y_{t-j} + e_t, \qquad (3.18)$$

may be applied. Even if the process is an ARMA(p,q) process, Said and Dickey (1984) found that the MA(q) portion of an ARMA (p,q) process under conditions of MA(q) parameter invertibility can be represented by an AR(p) process of the kind in Eq. (3.18) when p gets large enough (Banerjee *et al.*, 1993). If the series is afflicted with higher order autocorrelation, successive orders of lagged differencing will be required to render the residuals white noise. Often, the number of lags required will not be known in advance. If the number of lags is too low, the model will be misspecified and the residuals will be contaminated with autocorrelation. It is advisable to set the number of lags high enough so that the autocorrelation will be removed from the residuals. Said and Dickey suggest that one less than the AR order of the model will do. If the number of lags is higher than needed, there may be cost in efficiency as the coefficients of the excess lagged terms lose significance. The augmented Dickey–Fuller equations are identical to the three foregoing Dickey–Fuller equations, except that they contain the higher order lags of the differenced dependent variable to take care of serial correlation before testing for the unit root. SAS provides the critical values for these coefficients in accordance with the number of lagged difference terms applied.

To test for random walk nonstationarity under conditions of serial correlation in the residuals, the augmented Dickey–Fuller (ADF) test requires estimating regression Eq. (3.18). If the series has a higher order serial correlation, higher order differencing will be required in order to transform the residuals into white noise disturbances. This preparation should be completed before the test for stationarity is performed. If three lagged orders of differenced dependent variables are necessary to remove the autocorrelation from the residuals, and if the series has a random walk plus drift, Eq. (3.19) might be employed to test for nonstationarity:

$$y_t = \alpha_0 + \rho_1 y_{t-1} + \sum_{j=2}^{3} \beta_j y_{t-j} + e_t$$

$$= \alpha_0 + \rho_1 y_{t-1} + \beta_2 y_{t-2} + \beta_3 y_{t-3} + e_t.$$

(3.19)

If the series has random walk plus drift around a stochastic trend, the Dickey–Fuller test can be constructed with the addition of a time trend variable, according to

$$y_t = \alpha_0 + \rho_1 y_{t-1} + \sum_{j=2}^{p} \beta_j y_{t-j} + bt + e_t.$$

(3.20)

The question of how many autoregressive lags or what order of model to use in the test may arise. A likelihood ratio test may be conducted to determine whether the addition of the extra lag significantly adds to the variance of the model. Cromwell *et al.* (1994), assuming normality of the residuals, give the formula for the Likelihood ratio test:

$$\mathrm{LR} = T \ln \left(\frac{\sigma_{k-1}^2}{\sigma_k^2} \right),$$

where

T is the size of sample,

σ_i^2 is the residual variance of model i.

$\mathrm{LR} \sim \chi^2$ with 1 df, for the test of

h_0, the model is of order $\mathrm{AR}(k-1)$, and

h_a, the model is of order $\mathrm{AR}(k)$.

(3.21)

When additional lags no longer improve the fit of the model, the order of the model has been determined. At this point, the Dickey–Fuller test for the $(\rho_1 - 1)/\mathrm{SE}$ is performed and the observed value can be compared to the critical value for the model employed. If the observed absolute t value is less than the critical value, the series is nonstationary. If the observed absolute t value is greater than the critical value, no simple differencing is required.

3.6.3. ASSUMPTIONS OF THE DICKEY–FULLER AND AUGMENTED DICKEY–FULLER TESTS

The Dickey–Fuller tests presume that the errors are independent of one another—that is, they are distributed as white noise—and are homogeneous. All of the autoregressive terms required to render the residuals white noise have to be in the augmented Dickey–Fuller model for it to be properly estimated. If there are moving average terms, the model must be amenable to inversion into an autoregressive process. If there are multiple roots, the series must be differenced to the order of the number of roots before subjecting it to the test. For example, if there are two roots, the series would have to be twice differenced to render it potentially stationary. Testing for white noise residuals can be performed with an autocorrelation function and a partial autocorrelation function. These functions will be elaborated in the next chapter.

3.6.4. PROGRAMMING THE DICKEY–FULLER TEST

Some statistical packages have built-in procedures that perform Dickey–Fuller and augmented Dickey–Fuller tests. Although SPSS currently has no procedure that automatically performs the Dickey–Fuller test, SAS version 6.12 contains the augmented Dickey–Fuller test as an option within the Identify subcommand of the PROC ARIMA. For pedagogical purposes, the natural log of gross domestic product (GDP) in billions of current dollars is used. This series requires first and fourth lag seasonal differencing to render it stationary. An annotated SAS program is given to show how the series is at first assessed for stationarity using the Dickey–Fuller tests, then the augmented Dickey–Fuller tests, and then with a test for a seasonal root at lag 4.

```
options ls=80;                    /* Limits output to 80 columns      */
title 'C3pgm1.sas ' ;             /* Program on disk                  */
title2 'Source: Bureau of Econ Analysis, Dept of Commerce';
title3 'National Accounts Data';
title4 'Annual data from Survey of Current Business, August 1997';
title5 'downloaded from http://www.bea.doc.gov/bea/dn1.htm July 9, 1998';

data grdopr;                      /* Defines data set GRDOPR          */
  infile 'c:statssasgdpcur.dat';  /* Reads in data from GDPCUR.dat    */
  input year gdpcur;              /* Defines variables and order of vars */
time = _n_;                       /* Construction of a trend variable */
lgdp = log(gdpcur);               /* Takes natural log of GDP in current $ */
lglgdp=lag(lgdp);                 /* Takes lag of ln(GDPcurrent)      */

label gdpcur='GDP in current $Billions'
      lgdp='LN of GDP in current $Billions'
      year='Year of Observation'
      lglgdp='Lag of LN(GDPcur)';

proc print;                                  /* Lists out data for checking */
run;
```

```
proc arima data=grdopr;
 identify var=lgdp
 stationarity=(adf=(0)) nlag=20;          /* Stationary subcommand invokes */
                                          /* Regular Dickey-Fuller Tests    */
title6 'Ln(GDP) in need of Differencing'; /* at 0 lag only                  */
run;

proc arima data=grdopr;
     identify var=lgdp
     stationarity=(adf=(0,1,2,3,4));      /* Augmented Dickey-Fuller Tests */
title7 'Augmented Dickey-Fuller Tests';  /* at lags 0 through 4            */
run;

proc arima data=grdopr;
   identify var=lgdp(1)                   /* Test of First Differenced series */
     stationarity=(adf=(0,1,2,3,4));      /* Augmented Dickey Fuller Test     */
title6 'ADF of Diff1[Ln(GDPcur)]';       /* @ lags 0 thru 4                  */
run;

proc arima data=grdopr;
   identify var=lgdp(1)
     stationarity=(adf=(0) Dlag=4);       /* Augmented Dickey Fuller Test */
title6 'ADF of Diff1[Ln(GDPcur)]';       /* Seasonal root @ lag 4 test   */
run;

proc arima data=grdopr;                   /* Test of Diffd Stationary series */
   identify var=lgdp(1,4)
     stationarity=(adf=(0));              /* Dickey Fuller Test AT LAG 0 */
title6 'ADF of Diff1,4[Ln(GDPcur)]';     /* OF DIFFERENCED SERIES       */
run;
```

The output indicates that the natural log of the series is in need of differencing and one that needs no further differencing to render it stationary. Closer inspection reveals that regular differencing at lag 1 and seasonal differencing at lag 4 is necessary to effect stationarity.

Users of SAS version 6.12 or higher will find that the Dickey–Fuller or augmented Dickey–Fuller tests may be programmed by the inclusion of a stationarity option in the ARIMA procedure's identify subcommand. To render the residuals white noise and amenable for analysis, the serial correlation is first eliminated by the inclusion of autoregressive orders within the model. The list of AR orders to be tested is included within the parentheses of adf=( ). If there were first-order autoregression and the test were to be done on such a series, then adf=(1) would be used. Once the serial correlation is eliminated, nonsignificant probabilities indicate that differencing is in order. Significant probabilities indicate that the series is stationary in the output of these tests.

In this example there was no reason to suspect residual autocorrelation in the series, so the order of lagged differenced terms was set to zero. If there is reason to suspect higher order autoregression in the series, the user may specify more than one (autoregressive order) in the parenthesis. Hence, the following command invokes a regular Dickey–Fuller test.

```
proc arima data=qgrdopr;
  identify var=lgdp stationarity=(adf=(0)) nlag=20;
title5 'Ln(GDP) in need of Differencing';
run;
```

In addition to the usual ARIMA, the output for this Dickey–Fuller test of the natural log of annual U.S. gross domestic product from 1946 through 1997 indicates that the series is in need of differencing. The series has a significant single mean but it lacks a deterministic trend. Therefore, the line of output used for analysis is the middle line, entitled Single Mean.

Augmented Dickey-Fuller Unit Root Tests

Type	Lags	RHO	Prob<RHO	T	Prob<T	F	Prob<F
Zero Mean	0	1.0430	0.9133	12.1151	0.9999	--	--
Single Mean	0	-0.0292	0.9528	0.1487	0.9380	138.3900	0.0010
Trend	0	-2.1522	0.9629	1.0000	0.9351	0.5017	0.9900

Reading from the middle line for the single mean model, the null hypothesis of nonstationarity is confirmed. Some differencing would be in order here.

In the first line, the model being tested is that of the random walk without drift and without trend. The ρ is the coefficient of the lagged response variable. The probability less than rho is a Dickey–Fuller probability. The t test for $(\rho-1)/SE_{\rho}$ is the test for whether the lagged endogenous term is significant, according to the SAS simulations of Dickey–Fuller probabilities for τ_i. The null hypothesis is that the response series is nonstationary. The alternative hypothesis is that the response series is stationary. If this discovered probability is greater or equal to .05, then the null hypothesis cannot be rejected and simple differencing is needed to render the series stationary. There is no joint test of the mean, α, and ρ here, for it is already assumed that $\alpha = 0$.

In the second line, the test is performed for the model of random walk with drift but without a deterministic trend. The t test for $\tau_{\alpha} = (\rho-1)/SE$ is the test of significance used when there is a constant in the model, and the probability less than t indicates that the null hypothesis that $\rho = 1$

cannot be rejected. Therefore, it is inferred that the series being tested is nonstationary. The *F* test is the simultaneous test of the null hypothesis that the intercept and mean both equal zero—that is, that $\alpha = 1$ and $\rho_1 = 1$.

In the third line, labeled Trend, these are the probabilities found when there is a random walk, with drift around a deterministic trend. If there were a time trend variable in the model and the model had a constant, the third line would be used for interpretation. The *t* test is the same Studentized *t* test described earlier, but in the larger context of a single mean, a random walk, and a deterministic trend. The *F* test for this model tests the null hypothesis that neither the deterministic trend nor the stochastic trend is significant; in other words, the joint test here tests the null hypothesis that $\beta=0$ and $\rho=1$. Because our model does not have a deterministic trend, this is not the model that we examine to test our series stationarity. For all three contexts, nonsignificance of the *T* or ρ probabilities indicates that the series is in need of differencing to render it stationary.

If lagged AR terms are included to eliminate autocorrelation in the residuals, then the augmented Dickey–Fuller test is performed. It is invoked by specifying the number of lagged difference terms in the ADF = ((list of lagged terms) DLAG=orders of seasonal lags)) subcommand. In program C3PGM1.SAS, there is an example of how a seasonal lag at $T = 4$ is tested. Once the logged series has been differenced at lags 1 and seasonally differenced at lag 4, it becomes a stationary white noise series. The output for the last Dickey–Fuller test in the preceding program on the regular and seasonally differenced series can be interpreted from the Single Mean line below. The significant probabilities for coefficients for rho and tau suggest that the series is now stationary and that no further differencing is necessary (Hamilton, 1994; Leonard and Woodward, 1997; Meyer, 1998).

```
               Augmented Dickey-Fuller Unit Root Tests
```

Type	Lags	RHO	Prob<RHO	T	Prob<T	F	Prob<F
Zero Mean	0	-50.0517	0.0001	-7.4820	0.0001	--	--
Single Mean	0	-50.2561	0.0004	-7.4452	0.0001	27.7303	0.0010
Trend	0	-50.4605	0.0001	-7.3603	0.0001	27.1855	0.0010

SAS has computed the Dickey–Fuller tests with Monte Carlo studies of more than 60,000 replications, so the Dickey–Fuller probabilities obtained from SAS are likely to be more accurate than those found in the regular tables (Meyer, 1998).

3.7. STABILIZING THE VARIANCE

Weak or covariance stationarity requires not only mean stationarity, but variance or homogeneous stationarity as well. Often, series exhibit volatility or fluctuating variance. An example is the series showing the growth in the total gross federal debt.

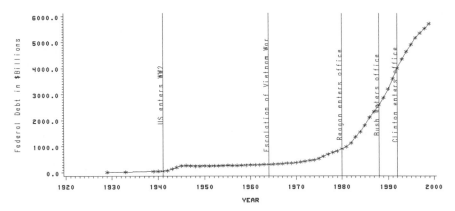

Figure 3.3   The growth of the gross total federal debt.

Graphing a series may reveal variance instability. If the variation in the series expands, contracts, or fluctuates with the passage of time, the change in variation will usually be apparent in a time plot. A simple graph of the series over time should reveal this volatility.

Once the researcher detects variance instability, he should consider variance stabilizing transformations. The natural log transformation, a power transformation, or a natural log of a series are examples of transformations that may stabilize the variance; a Box–Cox transformation (Eq. 3.23) is another common option. Examples of the power transformation already mentioned are the cube, square, square root, cube root, or fourth root of the original series. When a series variance is proportional to the level of a series or an exponential form of it, taking the log of the series may be another way of rendering its variance more stable.

To determine whether the natural log of a process is an appropriate transformation, one can test it with the SAS %LOGTEST macro. This macro estimates the process, with a chosen order of differencing and autoregression. It compares the fit of the maximum likelihood estimated model of the original series with that of its natural log. The test criteria are several measures of goodness of fit of the model. Optional criteria have different penalties for the number of degrees of freedom in the model. The penalties for more degrees of freedom increase according to whether the analyst employs the mean square error, the Akaike Information Criterion, or the Schwartz Information Criterion (Diebold, 1998; Ege *et al.,* 1993):

$$\text{Mean square error} = \sum_{t=1}^{T} \frac{e_t^2}{T}$$

$$\text{Akaike Information Criterion} = \exp(2k/T) \sum_{t=1}^{T} \frac{e_t^2}{T} \qquad (3.22)$$

$$\text{Schwartz Bayesian Criterion} = T^{(k)} \sum_{t=1}^{T} \frac{e_t^2}{T}.$$

Actually, SAS computes the natural log of these criteria and seeks the value of that transformation. If, for example, the natural log transformation produces a significant improvement of fit as indicated by a lower AIC, then a log transformation for the original series is recommended for variance stabilization.

The syntax of the SAS %LOGTEST macro begins with a percent sign, indicating the beginning of a macro and the macro name, LOGTEST. The arguments of the macro are embraced by the parentheses. Among the arguments are the data set name, the variable under consideration, the specification of the output data set, and the print command. The macro command, %LOGTEST(data set, variable, OUT=TRANS, PRINT= YES), terminated with a semicolon, produces the output:

TRANS	LOGLIK	RMSE	AIC	SBC
NONE	-259.390	327185.77	530.780	542.488
LOG	-233.079	88801.13	478.159	489.866

The comparison of the untransformed series, called NONE, and the natural log transformed series, called LOG, gives the root mean square residual, the Akaike Information Criterion, and the Schwartz Bayesian Criterion for the two series. Clearly, the natural log transformation of the total gross federal debt improves the variance stationarity of the series, as indicated by the lower RMSE, AIC, and SBC. Other options that may be added are a constant with the CONST= option, the AR=n, to specify the order of the AR model to be fit, and the DIF= option, which specifies the differencing to be applied before the test.

Because they are functions of the accumulation of error in the stochastic trend process, the variance and covariance are functions of time in the trend stationary process. Whether one applies the natural log of a series or a power transformation, all of these transformations are members of a family of Box–Cox transformations:

$$y = \frac{(X_t + C)^\lambda - 1}{\lambda} \qquad \text{if } 0 < \lambda \leq 1$$

$$(3.23)$$

$$y = \ln X_t + C \qquad \text{if } \lambda = 0,$$

where C is a constant, and λ is the shape parameter. In all cases, λ is a real number. When $\lambda = \frac{1}{2}$, the Box–Cox transformation reduces to a square root transformation. If the variance rises along with the level, then λ should be less than unity. If the variance declines as the level of the series increases, then λ should be set to some number greater than unity to stabilize the variance (Pankratz, 1991). Such series must be transformed into stationarity before the Box–Jenkins methodology can be applied.

3.8. STRUCTURAL OR REGIME STABILITY

If a series is covariance stationary, it has homogeneous variance. The condition of covariance stationarity implies a stable regime, in which the parameters of a model remain constant. Parameter constancy means that the parameters of a model fit the data equally well across the whole series: there is no significant difference between the residual variance from one part of the data set to another. Structural stability may be tested by the joint F (Chow) test that is part of an analysis of variance. Assuming that the series is long enough, the errors of the models are normally distributed, and those errors have equal variance, the researcher divides the sample into two subsets or segments, separated by a break-point. Three models may be developed. One model (M_{1+2}) may be formulated on the basis of the whole data set. From each of the two subsets of data, a model may be formulated. Models M_1 can be formed from the first segment and model M_2 can be formed from the second segment. Each model has a residual variance that makes up part of the test of structural stability. If there is no difference between the whole data set and the sum of its two parts, the residual variances of the whole model should be equal to those of the sum of the subset models and the series would reveal no significant break-points.

The joint F test for structural stability is a ratio of two residual variances. Each variance is itself a ratio of sums of squares to its degrees of freedom. The composite numerator variance consists, on the one hand, of the pooled models residual sums of squares subtracted from the sums of squares of the residuals based on the whole data set ($RSS_{1+2} - [RSS_1 + RSS_2]$). To provide the numerator variance, the resulting sum of squared residuals is then divided by its degrees of freedom, which $df_{num} = (n_1 + n_2 - p - 1) - (n_1 + n_2 - 2p - 2) = p + 1$, where p is the number of estimated parameters. The denominator variance consists of the ratio of the sum of squared residuals of the model based on the pooled data (RSS_{1+2}) to its respective degrees of freedom ($n_1 + n_2 - 2p - 2$). The joint F test is therefore:

$$F = \frac{\left(\dfrac{(RSS_{1+2} - [RSS_1 + RSS_2])}{(p + 1)}\right)}{\left(\dfrac{(RSS_1 + RSS_2)}{(n_1 + n_2 - 2p - 2)}\right)} \tag{3.24}$$

where RSS_{1+2} is the residual sum of squares of whole dataset, RSS_1 is the residual sum of squares of first dataset, RSS_2 is the residual sum of squares of second dataset, p is the number of parameters estimated. If the residual variances are constant and approximately equal, then the residuals are additive, and the joint F test yields a nonsignificant result. When this joint F test is nonsignificant, the series has structural stability across the break-point. When this joint F test is significant, the series lacks structural stability (Kennedy, 1992; Gujarati, 1995; Maddala, 1992).

3.9. STRICT STATIONARITY

In addition to all of the foregoing conditions holding for weak stationarity, strict stationarity requires normality of the distribution as well. Box–Jenkins time series analysis does require weak stationarity, but it does not require strict stationarity. Nonetheless, if one wishes to test his series for strict stationarity, he may analyze skewness ratio and the kurtosis coefficient of the distribution. If Y_t is a random variable with mean α, then the rth central moment may be defined as follows:

$$\mu_r = T^{-1} \sum_t [y_t - \alpha]^r,$$

for

r = level of moment, and

$t = 1, 2, \ldots, T.$

$$\tag{3.25}$$

From this formulation, the skewness ratio $(\beta_1)^{1/2} = (\mu_3) / (\mu_2)^{3/2}$ and the kurtosis coefficient $\beta_2 = (\mu_4) / (\mu_2)^2$ may be derived. The latter is normally distributed with mean 0 and standard error $= (24/T)^{1/2}$. If the random variable is normally distributed, the skewness ratio is 0 and the kurtosis coefficient is 3 (Cromwell *et al.*, 1994). It is more important that the series be rendered weakly stationary before the analysis begins.

3.10. IMPLICATIONS OF STATIONARITY

When a series exhibits weak stationarity, Box–Jenkins analysis becomes feasible. Whether the series is characterized by an autoregressive or moving average process, weak stationarity renders parameter values of the realization of the data-generating process stable in time (Granger and Newbold, 1986). For the purposes of this discussion, it is assumed that the series has already been transformed to a condition of stationarity through appropriate transformation.

3.10.1. FOR AUTOREGRESSION

A first-order autoregression obtains when a series whose current observation is a function of the immediately previous observation plus some innovation or random shock. This process has been formulated earlier in Eq. (3.3). An autoregression may be expanded as

$$
\begin{aligned}
Y_0 &= e_0 \\
Y_1 &= \phi_1 y_0 + e_1 \\
    &= \phi_1 e_0 + e_1 \\
Y_2 &= \phi_1 Y_1 + e_2 \\
    &= \phi_1(\phi_1 e_0 + e_1) + e_2 \\
    &= \phi_1^2 e_0 + \phi_1 e_1 + e_2
\end{aligned}
\tag{3.26}
$$

$$
\cdot
$$
$$
\cdot
$$

$$
Y_t = \phi_1^t e_0 + \phi_2^{t-1} e_1 + \cdots + \phi_1 e_{t-1} + e_t
$$

to show that it is a function of multiple lags. McCleary *et al.* (1980) have described this phenomenon as "tracking a shock through time" and have formulated it as follows: When one notes that $\varphi < 1$, then the power series of φ is one that diminishes over time. According to the preceding equation, if $\varphi^2 = .25$, then $\varphi^3 = .125$ and φ^4 is .0625. The farther back in time the analyst looks, the smaller the coefficient of the shock to the system in an autoregressive model. If this model is construed as an input–output system, the diminution of this coefficient may be interpreted as a leakage of effect from the system. Table 3.1 shows the leakage from the system at each time period. The autoregression and its corresponding leakage that characterizes this input–output system can be expressed by standard formulas:

Autoregression within the system is

$$\sum_{i=1}^{\infty} \phi_1^i e_{t-i}$$

(3.27)

and leakage from the system is

$$\sum_{i=1}^{\infty} (1 - \phi_1)^i e_{t_i}$$

where i is the number of past time periods. After a few time periods the effects of the previous shocks to the system are so small that they may be discarded as insignificant. For this attenuation to take place, however, the value of φ must be between -1 and $+1$. The evanescence of the effect is indicated by the amount less than 1 or greater than -1. The persistence of the effect is measured by its closeness to the value of 1 or -1. The closer the effect is to 1 or to -1, the longer the effect persists. If the value of $\varphi = 1$ or $\varphi = -1$, then this diminution of effect does not occur. The bounds of 1 and -1 for the autoregressive parameter are known as the bounds of stationarity. When φ equals 1 or -1, the process is no longer stationary. If the process is not stationary, it would need to be transformed to stationarity in order for it to be amenable to convergence or attenuation, which is required for the process to be analytically manageable. In sum, it is necessary therefore to have autoregressive coefficients of φ whose absolute value is less than 1.

With an autoregressive process, Vandaele (1983) notes that the variance of the process may be expressed as

$$\text{Var}(Y_t) = E(\phi_1 Y_{t-1} + e_t)^2$$

$$= E(\phi_1 Y_{t-1}^2 + 2\phi_1 Y_{t-1} e_t + e_t^2)$$

and because $E(Y_{t-1} e_t) = 0$,

(3.28)

$$\text{Var}(Y_t) = \phi_1^2 \text{Var}(Y_{t-1}) + 0 + E(e_t^2)$$

$$= \phi_1^2 \text{Var}(Y_{t-1}) + \sigma_e^2.$$

Table 3.1

Leakage from Autoregression

Time	Portion remaining	Leakage
$t = 0$	e_0	$\cdots$
$t = 1$	$\phi_1 e_0$	$(1 - \phi_1)e_0$
$t = 2$	$\phi_1^2 e_0$	$(1 - \phi_1^2)e_0$
.	.	.
$t = t$	$\phi_1^t e_0 = 0$	$(1 - \phi_1^t)e_0$

Under conditions of constant variance,

$$\text{Var}(Y_t) - \phi_1^2\text{Var}(Y_t) = \sigma_e^2$$

$$(1 - \phi_1^2)\text{Var}(Y_t) = \sigma_e^2 \tag{3.29}$$

$$\text{Var}(Y_t) = \frac{\sigma_e^2}{1 - \phi_1^2}.$$

If $\varphi = 1$, the process is unstable. The variance of the error becomes infinite as the time increases and the following process obtains:

$$(1 - \phi_1 L)Y_t = e_t$$

$$\text{if } \phi_1 = 1$$

$$(1 - L)Y_t = e_t \tag{3.30}$$

$$Y_t = Y_{t-1} + e_t$$

which leads to that nonstationary accumulation of random shocks, $Y_t = Y_0 + e_1 + e_2 + \cdots + e_t$. This is a nonstationary process. Hence, for this process to be stable, it is necessary that $\varphi < 1$ (Vandaele, 1983). When the autoregressive coefficient remains within the **bounds of stationarity,** the first-order autoregressive process may converge and be modeled.

In the second-order autoregressive process

$$Y_1 = \phi_1 Y_{t-1} + \phi_2 Y_{t-2} + e_t$$

$$\text{or} \tag{3.31}$$

$$(1 - \phi_1 L - \phi_2 L^2)Y_t = e_t.$$

Vandaele (1983) suggests that in autoregressive models the constant of the model may be parameterized in terms of its mean.

If the constant, C, is nonzero, then

$$(1 - \varphi_1 L - \varphi_2 L^2)(y_t - \mu) = C + e_t,$$

and if the mean, μ, is constant,

$$C = (1 - \varphi_1 - \varphi_2)\mu, \tag{3.32}$$

where

φ_i is an autoregressive parameter

C is a nonzero constant

μ is the mean.

where ϕ_i is an autoregressive parameter, C is a constant, μ is the mean. For this process to be convergent, new bounds of stationarity would have to hold:

Autoregressive model of order 2

Bounds of stationarity:

$$-1 < \phi_2 < 1 \qquad\qquad (3.33)$$

$$\phi_1 + \phi_2 < +1$$

$$\phi_2 - \phi_1 < +1.$$

Heretofore, we have considered autoregressive processes of order 1 and 2. We may also consider pth-order autoregressive processes. The bounds of stationarity may be elaborated for those also. Most of the time in the social sciences, data-generating processes will be explainable in terms of orders of 1 or 2. When the parameters of the data-generating process lie within the bounds of stationarity, the process becomes convergent and manageable. The implications of stationary extend beyond those of auto-regressive processes.

3.10.2. Implications of Stationarity for Moving Average Processes

The implications of stationarity extend to moving average processes as well. Indeed, according to Wold in 1938, a series may be explained in terms of an infinite linear combination of weighted innovations or random shocks. Such a series may be interpreted as an infinite moving average of innovations or shocks. More often than not, moving average models may be conceived of as a finite, rather than infinite, order of weighted past shocks. Equations (3.1) and (3.2) exemplify first- and second-order moving average processes. Most moving average models are first- or second-order. First-order moving average models (MA(1)) tend to be more common than second-order (MA(2)) models. Higher order moving average models tend to be more rare; often, they may be reformulated as lower order moving average models. In general, in a moving average process, a shock to the system enters the system and persists only for q time periods, after which it disappears completely.

Consider the MA(1) model. An MA(1) model is invertible to an infinite order AR(1) model, as shown in equation set (3.34):

$$Y = e_t - \theta_1 e_{t-1}$$

$$Y_t = (1 - \theta_1 L)e_t$$

$$\frac{Y_t}{(1 - \theta_1 L)} = e_t \tag{3.34}$$

$$(1 + \theta_1 L + \theta_1^2 L^2 + \cdots)Y_t = e_t$$

$$Y_t + \theta_1 Y_{t-1} + \theta_1^2 Y_{t-2} + \cdots + = e_t.$$

For this inversion to be effected, the parameter θ_1 must conform to certain bounds of invertibility. That is, in magnitude, the bounds of invertibility for an MA(1) model are defined by the inequality of $|\phi_1| < 1$. If $|\theta_1| = 1$, then the process would be unstable. Instead of converging, the process would be a nonstationary random walk. If it were a random walk, the process would require the integrated accumulation of outcomes from one shock. The effect would hardly be tractable. If $|\phi_1| > 1$, the process would not converge; rather, it would explode. Differencing would be required before stationarity could be attained.

In an MA(2) process, the shock lasts for two periods and then the impact it has on the model dies.

For the MA(2) case

$$Y_t = e_t - \theta_1 e_{t-1} - \theta_2 e_{t-2}$$

or

$$\frac{Y_t}{(1 - \theta_1 L - \theta_2 L^2)} = e_t. \tag{35}$$

This can also be expanded into another autoregressive series,

$$Y_t + \theta_1 Y_{t-1} + \theta_1^2 Y_{t-2} + \theta_1^3 Y_{t-3} + \cdots$$
$$+ \theta_2 Y_{t-2} + \theta_2^2 Y_{t-4} + \theta_2^3 Y_{t-6} + \cdots$$
$$+ 2\theta_1 \theta_2 Y_{t-3} + 3\theta_1^2 \theta_2 Y_{t-4} + 3\theta_1 \theta_2^2 Y_{t-5} + 4\theta_1^3 \theta_2 Y_{t-5} + \cdots = e_t.$$

For an MA(q) process, the effect of the shock persists for q lags and then desists. For this MA(2) process to be stationary, it must conform to the following boundary conditions of invertibility:

Bounds of Invertibility

For ARIMA(0,0,1) $-1 < \theta_1 < +1$

For ARIMA(0,0,2) $-1 < \theta_2 < +1$ \hfill (3.36)

$$\text{For ARIMA}(0,0,2) \quad \theta_1 + \theta_2 < +1$$
$$\text{For ARIMA}(0,0,2) \quad \theta_2 - \theta_1 < +1.$$

For an MA(2) model, similar conditions, also expressed in Eq. (3.36), must obtain for the process to be stable.

Given these conditions, the series formed can converge to a solution. From Eq. 3.34, it can be seen that an ARIMA (0,0,1) process can be converted into an infinite series of weighted past observations of the data-generating process. For this process to be tractable and stable, the parameters must reside within the **bounds of invertibility** for φ. If the parameters for φ equal or exceed the bounds of invertibility, one may assume that the series is nonstationary and should be differenced (McCleary *et al.*, 1980).

In the next chapter, the theory of the Box–Jenkins ARIMA models is discussed in greater detail. Derivation and use of the autocorrelation and partial autocorrelation functions are developed for identifying and analyzing time series. The characteristic patterns of the autocorrelation and partial autocorrelation functions for different types of models are reviewed, and programming of this identification procedure with SAS and SPSS is addressed.

REFERENCES

Banerjee, A., Dolado, J., Galbraith, J. W., and Hendry, D. F. (1993) *Co-Integration, Error Correction, and the Econometric Analysis of Non-Stationary Data.* New York: Oxford University Press, pp. 85–86, 100–102, 106–109.

Box, G. E. P. and Jenkins, G. M. (1976). *Time Series Analysis: Forecasting and Control.* San Francisco: Holden Day, p. 18.

Box, G. E. P., Jenkins, G. M., and Reinsel, G. C. (1994). *Time Series Analysis: Forecasting and Control,* 3rd ed. Englewood Cliffs: Prentice Hall. Chapter 3.

Clinton, William J. (1998). *Economic Report of the President.* Washington, DC: U.S. Government Printing Office. Table B-78, http://www.gpo.ucop.edu/catalog/erp_98_appen_b.html, data downloaded, October 31, 1998.

Cromwell, J. B., Labys, W. C., and Terraza, M. (1994). *Univariate Tests for Time Series Models.* Thousand Oaks, CA: Sage Publications, pp. 14, 19–24.

Diebold, F. (1998). *Elements of Forecasting.* South-Western College Publishing, pp. 87–91.

Dickey, D. and Fuller, W. (1979). "Distribution of the Estimators for Autoregressive Time Series with a Unit Root," *Journal of the American Statistical Association,* **74,** pp. 427–431.

Ege, G., (1993). *SAS ETS/User's Guide* 2nd ed. Cary, NC: SAS Institute, Inc., p. 143.

Enders, W. (1995). *Applied Econometric Time Series.* New York: John Wiley & Sons, pp. 181–185.

Farnum, N. R. and Stanton, L. W. (1989). *Quantitative Forecasting Methods.* Boston: PWS-Kent Publishing Co., p. 50.

Fuller, W. (1996). *The Statistical Analysis of Time Series,* 2nd ed. New York: John Wiley and

Sons, Table 10.A.2 of Appendix A, p. 642, constructed by David A. Dickey, is reprinted with permission of Wayne Fuller and John Wiley and Sons, publisher.

Gottman, J. M. (1981). *Time-Series Analysis. A Comprehensive Introduction for Social Scientists.* New York: Cambridge University Press.

Granger, C. W. J. and Newbold, P. (1986). *Forecasting Economic Time Series,* 2nd ed. San Diego: Academic Press, p. 4.

Greene, W. H. (1997). *Econometric Analysis.* Englewood Cliffs: Prentice Hall, pp. 847–851.

Hamilton, J. D. (1994). *Time Series Analysis.* Princeton: Princeton University Press. pp. 514–528.

Harvey, A. C. (1991). *The Econometric Analysis of Time Series.* Cambridge: MIT Press, pp. 23–24.

Harvey, A. C. (1993). *Time Series Models.* Cambridge: MIT Press., p. 11.

Hendry, D. F. (1995). *Dynamic Econometrics.* London: Oxford University Press, p. 22.

Kennedy, P. (1992). *Guide to Econometrics.* Cambridge: MIT Press, p. 57.

Gujarati, D. N. (1995). *Basic Econometrics.* New York: McGraw Hill, pp. 263–264.

Leonard, M. & Woodward, D. (1997) were helpful in interpretation of Dickey–Fuller tests. Cary, NC: SAS Institute, Inc. Personal communication.

Maddala, G. S. (1992). *Introduction to Econometrics.* 2nd ed. New York: Macmillan, p. 174.

McCleary, R. and Hay, Jr., R. with Meidinger, E. and McDowell, D. (1980). *Applied Time Series Analysis for the Social Sciences.* Beverly Hills: Sage, p. 20.

Meyer, K. (1998). Personal Communication. Cary, NC: SAS Institute, Inc. July 9, 1998. I am very grateful for Kevin Meyer's assistance provided in the interpretation of SAS Dickey–Fuller and Augmented Dickey–Fuller output.

Mills, T. C. (1990). *Time Series Techniques for Economists.* New York: Cambridge University Press, p. 64.

Mills, T. C. (1993). *The Econometric Modeling of Financial Time Series.* New York: Cambridge University Press, pp. 8–10.

Nelson, C. R. and Plosser, C. I. "Trends and Random Walks in Macroeconomic Time Series." *Journal of Monetary Economics* **10**, p. 143, 1982.

Pankratz, A. (1991). *Forecasting with Dynamic Regression Models.* New York: John Wiley & Sons, Inc., pp. 28–29.

Said, S. E. & Dickey, D. A. (1984). Testing for Unit Roots in Autoregressive-Moving Average Models of Unknown Order." *Econometric Theory,* pp. 7, 1–21.

SAS Institute (1995a). *SAS/ETS Software: Changes and Enhancements for Release 6.11* Cary, N.C.: SAS Institute, Inc., pp. 74, 78–79.

SAS Institute (1995b). *SAS/ETS Software: Time Series Forecasting System.* Cary, NC: SAS Institute, Inc., pp. 244–245.

Vandaele, W. (1983). *Applied Time Series and Box–Jenkins Models.* San Diego: Academic Press, pp. 33,55.

Wold, H. O. A. (1938). *A Study in the Analysis of Stationary Time Series.* Stockholm: Almqvist and Wicksell. (2nd ed., Uppsala, 1954).

Woodward, D. (1997). Cary, NC: SAS Institute, Inc. Personal communication, June 3, 1997.

Chapter 4

The Basic ARIMA Model

4.1. INTRODUCTION TO ARIMA

This chapter examines basic Box–Jenkins time series analysis. It reviews time sequence graphs and explains how inspection of these plots enables the analyst to examine the series for outliers, missing data, and stationarity. It expounds graphical examination of the effect of smoothing, missing data replacement, and/or transformations to stationarity. Correlogram review also permits the analyst to employ other basic analytical techniques, allowing identification of the type of series under consideration.

Two of these basic analytical tools, the sample autocorrelation function (ACF) and the sample partial autocorrelation function (PACF), are theoretically defined and derived. Their significance tests are given. Graphical characteristic patterns of the ACFs and PACFs are discussed. Once the characteristic ACF and PACF patterns of different types of models are understood and catalogued, they can be used to match and identify the nature of unknown data generating processes. To demonstrate application of these functions, we utilize ACF and PACF graphs (of the correlation over time), called correlograms. The characteristic ACF of nonstationary series is compared to the characteristic pattern after transformation to stationarity. The chapter then expounds the implications of bounds of stationarity and invertibility for correlograms, and derives and explains characteristic ACF patterns of the autoregressive processes, moving average, and ARMA processes. For the PACF, the characteristic patterns of those same processes are also distinguished and identified. A discussion of more complex ARMA processes and their patterns follows. Aspects of ARMA model order identification are also addressed with the corner method, and then the researcher is introduced to the integrated, ARIMA models as well.

Other types of autocorrelation functions are also discussed. The chapter briefly mentions the inverse autocorrelation function (IACF) and the sample extended autocorrelation function (EACF). Along with the discussion of the sample EACF is an explanation of the corner method for identifying the order of ARMA models. For preliminary graphing and plotting of the ACF and PACF plots, some SAS and SPSS programming syntax is provided. In sum, this chapter introduces the reader to basic theoretical and graphical identification of the basic ARIMA models, before addressing seasonal models in the following chapter.

4.2. GRAPHICAL ANALYSIS OF TIME SERIES DATA

4.2.1. TIME SEQUENCE GRAPHS

After undertaking background research regarding the series of interest and possible influences on it, the researcher first visually examines the data. He plots the series data against time in order to inspect it for outliers, missing data, or elements of nonstationarity. If any element of nonstationarity—such as a sudden sharp singular change; a random walk; a random walk plus drift, which is such random fluctuation around a nonzero intercept term (Enders, 1995); a random walk plus drift around a deterministic or stochastic trend (Cromwell et al., 1994); or even variance instability—is evident in the data, then the patterns of nonstationarity generally become apparent in either a time sequence plot or a correlogram. Those sudden drastic changes in the series could be evidence of outliers. The analyst may observe a random walk. He may perceive that the series drifts in one direction or another. He may discern a linear or polynomial trend. He may observe the unstable variation in the series. What do these nonstationary characteristics look like?

There are characteristic patterns of these components of nonstationarity. A white noise series has mere random variation. There is no discernible pattern in its representation, as can be seen in Fig. 4.1. In a white noise series, there is no upward or downward trend of observed data values. The temporal distribution of these values appears to be erratic or random. These series exhibit no drift, and no growth or diminution of variance. Moreover, no autocorrelation is apparent within the series.

When stationarity does not exist, there may be pure random variation around a zero mean or random walk about a previous nonzero level, called random walk with drift. This series may appear to randomly drift upward. The irregular change in mean signifies trend nonstationarity. An example of random walk with drift is the annual U.S. unemployment from 1954 through 1994, the upper of the two series shown in the SPSS chart in Fig. 4.2. These data are taken from Table B-42 of the 1998 *Economic Report of the (US) President.* The phenomenon observed appears to be random walk around a mean level of 5.75 percent. The existence of this mean is what enables us to use the term drift. The question arises whether the series should be centered before analysis. Although it is not necessary, the researcher may opt for pre-analysis centering. For simple analysis, centering is unnecessary, and not using it forces the student to learn the difference between the mean and the constant in time series analysis. For more complicated modeling—especially where intervention or transfer functions are involved—centering is recom-

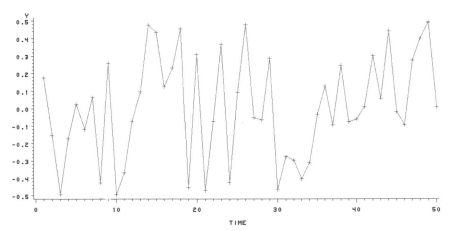

Figure 4.1 White noise simulation.

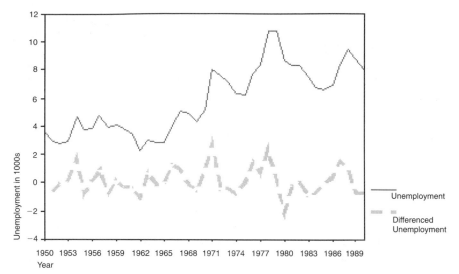

Figure 4.2   U.S. percent unemployment series.

mended. First differencing transforms the series into a condition of stationarity, and this differenced series is presented in the lower part of Fig. 4.2.

Another example of nonstationarity may be a trend, a more or less long-run tendency of increasing or decreasing mean. We can obtain an example of an SAS graph of a series exhibiting a linear trend from Federal Reserve

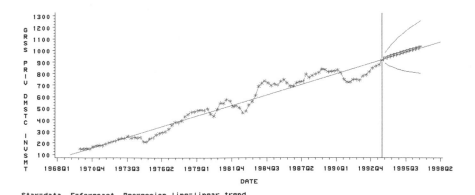

Star=data   F=forecast   Regression line=linear trend

Figure 4.3   Gross private domestic fixed investment by date, 1970 Q1 to 1993 Q4. Billions of dollars. Seasonally adjusted at annual rate. Source: Bureau of Economic Analysis, Survey of Current Business. Forecast of model with lead of 12 for forecast horizon.

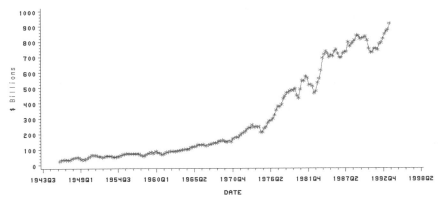

Figure 4.4 Gross private domestic fixed investment, 1946 Q1 to 1993 Q4, in billions of dollars. Seasonally adjusted at annual rate. Source: Bureau of Economic Analysis, Survey of Current Business. A possible quadratic trend.

Bank of Chicago, National Income and Products Accounts data archive. Culled from 1970 through 1993, the gross domestic private investment is a linear function of time and therefore exhibits a distinct linear trend Fig. 4.3. When gross domestic private investment is regressed against time, there is a significant positive linear component. When the series is examined over a longer time span, it may reveal an example of a quadratic trend, as can be seen in Fig. 4.4. It is often helpful to couple the graphical examination of the data with an objective statistical test.

In SAS, ASCII time plots or high-resolution graphic plots can be employed to display the series. The analyst may invoke the SPSS Time Sequence Plot or the SAS GPLOT procedure to obtain high-resolution graphical representation of the data. The SAS syntax for a graphical time sequence plot, where the series under consideration is the percent of civilian unemployment from Table B-42 of the 1998 *Economic Report of the President,* is shown in Fig. 4.5. The series is designated by the variable name UNEM P_RA, and the year is designated by the variable name YEAR (Bowerman and O'Connell, 1993; Ege *et al.,* 1993; Brocklebank and Dickey, 1994). The SAS command syntax to produce Fig. 4.5 follows:

```
symbol1 i=join c=blue;
axis1 label=(a=90 'Percent Unemployment');
proc gplot;
 plot unemp_ra * year/vaxis=axis1;
title justify=L 'Figure 4.5 U.S. Civilian Unemployment rate 1950-
97';
```

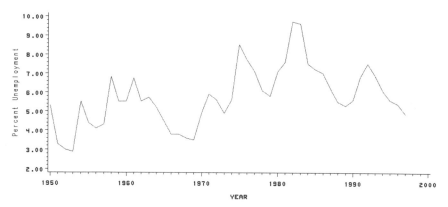

Figure 4.5   U.S. civilian unemployment rate, 1950–1997. Seasonally adjusted for all civilian workers. Source: 1998 Economic Report of the President, Table B-42.

```
title2 'Seasonally adjusted for All Civilian Workers';
title3 'Source: 1998 Economic Report of the President Table B-42';
title4 'Data from Government Printing Office On-line Services';
title5 'http://www.gpo.ucop.edu/catalog/erp98_appen_b.html on 2/12/98';
run;
```

This syntax is appended to the SAS program (*SAS/ETS Software: Applications Guide*, 1992) in order to generate a graphical time plot of the unemployment series. The SPSS syntax for a similar time plot may be entered in a syntax window once the data are already entered. This syntax can then be "selected" or "marked" and then the selection can be executed or run. The SPSS® Time Sequence Plot command syntax is given as follows (*SPSS-X Trends*, 1988; *SPSS Trends Release 6.0*, 1993):

```
TSPLOT VARIABLES= unempl
 /ID= date_
 /NOLOG
 /FORMAT NOFILL REFERENCE.
```

4.2.2. CORRELOGRAMS AND STATIONARITY

Correlograms for stationary processes exhibit characteristic patterns. The autoregressive parameters of a stationary process must reside within the bounds of stability. That is, the absolute values of the parameter estimates have to be less than unity. Only when the estimated values of these parameters adhere to this criterion will the process converge and the correlogram reveal rapid attenuation of the magnitude of the ACF. Rapid attenuation suggests that the magnitude of the ACF drops below the level of significance after a few lags. Because the autoregressive parameters of nonstationary processes may not be less than unity, the autocorrelations inherent in those processes may not rapidly dampen. Instead, they may very slowly decline, even undulate, or increase. Conversely, very gradual attenuation or wild fluctuation of the ACF before it drops below the level of significance is usually evidence of nonstationarity.

Similarly, the PACF of the moving average process exhibits rapid attenuation. For a moving average process to be stationary, the estimated values of the parameters must reside within the bounds of invertibility. Only then will this process converge and only then will the PACF of the moving average process attenuate rapidly. If the PACF of the moving average process does not rapidly dampen, the PACF will not attenuate rapidly, and that will be evidence of nonstationarity.

Often, after detection of nonstationary evidence, diagnosis of nonstationarity is helpful. Determination of whether the problem stems from a deterministic or a stochastic trend is in order. The diagnosis decomposes the components of nonstationary so that the series may be appropriately and effectively transformed. This diagnosis can be accomplished with the help of the Dickey–Fuller or augmented Dickey–Fuller tests, in cases of serial correlation, described in the previous chapter. A comparison of the second and third Dickey–Fuller or augmented Dickey–Fuller regressions will reveal whether trend stationarity exists. Once we know the precise nature of the nonstationarity, we can consider the appropriate corrective transformations: regression detrending for series with trend stationarity and differencing for series with stochastic trends. We also must determine the order of integration and undertake the appropriate transformation to effect stationarity. A linear or polynomial time trend may be used if the series has trend stationarity, and first or higher order differencing may be used if the series has stochastic trend. If there is heteroskedasticity in the series, it may be necessary to subject the series to a Box–Cox transformation or a log transformation to bring about variance stability. We can compare the AIC of the log transformation of the series with that of the raw series to see whether a natural log transformation is worth applying. Graphical inspection of the data should be coupled with a particular test of the series for nonstationarity to confirm the results of those tests (Mills, 1990).

4.3. BASIC FORMULATION OF THE AUTOREGRESSIVE INTEGRATED MOVING AVERAGE MODEL

The basic processes of the Box–Jenkins ARIMA (p,d,q) model include the autoregressive process, the integrated process, and the moving average process. As part of the orientation of the reader, fundamental definitions and notational conventions are specified, clarifying the mean and constant as well as the convention of the sign of moving average components. Our attention is then turned to the order of integration of the model, which is indicated by the $I(d)$ distribution designation. If a series is $I(0)$, then it is stationary and has an ARIMA$(p,0,q)$ designation. If a series requires first differencing to render it stationary, then $d=1$ and it is distributed as $I(1)$ and is given an ARIMA$(p,1,q)$ designation. Once the process has been transformed into stationarity, we can proceed with the analysis.

The series is then examined for autoregressive or moving average components. First, we have to consider centering and the difference between the mean and the constant. Consider the autoregressive process first. The parameter μ is the level of the process. In this text, the convention that Y_t is centered is employed here—such that $Y_t = y_t - \mu$. When the terms of constant and mean are not used interchangeably, it is helpful to distinguish between them. In the autoregressive process, where a series y_t is represented as

$$y_t - \mu + \phi_1(y_{t-1} - \mu) + e_t$$
$$(1 - \phi_1 L)y_t = (1 - \phi_1 L)\mu + e_t$$
$$(1 - \phi_1 L)Y_t = C + e_t \tag{4.1}$$
$$C = (1 - \phi_1)\mu \text{ for an ARIMA}(1,0,0) \text{ model,}$$

where the mean of the series is μ and the constant estimate of the autoregressive model is C (Ege et al., 1993; Vandaele, 1983; Babinec, 1996; Bresler, et al., 1992; Brocklebank and Dickey, 1994). The first-order autoregressive coefficient is designated ϕ_1. If the autoregressive process were a second-order process, then the mean-centered series could be represented by

$$(y_t - \mu) = \phi_1(y_{t-1} - \mu) + \phi_2(y_{t-2} - \mu) + e_t$$
$$y_t - \phi_1 y_{t-1} - \phi_2 y_{t-2} = \mu(1 - \phi_1 - \phi_2) + e_t$$
$$(1 - \phi_1 L - \phi_2 L^2)y_t = \mu(1 - \phi_1 - \phi_2) + e_t \tag{4.2}$$
$$(1 - \phi_1 L - \phi_2 L^2)y_t = C + e_t.$$

Hence, for an ARIMA(2,0,0) model,

$$C = \mu(1 - \phi_1 - \phi_2).$$

Higher order autoregressive processes would have three or more lags of the series. The autoregressive process is sometimes represented by [ARIMA(p,0,0)], where p is the order (the highest number of significant lags) of this process. For higher order autoregressive processes with a mean term in the model, the constant estimate is $C=\mu(1 - \phi_1 - \phi_2 - \cdots - \phi_p)$. Whereas the autoregressive process is a function of previous observations in the series, the moving average process is a function of the series innovations. When these series are stationary, the process remains in equilibrium around a constant level (Babinee, 1996; Zang, 1996).

Moving average processes are functions of current and past shocks around some mean. A first-order moving average process may be represented by kind of linear filter,

$$y_t = \mu + e_t - \theta_1 e_{t-1}$$
$$Y_t = e_t - \theta_1 e_{t-1},$$

(4.3)

where μ is the mean or constant estimate of this model, e_t is the current innovation or shock, θ_1 is the first-order moving average coefficient, and e_{t-1} is the previous shock to the system. A second-order moving average process is represented by

$$y_t = \mu + e_t - \theta_1 e_{t-1} - \theta_2 e_{t-2}$$
$$Y_t = e_t - \theta_1 e_{t-1} - \theta_2 e_{t-2}.$$

(4.4)

In the case of the second order moving average process, the current observation is a function of some mean or intercept, the current innovation, and two past innovations—one at lag 1 and the other at lag 2. Although some scholars use a plus rather than a minus sign parameterization of the previous moving average components, this amounts to mere convention of what value one assigns to $-\theta_{t-i}$, the weight of an innovation, at a particular time $t-i$ (Granger and Newbold, 1986; Harvey, 1993). The original parameterization employed by Box and Jenkins is the one used in this text. In this process $-\theta_2$ is the coefficient of the shock two time periods prior to the current time period. The moving average process is sometimes represented by [ARIMA(0,0,q)], where q represents the order of the process.

A combination of these processes is called the autoregressive moving average [ARMA(p,q)], sometimes referred to as an [ARIMA(p,0,q)] model. In this notation, the p is the order of the autoregressive process and the q is the order of the moving average process. With this stationary model, a zero in the middle position signifies the order of differencing required. If there are autoregressive and moving average components to the differenced series, such a series may be modeled as an ARIMA(p,d,q) model, where d is the order of differencing that is required to render the series stationary.

In sum, autoregression is the extent to which current output observation is a function of past outputs of the system. The order of autoregression signifies the number of previous observations of which the series is a significant function. Autoregression coefficients of higher orders would not be significant. The autoregressive process tends to have a longer memory; that of the moving average process is comparatively finite. The moving average process is only a function of a finite number of past shocks to the system. When the process under consideration contains both the autoregressive and the moving average component, it is referred to as a mixed autoregressive–moving average ARMA model. The model of the regular AR-IMA(1,1,1) process is

$$
\begin{aligned}
(1 - L)(y_t - \mu) &= \varphi_1(y_{t-1} - \mu) + e_t - \theta_1 e_{t-1} \\
(1 - L)(1 - \varphi_1 L)y_t &= (1 - \varphi_1)\mu + (1 - \theta_1 L)e_t
\end{aligned}
\tag{4.5}
$$

where y_t is the current output observation, e_t is the current shock to system θ_1 is the moving average parameter, μ = mean of the series, and φ_1 is the autoregressive parameter. The mean of the series is designated by μ. When the θ precedes the L in this set of parentheses, that is the first-order autoregressive parameter, and when the parameter preceding the L is the θ, that is the first-order moving average parameter. Given this notation, the model will be expounded in light of particular analytical functions.

4.4. THE SAMPLE AUTOCORRELATION FUNCTION

When we analyze the ARIMA process, we find several functions that are of considerable analytical utility. The first of these functions is the autocovariance function (ACV). This function shows the covariance in a series between one observation and another observation in the same series k lags away. The autocovariance at lag k is the autocovariance between a series Y_t at time t and the same series Y_{t-k}, lagged by k time periods. It may be formulated as

$$
\text{ACV}(k) = E(Y_t, Y_{t-k}) = \sum_{t=1}^{n-k} (Y_t - \bar{Y})(Y_{t-k} - \bar{Y}).
\tag{4.6}
$$

The autocorrelation function, ACF(k), may be construed as a standardization of the autocovariance function. The standardization is performed by dividing the autocovariance (with a distance of k lags between observations) by a quantity equal to the variance—that is, the product of the standard deviation at lag 0 and the standard deviation at lag 0. This is analogous to computing the Pearson product moment correlation of the series by dividing the covariance of a series and its lagged form by the product of the standard deviation of the series times itself. Because of covariance stationarity, it does not matter whether the k is a lead or a lag from reference time t: The autocorrelation would be the same, regardless of where the reference point is in the series, as long as the time lag (or lead) between the two time points is the same.

$$\text{ACF}(k) = \frac{\text{ACV}(Y_t Y_{t-k})}{\text{std dev } Y_t * \text{std dev } Y_t}$$

$$= \frac{\sum_{t=1}^{n-k}(Y_t - \bar{Y})(Y_{t-k} - \bar{Y})/(n-k)}{\sum_{t=1}^{n}(Y_t - \bar{Y})^2/n} \tag{4.7}$$

$$= \frac{E(Y_t Y_{t-k})}{\sigma_y^2}.$$

The expected value of the autocorrelation function for lag 1 (where $k = 1$) is derived by dividing Eq. (4.6) by the output variance (which is the square root of the variance of Y at time t times the square root of the variance of Y at the same time period). Given this definition of the autocorrelation function, shown in Eq. (4.7), different characteristic patterns emerge for various autoregressive and moving average autoregressive processes. To these patterns, we now turn our attention.

It is instructive to examine the characteristic differences between the ACFs of those two processes. The first-order autoregressive process, sometimes referred to as ARIMA(1,0,0) or AR(1), may be represented by the formula in equation set (4.8). We can use this equation to illustrate the formulation of the autocorrelation function. The characteristic pattern of the autoregressive process is seen to be one of gradual attenuation of the magnitude of the autocorrelation. The autocorrelation function for such a process is computed with the autoregression parameter, ϕ.

The ARIMA(1,0,0) process can be written

$$(1 - \varphi_1 L)y_t = \mu + e_t;$$

therefore;

$$y_t = \mu + \varphi_1 y_{t-1} + e_t, \tag{4.8}$$

and with the autocovariance for lag 1,

$$\text{ACV}(1) = E[(Y_t)(Y_{t-1})]$$

$$= E[(Y_{t-1})(\varphi_1 Y_{t-1} + e_{t-1})]$$

$$= E(\varphi_t Y_{t-1}^2 + Y_{t-1} e_{t-1}) \tag{4.9}$$

$$= \varphi_1 Y_{t-1}^2 \text{ and because of stationarity}$$

$$= \varphi_1 \sigma_y^2.$$

Assume that the series is centered, so that $Y_t = y_t - \mu$ and that $Y_{t-1} = y_{t-1} - \mu$. We may take the covariance of the series and its lag at time $t - 1$ in Eq. (4.9). This result is the first-order autocovariance of an autoregressive model. Because Y_t and Y_{t-1} are independent

$$E(Y_1 e_{t-1}) = 0,$$

and because of homogeneity

$$E(Y_{t-1}^2) = \sigma_y^2. \tag{4.10}$$

Therefore,

$$\text{ACV}(1) = \varphi_1 E Y_t^2 = \varphi_1 \sigma_y^2.$$

The autocorrelation, $E[\text{ACF}(1)]$, can be computed by dividing the covariance by the variance:

$$\text{ACF}(1) = \frac{\text{Cov}(Y_t, Y_{t-1})}{\text{Var}(Y_t)}$$

$$= \varphi_1 \frac{\sigma_y^2}{\sigma_y^2} \tag{4.11}$$

$$= \varphi_1.$$

This autocorrelation is that for the first-order autoregressive process. If the process is second-order, ARIMA(2,0,0), then the manifestation is

$$\text{ACF}(2) = \frac{\text{ACV}(2)}{\text{Variance}}$$

$$\text{ACV}(2) = E(Y_t Y_{t-2})$$

$$= E[(\varphi_1 Y_{t-1} + e_t)(Y_{t-2})]$$

$$= E[\varphi_1(\varphi_1 Y_{t-2} + e_{t-1}) + e_t)(Y_{t-2})] \tag{4.12}$$

$$= E[(\varphi_1^2 Y_{t-2}^2 + \varphi_1 Y_{t-2} e_{t-1} + Y_t e_{t-2})]$$

$$= \varphi_1^2 E(Y_{t-2}^2) = \varphi_1^2 \sigma_y^2;$$

so

$$\text{ACF}(2) = \frac{\varphi_1^2 \sigma_y^2}{\sigma_y^2} = \varphi_1^2.$$

The autocorrelation function defines the autoregressive process as the expectation of the current observation times that of the previous observation. By mathematical induction, the more general case can be inferred. In Eq. (4.13), this AR process is generalized in this way to the kth power for the ARIMA(1,0,0) or AR(1) model:

$$\text{ACF}(3) = \varphi_1^3$$

$$\cdot$$

$$\cdot \qquad\qquad (4.13)$$

$$\cdot$$

$$\text{ACF}(k) = \varphi_1^k.$$

Therefore, the strength of the autocorrelation of the stationary autoregressive process exponentially diminishes over time, as long as the magnitude of the autoregressive parameter remains less than 1. With this exponential attenuation, the decline in magnitude approaches zero as the time lag becomes infinite. This exponential decline in magnitude of the parameter forms the characteristic pattern of the ACF for the autoregressive process. The autocorrelation function has different implications for the moving average process.

If the magnitude of the autoregressive parameter equals unity, then the process becomes a nonstationary ARIMA(0,1,0) process. If $Y_t = \varphi_1 Y_{t-1} + e_t$, then $Y_t = e_t/(1 - \varphi_1 L) = e_t(1 + \varphi_1 L + \varphi_1 L^2 + \varphi_1 L^3 + \cdots + \varphi_1^p L^p)$. In other words, if $\varphi_1 = 1$, then this process represents a random walk. But if $\theta_1 > 1$, then the process exhibits a nonstationary stochastic trend and/or goes out of control. Therefore, stationarity requires that the autoregressive parameter remain within particular limits.

A moving average process exhibits a different characteristic autocorrelation function pattern. The characteristic pattern consists of sharp spikes up to and including the lag, indicating the order of the MA(q) process under consideration. Consider the case of the first-order moving average process, sometimes referred to as an ARIMA(0,0,1) or MA(1), and represented by the expected value of the series at time, t_{t-1}. For this process the autocovariance, ACV(1) is

$$\begin{aligned}
E(y_t, y_{t-1}) &= E[(e_t - \theta_1 e_{t-1})(e_{t-1} - \theta_1 e_{t-2})] \\
&= E[e_t e_{t-1} - \theta_1 e_{t-1}^2 - \theta_1 e_t e_{t-2} - \theta_1^2 e_{t-1} e_{t-2}] \\
&= -\theta_1 E e_{t-1}^2 \\
&= -\theta_1 \sigma_e^2.
\end{aligned} \qquad (4.14)$$

In this first-order moving average process, the autocovariance equals minus the magnitude of the shock, θ, times the variance of the shock at the first time lag. The autocorrelation function is formed from the autocovariance and the process variance. The process variance of the first-order moving average is given by

$$
\begin{aligned}
\text{Variance} = E(y_t^2) &= E(e_t - \theta_1 e_{t-1})^2 \\
&= E(e_t^2 - 2\theta_1 e_1 e_{t-1} + \theta_1^2 e_{t-1}^2) \\
&= E(e_t^2 - 2\theta_1 E e_1 e_{t-1} + \theta_1^2 E e_{t-1}^2) \\
&= \sigma_e^2(1 + \theta_1^2).
\end{aligned}
\tag{4.15}
$$

The autocorrelation is equal to the covariance divided by the process variance. For the first-order moving average process, the ACF at lag 1 equals

$$
E(\text{ACF}(1)) = \frac{-\theta_1}{1 + \theta_1^2}.
\tag{4.16}
$$

If the ACV for the first-order moving average is calculated at lag 2 (two lags difference between two first-order moving averages), the numerator and hence the ACF(2) is found to disappear completely:

$$
\begin{aligned}
E(y_t y_{t-2}) \\
= E(e_t - \theta_1 e_{t-1})(e_{t-2} - \theta_1 e_{t-3}) \\
= E[(e_t)(e_{t-2}) + \theta_1^2(e_t)(e_{t-3}) - \theta_1 e_t e_{t-3} - \theta_1(e_t)(e_{t-2})] \\
= 0.
\end{aligned}
\tag{4.17}
$$

Hence, the moving average is shown to spike at the lag of its order and then drop to zero:

$$
\text{ACF}(1) = \frac{0}{1 + \theta_1^2} = 0.
\tag{4.18}
$$

At higher orders, the ACF, say from ACF(3) to ACF(q), where $q > 1$, equals zero as well. Therefore, the ACF(1) of the first-order moving average is shown to have finite memory: After the time period of that shock, its autocorrelation drops to zero and disappears.

It is possible to compute the ACF(1) for a second-order moving average process using the same method. From a derivation of the equations, it may be seen that the ACF for a second-order moving average will have negative values for ACF(1) and ACF(2) but zero values for higher lags. Consider the first-order autocovariance, ACV(1). The multiplication proceeds by multiplying the first, the outside, and the inside terms that result in squared components of the same kind, and finally the rest of the inside terms in sequence:

$$
\begin{aligned}
E(Y_t Y_{t-1}) &= E[(e_t - \theta_1 e_{t-1} - \theta_2 e_{t-2})(e_{t-1} - \theta_1 e_{t-2} - \theta_2 e_{t-3})] \\
&= E[(e_t e_{t-1} - \theta_2 e_t e_{t-3} \\
&\quad - \theta_1 e_{t-1}^2 + \theta_1 \theta_2 e_{t-2}^2 \\
&\quad - \theta_1 e_t e_{t-2} + \theta_1^2 e_{t-1} e_{t-2} \\
&\quad - \theta_1 \theta_2 e_{t-1} e_{t-3} + \theta_2 e_{t-1} e_{t-2} \\
&\quad + \theta_1^2 e_{t-2} e_{t-3})] \\
&= -\theta_1 E e_{t-1}^2 + \theta_1 \theta_2 E e_{t-2}^2 \\
&= -\sigma_e^2 \theta_1 (1 - \theta_2).
\end{aligned}
\tag{4.19}
$$

The output variance is

$$
\begin{aligned}
E(Y_t^2) &= E[(e_t - \theta_1 e_{t-1} - \theta_2 e_{t-2})]^2 \\
&= E[(e_t^2 - 2\theta_1 e_t e_{t-1} - 2\theta_2 e_t e_{t-2} \\
&\quad + \theta_1^2 e_{t-1}^2 + 2\theta_1 \theta_2 e_{t-1} e_{t-2} \\
&\quad + \theta_2^2 e_{t-2}^2)] \\
&= E e_t^2 + \theta_1^2 E e_{t-1}^2 + \theta_2^2 E_{t-2}^2 \\
&= \sigma_e^2 (1 + \theta_1^2 + \theta_2^2).
\end{aligned}
\tag{4.20}
$$

To obtain the ACF(1) for the second-order moving average process, the autocovariance is divided by the process variance. The expected value of ACF(2) = ACV(2)/Var is

$$
\begin{aligned}
E(\mathrm{ACF}(2)) &= \frac{-\sigma_e^2 \theta_1 (1 - \theta_2)}{\sigma_e^2 (1 + \theta_1^2 + \theta_2^2)} \\
&= \frac{-\theta_1 (1 - \theta_2)}{1 + \theta_1^2 + \theta_2^2}.
\end{aligned}
\tag{4.21}
$$

From Eq. (4.21), it can be seen that for positive innovations, there will be two negative spikes on the ACF(2) from the parameters in the numerator. For a second-order moving average process, we may also compute ACF(2). We can compute the autocovariance using

$$
\begin{aligned}
E(Y_t Y_{t-2}) &= E(e_t - \theta_1 e_{t-1} - \theta_2 e_{t-2})(e_{t-2} - \theta_1 e_{t-3} - \theta_2 e_{t-4}) \\
&= -\theta_2 E e_{t-2}^2 \\
&= -\theta_2 \sigma_e^2.
\end{aligned}
\tag{4.22}
$$

When we divide the variance into the autocovariance, we obtain ACF(2):

$$\text{ACF}(2) = \frac{-\theta_2 \sigma_e^2}{\sigma_e^2(1 + \theta_1^2 + \theta_2^2)}$$

$$= \frac{-\theta_2}{1 + \theta_1^2 + \theta_2^2}.$$

(4.23)

These autocorrelation formulations are for moving averages. For a second-order moving average, if there is a positive innovation at lag 2, then this will appear as a spike on the ACF at lag 2. We can similarly show that the ACF(q), where $q > 2$ for a second-order moving average process, is 0. That is, the ACF(3) for the MA(2) model may be shown in Eq. (4.24) to equal zero:

$$E(y_t, y_{t-3})$$

$$= (e_t - \theta_1 e_{t-1} - \theta_2 e_{t-2})(e_{t-3} - \theta_1 e_{t-4} - \theta_2 e_{t-5})$$

$$= 0.$$

(4.24)

Because there are no identical product terms, the ACF for the MA process drops off immediately after the order of its time lag of the process has transpired.

The autocorrelation at lag k, ACF(k), for an MA(q) process can be expanded by mathematical induction to show that ACV(k)/Var is

$$\text{ACF}(k) = \left(\frac{-\theta_k + \theta_1 \theta_{k+1} + \theta_2 \theta_{k+2} + \cdots + \theta_{q-k}\theta_q}{1 + \theta_2^2 + \cdots + \theta_q^2} \right)$$

for $k = 1,2,3,\ldots,q$

if $k = < q$, ACF(k) $\neq 0$

and if $k > q$, ACF(k) $= 0$,

(4.25)

where k is the order of correlation and q is the order of moving average process. Unlike the exponential attenuation of the ACF of the autoregressive process, the characteristic pattern of the moving average process is delimited by the order of the process and drops to zero immediately thereafter. Consequently, the memory of the moving average process is finite and limited to the order of its process.

A more complex situation is that of the mixed AR and MA process. This kind of process is commonly referred to as an ARIMA(1,0,1) or ARMA process. The implications for the ACF in the ARMA process are interesting. With a centered series, the ARMA process possesses the autoregressive component on the left-hand side of the lower equation of (4.26) and the moving average component on the right-hand side:

$$Y_t = \varphi_1 Y_{t-1} + e_t - \theta_1 e_{t-1}$$

$$(1 - \varphi_1 L)Y_t = (1 - \theta_1 L)e_t.$$

(4.26)

Therefore

$$Y_t = (1 - \varphi_1 L)^{-1}(1 - \theta_1 L)e_t.$$

From Eq. (4.25), we can see that if $\phi_1 = \theta_1$, then the ARMA reduces to an ARIMA(0,0,0), a white noise process. Another way of expressing Eq. (4.26) is

$$
\begin{aligned}
Y_t &= (1 + \varphi_1 L + \varphi_1^2 L^2 + \cdots + \varphi_1^p L^p)(1 - \theta_1 L)e_t \\
&= (1 + \varphi_1 L + \varphi_1^2 L^2 + \cdots + \varphi_1^p L^p) \\
&\quad - (\theta_1 L + \theta_1 \varphi_1 L^2 + \theta_1 \theta_1^2 L^3 + \cdots + \theta_1 \varphi_1^{p-1} L^p)e_t \\
&= (1 + (\varphi_1 - \theta_1)L + (\varphi_1 - \theta_1)\varphi_1 L^2 + \cdots + (\varphi_1 - \theta_1)\varphi_1^{p-1} L^p)e_t \\
&= (1 - (\theta_1 - \theta_1)L - (\theta_1 - \varphi_1)\varphi_1 L^2 - \cdots - (\theta_1 - \varphi_1)\varphi_1^{p-1} L^p)e_t.
\end{aligned}
\tag{4.27}
$$

If there is a small difference between the autoregressive parameter ϕ_1 and the moving average parameter θ_1, and that difference is called v, then $\phi_1 = \theta_1 + v$. In this case each $\phi_1 - \theta_1 = v$, and the equation in (4.26) reduces to an autoregressive model of order $(p - 1)$ in the penultimate equation of equation set (4.27) or a kind of MA model, as revealed in the final equation of that set. Of course, the absolute values of such AR or MA parameters must lie within bounds permitting series convergence.

Consider the first order ACF(1) for the ARMA. First the variance and then the autocovariance are computed. Because $E(Y_{t-1} e_{t-1}) = E(\phi_1 Y_{t-2}, e_{t-1} + e_{t-1}^2 - \theta_1 e_{t-2} e_{t-1}) = \sigma_e^2$, the variance for the ARMA(1,1) is:

$$
\begin{aligned}
\text{Var}(Y_t) &= E(\varphi_1 Y_{t-1} + e_t - \theta_1 e_{t-1})^2 \\
&= \varphi_1^2 Y_{t-1}^2 + \varphi_1 Y_{t-1} e_t - \varphi_1 \theta_1 Y_{t-1} e_{t-1} \\
&\quad + \varphi_1 Y_{t-1} e_t + e_t^2 - \theta_1 e_t e_{t-1} \\
&\quad - \varphi_1 \theta_1 Y_{t-1} e_{t-1} - \theta_1 e_t e_{t-1} + \theta_1^2 e_{t-1}^2 \\
&= \varphi_1^2 \sigma_y^2 + (1 - 2\varphi_1 \theta_1 + \theta_1^2)\sigma_e^2 \\
&= \varphi_1^2 \,\text{Var}(Y_t) + (1 - 2\varphi_1 \theta_1 + \theta_1^2)\sigma_e^2.
\end{aligned}
\tag{4.28}
$$

Therefore,

$$
\text{Var}(Y_t) = \frac{(1 + \theta_1^2 - 2\,\varphi_1 \theta_1)\sigma_e^2}{(1 - \varphi_1^2)}.
$$

Wei (1990) computes the autocovariance as follows:

$$
\begin{aligned}
\text{ACV}(k) &= E(Y_{t-k} Y_t) = E[(\varphi_1 Y_{t-k} Y_{t-1}) + Y_{t-k} e_t - \theta_1 Y_{t-k} e_{t-1})] \\
&= E[\varphi_1 Y_{t-k} Y_t] + E(Y_{t-k} e_t) - \theta_1 E(Y_{t-k} e_{t-1}).
\end{aligned}
$$

If $k = 0$, then $\text{Var}(Y_t) = \text{ACV}(0)$, and

$$\text{Var}(Y_t) = \varphi_1 \text{ACV}(1) + \sigma_e^2 - \theta_{t-1} E(Y_t e_{t-1}).$$

Because

$$E(Y_t e_{t-1}) = \varphi_1 E(Y_{t-1} e_{t-1}) - \theta_1 E(e_{t-1})^2 = (\varphi_1 - \theta_1)\sigma_e^2$$

$$\text{Var}(Y_t) = \varphi_1 \text{ACV}(1) + \sigma_e^2 - \theta_1(\varphi_1 - \theta_1)\sigma_e^2. \tag{4.29}$$

If $k = 1$, then

$$\text{ACV}(1) = \varphi_1 \text{Var}(Y_t) - \theta_1 \sigma_e^2.$$

From Eq. (4.28), we obtain $\text{Var}(Y_t)$, and by substitution, obtain

$$\text{ACV}(1) = \frac{\varphi_1(1 + \theta_1^2 - 2\varphi_1\theta_1)\sigma_e^2}{1 - \varphi_1^2} - \theta_1 \sigma_e^2.$$

After the rightmost term is multiplied by $1 - \theta^2$, the numerator terms can be collected and factored. Then the ACF(1) for the ARMA is computed by dividing the autocovariance by the variance:

$$\text{ACF}(1) = \frac{\text{ACV}(1)}{\text{Variance}}$$

$$= \frac{(1 - \varphi_1\theta_1)(\varphi_1 - \theta_1)}{(1 - 2\varphi_1\theta_1 + \theta_1^2)\sigma_e^2}. \tag{4.30}$$

We have an exponentially attenuating ACF. The magnitude is modulated by the order of the theta in the denominator. Similarly, the ACF(k) of the ARMA(1,1) or ARIMA(1,0,1) is equal to ACF(k) = ϕ_1ACF($k - 1$) for $k \geq 2$ (Box *et al.*, 1994; Griffiths et al., 1993; Vandaele, 1983; Wei, 1990). Therefore, these models may have ACFs that taper off. For the most part, most complex models may be reduced to small-order AR, MA, or ARMA processes. Clearly, the ACF is a valuable instrument for identification of the nature of the data-generating process.

4.5. THE STANDARD ERROR OF THE ACF

Although the magnitude and relative magnitude of the ACF are important, the standard error and confidence interval are essential for proper inference. Unless we know the confidence limits, it is hard to tell below what magnitude of the ACF may be attributable to normal error variation within the series and above what magnitude of the ACF may be clearly statistically significant. Once we know the magnitude of the standard error of the ACF, we can estimate the confidence limits formed by ± 2 standard errors. ACFs with magnitudes beyond the confidence limits are those worthy of attention.

The standard error of the ACF has been derived. Box and Jenkins (1976) use Bartlett's approximation of the variance of the ACF to obtain the standard error of the ACF. They maintain that if the samples are large and the series is completely random, the variance (ACF) approximately equals the inverse of the sample size:

Var(ACF) $\cong$ $1/T$, where T = number of observations in data set.   (4.31)

The standard error then is the square root of this variance:

$$\text{Standard error(ACF)} = 1/\sqrt{T} \text{ for random series, and}$$

$$\text{Standard error(ACF)} = \frac{1}{\sqrt{T}}\left(1 + 2\sum_{k=1}^{q} r_k^2\right) \text{ for } MA(q). \tag{4.32}$$

Therefore, the confidence limits are formed from $\pm\, 1.96/\sqrt{T}$ or $\pm\, 2/\sqrt{T}$. If the process is an MA process, SPSS makes a slight adjustment by adding 2 times the sum of the autocorrelations (*SPSS 7.0 Statistical Algorithms*, 1996).

The significance of the autocorrelation coefficient can also be determined by either the Box–Pierce portmanteau Q statistic or the modified Ljung–Box Q statistic (Box et al., 1994; Cromwell et al., 1994):

$$\text{Box–Pierce } Q \text{ statistic}_{(df=k-p-q-1)} = T\sum_{k=1}^{m} r_k^2, \text{ and}$$

$$\text{Ljung–Box } Q \text{ Statistic}_{(df-k-p-q-1)} = T(T+2)\sum_{m=1}^{m} \frac{r_k^2}{T-k}, \tag{4.33}$$

where m is any positive maximum lag, T is the number of observations, k is the lag of autocorrelation, r_k is the autocorrelation for lag k, SAS and SPSS use the modified Box–Ljung Q statistic to test the significance of autocorrelations and partial autocorrelations. Given the degrees of freedom, the Box–Ljung Q is known to provide better chi-square significance tests at lower sample sizes than the earlier Box–Pierce Q statistic (Mills, 1990).

4.6. THE BOUNDS OF STATIONARITY AND INVERTIBILITY

Certain conditions must hold for these processes, which consist in part of series, to be asymptotically convergent and hence stable. According to Wold's decomposition theorem, these stationary processes may be expressed as a series of an infinite number of weighted random shocks, with the sum of the absolute values of the weights being less than infinity. By inverting the autoregressive component of a first-order autoregressive process, so that $Y_t - \mu = (1 - \phi_1 L)^{-1} e_t = 1 + \phi_1 L + \phi_1^2 L^2 + \cdots) e_t$, one obtains an infinite sequence of moving average shocks. For these infinite series to converge rather than randomly walk or even diverge, the component roots have to lie within certain limits. The roots of the equation for the process to be stationary have to reside outside the unit circle. For a first-order autoregression equation, if the $|\phi_1| < 1$, then the series at the bottom of equation sets (4.34 and 4.35) converges. If $|\phi_1| = 1$, there is a unit root in Eq. (4.8) and the process becomes a nonstationary random walk that does not stabilize. When the process is a random walk, the process after inversion becomes an infinite, random sum of shocks. The series may drift about, and in so doing, fails to converge. If the series is nonstationary and $|\phi_1| > 1$, the series goes out of control. The summation of shocks endows it with an exponential stochastic trend while it fails to converge. Alternatively, a nonstationary variance may be unstable and increase to infinity as time progresses. This condition begins as the process goes beyond the unit bound of stationarity. Unless roots of the equation lie within the bounds mentioned, the first-order autoregressive process will not converge and will be characterized by asymptotic instability. In the correlogram, a nonstationary autoregressive process exhibits a slow rather than an exponential diminution in magnitude of autocorrelation. Therefore, these parameter limits are called the **bounds of stationarity** for autoregressive processes.

Higher order AR(p) models have bounds of stationarity as well. For example, the AR(2) model, $y_t = \mu + \phi_1 y_{t-1} + \phi_2 y_{t-2} + e_t$, has three sets of boundaries of stationarity: $\phi_1 + \phi_2 < 1$, $\phi_2 - \phi_1 < 1$, and $\phi_2 < 1$. For a nonlinear model with two roots, the roots of the equation, $Y_t (1 - \phi_1 L - \phi^2 L^2) = e_t$, must lie within limits that make solution of the equation possible. Wei (1990) and Enders (1993) show that these limits are partly determined by the discriminant, $\sqrt{\phi_1^2 + 4\phi_2}$, of the solution equation for the roots and provide a good detailed exposition of the characteristic root derivation of these parameters. If the discriminant is positive, the parameters remaining within the bounds of stationarity guarantee that the process will converge. If the discriminant is negative, the formulation is converted to a cosine function and under specific conditions this cycles or undulates with some attenuation. In short, for the process to remain stationary, the characteristic roots must lie outside the unit circle, and higher order models have similar constraints. These limits constitute the bounds of stationarity for autoregressive processes.

The MA models have similar limits within which they remain stable. These boundaries are referred to in MA models as **bounds of invertibility.**

The MA(1) model, $y_t = \mu + e_t - \theta_1 e_{t-1}$, has its bounds of invertibility. That is, $y_t - \mu = e_t - \theta_1 e_{t-1} = (1 - \theta_1 L) e_t$. Another way of expressing this is $Y_t/(1 - \theta_1 L) = Y_t + \theta_1 L + \theta_1^2 L^2 + \ldots + \theta_1^q L^q = e_t$. When $\theta_1 = -\phi_1$, as long as this invertibility obtains, the moving average process is another expression of an infinite autoregressive process. This condition exhibits the duality of autoregression and moving average processes. This condition exists for an MA(1) model as long as $\theta_1 < 1$. For an MA(1) model, this inequality defines the bounds of invertibility.

Consider the first-order moving average process,

$$Y_t = \phi_1 Y_{t-1} + e_t$$

$$(1 - \phi_1 L)Y_t = e_t$$

$$
\begin{aligned}
Y_t &= \frac{e_t}{(1 - \phi_1 L)} \\
&= (1 - \phi_1 L)^{-1} e_t \\
&= (1 + \phi_1 L + \phi_1^2 L^2 + \ldots + \phi_1^p L^p) e_t \\
&= e_t + \phi_1 e_{t-1} + \phi_1^2 e_{t-2} + \ldots + \phi_1^p e_{t-p}.
\end{aligned}
$$

(4.34)

This process is one which can be extended as follows (McCleary *et al.*, 1980):

If $Y_t = e_t - \theta_1 e_{t-1}$

and

$$e_{t-1} \cdot Y_{t-1} \left(- \theta_1 e_{t-2}\right),$$

and

$$
\begin{aligned}
Y_t &= e_t - \theta_1 (Y_{t-1} - \theta_1 e_{t-2}) \\
&= e_t - \theta_1 Y_{t-1} + \theta_1^2 e_{t-2}.
\end{aligned}
$$

(4.35)

By extension;

$$e_{t-2} = Y_{t-2} - \theta_1 e_{t-3},$$

for which reason

$$
\begin{aligned}
Y_t &= e_t - \theta_1 Y_{t-1} - \theta_1^2 e_{t-2} \\
&= e_t - \theta_1 Y_{t-1} - \theta_1^2 Y_{t-2} - \theta_1^3 e_{t-3} \\
&= e_t - \sum_{i=1}^{\infty} \theta_1^i y_{t-i}.
\end{aligned}
$$

The moving average process in Eq. (4.35) is expressed as a function of the sum of a current and an infinite series of past observations. If we transfer the sum portion of the formula to the left-hand side of the equation, we can conceive of the above formula as a lag function of Y_t in which the Y_t portion may be divided by the lag function $(1 - \Sigma \theta L^i)$. This division renders the process invertible. Yet this series converges only if $|\theta_1| < 1$. If $|\theta_1| = 1$, then the series becomes unstable and nonstationary. If $|\theta_1| > 1$, then its magnitude grows beyond limit and the series becomes unstable. Only if $-1 < \theta_1 < 1$ does the process asymptotically converge to a limit. For this reason these bounds of a moving average process are called the bounds of invertibility. For an MA(2) model, the following bounds of invertibility hold: $\theta_1 + \theta_2 < 1$, $\theta_2 - \theta_1 < 1$, and $\theta_2 < 1$. Prior to modeling, series have to be tested for stability and convergence. One way of doing this is to test for a unit root. Because most series are AR(1), AR(2), MA(1), MA(2), or some combination thereof, the limits discussed in this section are used to test the bounds of stability or invertibility. If these conditions do not hold, we can transform the series so that the roots lie within those boundaries. Similar conditions have to hold for the moving average processes to be asymptotically stable.

4.7. THE SAMPLE PARTIAL AUTOCORRELATION FUNCTION

The other analytical function that serves as a fundamental tool of Box–Jenkins time series analysis is the sample partial autocorrelation function (PACF). This partial autocorrelation function, used in conjunction with the autocorrelation function, can be used to distinguish a first-order from a higher order autoregressive process. It works in much the same way as a partial correlation. This function, when working at k lags, controls for the confounding autocorrelations in the intermediate lags. The effect is to partial out those autocorrelations, leaving only the autocorrelation between the current and kth observation.

It is helpful to derive the partial autocorrelation function in order to understand its source and meaning. Consider the first-order autoregression process:

$$Y_t = \varphi_1 Y_{t-1} + e_t$$

$$Y_t Y_{t-1} = \varphi_1 Y_{t-1} Y_{t-1} + e_t Y_{t-1}$$

$$E(Y_t Y_{t-1}) = \varphi_1 E(Y_{t-1} Y_{t-1}) + E(e_t Y_{t-1}). \qquad (4.36)$$

With $\gamma_1 =$ autocovariance $(\mathrm{ACV}(Y_t))$
and $\gamma_0 =$ Variance (Y_t), then

$$\gamma_1 = \varphi_1 \gamma_0.$$

When we divide both sides of the last equation in (4.36) by γ_0, we obtain the result in Eq. (4.37). From this we observe that the first partial autocorrelation is equal to the first autocorrelation:

$$\rho_1 = \varphi_1. \tag{4.37}$$

Furthermore, to obtain the PACF for the second parameter, the first-order autocorrelation should be controlled. Yet the autocorrelation at lag k is a function of the intervening lags:

$$\rho_k = \varphi_1^k. \tag{4.38}$$

Just as the ACF of lag *3* is correlated with the ACF of the previous lag, the ACF of lag k is correlated with the intervening ACFs. To ascertain the partial autocorrelation of lag *1* with lag *3* controlling for the autocorrelation at lag *2*, it is possible to apply the ordinary formula for partial correlation. Recalling that

$$r_{xz.y} = \frac{r_{xz} - r_{xy}r_{yz}}{\sqrt{(1 - r_{xy}^2)(1 - r_{xz}^2)}}$$

$$x = Y_t$$

$$y = Y_{t-1}$$

$$z = Y_{t-2}$$

$$\tag{4.39}$$

$$\text{PACF}(2) = \rho_{13.2} = \frac{\rho_2 - \rho_1\rho_1}{\sqrt{(1 - \rho_1^2)(1 - \rho_1^2)}} = \frac{\begin{vmatrix} 1 & \rho_1 \\ \rho_1 & \rho_2 \end{vmatrix}}{\begin{vmatrix} 1 & \rho_1 \\ \rho_1 & 1 \end{vmatrix}}$$

$$= \frac{\rho_2 - \rho_1^2}{1 - \rho_1^2}.$$

The derivation of PACF(k) is a little more complicated. Consider an autoregressive process of order k. The partial correlation is generally derived by the Cramer's rule solution to the Yule–Walker equations (Pandit and Wu, 1993):

$$(\text{PACF}(3)) = \frac{\begin{vmatrix} 1 & \rho_1 & \rho_1 \\ \rho_1 & 1 & \rho_2 \\ \rho_2 & \rho_1 & \rho_3 \end{vmatrix}}{\begin{vmatrix} 1 & \rho_1 & \rho_2 \\ \rho_1 & 1 & \rho_1 \\ \rho_2 & \rho_1 & 1 \end{vmatrix}} \tag{4.40}$$

This formula can be extended further. In general, the Yule–Walker equations, explaining the derivation of the partial autocorrelation function from the autocorrelations, are

$$\rho_1 = \varphi_1 + \quad \varphi_2 \rho_1 + \cdots + \quad \varphi_p \rho_{p-1}$$

$$\rho_2 = \varphi_1 + \quad \varphi_2 + \cdots + \quad \varphi_p \rho_{p-2}$$

$$\vdots$$

$$\rho_p = \varphi_1 \rho_{p-1} + \varphi_2 \rho_{p-2} + \cdots + \varphi_p.$$

(4.41)

Expressed in matrix form, they are

$$\varphi = \mathbf{P}_p^{-1} \rho_p.$$

For an autoregressive process, the partial autocorrelation function exhibits diminishing spikes through the lag of the process, after which those spikes disappear. In an AR(1) model, there will be one spike in the PACF. If the autocorrelation is positive, the partial autocorrelation function will exhibit positive spikes. If the autocorrelation is negative, then the PACF for the AR(1) model will exhibit negative spikes. Because the model is only that of an AR(1) process, there will be no partial spikes at higher lags. Similarly, in an AR(2) model, there will be two PACF spikes with the same sign as those of the autocorrelation. No PACF spikes will appear at higher lags. Therefore, the PACF very clearly indicates the order of the autoregressive process.

The PACF is not as useful in identifying the order of the moving average process as it is in identifying the order of the autoregressive process. For a moving average ARIMA(0,0,1) model, the ACF(1) and therefore PACF(1) was derived from Eqs. (4.15)–(4.18) to be equal to the first equation in the following set. For the MA(1) process, the PACF at the kth lag equals the third equation in set (4.43):

$$\text{PACF}(1) = -\frac{\theta_1}{1 + \theta_1^2}$$

$$\text{PACF}(2) = -\frac{\theta_1^2}{1 + \theta_1^2 + \theta_1^4}$$

(4.43)

$$\text{PACF}(k) = -\frac{\theta_1^k (1 - \theta_1^2)}{1 - \theta_1^{2(k+1)}}, \text{ where } k > 1.$$

The implications of this formulation are several. For the first-order moving average model, the PACF gradually attenuates as time passes. If the shock is positive, then the PACF will be negative in sign and will be exponential in decline of size. If the innovation is negative, then the PACF will be positive and exponentially diminish in size.

If the data-generating process under consideration is MA(2), then the ACF and PACF are

$$ACF(1) = \frac{-\theta_1(1 - \theta_2)}{1 + \theta_1^2 + \theta_2^2}$$

$$ACF(2) = \frac{-\theta_2}{1 + \theta_1^2 + \theta_2^2}$$

$$ACF(k > 2) = 0$$

and for higher order MA(q) processes,

$$ACF(k) = \frac{-\theta_k + \theta_1\theta_{k+1} + \cdots + \theta_{q-k}\theta_q}{1 + \theta_1^2 + \cdots + \theta_q^2}$$

$$ACF(k > q) = 0.$$

For the MA(2) process

the PACF(1) = ρ_1,

$$PACF(2) = \frac{\rho_2 - \rho_1^2}{1 - \rho_1^2},$$

$$PACF(3) = \frac{\rho_1^3 - 2\rho_1\rho_2 + \rho_1\rho_1^2}{1 - \rho_2^2 - 2\rho_1^2 + 2\rho_1^2\rho_2},$$

.
.
.

(4.44)

The ACF will indicate the order of the model. There will be as many significant spikes as the model order. As for the PACF for the MA(2) model, as long as the roots are real and positive the PACF of an MA(q) process is that of a dampened exponential, but as long as those roots are complex, then the PACF is one of attenuated undulation (Box *et al.*, 1994).

4.7.1. STANDARD ERROR OF THE PACF

The estimated standard errors of the partial autocorrelation are the same as those of the autocorrelation. They approximately are equal to the inverse of the square root of the sample size.

4.8. BOUNDS OF STATIONARITY AND INVERTIBILITY REVIEWED

The autoregressive function can be formulated as an infinite series:

$$Y_t = \varphi_1 Y_{t-1} + \varphi_2 Y_{t-2} + \cdots + \varphi_p Y_{t-p} + e_t$$

$$= (1 - \varphi_1 L - \varphi_2 L^2 - \cdots - \varphi_p L^p)Y_t + e_t$$

and

$$Y_0 = e_0$$

$$Y_1 = \varphi_1 Y_0 + e_1$$

$$= \varphi_1 e_0 + e_1$$

$$Y_2 = \varphi_1 Y_1 + e_2$$

$$= \varphi_1(\phi_1 e_0 + e_1) + e_2 \qquad (4.45)$$

$$= \varphi_1^2 e_0 + \varphi_1 e_1 + e_2$$

.

.

$$Y_t = \varphi_1^t e_0 + \varphi_1^{t-1} e_0 + \ldots + \varphi_1 e_{t-1} + e_t.$$

If ϕ_1 were greater than 1, then the series would lead to uncontrolled explo-
sion of output y_t. If θ_1 were equal to 1, there would be trend (nonstationarity)
in the series, and either regression detrending or differencing would be
required to eliminate it. For the series to be stationary, θ_1 must be less
than $+1$ and more than -1. That is, $|\phi_1|$ must be less than 1, if the parameter
estimate is to remain within the bounds of stationarity. If the series converges, an infinite-order autoregressive process is equivalent to a first-order
moving average process. Similarly, a first-order autoregressive process is
equivalent to an infinite-order moving average process by dint of $1/(1-
L) = 1 + L + L^2 + L^3 + \ldots$. That is to say, $(1 - \varphi_1 L)Y_t = e_t$. In other
words, $Y_t = e_t/(1 - \varphi_1 L) = (1 + \pi_1 L + \pi_1 L^2 + \pi_1 L^3 + \cdots) = e_t$. Because
$\pi_1 = -\theta_1$, $Y_t = (1 - \theta_1 L - \theta_1 L^2 - \theta_1 L^3 - \cdots)e_t$. In these respects, there
is a duality between the autoregressive and the moving average process
(Gottman, 1981). Yet for these AR and MA processes to be invertible and
hence stable, then the bounds of stationarity and bounds of invertibility
must obtain.

4.9. OTHER SAMPLE AUTOCORRELATION FUNCTIONS

Other correlation functions have been found to be useful in identifying
univariate time series models. These are the Inverse Autocorrelation Function (IACF) and the Extended Sample Autocorrelation Function (EACF
or ESACF). Ege *et al.* (1993) explained that when the usual invertible
model, $\phi(L)W_t = \theta(L)e_t$, was reparameterized as $\theta(L)Z_t = \theta(L)e_t$, the ACF
of the reparameterized model is really the IACF of the initial model. They
note that the IACF for an overdifferenced model has the appearance of a

stationary sample ACF, but that an IACF of a nonstationary model has
the appearance of a noninvertible moving average. (See also Abraham and
Ledolter, 1984; Cleveland, 1972; Chatfield, 1980; and Wei, 1990, for further
discussion of this function.) For ARMA models, the ACF, IACF, and
PACF all exhibit tapering-off correlations, for which reason it is not always
easy to identify the orders of the ARMA model by the usual ACF and
PACF. Although the extended autocorrelation function has been found
very helpful for identification of these ARMA(p,q) processes, as of this
writing the EACF is contained in the SAS and not the SPSS package.
Because the EACF is so useful in this identification process, the general
theory of the sample EACF is presented here (Liu *et al.*, 1986).

The sample extended autocorrelation function is presented in the form of
a table. Consider an ARMA(p,d,q) model. A tabular matrix is constructed.
The structure of the matrix is determined by the orders of the possible ARMA
models. If these data are being analyzed before differencing, the matrix would
have $p + d + 1$ rows, where d is the order of differencing. If no differencing
is required, this matrix has $p + 1$ rows and $q + 1$ matrix columns. In this case,
there are $p + 1$ rows, extending from 0 to $p + 1$ rows, and $q + 1$ columns
extending from 0 to q columns, in this table. Iterated regression analysis is
employed to yield ACF parameters to fill the contents of the cells of the matrix.
An ARMA model is run for each column of the table. An mth-order autoregression of Z_t of the matrix for the jth moving average order of the matrix
determines what is placed in the cell. More precisely, where $w_t(j)$ refers to a
mean centered stationary series and $w_t(j) = (1 - \phi_1^{(j)}L - \ldots - \phi_m^{(mj)}L^m)Z_t$,
the significance of the $\theta_t^{(mj)}$ sample autocorrelation determines what is placed
in the cell of the matrix. In this case, m refers to the pth autoregressive order
of the model and j refers to the $q+1$ moving average order of the matrix. If
the sample autocorrelation coefficient is significant, an "X" is placed in the
cell. Where the coefficient is not significant, a "0" is placed in the cell. The
matrix of X's and 0's usually displays a triangular shape of zeroes, the upper
left-hand vertex of which indicates the order of the ARMA(p,q) model.

By way of illustration, Wei (1990) presents the iterated regressions:

$$Z_t = \sum_{i=1}^{p} \varphi_i^{(1)} Z_{t-i}$$

$$Z_t = \sum_{i=1}^{p} \varphi_i^{(1)} Z_{t-i} + \beta_1^{(1)} \hat{e}_{t-1}^{(1)} + e_t^{(1)} \qquad \text{where } t = p + 2, \ldots, n$$

$$Z_t = \sum_{i=1}^{p} \varphi_i^{(2)} Z_{t-i} + \beta_1^{(2)} \hat{e}_{t-1}^{(1)} + \beta_2^{(2)} \hat{e}_{t-2}^{(0)} + e_t^{(2)} \qquad \text{where } t = p + 3, \ldots, n$$

$$Z_t = \sum_{i=1}^{p} \varphi_i^{(q)} Z_{t-i} + \sum_{i=1}^{p} \beta_i^{(q)} \hat{e}_{t-i}^{(q-i)} + e_t^{(q)} \qquad \text{where } t = p + q + 1, \ldots, n.$$

Table 4.1

```
Command ===>
              Figure 4.6 Seasonally Adjusted Unemployment Rate of Civilian Labor Force
                   Labor Force Statistics from the Current Population Survey    20:25 Sunday, F
                                      16 Years And Older (null)
                           Data extracted on: September 17, 1996 (11:29 AM)
                                  Bureau of Labor Statistics Data

                                         ARIMA Procedure


                                  Name of variable = PCTUNEMP.

                                  Mean of working series = 5.749486
                                  Standard deviation     = 1.599069
                                  Number of observations =      584

                                         Autocorrelations

 Lag Covariance Correlation -1 9 8 7 6 5 4 3 2 1 0 1 2 3 4 5 6 7 8 9 1           Std
   0   2.557020    1.00000  |                    |********************|          0
   1   2.525555    0.98769  |                  . |********************|   0.041380
   2   2.490843    0.97412  |                 .  |*******************  |   0.071086
   3   2.441346    0.95476  |                .   |*******************  |   0.091120
   4   2.381541    0.93137  |                .   |*******************  |   0.106887
   5   2.310623    0.90364  |               .    |******************   |   0.119981
   6   2.230860    0.87245  |               .    |*****************    |   0.131118
   7   2.147048    0.83967  |              .     |*****************    |   0.140708
   8   2.063304    0.80692  |              .     |****************     |   0.149041
   9   1.976829    0.77310  |              .     |***************      |   0.156343
  10   1.889814    0.73907  |             .      |***************      |   0.162757
  11   1.808297    0.70719  |             .      |**************       |   0.168406
  12   1.727678    0.67566  |             .      |**************       |   0.173416
  13   1.657529    0.64823  |             .      |*************        |   0.177867
  14   1.592406    0.62276  |             .      |************         |   0.181867
  15   1.532676    0.59940  |             .      |************         |   0.185483
  16   1.478041    0.57803  |            .       |************         |   0.188770
  17   1.429423    0.55902  |            .       |***********          |   0.191777
  18   1.385592    0.54188  |            .       |***********          |   0.194547
  19   1.346589    0.52662  |            .       |***********          |   0.197115
  20   1.308561    0.51175  |            .       |**********           |   0.199510
  21   1.269944    0.49665  |            .       |**********           |   0.201745
```

Figure 4.6

```
Command ===>
                     Figure 4.7 Seasonally Adjusted Civilian Unemployment Rate
                   Labor Force Statistics from the Current Population Survey    20:37 Sunday, F
                                      16 Years And Older (null)
                           Data extracted on: September 17, 1996 (11:29 AM)
                                  Bureau of Labor Statistics Data

                                         ARIMA Procedure


                                     Inverse Autocorrelations

            Lag Correlation -1 9 8 7 6 5 4 3 2 1 0 1 2 3 4 5 6 7 8 9 1
              1    -0.40013  |          ********| .                   |
              2    -0.10889  |                **| .                   |
              3     0.02399  |                 .| .                   |
              4    -0.03997  |                .*| .                   |
              5    -0.02379  |                 .| .                   |
              6     0.03753  |                 .|*.                   |
              7     0.05694  |                 .|*.                   |
              8    -0.06645  |                .*| .                   |
              9    -0.00234  |                 .| .                   |
             10     0.02391  |                 .| .                   |
             11    -0.04957  |                .*| .                   |
             12     0.08521  |                 .|**                   |
             13    -0.02140  |                 .| .                   |
             14    -0.00678  |                 .| .                   |
             15    -0.01630  |                 .| .                   |
             16     0.00799  |                 .| .                   |
             17     0.00220  |                 .| .                   |
             18     0.03482  |                 .|*.                   |
             19    -0.04516  |                .*| .                   |
             20    -0.06347  |                .*| .                   |
             21     0.14898  |                 .|***                  |
             22    -0.12097  |                **| .                   |
             23     0.05663  |                 .|*.                   |
             24    -0.01290  |                 .| .                   |
```

Figure 4.7

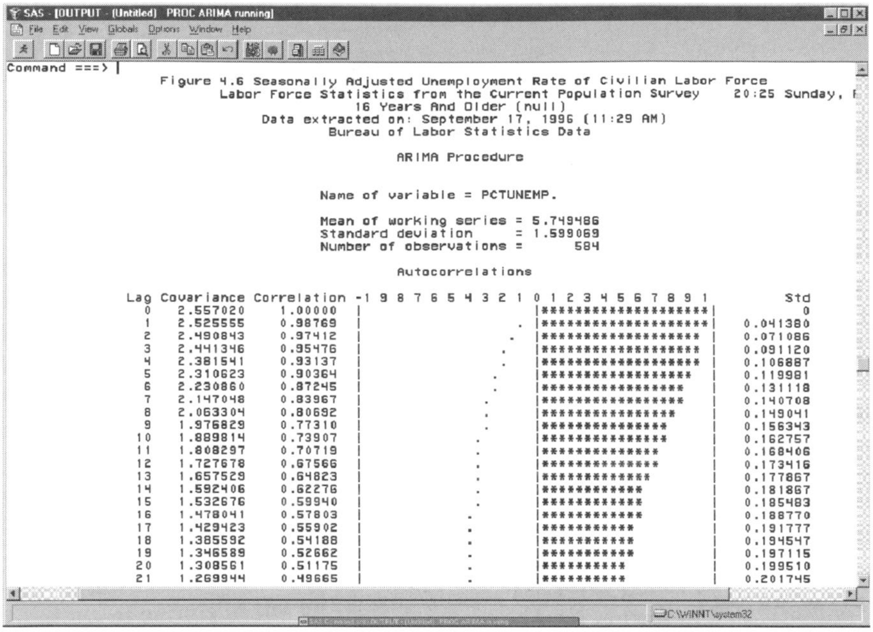

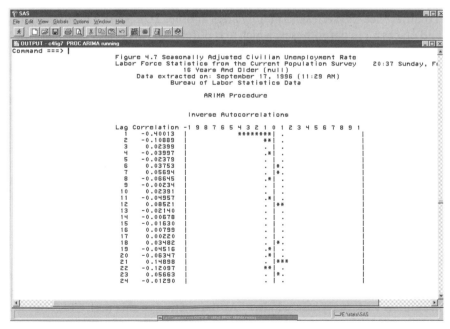

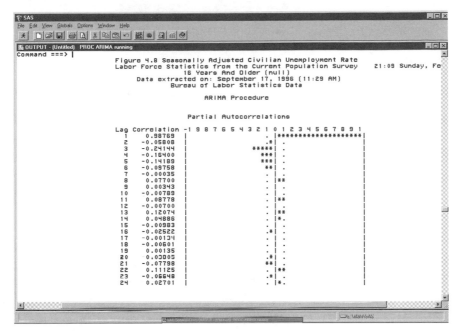

Figure 4.8

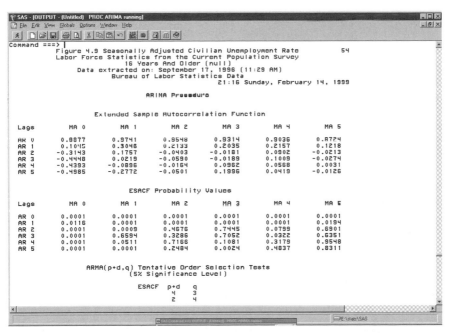

Figure 4.9

ACF command syntax employs the DIFF=1 subcommand for identifying this differenced series:

```
title 'Seasonally Adjstd Unemploymt Rate of Civilian Labor Force'.
subtitle 'Labor Force Stats from Current Population Survey'.
data list / pctunemp 1-4 (1).
begin data.
3.4
3.8
..
..
5.4
5.1
end data.
list variables=all.
execute.

Date year 1948 month 1 12.
execute.
*Sequence Charts .
title 'US Unemployment Rate'.
subtitle 'source: Bureau of Labor Statistics'.
TSPLOT VARIABLES= pctunemp
 /ID= year_
 /NOLOG
 /FORMAT NOFILL NOREFERENCE.
title 'Identification of Series'.
ACF
VARIABLES= pctunemp
 /NOLOG
 /DIFF=1
 /SERROR=IND
 /PACF.
```

The SPSS ACF and PACF output for the differenced series can be found in Figs. 4.10 and 4.11, respectively.

4.10.2. STATIONARITY ASSESSMENT

When the researcher attempts to analyze the data, he first graphs the series. If there is sufficient evidence that the series is nonstationary, he attempts to identify that evidence, confirms it, and transforms the series to stationarity. Preliminary evidence is gathered by graphing the data. If the series exhibits either a deterministic or stochastic trend upward, the researcher has reason to suspect nonstationarity. If the ACF attenuates very slowly, that is evidence of nonstationarity. In the seasonally adjusted civilian labor force U.S. percent unemployment series, extending from

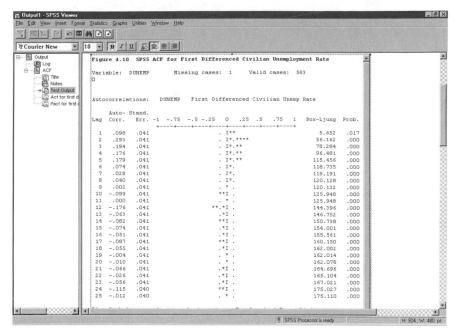

Figure 4.10

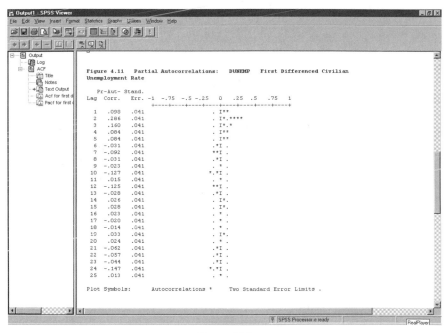

Figure 4.11

January 1948 through August 1996, it can be seen from the sample ACF in Fig. 4.6 that for 24 lags, all of the spikes exceed the confidence limit dots. This ACF is an example of a series exhibiting nonstationarity.

Statistical tests for nonstationarity, such as the Dickey–Fuller test, and first differencing of the series, may be applied. If the residuals from that first differencing are stationary, then the series has been rendered stationary by this transformation. If the residuals are not yet stationary, then second differencing may be in order. This is first differencing of the first differenced series. Generalized differencing is usually performed until stationarity is attained. If the series possess autocorrelation, the augmented Dickey–Fuller tests may be employed to determine when no further differencing is needed.

4.10.3. IDENTIFYING AUTOREGRESSIVE MODELS

Once the series has been rendered stationary, the ACF and PACF are examined to determine the type and order of the model. During the discussion of the nature of the autoregressive models, with their current observations being functions of earlier observations plus errors, these models were found to have a gradually attenuating ACF on the one hand, and a PACF that spikes at the order of the autoregressive model on the other. A first-order autoregressive process, an AR(1) model, would have an exponentially declining function as the lag k increases. The magnitude of the ACF is equal to θ^k. Consider a characteristic form of the AR(1) model. The ACF and the PACF of the AR(1) process have the following shown in Fig. 4.12 if $\theta_1 > 0$. If, on the other hand, the $\theta_1 < 0$, then the ACF and PACF would have the general appearance of Fig. 4.13. An actual example of an ARIMA (1,0,0) or AR(1) model is the Gallup Poll Index (Gallup Poll Index, 1996) of public approval of President William J. Clinton's job performance. These polls are taken one or more times per month and ask, "Do you approve of the way that the President is handling his job?" Because the intervals are supposed to be equally spaced and these polls are not temporally equidistant, the average monthly approval percentage is computed and a working assumption that these averages are equally spaced, though in fact the polls were not, is made. The characteristic ACF and PACF patterns of this series can be seen in Figs. 4.14 and 4.15, respectively. It may appear that the number of significant spikes is 2 in the PACF, but the most parsimonious model is an AR(1).

The second-order autoregressive process, the ARIMA(2,0,0) model, has an appearance similar to that of the AR(1) model. The AR(2) process has a longer memory than the AR(1) process in that the AR(2) process is a

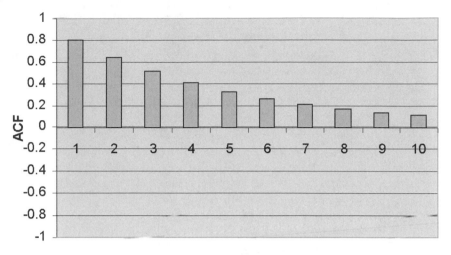

Figure 4.12 ACF of AR(1) series where $\phi_1 > 0$.

function of the previous two observations plus an error term. The ACF of the AR(2) process gradually attenuates as the AR(1) model does, except that the attenuation begins after the second lag rather than after the first lag. The PACF of the AR(2) model is what differs from that of the AR(1) model. The PACF of the AR(2) process clearly has two significant spikes,

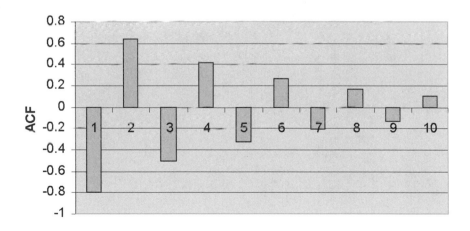

Figure 4.13 ACF of AR(1) series where $\phi_1 < 0$.

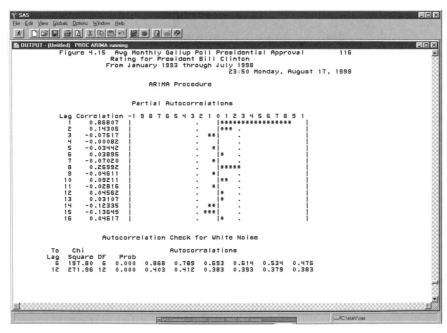

```
SAS                                                                        _ B X
File  Edit  View  Globals  Options  Window  Help
 *  D|&|H| |@| |X|&|&|-| |#|@| |&||
OUTPUT - (Untitled)  PROC ARIMA running                                    _ D X
        Figure 4.14   Avg Monthly Gallup Poll Presidential Approval        90
                      Rating for President Bill Clinton
                  From January 1993 through July 1998
                                             23:50 Monday, August 17, 1998

                           ARIMA Procedure


                     Name of variable = APPROV_1.

                     Mean of working series = 52.57783
                     Standard deviation     = 6.581291
                     Number of observations =       67

                            Autocorrelations

Lag Covariance Correlation -1 9 8 7 6 5 4 3 2 1 0 1 2 3 4 5 6 7 8 9 1
  0  43.313392   1.00000    |                    |********************|
  1  37.598897   0.86807    |              .     |*****************   |
  2  34.165445   0.78880    |            .       |****************    |
  3  30.007841   0.69281    |          .         |**************      |
  4  26.608257   0.61432    |         .          |***********         |
  5  23.117868   0.53373    |        .           |**********          |
  6  20.612191   0.47588    |       .            |********* .         |
  7  17.452506   0.40294    |      .             |********            |
  8  17.834016   0.41174    |      .             |********  .         |
  9  16.602554   0.38331    |      .             |********  .         |
 10  17.009389   0.39271    |      .             |********  .         |
 11  16.394640   0.37851    |      .             |*******   .         |
 12  16.577368   0.38273    |      .             |********  .         |
 13  16.380727   0.37819    |      .             |********  .         |
 14  15.293739   0.35309    |       .            |*******   .         |
 15  12.232544   0.28242    |       .            |*****     .         |
 16  10.991609   0.25377    |        .           |*****     .         |
                              "." marks two standard errors
```

Figure 4.14

```
SAS                                                                        _ B X
File  Edit  View  Globals  Options  Window  Help
 *  D|&|H| |@| |X|&|&|-| |#|@| |&||
OUTPUT - (Untitled)  PROC ARIMA running                                    _ D X
        Figure 4.15   Avg Monthly Gallup Poll Presidential Approval        116
                      Rating for President Bill Clinton
                  From January 1993 through July 1998
                                             23:50 Monday, August 17, 1998

                           ARIMA Procedure


                       Partial Autocorrelations

     Lag Correlation -1 9 8 7 6 5 4 3 2 1 0 1 2 3 4 5 6 7 8 9 1
      1   0.86807    |                    .   |****************    |
      2   0.14305    |                        |***  .              |
      3  -0.07517    |                     . **|    .              |
      4  -0.00082    |                        |    .              |
      5  -0.03442    |                     .  *|    .              |
      6   0.03895    |                     .   |*   .              |
      7  -0.07020    |                     . * |    .              |
      8   0.26992    |                     .   |*****              |
      9  -0.04611    |                     . * |    .              |
     10   0.09211    |                     .   |**  .              |
     11  -0.02816    |                     . * |    .              |
     12   0.04562    |                     .   |*   .              |
     13   0.03107    |                     .   |*   .              |
     14  -0.12335    |                     .**|    .              |
     15  -0.13649    |                     .***|    .              |
     16   0.04617    |                     .   |*   .              |


                 Autocorrelation Check for White Noise

      To   Chi                   Autocorrelations
     Lag  Square DF  Prob
      6  197.80   6  0.000  0.868  0.789  0.693  0.614  0.534  0.476
     12  271.96  12  0.000  0.403  0.412  0.383  0.393  0.379  0.383
```

Figure 4.15

whereas that of the AR(1) process has only one significant spike. If the ACF of the AR(2) process is positive, then the PACF spikes will be positive. If the ACF is negative, then the PACF spikes will be negative. The ACF in Fig. 4.16 and the PACF in Fig. 4.17 of the Chicago Hyde Park purse snatching series extending from January 1969 to September 1973 collected by Reed to evaluate Operation Whistlestop (reported in McCleary *et al.*, 1980) represent an underlying AR(2) process. In general, the autoregressive process is identified by the characteristic patterns of its ACF and PACF. The ACF has a gradual attenuation and the PACF possesses the same number of spikes as the order of the model. (Makidakas *et al.*, 1983; Bresler *et al.*, 1991).

4.10.4. IDENTIFYING MOVING AVERAGE MODELS

Unlike the autoregressive processes, moving average processes have short-term, finite memories. These processes are functions of the error terms. In the first-order moving average process, the MA(1) process, is a function of the current error and the previous error. Consequently, the

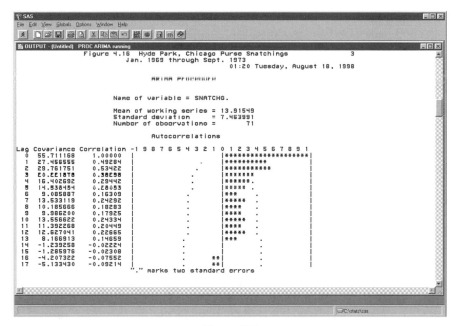

Figure 4.16

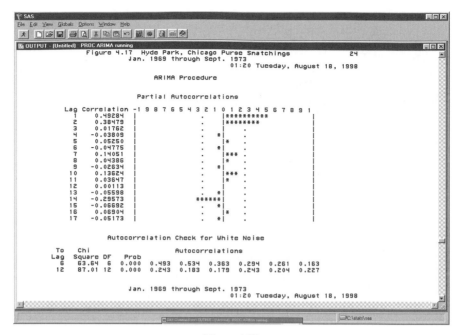

Figure 4.17

ACF of the MA(1) process usually has only one significant spike, whereas the PACF of the MA models generally exhibits gradual attenuation. Owing to the SPSS and SAS parameterization of the model, the ACF and PACF of the first-order moving average happens to be negative when the $\theta_1 > 0$. Examples of the negative spikes (with a positive θ_1) of an MA(1) ACF are shown in Fig. 4.18 and those of PACF in Fig. 4.19. In contrast, a first-order MA(1) model with a positive θ_1 has negative spikes in its respective ACF and PACF.

Second-order MA(2) models commonly have two significant spikes in the ACF followed by no subsequent significant spikes. MA(2) models, owing to the invertibility of the AR with the MA models, have gradual attenuation in the PACF. For stationarity to obtain, the roots have to be real and lie outside the unit circle. If the θ_1 and θ_2 are positive and real, these spikes in the ACF and PACF are negative. If the θ_1 and θ_2 are negative, with complex roots, then the spikes of the ACF function are positive, as shown in Fig. 4.20. The PACF characteristic pattern of the MA(2) models, with negative spikes, is displayed in Fig. 4.21. In reality, the models are more mixed than these ideal types.

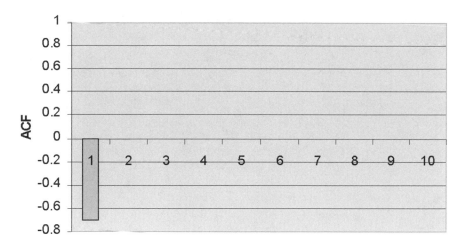

Figure 4.18   ACF of MA(1) series, $\theta_1 > 0$.

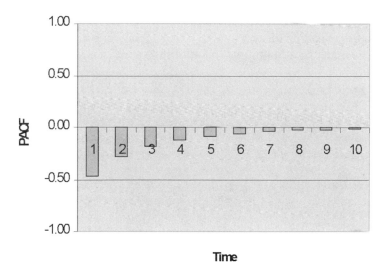

Figure 4.19   PACF of MA(1) series, $\theta_1 > 0$.

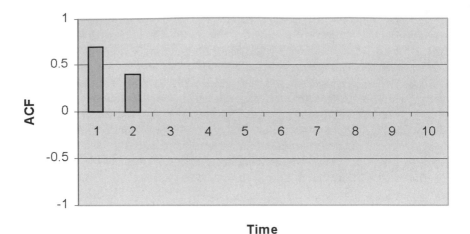

Figure 4.20 ACF of MA(2) series, θ_1, $\theta_2 < 0$.

It is helpful to examine a less than pure example of an MA(2) model. The Democratic percentage of seats in the U.S. House of Representatives, when analyzed from 1896 through 1992, reveals an MA(2) moving average process. U.S. House of Representatives, 1998; Stanley and Neimi, 1995). The ACF for this series is found in Fig. 4.22, and the PACF is found in Fig. 4.23. While the ACF clearly indicates an MA(2) series, the PACF is more ambiguous. For this particular PACF, there is less of a gradual and more of an irregular attenuation, like that shown in Fig. 4.21. At first glance,

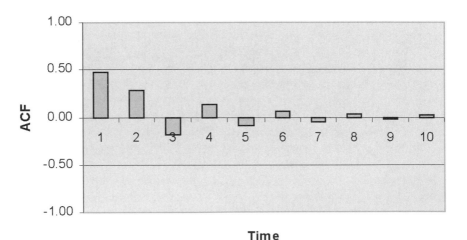

Figure 4.21 PACF of MA(2) series, $\theta_1 < 0$, $\theta_2 < 0$.

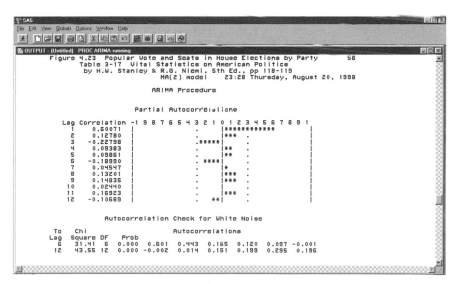

Figure 4.22

Figure 4.23

the series ACF and PACE appear to suggest that the underlying data generating process is an AR(1), but a comparison of SBC for estimated AR(1), MA(1), and MA(2) models supports the conclusion that the underlying process is really MA(2). In this instance the θ_1 and θ_2 are both negative, so the significant spikes in the ACF are both positive (Makridakis *et al.*, 1983; Bresler *et al.*, 1991).

4.10.5. IDENTIFYING MIXED AUTOREGRESSIVE–MOVING AVERAGE MODELS

In reality, many models are mixed models. Some models require differencing before they can be analyzed. These models are nonstationary before differencing but after differencing may be modeled as either AR or MA models. Still, other models are mixed autoregressive moving average processes that requires no prior differencing. Mixed ARMA(1,1) models have at least four different characteristic patterns, classified on the basis of combinations of different signs of the autoregressive and moving average parameters.

Consider first an ARMA(1,1) model with both autoregressive and moving average parameters greater than zero. The characteristic ACF and PACF patterns (Figs. 4.24 and 4.25) exhibit beginning spikes that are positive in sign. The ACF spikes gradually taper off in the correlogram. The gradual decay is not exponential in that it seems to drop, level off, then drop, level off and so on. This ACF pattern continues until the ACF spikes drop below significance. The PACF for the ARMA(1,1) exhibits significant spikes at lags 1, 3, 10, and 14. Stepdown ACF attenuation combined with a positive PACF spike can suggest an ARMA(1,1) model.

Another ARMA(1,1) model is characterized by both the autoregressive component and the moving average being negative. A different characteristic identification pattern is exhibited by this mixed ARMA(1,1) model and can be seen in Figs. 4.26 and 4.27. The ACF exhibits alternating spikes that begin on the negative side and alternate to the positive side. As can be seen in Fig. 4.26 the magnitude of the spikes dampens exponentially. The PACF for this model reveals a single negative spike at lag 1, which can be seen in Fig. 4.27. The significant spike at lag 11 is ignored as the usual one out of 20 tolerable errors.

Another ARMA(1,1) model has an autoregressive component that is negative and a moving average component that is positive. In the ACF shown in Fig. 4.28, the negative autoregressive parameter yields first a

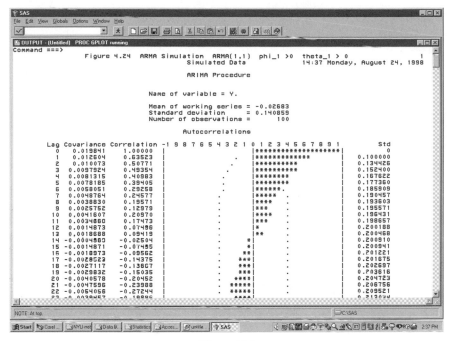

Figure 4.24

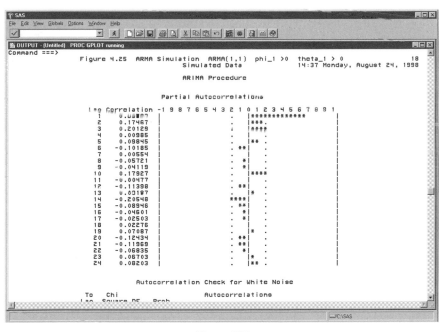

Figure 4.25

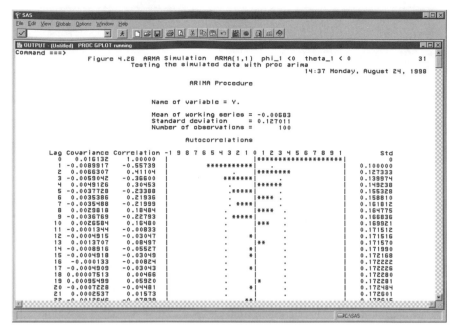

Figure 4.26

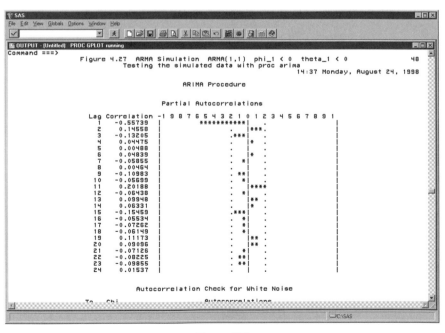

Figure 4.27

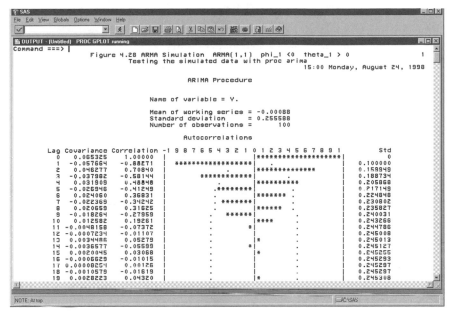

Figure 4.28

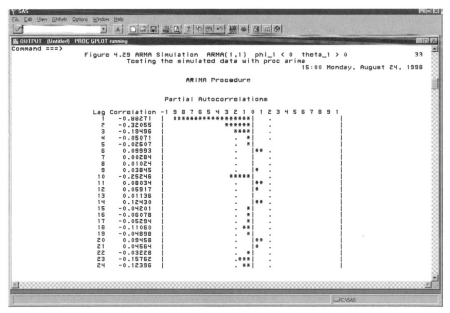

Figure 4.29

negative and then a positive spike. The signs of the spikes continue to alternate and the magnitude of the spikes decays in the characteristic ACF pattern. The PACF for this ARMA(1,1) model remains negative but gradually tapers off into nonsignificance, notwithstanding a significant spike at lag 10 (Fig. 4.29).

Finally, the characteristic correlograms for the ARMA(1,1) model with the positive AR parameter and the negative MA parameter produces a slightly different characteristic pattern. The ACF for this series is one with positive and very gradual (not exponential) declining magnitude, exemplified in Fig. 4.30. In Fig. 4.31, the PACF exhibits pronounced significant negative spikes at lags 1 and 14.

From these characteristic patterns, the analyst would identify the nature of the mixed autoregressive moving average model. When the ARMA model is characterized by ambiguity owing to all of the correlograms tailing off, the researcher may have recourse to the EACF (ESACF). To be sure of the order of an ARMA, the researcher identifies the upper left vertex of the triangle of zeroes in the EACF and uses the location of intersection of marked rows and columns as indication of the order of the ARMA model.

In concluding this chapter, it is helpful to be able to examine an identifi-

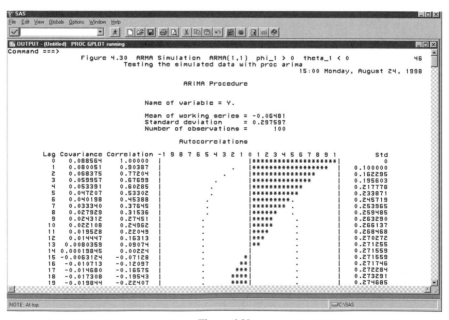

Figure 4.30

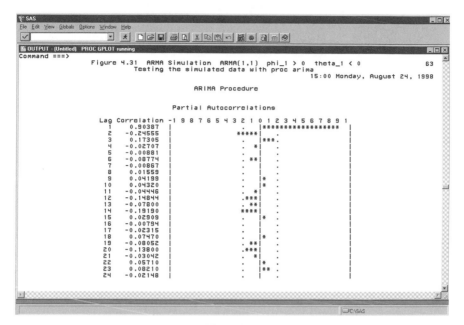

Figure 4.31

cation table. Table 4.2 contains the characteristic form of the integrated, autoregressive, moving average, and mixed models to facilitate identification of the characteristic patterns by examination of the ACF and PACF. The researcher first tests the series for nonstationarity. If it appears to be nonstationary in mean or variance, he applies an appropriate transformation to render it stationary. The ACF and PACF are examined to determine the nature of the series. If the ACF slowly attenuates, the series may require further differencing. Once stationarity is achieved, the series characteristics may be examined further. If the underlying process is autoregressive, moving average, or mixed, the correlograms will exhibit the forms described in Table 4.2. After the researcher identifies the nature of the series, he can estimate the parameters. To assess accuracy of estimation, he examines the residuals of the estimation. If they are white noise, random error devoid of tell-tale residual pattern, he assumes that the estimation is correct. If a pattern of significant spikes persists in the residuals, alternative identification and estimation may be in order. When the characteristic patterns of spikes have been properly identified and estimated in the model, the residuals of the estimation process will resemble random insignificant white noise and that stage of the modeling will have been completed (Cook and Campbell, 1979).

Table 4.2
Identification of ARIMA Processes

Process	ACF	PACF
White noise process		
ARIMA(0,0,0)	no significant spikes	no significant spikes
Integrated process		
ARIMA(0,1,0) $d = 1$	slow attenuation	1 spike at order of differencing
Autoregressive processes		
ARIMA(1,0,0) $\varphi_1 > 0$	exponential decay, positive spikes	1 positive spike at lag 1
ARIMA(1,0,0) $\phi_1 < 0$	oscillating decay, begins with negative spike	1 negative spike at lag 1
ARIMA(2,0,0) $\varphi_1, \varphi_2 > 0$	exponential decay, positive spikes	2 positive spikes at lags 1 and 2
ARIMA(2,0,0) $\varphi_1 < 0, \varphi_2 > 0$	oscillating exponential decay	1 negative spike at lag 1, 1 positive spike at lag 2
Moving average processes		
ARIMA(0,0,1) $\theta_1 > 0$	1 negative spike at lag 1	exponential decay of negative spikes
ARIMA(0,0,1) $\theta_1 < 0$	1 positive spike at lag 1	oscillating decay of positive and negative spikes
ARIMA(0,0,2) $\theta_1, \theta_2 > 0$	2 negative spikes at lags 1 and 2	exponential decay of negative spikes
ARIMA(0,0,2) $\theta_1, \theta_2 < 0$	2 positive spikes at lags 1 and 2	oscillating decay of positive and negative spikes
Mixed processes		
ARIMA(1,0,1) $\varphi_1 > 0, \theta_1 > 0$	exponential decay of positive spikes	exponential decay of positive spikes
ARIMA(1,0,1) $\varphi_1 > 0, \theta_1 < 0$	exponential decay of positive spikes	oscillating decay of positive and negative spikes
ARIMA(1,0,1) $\varphi_1 < 0, \theta_1 > 0$	oscillating decay	exponential decay of negative spikes
ARIMA(1,0,1) $\varphi_1 < 0, \theta_1 < 0$	oscillating decay of negative and positive spikes	oscillating decay of negative and positive spikes

This chapter has presented the basis for identification of ARIMA processes. We have discussed various series, their nonstationarity, their transformations to stationary, and their autoregressive or moving average characteristics, as well as different types of these models. We explained and illustrated the characteristics by which they can be identified. In this way, we have elaborated the additive Box–Jenkins models. Computer program syntax and data sets are available on the World Wide Web by which they may be tested.

REFERENCES

Abraham, B., and Ledolter, J. (1984). "A Note on Inverse Autocorrelations," *Biometrika*, **71**, pp. 609–614.

Babinec, T., Aug, 13–14, 1996. Parameterization of constant in time series models, personal communication. Thanks must also go to Wei Zang at SPSS for his commentary.

Box, G. E. P., and Jenkins, G. M. (1976). *Time Series Analysis: Forecasting and Control.* 2nd ed. San Francisco: Holden-Day., p. 34.

Box, G. E. P., Jenkins, G. M., and Reinsel, G. C. (1994). *Time Series Analysis Forecasting and Control,* 3rd. ed. Englewood Cliffs, NJ: Prentice Hall, pp. 32–33, 66, 68, 70–75, 188, 314–315, 547.

Bowerman, B., and O'Connell, R. T. (1993). *Forecasting and Time Series.* Belmont, CA: Wadsworth, pp. 437–486, 489.

Bresler, L., Cohen, B. L., Ginn, J. M., Lopes, J., Meek, G. R., and Weeks, H. (1991). *SAS/ETS Software: Applications Guide 1.* Version 6, 1st ed. Cary, NC: SAS Institute, p. 36.

Brocklebank, J., and Dickey, D. (1994). *Forecasting Techniques Using SAS/ETS Software: Course Notes.* Cary, NC: SAS Institute, Inc., pp. 66, 71–72, 163–175, 176–177.

Chatfield, C. (1980). "Inverse Autocorrelations." *Journal of the Royal Statistical Society,* **A142,** pp. 363–377.

Cleveland, W. S. (1972). "The Inverse Autocorrelations of a Time Series and their Applications." *Technometrics,* **14,** pp. 277–293.

Clinton, President W. J. and Council of Economic Advisors. (1995). *The Economic Report of the President.* Washington, DC: Government Printing Office, p. 314.

Cook, T. D., and Campbell, D. T. (1979). *Quasi-Experimentation: Design and Analysis Issues for Field Settings.* Boston: Houghton-Mifflin, pp. 233–252.

Cromwell, J. B., Labys, W.C., and Terraza, M. (1994). *Univariate Tests for Time Series Models.* Thousand Oaks, CA; Sage Publications, pp. 10–19, 25–27.

Ege, G., Erdman, D. J., Killam, B., Kim, M., Lin, C. C., Little, M., Narter, M. A., and Park, H. J. (1993). *SAS/ETS User's Guide. Version 6,* 2nd ed. Cary, NC: SAS Institute, Inc., pp. 100–180, pp. 136–137.

Enders, W. (1995). *Applied Econometric Time Series.* New York: John Wiley & Sons, pp. 221–227, 300.

Gallup Poll Presidential Approval Index, Gallup Poll, Inc., World Wide Web, http://www.gallup.com/polltrends/jobapp.htm, Aug–Nov, 1996. Data are analyzed and posted with permission of the Gallup Organization, Inc., 47 Hulfish Street, Princeton, N.J. 08542.

Gottman, J. M. (1981). *Time Series Analysis: A Comprehensive Introduction for Social Scientists.* New York: Cambridge University Press, pp. 153ff, 174–177.

Granger, C. W. J. (1989). *Forecasting in Business and Economics,* 2nd ed. San Diego: Academic Press, p. 40.

Granger, C. W. J. and Newbold, P. (1993). *Forecasting Economic Time Series,* 2nd ed. San Diego: Academic Press, pp. 7–120.

Griffiths, W. E., Hill, R. C., and Judge, G. G. (1993). *Learning and Practicing Econometrics.* New York: Wiley, p. 662.

Harvey, A. C. (1993). *Time Series Models,* 2nd ed. Boston: MIT Press, pp. 1–33.

Liu, L. M., Hudak, G. B., Box, G. P., Muller, M. E., and Tiao, G. C. (1986). *The SCA Statistical*

System: Reference Manual for Forecasting and Time Series Analysis, Version III. DeKalb, IL: Scientific Computing Associates, pp. 3–19.

Makridakis, S., Wheelwright, S. C., and McGee, V. E. (1983). Forecasting: Methods and Applications, 2nd ed. New York: Wiley, pp. 421–422, 442–443.

McCleary, R., and Hay, Jr., R. with Meidinger, E. and McDowell, D. (1980). Applied Time Series Analysis for the Social Sciences. Beverly Hills: Sage, pp. 18–83, 315–316. Data used with permission of author.

Mills, T. C. (1990). Time Series Techniques for Economists. Cambridge: Cambridge University Press, pp. 116–180.

Pandit, S. M., and Wu, S. (1993). Time Series and System Analysis with Applications. Malabar, FL: Krieger Publishing, p. 131.

SAS Institute, Inc. (1992). SAS /ETS Software: Applications Guide. Time Series Modeling and Forecasting, Financial Reporting, and Loan Analysis. Version 6, 1st ed. Cary, NC: SAS Institute, Inc., pp. 35–108.

SPSS, Inc. (1996). SPSS 7.0 Statistical Algorithms, Chicago: SPSS, Inc. Various drafts of these algorithms were generously provided by Tony Babinec, Director of Business Management, and David Nichols, Senior Support Statistician, pp. 3–7, 44–51.

SPSS, Inc. (1988). SPSS-X Trends. Chicago: SPSS, Inc.: SPSS, Inc., pp. B-29–B-154.

SPSS, Inc. (1993). SPSS for Windows Trends Release 6.0. Chicago: SPSS, Inc. pp. 264–271.

Stanley, H. W., and Neimi, R. G. (1995). Vital Statistics on American Politics, 5th ed. Washington: CQ Press, Inc., 118–119.

U.S. House of Representatives (1998). 'Political Divisions of the U.S. Senate and House of Representatives on Opening Day 1855 to the Present'; World Wide Web: http://clerkweb.house.gov/histrecs/history/elections/political/divisions.htm. Retrieved 1998.

Vandaele, W. (1983). Applied Time Series and Box–Jenkins Models. Orlando: Academic Press, pp. 39, 46–47.

Wei, W. (1990). Time Series Analysis. Univariate and Multivariate Methods. Redwood City, CA: Addison-Wesley Publishing Co., pp. 21, 39–40, 58–59, 123–132.

Zang, W. (1996). SPSS, Inc. See Babinec, T. (1996). Personal communication.

Chapter 5

Seasonal ARIMA Models

5.1. CYCLICITY

Cyclicity can be defined as long wave swings, whereas seasonality is generally defined as annual periodicity within a time series (Granger, 1989). Cycles involve deviations from trends or equilibrium levels. They may assume the likeness of a sinc wave. They are characterized by phases and turning points in the series. There are several different classifications of phases of a cycle. The cycle may be described by four basic phases. The reference point is an equilibrium level or trend line. In the upswing phase, the series value increases. When the series reaches a maximum, the turning point is called the peak of the cycle. The cycle then enters the downswing or contraction phase. When the series value reaches the equilibrium or trend line, a point of inflection (where the concavity of the cycle changes) has been reached. After the point of inflection, the value of the series goes negative. The series value eventually reaches a minimum value, the turning point at which is called the trough of the cycle. The upswing phase is resumed until the point of inflection is reached again, which completes the cycle.

The cycle is measured according to its frequency, duration, amplitude, and phase shift. The frequency pertains to the number of cycles per span of some standard number of time periods. The duration (wavelength) of the cycle refers to the number of time periods the cycle spans. The amplitude of the cycle refers to the magnitude of the distance between minimum to maximum series values during the cycle. The phase shift refers to horizontal displacement of the cycle—measured by the angle (usually measured in radians) added to the equilibrium level to create an intercept for the beginning of the cycle.

More complex classifications of the phases of business cycles have been propounded by several prominent economists, a number of whom formed the National Bureau of Economic Research in 1920. Among these economists were Arthur Burns and Wesley Mitchell, who focused on the sequence of changes delineated by turning points: expansion, recession, and recovery. In another classification, proposed by Joseph Schumpeter, there are upswing, recession, depression, and revival phases. At the peak of the cycle, the upswing turns into the recession phase. At the point of inflection, or equilibrium level, the recession turns into the depression phase. At the lower turning point, or trough, of the cycle, the depression turns into the upswing again. This upswing has been called the expansion, recovery, or revival phase (Neimira and Klein, 1994).

Not only have economists sought to determine the durations of expansion and contraction phases along with the diffusion of effect of these cycles on related series, they have also endeavored to identify leading, coincident, and lagging indicators of business cycle turning points. The methodology

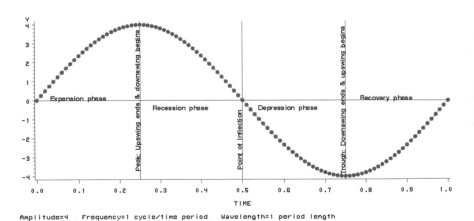

Figure 5.1 Schumpeter's business cycle: fluctuation around an equilibrium level or trend line.

of leading indicators is dealt with in Chapter 9 in the discussion of transfer function models.

Many important cycles are seasonal. They have annual periodicity. A crop price cycle is associated with varying yield from the annual harvests. When there is developing abundance of supply, the price tends to decrease. As the supply becomes depleted, the price gradually rises. The recurrent decline and rise in produce price is associated with a seasonal pattern. Depending on the goods or service under consideration, the span of seasonal activity may extend over a month, a quarter, a half-year, or a year. Examples of activities associated with a particular season include the summer purchase of swim suits, summer flooding the unemployed with out-of-school young adults, autumn return-to-school purchases, or purchase of winter sporting equipment. Purchasing during the Thanksgiving–Christmas holiday season is another example. Many series that have not been descasonalized are riven with variation that demands special attention, whether for modeling the series or for forecasting its values.

To model seasonality, the length of the series must exceed the length of the span of the seasonality (Enders, 1995). Incomplete spans of seasonality may add error to the analysis. Enders writes that when seasonal variation predominates, much of the error in the forecast may derive from this variation. Therefore, we should remove or model seasonality to whatever extent is possible before forecasting. There are several methods for adjusting for or modeling seasonality. Ratio-to-moving averages, Winter's exponential smoothing, or the Census X-11 methods, which were discussed in Chapter 2, have been used to model or extract seasonality. Because these methods were discussed earlier, they will not be reviewed here. Seasonal dummy or trigonometric function variables may be employed with autoregression methods, which are discussed later, to model deterministic seasonality or cyclicity.

Especially when a series is being used for forecasting, the seasonality, which contributes to error variance, should be removed. When that is done the series is called seasonally adjusted. If the series is not seasonally adjusted first, seasonality can be modeled in the Box–Jenkins approach by employing seasonal components alone or mixing these seasonal with regular nonseasonal components to construct multiplicative Box–Jenkins models. Within Box–Jenkins models, seasonality may refer to any repetition of pattern of activity (McCleary et al., 1980; Makridakis et al., 1983). Seasonal variation has an order to it. By convention, the order of seasonality is the number of seasons in an annual period. Quarterly seasonal peaks in data indicate a seasonal order of 4. If the data are measured daily but monthly seasonality is present, then the order of the seasonality is 12 (Bowerman and O'Connell, 1993). In order to approach the basics of seasonal modeling, we turn first to the subject of seasonal stationarity and its complement, seasonal nonstationarity.

5.2. SEASONAL NONSTATIONARITY

If the series under consideration exhibits annual patterns of nonstationarity, these characteristics need to be identified, removed, or modeled before further analysis can proceed. Annual patterns may manifest themselves as quarterly shifts in the mean level of the series. Alternatively, they may appear as monthly fluctuations in variance. By controlling for such types of seasonal nonstationarity, we can identify the confounding effects of seasonal features and set them aside in or completely remove them from the model. How the researcher proceeds usually depends on the kind of nonstationarity detected. Graphical review, by time sequence plot or correlogram, of the time series enables the analyst to find and identify different types of seasonal nonstationarity.

One type of nonstationarity to look for would be a seasonal shift in the mean level of the series. Sudden changes in the mean level of a series may follow from a shift in deterministic regime or a local trend. The series might reach a threshold level or experience a delayed reaction to other influences that may bring about a sudden shift of level. Enders (1995) discusses how within the periods there may be stationarity, but between periods there may be nonstationarity. He proceeds to give an example where these step or ramp shifts in mean might artificially produce the appearance of an overall significant positive (or negative) trend. If this characteristic trend is not removed from the data or controlled by modeling, the nonstationarity may preclude proper analysis.

The series in Fig. 5.2 is an example of nonstationarity brought about by sudden shifts in level. The series in this graph has three distinct different levels of mean. McCleary *et al.* (1980) also give an example of this kind of nonstationarity. The first level proceeds around a mean of 3.4 or so for four time periods before rising to a new level. The second level hovers around 5.5 for four time periods before abruptly increasing again. One time period later the series assumes a new level around 7.4 and hovers there for four more time periods. The series is characterized by three different levels within the time horizon of our data capture. Each time the series shifts level, it does so within a time span of one period. The time-dependent mean shift therefore is rather abrupt. It is the graphical review that brings this aspect of the structural change to the fore.

We can detect sudden time-dependent mean shifts with regression analysis. The researcher can test for significant regime shifts with dummy variable

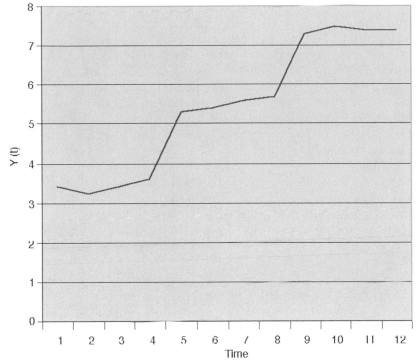

Figure 5.2 Seasonal nonstationarity with shifts of level.

regression analysis. With the appearance of three distinct levels in the series, it would be necessary to construct two dummy variables. Each dummy variable would be constructed to indicate a change of level from a reference or equilibrium level. The dummy variable is given a value of 0 before the change and a value of 1 after the change. If the dummy variable regression coefficient is found to be significant, the regime level of the model is changed by the magnitude of the significant coefficient. If the dummy variable is not significant and there are enough observations in this part of the series, the apparent shift in regime level is not distinguishable from ordinary error variance. If a significant change in regime takes place haphazardly, then there is simple mean nonstationarity. A seasonal pulse could be a source of seasonal nonstationarity as well. Similarly, a dummy variate could be used to model such a pulse (Reilly, 1999). If the mean level changes significantly at annual intervals, then seasonal mean nonstationarity obtains. The residuals may be used to model the remainder of an additive series.

A seasonal shift in slope of a trend may be another kind of seasonal nonstationarity. As displayed in Fig. 5.3, with each shift in level, the trend increases. This piecewise increase in inclination has the appearance of a ramp function, as opposed to the previously displayed step function. Ramp functions represent a series that eventually exhibits a gradual increase in the level of a series. This change in slope takes place often in response to some external event or influence. We can similarly test these functions with dummy variable regression analysis. The use of dummy variables to denote the segment of this series where the level substantially increases will yield significant regression coefficients. The residuals can be used for further ARIMA analysis.

Repeated wavelike patterns that span periods longer than 1 year are called cycles. If these cycles are not removed beforehand, the researcher can model them with multiplicative Box–Jenkins models. In Fig. 5.4, the well-known Wolfer sunspot data from 1770 through 1869 from Box *et al.* (1994) has been graphed. Figure 5.4 displays these cycles as possessing an 11-year span.

The Wolfer sunspot time series in Fig. 5.4, exhibits this kind of nonstationarity. When the seasonality in the series derives from annual periodic

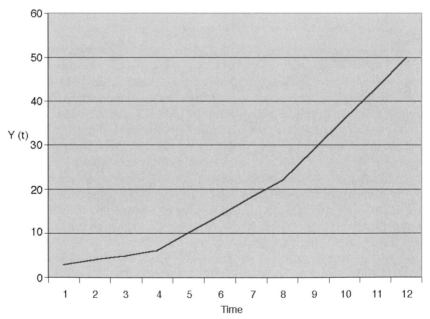

Figure 5.3   Seasonal nonstationarity with shifts of trend.

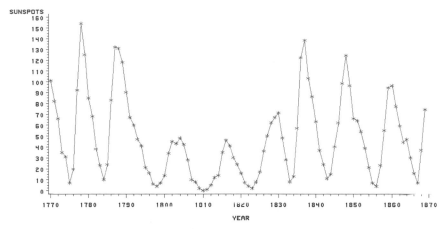

Figure 5.4   Wolfer sunspot data.

fluctuation, such as in the Sutter County, California, labor force data, presented and analyzed in McCleary *et al.* (1980), the analyst can model this seasonality as well. In the case of the Sutter County data, nonstationarity flows from trend as well as seasonality. Seasonal changes may appear as periodic peaks or valleys in the series. One of the seasonal examples mentioned is the growth in the workforce during the summer, when migrant laborers and students enter the work force. That seasonality can be observed in the time sequence plot of the Sutter County work force data in Fig. 5.5. In this graph the annual periods clearly indicate seasonality in the work force size.

Seasonal nonstationarity may also be detected by correlograms as well. Consider the monthly time series data from the Sutter County, California, labor force size. There appear to be annual or 12-month spikes in the ACF and PACF correlograms. The ACF in Fig. 5.6 clearly exhibits this prima facie evidence of seasonal nonstationarity. The PACF in Fig. 5.7 reveals the seasonal spikes as well. Slow attenuation of the seasonal peaks in the Fig. 5.6 ACF signifies seasonal nonstationarity. The 12-month ACF periodicity can be seen in the periodic peaks at lags 12 and 24, suggestive of seasonal differencing at lag 12. The sample PACF of seasonal models is often difficult to interpret. When the parameterization of the seasonal model is discussed, we will see that the multiplication of the nonseasonal by the seasonal factors can produce significant interaction terms, which render analysis of the individual lag structure somewhat complicated. For this reason, it is the ACF, and not the PACF, that is used as the principal

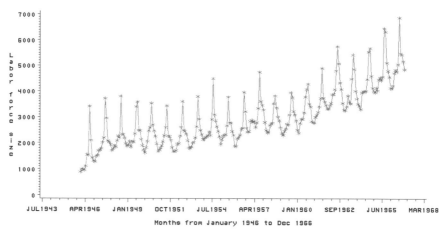

Figure 5.5   Labor force size, Sutter County, California, Jan 1946 through Dec 1966.

Figure 5.6

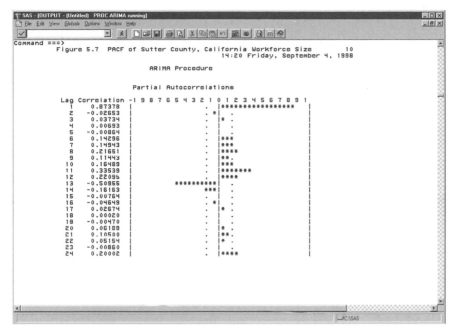

Figure 5.7

guide to seasonal model analysis. As an illustration, the sample PACF in Fig. 5.7 shows significant positive spikes at lags 11 and 12, a significant negative spike at lag 14, and a large negative spike at lag 13, suggestive of a multiplicative model that we will soon examine in detail. However, the PACFs tip off the analyst to the multiplicative nature of the series, as they reveal statistical significance of the interaction product spikes. Together with the time sequence plot, the sample ACF and PACF suggest seasonal nonstationarity.

Seasonal nonstationarity may be detected by unit root tests. When a series is nonstationary, it requires a transformation to render it stationary. If there is seasonal trend nonstationarity and this trend is deterministic, a regression of the response against a linear measure of time will not only control for the trend; it will also yield residuals amenable to further analysis. If the series is one of stochastic trend, with an accumulation of moving average shocks leading to the movement away from a starting point, then the series may be difference stationary, in which case differencing will render the series stationary.

For the most part, the parameterization of the augmented Dickey–Fuller test was discussed earlier in Chapter 3. To test for a seasonal root at lag

12, the maximum testable seasonal lag, under the current SAS system, at the time of this writing, the DLAG option is used to specify the order of the seasonal lag, while the ADF = 0 option specifies the autoregressive lag used to effect the white noise condition necessary for proper testing of seasonal nonstationarity. To properly neutralize the contamination of the autocorrelation in the testing of the series, it is recommended that the ADF option = $(p - 1)$, where p equals the autoregressive order of the process to be tested (Meyer, 1999). If there is no autocorrelation in the series, the following programming syntax is used:

```
PROC ARIMA;
IDENTIFY VAR=XX Stationarity=(adf=0 DLAG=12);
Run;
```

From this command syntax the following output is obtained.

```
   Seasonal Dickey-Fuller Unit Root Tests

   Type         Lags RHO          Prob<RHO T         Prob<T

Zero Mean     0     -111.044 0.0001    -7.9546 0.0001


Single Mean 0       -111.043 0.0001    -7.9370 0.0001
```

Assuming that the series contains within it enough observations, a significant ρ or T test indicates that the series is seasonally stationary and no

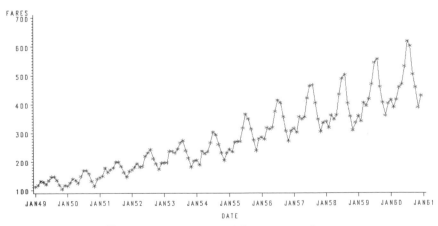

Figure 5.8 International airline passenger fares.

seasonal differencing is required. A nonsignificant ρ or T test indicates that the series is seasonally nonstationary in its tested form and is in need of differencing before further modeling is attempted. When there is a seasonal unit root, the model requires more differencing at a seasonal lag prior to further analysis. Although SAS has the test for the seasonal unit root, SPSS currently has no such procedure.

There may be seasonal variations in volatility or variance. Figure 5.8 illustrates growing variation in annual international airline ticket sales data (Box *et al.*, 1994). When a change in volatility cannot be handled by differencing alone, then a Box-Cox, log, power, arcsine, or square root transformation may be in order to stabilize the variance along with the seasonal differencing.

5.3. SEASONAL DIFFERENCING

The first order of business is to detect and eliminate the nonstationarity. As stochastic nonseasonal nonstationarity is commonly eliminated by first or second differencing, we can often eliminate stochastic seasonal nonstationarity by seasonal differencing. Seasonal differencing means differencing by the order of seasonal periodicity. When a series exhibits regular patterns of behavior within an annual period, the order of periodicity signifies the number of seasons within a year. Quarterly data would have a periodicity of 4, since there are four seasons within the year. If data are monthly, and their time plot or sample ACF reveals quarterly peaks, then quarterly nonstationarity exists. Seasonal differencing with an order of 4 could be used to resolve the problem. If the data are recorded daily with monthly peaks (or valleys), then differencing with an order of 12 might eliminate the seasonal nonstationarity. If fluctuations in the series are also annual, as in the Sutter County work force data series, a regular first difference and a seasonal 12th difference could transform the series into stationarity. The seasonal difference has a smaller variance than the nonseasonal difference, and taking seasonal differences has the effect of rendering seasonal variation in the series stationary. Modeling seasonality in Box–Jenkins methodology takes place in much the same way as modeling of regular nonseasonal series.

The formulation of seasonal differencing may be additive or multiplicative. On the one hand, there is an additive formulation of a difference at a seasonal lag–such as $(1 - L^s)$—as in $(1 - L^4)y_t = C + e_t$. On the other hand, a model with multiplicative seasonal differencing entails the multiplication of the nonseasonal by the seasonal differencing factors. When the product of a regular difference and a seasonal difference is required to render a series stationary, one multiplies the first regular factor by the seasonal factor to obtain the differencing for the series. The multiplicative seasonal ARIMA model has multiplicative differencing, such as $(1 - L^d)$ $(1 - L^s)y_t = C + e_t$, where Y_t is the undifferenced series variable, d is the order of regular differencing and s is the order of seasonal differencing. In the differenced model, z_t is used to indicate previous differencing of Y_t. For the multiplicative model, w_t is used by convention to indicate seasonal differencing of the regularly differenced model, $w_t = \mu + (1 - L^s)z_t$. Consequently, the centered form that a series with regular and seasonal differencing takes is $W_t = (1 - L^d)(1 - L^s)Y_t$.

An example of a seasonal model with differencing at a seasonal lag indicated by the length of the seasonal period is the Sutter County, California, labor force data (McCleary *et al.* 1980). The sample ACF of this series exhibits a seasonal nonstationarity of order 12. Under these circumstances, we would try seasonal differencing of order 12 as a means of effecting stationarity. To complete the transformation to stationarity, this model also requires regular differencing at the first lag. In sum, it involves first-order regular and 12th-order seasonal differencing before the series is sufficiently stationary to be amenable to further analysis. When one multiplies the regular and the seasonal factors in the Sutter County work force series, the result is a simple product of the lag factors:

$$
\begin{aligned}
W_t &= (1 - L)(1 - L^{12})\text{Workforce}_t \\
&= \text{Workforce}_t - \text{Workforce}_{t-1} - \text{Workforce}_{t-12} \qquad (5.1) \\
&\quad + \text{Workforce}_{t-13}.
\end{aligned}
$$

The net effect of the multiplication of the lag factors is to transform the nonstationary lag structures of the workforce series into a stationary one, which may be analyzed further into regular and seasonal ARIMA components. The final lag structure of the transformed series is simply the product of the lags resulting from the multiplication of regular by seasonal differencing factors.

5.4. MULTIPLICATIVE SEASONAL MODELS

Just as ARIMA notation has regular parameters, it may also have seasonal parameters. The regular parameters of the ARIMA model are denoted in the formulation of ARIMA(p,d,q) by uncapitalized letters respectively representing the regular autoregressive, integration, and moving average orders of the model. Similarly, the seasonal components of the ARIMA model are denoted by ARIMA(P,D,Q)$_s$, where capitalized letters respectively represent the seasonal components of the model and the s

indicates the order of periodicity or seasonality. Seasonal ARIMA models are sometimes called SARIMA models. A full formulation of a multiplicative SARIMA model has the general form ARIMA$(p,d,q)(P,D,Q)_s$. The parentheses enclose the nonseasonal and the seasonal factors, respectively. The parameters enclosed indicate the order of the model. An example of a model with monthly data characterized by regular as well as seasonal random walk is an ARIMA$(0,1,0)(0,1,0)_{12}$ model. The differencing required is that of first-order nonseasonal and seasonal differencing, with the seasonal differencing performed at a lag of 12 months. Seasonal models formulate the between-season periodic variation, whereas nonseasonal models formulate the within-season variation (Wei, 1990; Box *et al.*, 1994). Multiplicative seasonal models consist of multiplication of nonseasonal and seasonal factors.

One may multiply the regular and the seasonal factors when they are expressed in terms of their lag operators. When the two factors are multiplied, the seasonal model assumes a more complicated form than with the simple additive models discussed in the previous chapter. In this process, the models are reduced to their lag factors and then multiplied to give the penultimate structure. The terms are then collected and redistributed to give the final equation. The transformed expansion of the Sutter County labor force differencing was just explained. The factors on the right-hand side of the model are similarly expanded. If the parameters of the series are small enough and if the sample size is small enough, when this multiplication takes place, the product or interaction term may turn out to be nonsignificant. In the Sutter County labor force model, the constant happens to equal 0 and therefore drops out of the model. An example of a series, where the positive interaction term with a magnitude of 0.12 at lag 13 happens to be nonsignificant, is

$$
\begin{aligned}
(1 - 0.4L)(1 - 0.3L^{12})Y_t &= (1 - 0.4L - 0.3L^{12} + 0.12L^{13})Y_t \\
&= Y_t - 0.4Y_{t-1} - 0.3Y_{t\ 12} + 0.12Y_{t-13}.
\end{aligned} \tag{5.2}
$$

Owing to the multiplication of the regular with the seasonal moving average term, there is a small positive differencing interaction at lag 13. In this case, the other lags have negative spikes while the 13th lag is a spike in the opposite direction. If the series is not long enough, this interaction term may turn out to be nonsignificant. Under these circumstances, a researcher identifying the model might treat it as if it did not contain its interaction term and might specify it as an additive subset model: $(1 - 0.4L - 0.3L^{12})$, requiring only first- and 12th-order regular differencing. The only difference between the additive and multiplicative model is the inclusion of such interactions. These interactions, with their opposite directions, may complicate the appearance of a sample PACF used for analysis of these models. It is this seasonal differencing $(1 - L^s)$ that Box and Jenkins (1976) called the simplifying operator, insofar as it renders the residual series stationary and amenable to further analysis.

5.4.1. SEASONAL AUTOREGRESSIVE MODELS

Apart from seasonal differencing, there are seasonal autoregressive models (SAR), seasonal moving average (SMA) models, and seasonal autoregressive moving average (SARMA) models. Seasonal autoregressive models contain autoregressive parameters at seasonal lags. A centered, purely seasonal autoregressive model, $(1 - \Phi_1 L^3)Y_t = e_t$, might be identified by exponentially declining ACF spikes at every third lag. The ACF at the first lag might equal 0.5, while the ACF at lag 3 would equal 0.25, and the ACF at lag six would equal 0.125, etc., as can be seen in Fig. 5.9. The PACF for a purely SAR model reveals a positive spike at the seasonal lag as shown in Fig. 5.10. Hence, either the time sequence plot, ACF, or PACF can be used as a primary instrument for identifying seasonal autoregressive models.

A multiplicative seasonal autoregressive model contains both nonseasonal and seasonal autoregressive factors. A simple example of a seasonal autoregressive model would be one with a regular first-order autoregressive term and a seasonal autoregressive term of order 12. A simple formulation of this $\text{ARIMA}(1,0,0)(1,0,0)_{12}$ is:

$$
\begin{aligned}
&y_t = \mu + e_t/(1 - \varphi_1 L)(1 - \Phi_{12} L^{12}) \\
&C = \mu - \varphi_1 \mu - \Phi_{12}\mu + \varphi_1 \Phi_{12}\mu \\
&(1 - \varphi_1 L)(1 - \Phi_{12} L^{12})(y_t - \mu) = e_t \\
&(1 - \varphi_1 L)(1 - \Phi_{12} L^{12})Y_t = e_t,
\end{aligned}
\tag{5.3}
$$

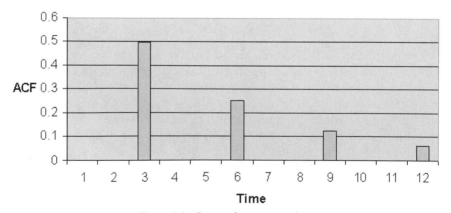

Figure 5.9 Seasonal autoregression.

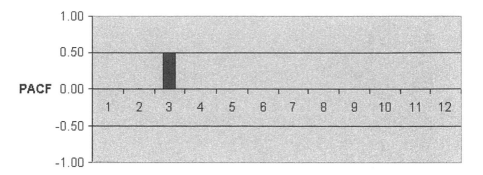

Figure 5.10 Seasonal autoregression.

where C is the constant, μ is the mean. We can expand this to

$$Y_t(1 - \varphi_1 L - \Phi_{12} L^{12} + \varphi_1 \Phi_{12} L^{13}) = e_t$$
or
$$Y_t - \varphi_1 Y_{t-1} - \Phi_{12} Y_{t-12} + \varphi_1 \Phi_{12} Y_{t-13} + e_t$$
or   (5.4)
$$Y_t = \varphi_1 Y_{t-1} + \Phi_{12} Y_{t-12} - \varphi_1 \Phi_{12} Y_{t-13} + e_t$$
or
$$(y_t - \mu) - \varphi_1(y_{t-1} - \mu) - \Phi_{12}(y_{t-12} - \mu) + \varphi_1 \Phi_{12}(y_{t-13} - \mu) = e_t.$$

Note that an interaction term is present in this model as well. The sign of that term is opposite that of the other autoregressive terms and it emerges at a lag one more than the highest seasonal autoregressive term. As the main effects of that interaction term have smaller parameters, the interaction term may disappear into insignificance.

If the model being analyzed required differencing, it becomes more elaborate. The differencing factors are multiplied by the autoregressive factors, rendering the result more complicated. In the following example, first-order regular differencing combined with regular and seasonal autoregression has a sample ACF characterized by regular first-order and seasonal 12th-order autoregressive characteristics in monthly data. The notation for such a model is $\text{ARIMA}(1,1,0)(1,0,0)_{12}$. A model of a series that exemplifies a model that contains regular and seasonal autoregressive parameters, along with the first differencing, is

$$(1 - L)(1 - \varphi_1 L)(1 - \Phi_{12} L^{12}) y_t = \mu + e_t. \qquad (5.5)$$

Expansion of this seasonal autoregressive series yields

$$
\begin{aligned}
(1 - L)(1 - \varphi_1 L - \Phi_{12}L^{12} + \varphi_1\Phi_{12}L^{13})y_t &= \mu + e_t \\
= (1 - \varphi_1 L - \Phi_{12}L^{12} + \varphi_1\Phi_{12}L^{13} - L + \varphi_1 L^2 + \Phi_{12}L^{13} \\
- \varphi_1\Phi_{12}L^{14})y_t &= \mu + e_t \\
= (1 - L - \varphi_1 L + \varphi_1 L^2 - \Phi_{12}L^{12} + \Phi_{12}L^{13} + \varphi_1\Phi_{12}L^{13} \\
- \varphi_1\Phi_{12}L^{14})y_t &= \mu + e_t \qquad (5.6) \\
= y_t - y_{t-1} - \varphi_1 y_{t-1} + \varphi_1 y_{t-2} - \Phi_{12}y_{t-12} + \Phi_{12}y_{t-13} \\
+ \varphi_1\Phi_{12}y_{t-13} - \varphi_1\Phi_{12}y_{t-14} &= \mu + e_t.
\end{aligned}
$$

A simple first difference multiplied by a first- and seasonal 12th-order autoregressive term yields an equation consisting of a first difference plus a second order regular autoregressive term along with seasonal 12th- and 13th-order autoregressive terms coupled with two interaction terms of 13th- and 14th-order. Although the seasonal ACF is not complicated, the seasonal PACF becomes more complex. It is the PACF that will reveal the opposite signed interaction term at the appropriate lag, signifying the presence of a multiplicative seasonal autoregressive model. The actual ACF and PACF of this model will be examined in detail when the problem of identifying seasonal models is addressed.

5.4.2. SEASONAL MOVING AVERAGE MODELS

The seasonal moving average model typically possesses a seasonal moving average component. Moreover, the seasonal moving average model is a common multiplicative model. Because the differencing factors are not multiplied directly by the moving average factors, these models tend to be a little simpler than the type just described. If the interactions are small or negligible, then the multiplicative moving average model may reduce to an additive model whose nonsignificant multiplicative components have been trimmed away. In other words, an ARIMA $(0,0,1)(0,0,1)_{12}$ model is a common seasonal moving average model that may reduce to an additive subset model if the interaction term is negligible. Although the seasonal multiplicative model has a regular moving average component at lag one and a seasonal component at lag 12 along with a reversed signed interaction component at lag 13, the additive subset may only have regular and seasonal moving average components at lags 1 and 12, lacking that reverse signed component at lag 13. In general, multiplicative models retain the significant interaction terms distinguishing the multiplicative from the nonseasonal or seasonal subset model. This full multiplicative model, with the mean term included, can be formulated as

$$
y_t = \mu + (1 - \theta_1 L)(1 - \Theta_{12}L^{12})e_t \qquad (5.7)
$$

and expanded to

$$y_t = \mu + e_t - \theta_1 e_{t-1} - \Theta_{12} e_{t-12} + \theta_1 \Theta_{12} e_{t-13}. \qquad (5.8)$$

Owing to the multiplication of factors, the interaction term with a sign opposite that of the other terms is present in the seasonal multiplicative model. That interaction term usually appears at a lag equal to the sum of the lags of the factors, unless the product of the first- and 12th-order moving average terms turns out to be nonsignificant.

Often, models require regular as well as seasonal differencing. A common model has an $ARIMA(0,1,1)(0,1,1)_{12}$ structure. Three models having this form of moving average structure are the Sutter County, California, work force size, depicted in Fig. 5.5, from January 1946 through December 1966; the international airline fares, depicted in Fig. 5.8, from January 1949 through December 1960; and the U.S. civilian unemployment rate of persons over 16, seasonally adjusted, during the period from January 1948 through August 1996, the last of which is shown in the SPSS chart in Fig. 5.11. The equation for this model is

$$(1 - L)(1 - L^{12})Y_t = (1 - \theta_1 L)(1 - \Theta_{12} L^{12}) e_t. \qquad (5.9)$$

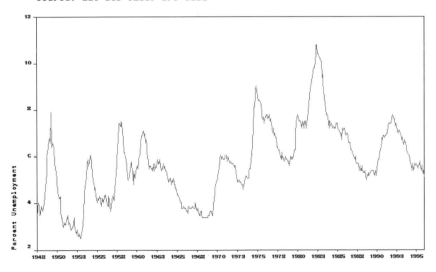

Figure 5.11 U.S. civilian unemployment rate of seasonally adjusted CPS data.

It reveals the regular and seasonal differencing at lags 1 and 12 as well as the moving average parameters at lags 1 and 12. When the equation is expanded it has the following formulation:

$$Y_t - Y_{t-1} - Y_{t-12} + Y_{t-13} = e_t - \theta_1 e_{t-1} - \Theta_{12} e_{t-12} + \theta_1 \Theta_{12} e_{t-13}$$
or   (5.10)
$$Y_t = Y_{t-1} + Y_{t-12} - Y_{t-13} + e_t - \theta_1 e_{t-1} - \Theta_{12} e_{t-12} + \theta_1 \Theta_{12} e_{t-13}.$$

This seasonal moving average model requiring a multiplicative formulation and is a common one among series.

5.4.3. SEASONAL AUTOREGRESSIVE MOVING AVERAGE MODELS

A slightly more complicated model would be the mixed, multiplicative, seasonal model. In addition to possible regular and seasonal differencing, this model contains both regular and seasonal autoregressive and seasonal moving average parameters. An example of this kind of series would be an ARIMA$(1,1,1)(1,1,1)_4$. Such a model possesses both regular and quarterly seasonal characteristics. In its factored form, the model, with a constant equal to 0, is

$$(1 - L)(1 - L^4)(1 - \varphi_1 L)(1 - \Phi_4 L^4)Y_t = (1 - \theta_1 L)(1 - \Theta_4 L^4)e_t$$
$$(1 - L - L^4 + L^5)(1 - \varphi_1 L - \Phi_4 L^4 + \varphi_1 \Phi_4 L^5)Y_t$$
$$= (1 - \theta_1 L - \Theta_4 L^4 + \theta_1 \Theta_4 L^5)e_t$$

With the left-hand side expanded, this equation becomes

$$(1 - L - L^4 + L^5 - \varphi_1 L + \varphi_1 L^2 + \varphi_1 L^5 - \varphi_1 L^6 - \Phi_4 L^4 + \Phi_4 L^5$$
$$+ \Phi_4 L^8 - \Phi_4 L^9 + \varphi_1 \Phi_4 L^5 - \varphi_1 \Phi_4 L^6 - \varphi_1 \Phi_4 L^9 + \varphi_1 \Phi_4 L^{10})Y_t \quad (5.11)$$
$$= (1 - \theta_1 L - \Theta_4 L^4 + \theta_1 \Theta_4 L^5)e_t.$$

Expanded, this equation is

$$Y_t - Y_{t-1} - Y_{t-4} + Y_{t-5} - \varphi_1 Y_{t-1} + \varphi_1 Y_{t-2} + \varphi_1 Y_{t-5}$$
$$- \varphi_1 Y_{t-6} - \Phi_4 Y_{t-4} + \Phi_4 Y_{t-5} + \Phi_4 Y_{t-8} - \Phi_4 Y_{t-9} \quad (5.12)$$
$$+ \varphi_1 \Phi_4 Y_{t-5} - \varphi_1 \Phi_4 Y_{t-6} - \varphi_1 \Phi_4 Y_{t-9} + \varphi_1 \Phi_4 Y_{t-10}$$
$$= e_t - \theta_1 e_{t-1} - \Theta_4 e_{t-4} + \theta_1 \Theta_4 e_{t-5}.$$

It is clear that the seasonal ARIMA models may become quite complicated, which is why they are usually expressed in factored terms.

5.5. THE AUTOCORRELATION STRUCTURE OF SEASONAL ARIMA MODELS

To be able to identify the seasonal ARIMA model, we must understand the basis of its autocorrelation structure. The autocorrelation function of the SARIMA model is computed in the same way as that of regular ARIMA models. The autocorrelation is the autocovariance divided by the variance, except in the multiplicative models the seasonal components are entered as factors. A brief discussion of some simple autoregression and moving average models is presented along with the rules for identification of the autocovariance structure.

The autocorrelation structure of the seasonal autoregression model is based on the ARIMA$(p,0,0)(P,0,0)_s$ model or some variant thereof. First-order structures are reported to be the most common. Second-order structures are reportedly less common and higher order structures are rather rare (McCleary and Hay, 1980). Consider the simple additive seasonal ARIMA$(1,0,0)_4$ model, $(1 - \Phi_4 L^4)Y_t = e_t$. Owing to the boundary requirements of stability and invertibility, which hold for seasonal and regular parameters in ARIMA models, this fourth-order seasonal autoregressive model can be expressed as an exponentially weighted moving average model, $Y_t = e_t/(1 - \Phi_4 L^4) = (1 + \Phi_4 L^4 + \Phi_4^2 L^8 + \Phi_4^3 L^{12} + \ldots)e_t$. As the exponentiation of the seasonal component increases, the magnitude of the seasonal autoregressive parameter, which has to be less than 1, decreases. If the process were a multiplicative one, where a regular autoregressive process is multiplied by the seasonal one, the seasonal pattern would be superimposed on the regular one. Seasonal autoregressive decay would periodically enhance regular autoregressive decay when the parameter values were of the same sign. It would periodically suppress the regular autoregressive decay when the parameter values were of opposite signs. The presence of multiplicative seasonal factors complicates the identification process somewhat. Examples of the characteristic patterns of autocorrelation will be provided in the next section dealing with identification. How the researcher can handle these complications will be addressed as well.

The autocovariance and autocorrelation structure of the seasonal moving average model is similar to that of nonseasonal moving average models. If, on the one hand, the model is a purely seasonal moving average model, the seasonal autocorrelation is formed by taking the lag 12 autocovariance and dividing it by the variance. The formula is the same except that the seasonal parameter would replace the nonseasonal parameter:

$$\rho_{12} = -\frac{\Theta_{12}}{1 + \Theta_{12}^2}. \tag{5.13}$$

However, a seasonal multiplicative moving average model has a more elaborately formed ACF. Series G in Box and Jenkins (1976), for example, is the natural log of the international airline ticket fares, and its model is an ARIMA$(0,1,1)(0,1,1)_{12}$. If one uses an already differenced series, that version can be formulated as $W_t = (1 - L)(1 - L^{12})Y_t$, where W_t is the regular and seasonally differenced natural log of airline fares. Box and Jenkins (1976) have computed the autocovariances for W_t:

$$
\begin{aligned}
\gamma_0 &= (1 + \theta^2)(1 + \Theta^2)\sigma_e^2 \\
\gamma_1 &= -\theta(1 + \Theta^2)\sigma_e^2 \\
\gamma_{11} &= \theta\Theta\sigma_e^2 \\
\gamma_{12} &= -\Theta(1 + \theta^2)\sigma_e^2 \\
\gamma_{13} &= \theta\Theta\sigma_e^2 \\
\gamma_j &= 0, \text{ in other cases.}
\end{aligned}
\tag{5.14}
$$

The autocorrelations are computed by dividing the autocovariances by the variance of W_t (Mills, 1990):

$$
\begin{aligned}
\rho_1 &= \gamma_1/\gamma_0 = -\theta_1/(1 + \theta_1^2) \\
\rho_{12} &= \gamma_{12}/\gamma_0 = -\Theta_{12}/(1 + \Theta_{12}^2) \\
\rho_1\rho_{12} &= \gamma_1\gamma_{12}/\gamma_0 = \theta_1\Theta_{12}/(1 + \theta_1^2)(1 + \Theta_{12}^2) \\
\rho_j &= 0, \text{ in other cases.}
\end{aligned}
\tag{5.15}
$$

If we know which of the sample ACFs and PACFs is significant, we can proceed to identify the different seasonal models.

5.6. STATIONARITY AND INVERTIBILITY OF SEASONAL ARIMA MODELS

Seasonal models as well as regular ARIMA models have parameters that must meet the bound of stationarity and invertibility. The seasonal autoregressive models ARIMA$(p,d,0)(P,D,0)s$ need to be stationary for analysis. For stationarity to exist, both the regular and the seasonal autoregressive parameters need to lie within the bounds of stationarity. That is,

$$
-1 < \varphi_p, \Phi_s < +1.
\tag{5.16}
$$

Autoregressive processes whose parameter estimates remain within these bounds are invertible. Consider the basic ARIMA$(1,0,0)(1,0,0)_{12}$ model:

$$
\begin{aligned}
(1 - \phi_1 L)(1 - \Phi_{12}L^{12})Y_t &= e_t \\
y_t - \mu = e_t/(1 - \phi_1 L)(1 - \Phi_{12}L^{12}) \\
&= (1 + \phi_1 L + \phi_1\Phi_{12}L^{13} + \cdots)e_t
\end{aligned}
\tag{5.17}
$$

If we find that these parameters lie within the bounds of stationarity, their products will also lie within the bounds of stationarity. As the regular and seasonal infinite series converge, so their products converge. Stability is thereby assured.

The bounds of invertibility similarly must hold for multiplicative seasonal moving average models. Hence, the series

$$W_t = (1 - \theta_1 L)(1 - \Theta_{12} L^{12}) e_t \qquad (5.18)$$

would have to possess regular and seasonal parameters that lie within the same bounds of invertibility ($|\theta_1|$, $|\theta_s| < +1$) for the mixed seasonal moving average model to be stationary. If the moving average parameters were confined to this range, the product of these factors would also be confined to these bounds. Only under such conditions would the series converge and remain stable.

5.7. A MODELING STRATEGY FOR THE SEASONAL ARIMA MODEL

5.7.1. IDENTIFICATION OF SEASONAL NONSTATIONARITY

After graphing the series, and viewing its sample correlograms, the analyst can look for seasonal patterns. To identify seasonality, the researcher searches for evidence of annual fluctuation in the data. The researcher must know what distinguishes seasonal nonstationarity, seasonal autoregression, and seasonal moving averages from nonseasonal patterns; he must also know how to distinguish these seasonal patterns from one another. To perform these analyses, he can rely primarily on the time sequence plot and the ACF to perform these analyses, pursuing a strategy of inquiry that maximizes opportunity for ascertaining the optimal model. When there is a seasonal unit root, the model requires more differencing at a seasonal lag prior to further analysis (Meyer, 1998). Although SAS has the test for the seasonal unit root, SPSS currently has no such procedure.

5.7.2. PURELY SEASONAL MODELS

The first part of the strategy entails graphing the series with a time plot as well as running the sample ACF and PACF correlograms. The first thing the researcher looks for is evidence of nonstationarity. He plots the series against time and checks for nonseasonal or seasonal changes that reveal nonstationarity. Peaks in a series every 12 months would indicate annual seasonality;

peaks in a series every 3 months would indicate quarterly seasonality. An example of seasonal nonstationarity would be a quarterly shift of the series level. After plotting the series against time, the analyst turns to analysis of the correlograms.

To confirm the nature of this periodicity, the researcher can examine ACFs and PACFs. The characteristic patterns of the seasonal ARSAR, MASMA, and SARSMA models reveal the nature of the seasonality in the model. Seasonal models have pronounced regular ACF and PACF patterns with a periodicity equal to the order of seasonality, the number of times per year that seasonal variation occurs. If the seasonality is annual, the prominent seasonal ACF spikes are heightened patterns at seasonal lags over and above the regular nonseasonal variation once per year. If the seasonality is quarterly, there will be prominent ACF spikes four times per year.

Purely SAR models have significant and pronounced ACFs, which exponentially attenuate at seasonal lags. If the decay of the seasonal ACF is very gradual, then the series remains seasonally nonstationary and in need of seasonal differencing as in Fig. 5.6. In other words, if the model is a purely seasonal autoregressive (SAR) model of order sP, then the seasonal autoregressive Φ_{t-s} parameter effect is observed at the $t - s$ lag, where P is the number of seasonal autoregressive parameters necessary to specify the model and s is the order of seasonality at which the influence of the previous value is experienced by the series. If the seasonality is quarterly, then s in the case of pure seasonality would equal 4, 8, 12, etc. SAR models have sP significant and pronounced autoregressive spikes in the IACF and PACF at $s, 2s, 3s, \ldots,$ Ps seasonal lags. The ACF of this model will exhibit exponentially declining spikes at seasonal lags, while the PACF of this model will exhibit as many spikes at seasonal lags as represent the order of the model.

Purely SMA models exhibit the pronounced MA pattern at seasonal lags. These models possess significant and pronounced ACF spikes every sQ lags, where Q is the number of seasonal moving average parameters necessary to specify the model and s is the order of the seasonality. In other words, pronounced, significant spikes are found at each seasonal lag up to Q seasonal lags. If the model is a purely seasonal MA (SMA) model of order sQ, then the ACF will exhibit as many Θ_{t-s} spikes at $t - s$ seasonal lags as is the order of the SMA model, whereas the PACF will exhibit exponential decline at $t - s$ seasonal lags. The prominent seasonal IACF and PACF spikes taper off gradually. That is to say, the prominent seasonal MA spikes tail off at multiples of the order of seasonality. If this tapering is very gradual in the SMA model, then the series remains seasonally nonstationary and in need of seasonal differencing. Once stationarity has been attained by regular and seasonal differencing, the model can be identified.

In other words, in purely seasonal models, the prominent spikes of the

ACF will be found between the periods rather than within them. If the model is a mixed SARSMA model of order (sP,sQ), then the seasonal lags will taper off exponentially after Q lags, whereas the seasonal lags of the PACF will taper off gradually after P lags (Bresler *et al.,* 1991). When models are purely seasonal, they may be modeled as additive models at seasonal lags. The spikes will be apparent between the seasonal periods. The purely seasonal models will show relatively few autoregressive or moving average patterns within those seasonal periods.

After each attempt at identification, the researcher estimates the components and examines the residuals. When all of the seasonal parameters are properly identified and estimated, the ACF and PACF of the residuals should resemble white noise.

Once the parameters have been identified, the modeling includes several other steps. The parameters identified have to be estimated by means discussed in the next chapter. The researcher then diagnoses and fine-tunes the fitting and may produce forecasts. With metadiagnosis, he compares alternative models and their forecasts. From this analysis, he may find an optimal solely seasonal model.

5.7.3. A MODELING STRATEGY FOR GENERAL MULTIPLICATIVE SEASONAL MODELS

Box–Jenkins modeling methodology for multiplicative models deals with nonseasonal and seasonal factors in the model. The time plot shows evidence of seasonal variation over and above regular patterns of variation, and in so doing indicates whether preliminary transformation is in order. Seasonal nonstationarity will confound the autocorrelations and make it hard to model the series and/or forecasting from that model. Testing for seasonal roots may be performed to determine whether seasonal differencing is in order (Frances, 1991; Meyer, 1998; Reilly, 1999). Seasonal differencing is performed to neutralize the seasonal nonstationarity and regular differencing may follow to render the remainder of the series stationary. If there is residual nonhomogeneity, a Box–Cox, natural log, or power transformation might be applied to achieve covariance stationary. As noted earlier, when the variance is not constant and the standard deviation is proportional to the mean, then the natural log may be the transformation of choice. To be sure that the seasonal and regular differencing is no longer needed, regular and seasonal stationarity may be tested with Dickey–Fuller and augmented Dickey–Fuller tests for nonseasonal as well as seasonal unit roots. These tests will indicate whether series require regular and seasonal differencing to bring about the stationarity necessary for identification of the series. Then the appropriate differencing is performed to effect covariance stationarity.

The parameters of the model are usually identified predominantly with the ACF, and, to a lesser extent, with the PACF, mainly because the PACFs of seasonal models become relatively complicated and difficult to interpret. The analyst looks for evidence of nonseasonal AR, MA, or ARMA patterns upon which are superimposed seasonal AR, MA, or ARMA patterns. For additive MA models with both nonseasonal and seasonal parameters, there will be q spikes before the seasonal spike, where q is the order of the nonseasonal MA parameter. For multiplicative MA models, the ACF has q spikes before and q spikes after the seasonal lag. During this phase of the modeling process, the analyst searches for evidence of interaction terms and hence multiplicative models. The seasonal components are identified first, and then the nonseasonal multiplicative factors. Testing these parameters for stationarity and invertibility can assure the analyst of a good model. Estimation of those parameters is undertaken by one of the selected computer algorithms, and then diagnosis of the residuals is performed to be sure that the variation is properly modeled. If the residuals are not white noise, the analyst examines them for telltale patterns and returns to the identification stage for either complete remodeling or fine-tuning of the existing model. Each time he estimates a model, he compares the fit of the models in a metadiagnosis to see which model has the better fit. This process is reiterated until the residuals are white noise and the optimal fit is attained.

5.7.3.1. Identification of Multiplicative Seasonal Components

Once the model has been rendered both nonseasonally and seasonally stationary, the Box–Jenkins method proceeds after the fashion of modeling with a nonseasonal series (Wei, 1990). If the model is a natural log multiplicative $ARIMA(0,1,1)(0,1,1)_{12}$ model, it will have to have been natural log transformed and then differenced at lags 1 and 12 before the remainder of the modeling takes place. The structure of the ACF and PACF depends on the relative direction and magnitude of the parameters.

In order to facilitate identification of seasonal models, the analyst must consider some of the characteristic patterns of common seasonal models. Simple additive seasonal models are examined first. The analyst begins by looking for purely seasonal nonstationarity. Over and above any other pattern, he searches for seasonal patterns of shifts in level or rise and fall in the time plot as well as periodic spikes in the sample ACFs and PACFs. In the correlograms, this phenomenon is characterized by periodic peaks, which in and of themselves decline in magnitude very gradually. The general ACF patterns of these nonstationary characteristics are presented in

Fig. 5.12. This kind of seasonal peaking and gradual attenuation is generally characteristic of seasonal ARIMA models as well. When the series requires seasonal differencing for stationarity, then the periodic peaks in the sample ACF in Fig. 5.12 may indicate the order of differencing required. Once the series is properly differenced, the sample ACF will decay rapidly (Granger and Newbold, 1986). After the transformation phase of the modeling is completed, the residuals will appear to be stationary.

The analyst then examines the series for the proper orders of regular and seasonal autoregression and/or moving average parameters. Seasonal autoregressive models may be identified by the outline of an autoregressive pattern at seasonal lags. The nonseasonal periods will manifest comparatively reduced ACFs. In these models, there is a gradual (but not too gradual) attenuation of the ACF and a significant seasonal spike at each of the lags indicating the order of the seasonal autoregression. That is, if there is only one significant seasonal PACF spike, then the model may be a seasonal AR(*1*), sometimes referred to as a SAR(*1*). If there are two significant seasonal PACF spikes, with more gradual attenuation of the seasonal ACF, then the model may be an SAR(*2*). Seasonal multiplicative models contain one or more interaction terms, whose sign equals the prod-

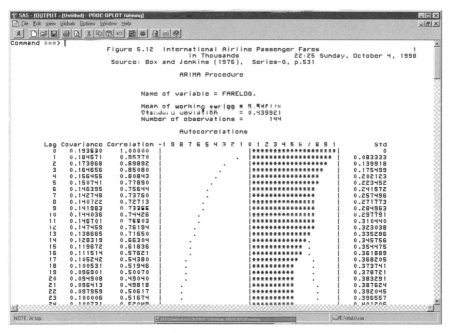

Figure 5.12

Figure 5.13

uct of the signs of its components. The general autocorrelation structure, after first differencing, of the $ARIMA(1,1,0)(1,0,0)_{12}$ series of the natural log of the Consumer Price Index for all urban consumers since 1980 is given in Figs. 5.13 and 5.14. In the first correlogram, the ACF for this model is presented, whereas in the second correlogram, a PACF of the model is shown. All of the seasonal autoregressive parameters are positive. If the φ parameter were negative, the sample PACF would exhibit a negative rather than a positive seasonal spike, while the seasonal ACF parameters would exhibit diminishing oscillations. If the seasonal spike were negative, there would appear to be negative seasonal spikes in the general patterns. Figures 5.15 and 5.16 depict correlograms where, following differencing, the superimposed seasonal pattern overlays the regular patterns, yielding an $ARIMA(2,1,0)(1,1,0)_{11}$ model.

Some multiplicative seasonal models are easy to identify. Figures 5.17 and 5.18 present the ACF and PACF for an $ARIMA(0,0,1)(0,0,1)_{12}$ seasonal moving average model. These models have ACFs with spikes at the first significant nonseasonal lag and several spikes around the seasonal lag for the model. The PACFs tend to gradually attenuate but exhibit reverse spikes after the seasonal first lag of the model.

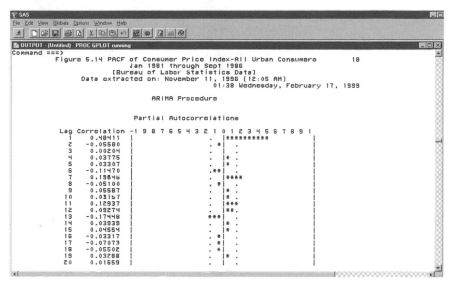

Figure 5.14

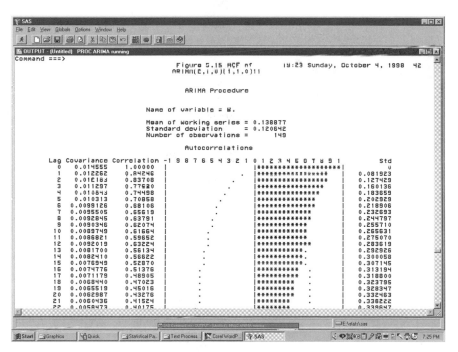

Figure 5.15

ARIMA(2,1,0)(1,1,0)11

ARIMA Procedure

Partial Autocorrelations

Lag	Correlation	-1 9 8 7 6 5 4 3 2 1 0 1 2 3 4 5 6 7 8 9 1
1	0.84246	. |******************
2	0.43873	. |*********
3	0.04599	. |* .
4	0.02490	. | .
5	0.02806	. |* .
6	0.03195	. |* .
7	0.03437	. |* .
8	0.04416	. |* .
9	0.03671	. |* .
10	0.07425	. |* .
11	0.00170	. | .
12	0.20455	. |****
13	-0.21623	****| .
14	-0.03061	. *| .
15	-0.00635	. | .
16	-0.00530	. | .
17	-0.00543	. | .
18	-0.00559	. | .
19	-0.00608	. | .
20	-0.00636	. | .
21	-0.00682	. | .
22	-0.00953	. | .
23	-0.00737	. | .
24	-0.01757	. | .
25	-0.00041	. | .
26	-0.00640	. | .
27	-0.00712	. | .
28	-0.00668	. | .
29	-0.00744	. | .

Figure 5.16

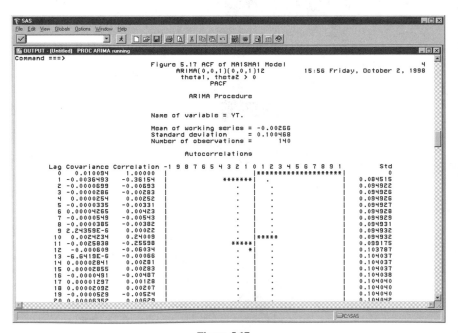

Figure 5.17

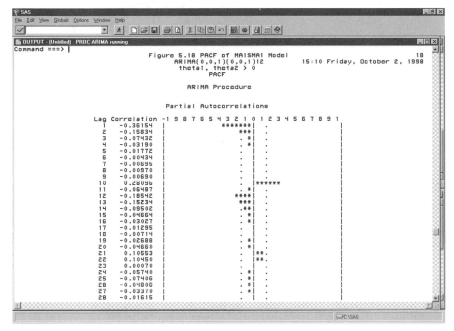

Figure 5.18

Sometimes testing for seasonal unit roots is required to be sure that the model is in fact seasonal and multiplicative. From a visual inspection of the correlograms in Figures 5.19 and 5.20, we might suspect the model to be an $ARIMA(0,0,2)(0,0,1)_{12}$ model. The sign-reversed moving average parameter at lag 13 shown in the PACF can indicate the multiplicative model. When we test for a seasonal unit root test, we find that the apparent seasonal unit root at lag 12 is not real. Consequently, we should reformulate the apparently multiplicative seasonal model as a nonseasonal moving average model with parameters at lags 1, 2, and 13. This MA(3) model yields pure white noise residuals, but a more parsimonious alternative might be an MA(2) model with moving average parameters at lags 1 and 2. Although the MA(2) model has a lower SBC than the MA(3) model, it lacks white noise residuals because its correlograms contain a residual significant spike at lag 13. The model selection decision therefore hinges on the tradeoff between the white noise residuals and the incremental parsimony.

The characteristic correlogram pattern of the multiplicative autoregression model can be complicated. There are basically two sets of patterns. There is the regular pattern within the seasonal period and there is the seasonal period pattern. The regular pattern within periods has been described in Chapter 3 and 4. An ACF of such a model exhibits a seasonal

```
SAS
File  Edit  View  Globals  Options  Window  Help
OUTPUT - (Untitled)  PROC ARIMA running
Command ===>
                          Figure 5.19 ACF of MA(2)SMA(1) Model                          4
                               ARIMA(0,0,2)(0,0,1)           14:31 Friday, October 2, 1998
                              theta_2, theta_12 > 0

                                ARIMA Procedure


                         Name of variable = YT.

                         Mean of working series = 0.046752
                         Standard deviation     = 4.000958
                         Number of observations =      140

                               Autocorrelations

Lag Covariance Correlation -1 9 8 7 6 5 4 3 2 1 0 1 2 3 4 5 6 7 8 9 1          Std
  0  16.007667    1.00000   |                    |********************|           0
  1  -3.323398   -0.20761   |                ****|  .                 |    0.084515
  2  -3.678868   -0.22982   |               *****|   .                |    0.088083
  3  -0.216544   -0.01353   |                .   |   .                |    0.092267
  4   0.199477    0.01246   |                .   |   .                |    0.092281
  5  -0.129410   -0.00808   |                .   |   .                |    0.092293
  6  -0.154151   -0.00963   |                .   |   .                |    0.092298
  7   0.174370    0.01089   |                .   |   .                |    0.092305
  8  -0.182105   -0.01138   |                .   |   .                |    0.092314
  9  -0.059401   -0.00371   |                .   |   .                |    0.092324
 10   0.298156    0.01863   |                .   |   .                |    0.092325
 11  -0.571547   -0.03570   |                .  *|   .                |    0.092352
 12  -1.696495   -0.10598   |                . **|   .                |    0.092451
 13   2.285170    0.14275   |                .   |***.                |    0.093314
 14   0.257913    0.01611   |                .   |   .                |    0.094862
 15   0.043291    0.00270   |                .   |   .                |    0.094881
 16  -0.255352   -0.01595   |                .   |   .                |    0.094882
 17   0.132111    0.00825   |                .   |   .                |    0.094901
 18   0.142170    0.00888   |                .   |   .                |    0.094906
 19  -0.297633   -0.01859   |                .   |   .                |    0.094912
 20   0.169383    0.01058   |                .   |   .                |    0.094938
 21  -0.080350   -0.00502   |                .   |   .                |    0.094946
 22   0.090812    0.00567   |                .   |   .                |    0.094948
C:\SAS
```

Figure 5.19

```
Command ===>
                          Figure 5.20 PACF of MA(2)SMA(1) Model                          6
                               ARIMA(0,0,2)(0,0,1)           14:29 Friday, October 2, 1998
                              theta_2, theta_12 > 0

                                ARIMA Procedure


                           Partial Autocorrelations

Lag Correlation -1 9 8 7 6 5 4 3 2 1 0 1 2 3 4 5 6 7 8 9 1
  1   -0.20761   |                ****|   .                |
  2   -0.28522   |              ******|   .                |
  3   -0.15251   |                 ***|   .                |
  4   -0.11449   |                 .**|   .                |
  5   -0.08957   |                 .**|   .                |
  6   -0.07408   |                 . *|   .                |
  7   -0.04388   |                 . *|   .                |
  8   -0.05093   |                 . *|   .                |
  9   -0.03811   |                 . *|   .                |
 10   -0.01185   |                 .  |   .                |
 11   -0.05532   |                 . *|   .                |
 12   -0.16066   |                 ***|   .                |
 13    0.04022   |                 .  |*  .                |
 14   -0.01256   |                 .  |   .                |
 15    0.04518   |                 .  |*  .                |
 16    0.01741   |                 .  |   .                |
 17    0.03615   |                 .  |*  .                |
 18    0.03730   |                 .  |*  .                |
 19    0.01679   |                 .  |   .                |
 20    0.02880   |                 .  |*  .                |
 21    0.01452   |                 .  |   .                |
 22    0.02685   |                 .  |*  .                |
 23    0.02597   |                 .  |*  .                |
 24    0.00630   |                 .  |   .                |
```

Figure 5.20

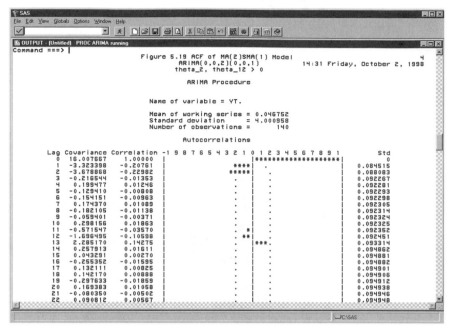

autoregressive pattern superimposed on a regular series, as if there were a seasonal modulation of the amplitude of the regular series with the annual periodicity being the order of seasonality of the model. The parameters of the within series autoregression are contained in the first factor, while those of the seasonal variation are contained within the second factor of the $(1,0,0)(1,0,0)_{12}$ model. In general, this kind of pattern was shown in Section 5.4.1. The PACF of such a series would reveal a spike at the lag of the interaction term. With higher order models, there would be a cluster of significant spikes around the lag of the interaction term.

The characteristic correlogram pattern of seasonal moving average models is easier to identify than those of autocorrelated models. The ACF reveals regular and the periodic seasonal moving average spikes. If the model has an MA(1) component, than there will only be one regular significant spike. If the model has an MA(2) component, then the model with have two regular significant spikes. The duration of the seasonal spiking will indicate the order of the seasonal component. Multiplicative models possess an interaction term, whose sign is equal to the sign of the product of its components. The signs of the moving average parameters are reversed. They are treated as negative when multiplied to yield the interaction product term. Therefore, if regular and seasonal moving average parameters are positive, the sign of the interaction term is negative. With higher order models, there is often a cluster of interaction terms in the PACF. The PACF of these models manifests periodic gradual attenuation. When seasonal moving average models are identified this way, it remains for the user to fit the model.

5.7.3.2. Estimation, Diagnosis, and Comparison of Multiplicative Models

Different combinations of multiplicative parameters can be estimated. To determine whether the identified parameters are statistically significant, we can examine t ratios of the parameters to their standard errors. Then we may have to consider some important modeling questions. If some of the parameters are nonsignificant, do they seem to be theoretically necessary? Is the retention of those parameters necessary for the residuals to approximate white noise? Does the model converge? These factors need to be considered as we evaluate the adequacy of the model. For pruning or fine-tuning the model, we take note of the mean square error, likelihood ratio, AIC, and/or SBC. The model may be simplified by trimming out the theoretically unimportant and statistically nonsignificant parameter(s). We note whether a significant reduction in the likelihood ratio, AIC, or SBC has occurred, and whether this constitutes model improvement. Thus, we can assess the fit of the model.

Another way of assessing the utility of the model is to compare the forecast accuracy of the models. Positive or negative forecast bias can occasionally be detected in the forecast profile graphs. The accuracy of the forecast can be seen in the narrowness of the confidence intervals around the point forecast. The accuracy can also be compared with the mean absolute percentage error over different time horizons. The details of forecasting are covered in Chapter 7 and the comparative evaluation of models and forecasts are covered in Chapter 11. In this way, we can diagnose and compare plausible alternative models.

5.7.3.3. Model Simplification

Once we have modeled these seasonal components, we can identify the regular components. Although a researcher begins with identifying a larger multiplicative model, he should then estimate, diagnose, and trim it down to a more parsimonious yet still adequate model. Maintaining explanatory power while simplifying the model furthers parsimony. When the researcher attempts to reduce the multiplicative model to a simpler additive model, by pruning interaction terms of borderline significance, he seeks to maximize parsimony. Sometimes the multiplicative seasonal model will reduce to a purely seasonal model. This is another type of additive seasonal model where the regular nonseasonal components turn out to be nonsignificant. Model comparison criteria—such as measures of fit, parsimony, and forecast accuracy—may be used to compare and contrast the multiplicative with the simpler model.

5.7.3.4. Metadiagnosis

After the seasonal models have been compared and simplified, the researcher may perform further metadiagnosis by comparing the models for fit, parsimony, reliability, and forecast accuracy. The criteria for fit (sum of squared errors), parsimony (minimum information criteria), and forecast accuracy (minimum squared forecast error, minimum absolute percentage error, etc.) will be discussed in detail in the chapter on metadiagnosis and forecasting. After the model is estimated on historical data, it should be tested on data set aside for validation, the hold-out sample. If the same model is an adequate description of the process on the validation data set, then the model is stable and reliable. It then passes a pretest of predictive validity. Each of the competing models should be tested in these ways in order to determine which among them is the optimal model. Metadiagnosis and forecasting will be covered in detail in Chapter 7.

5.8. PROGRAMMING SEASONAL MULTIPLICATIVE BOX–JENKINS MODELS

5.8.1. SAS PROGRAMMING SYNTAX

The programming of seasonal models using SAS and SPSS is very simple. First, there is the question of differencing to effect seasonal stationarity. In SAS, the programming of tests for seasonal unit roots have been discussed in section 5.2. Differencing is specified in the SAS ARIMA procedure within the IDENTIFY subcommand. For a purely seasonal model, a seasonal differencing of 4 might be required, and this could be accomplished with the following option in the identify subcommand, coupled with an NLAG option for the number of lags in the correlogram to examine.

```
PROC ARIMA;
IDENTIFY VAR=Y (4) NLAG=30;
RUN;
```

A seasonal difference of 12 accompanying a regular first difference would be modeled by

```
PROC ARIMA;
IDENTIFY VAR- Y(12) NLAG=35;
RUN;
```

In the event that nonseasonal as well as seasonal differencing are required in the same model, then the following option may be implemented:

```
PROC ARIMA;
IDENTIFY Var=Y(1,12) NLAG=35;
RUN:
```

Specifying a seasonal model involves more than the differencing needed to bring about stationarity. It involves the specification of the SAR or SMA parameters as well. Consider the purely seasonal model. This model may be a subset of a larger multiplicative model. In order to specify the purely seasonal model,

$$(1 - L^4)y_t = \mu + (1 - \Theta_1 L^4)e_t$$
$$y_t = y_{t-4} + \mu + e_t - \Theta_1 e_{t-4}, \tag{5.19}$$

which is an ARIMA(0,1,1)4 model, the user would include the parameters in the ESTIMATE subcommand of the SAS ARIMA procedure. The way to specify only at the fourth lag and to prevent all of the earlier lags from being specified is to place the 4 in parentheses in the SAS ESTIMATE subcommand.

```
PROC ARIMA;
IDENTIFY Var=Y(4) NLAG=30;
ESTIMATE Q=(4) PRINTALL PLOT;
RUN;
```

The ESTIMATE subcommand generates the parameter estimation, including the constant, with the t tests. It also generates a variance and standard error of the residuals, along with the AIC and SBC criteria for the model, plus Q statistics for autocorrelation in the residuals. The PRINTALL option generates the optimization summary or iteration history of the model. The PLOT option invokes and ACF, IACF, and PACF of the residuals in addition to the Q statistics.

If the user wishes to estimate an ARIMA(0,1,1)(0,1,1)$_{12}$ multiplicative model,

$$(1 - L)(1 - L^{12})y_t = \mu + (1 - \theta_1 L)(1 - \Theta_{12}L^{12})e_t, \qquad (5.20)$$

he may use the differencing option to specify the orders of the nonseasonal and the seasonal differencing, and then he may invoke the ESTIMATE subcommand and define his factored multiplicative model in the ESTIMATE subcommand as the product of two factors:

```
PROC ARIMA;
IDENTIFY VAR=Y(1,12);
ESTIMATE Q=(1)(12) PRINTALL PLOT;
RUN;
```

If, however, the interaction "product" term does not turn out to be significant, the researcher may wish to run a subset model. The ESTIMATE subcommand in this case employs only the significant components that comprise the additive subset of the factored model. They are combined in one factor's parentheses.

```
PROC ARIMA;
IDENTIFY VAR=Y(1,12);
ESTIMATE Q=(1 12) PRINTALL PLOT;
RUN;
```

If the user wished to estimate a seasonal multiplicative ARIMA$(1,1,0)(1,1,1)_{12}$ model, with regular and seasonal differencing, the formulation becomes more complex:

$$(1 - L)(1 - L^{12})(1 - \varphi_1 L)(1 - \Phi_{12}L^{12})y_t = \mu + (1 - \Theta_{12}L^{12})e_t. \qquad (5.21)$$

The differencing accounted for by the first two factors on the left-hand side of the equation is taken care of in the IDENTIFY subcommand, while the AR and SAR factors are specified with the P= (1) (12) options in the ESTIMATE subcommand. The seasonal moving average is specified with the Q= (12) option. The parentheses guarantee a purely seasonal specification here.

```
PROC ARIMA;
   IDENTIFY=Y(1,12);
   ESTIMATE P=(1)(1) Q=(12) PRINTALL PLOT;
   RUN;
```

5.8.2. SPSS Programming Syntax

In SPSS, similar syntax may be employed to generate either a purely seasonal model or a multiplicative seasonal model. For the purely seasonal model specified in Eq. (5.21), the SPSS syntax required is as follows:

```
ARIMA Y/
   MODEL=CONSTANT/
   SD=4/
   SQ=(4)/
   MXITER=10/
       PAREPS .001/
       SSQPCT .001/
       FORECAST EXACT.
```

For the multiplicative ARIMA$(0,1,1)(0,1,1)_{12}$ model, the following syntax may be used:

```
ARIMA Y/
   MODEL=(0,1,1)(0,1,1) 12 CONSTANT/
   MXITER=10/
   PAREPS .001/
   SSQPCT .001/
   FORECAST EXACT.
```

The ARIMA$(1,1,0)(1,1,1)_{12}$ model is programmed by reformulating the ARIMA specification in the second line to read

```
ARIMA Y/
  MODEL=(1,1,0)(1,1,1)12 CONSTANT/
  MXITER=10/
  PAREPS .001/
  SSQPCT .001/
  FORECAST EXACT.
```

In these ways, both SAS and SPSS can be used to program purely (additive) seasonaland multiplicative seasonal models.

5.9. ALTERNATIVE METHODS OF MODELING SEASONALITY

Suppose that the scientist discovers that a seasonal shift in level occurs within the series. This shift may be either deterministic or stochastic. If the shift is well defined and well behaved throughout the series, it may be modeled by traditional methods of analysis. Moving average, Winters exponential smoothing, decomposition, and regression techniques may be used to control for deterministic trend and/or seasonality. A Winters exponential smoothing model can handle either additive or multiplicative seasonality. An X-11 or X-12 decomposition may extract the trend-cycle as well as seasonal components, leaving the stochastic residuals for subsequent analysis. If a consistent deterministic linear or polynomial trend is discovered, then a linear or polynomial regression analysis may be used to control for this trend. If sales are found to vary systematically according to the four seasons of the year, a multiple linear regression analysis may model the seasonal part of the process. If there are four shifts per year, the researcher may need three seasonal dummy independent variables to model the seasonal changes. The seasonal dummies, named *time,* in Eq. (5.22) are all coded 1 and 0, depending upon whether the observation respectively takes place during that season or during another. All measures are implicitly coded in comparison with the reference season. In this case, the reference category is the autumn sales season. The residuals may be used for further modeling in accordance with the Wold decomposition theorem, which maintains that any series is a combination of deterministic and stochastic processes.

At other times, we can employ a trigonometric function—such as a sine or cosine—in such a regression model to represent and control for this annual periodicity. Especially when series data are more or less continuous

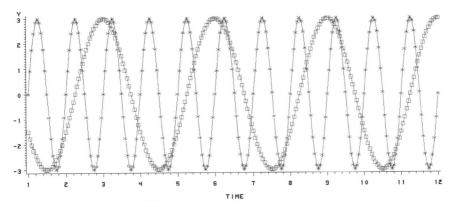

Figure 5.21 Deterministic trigonometric function modeling monthly sine and quarterly cosine seasonality.

and contain seasonal variation of the kind shown in Fig. 5.21, trigonometric predictor variables may be constructed out of these functions and employed on the right-hand side of a regression model explaining seasonal variation in series Y_t [Eq (5.23)]. These functions may be adapted to model deterministic long-wave cyclical variation as well.

$$y_t = \mu + \text{winter} * \text{time} + \text{spring} * \text{time} + \text{summer} * \text{time} + e_t \quad (5.22)$$

$$Y_t = C + \sum_{i=1}^{\text{period}/2} (X_{1t} + X_{2t}) + e_t,$$

where

$$X_{1t} = b_1 \sin\left(\frac{(2\pi \text{ freq Time})}{\text{periodicity}} + (\text{phase shift})\right)$$

and

$$X_{2t} = b_2 \cos\left(\frac{(2\pi \text{ freq Time})}{\text{periodicity}} + (\text{phase shift})\right) \quad (5.23)$$

and where

$$C = \text{constant}$$

$$\text{periodicity} = \text{order of seasonality}$$

$$b_i = \text{amplitude}.$$

Regression models with time-varying covariates are useful in modeling effects other than seasonality. In addition to modeling deterministic cyclicity with trigonometric functions and seasonality with seasonal dummy variables, regression analyses with holiday dummy variables can also be used to model holiday effects—such as those of Thanksgiving or Easter. Such variables are coded 0 when the holiday is not in effect and 1 if the holiday is in effect. Trading day variables may be included as well. For each month the trading day variable may have a value equal to the number of trading days in that month. With the inclusion of seasonal dummies, holiday dummies, and trading day covariates, the regression model that regresses the value of a series on time may account for a variety of time series effects. An example of an autoregression model of monthly GDPPC at a particular monthly time t might be

$$
\begin{aligned}
\text{GDPPC}_t = C &+ \varphi_1 \text{GDPPC}_{t-1} + \beta_{2T} \text{Time} \\
&+ \beta_{3t} \text{Winter} + \beta_4 \text{Spring} + \beta_{5t} \text{Summer} \\
&+ \beta_{6t} \text{Holiday} + \beta_{7t} \text{Tradingdays} + e_t, \qquad (5.24)
\end{aligned}
$$
where GDPPC_t = gross domestic product per capita.

In this model, there are three seasonal dummy variables—namely, Winter, Spring, and Summer—while there is a Holiday dummy coded 1 for holidays and 0 for all other times. There is a time varying covariate—that is, TRADINGDAYS—which contains the number of trading days for the months surveyed (Diebold, 1998).

The residuals, of course, may be saved and used in combination with Box–Jenkins ARIMA modeling. Such combined autoregression and ARIMA models have been found to be effective in forecast competitions. When different models or combinations of them fit, plausible alternative models need to be formulated. Model comparison criteria relating to measures of model fit (sum of squared residuals), parsimony (minimum information criteria), or forecast accuracy (sum of squared forecast errors, etc.), which will be covered in the chapter on metadiagnosis and forecasting, may be used to determine which is the optimal model.

5.10. THE QUESTION OF DETERMINISTIC OR STOCHASTIC SEASONALITY

The nature of seasonality is important to proper specification of the model. Tests for seasonal nonstationarity may be in order (Frances, 1991; Meyer, 1998). When these tests indicate that adjustments for seasonality are in order, then the question arises as to how to control for seasonality.

If, on the one hand, seasonal factors follow a precise deterministic functional form, then seasonality may be modeled by dummy or trigonometric variables in a model (Granger and Newbold, 1986; Diebold, 1998; Reilly, 1999). If, on the other, the seasonality, depending on the memory of the series, may follow a more stochastic form—where the seasonality may be more or less recently emergent in the series, seasonal differencing and a seasonal multiplicative Box–Jenkins modeling may be preferred. If the researcher is unclear as to the nature of the seasonality, he may apply an appropriate seasonal unit root test—for example, the augmented Dickey–Fuller or Phillips–Perron Test—to help ascertain the nature of the seasonal variation. He can then model the seasonal variation in different ways and compare the models for goodness of fit as well as forecast accuracy with a view toward choosing the optimal model.

REFERENCES

Bowerman, B. L., and O'Connell, R. T. (1993). *Forecasting and Time Series: An Applied Approach,* 3rd ed. Belmont, CA: Duxbury Press, p. 526.

Box, G. E. P., and Jenkins, G. M. (1976). *Time Series Analysis: Forecasting and Control.* Oakland, CA: Holden-Day, pp. 303, 313.

Box, G. E. P., Jenkins, G. M., and Reinsel, G. C. (1994). *Time Series Analysis: Forecasting and Control,* 3rd ed. Englewood Cliffs, NJ: Prentice Hall, pp. 352, 366–369, 542–543, 546–547. Data are used with permission of author and publisher.

Bresler, L. E., Cohen, B. L., Ginn, J. M., Lopes, J., Meek, G. R., Weeks, H. (1991). *SAS/ ETS®* Software: Applications Guide 1: Time Series Modeling and Forecasting, Financial Reporting, and Loan Analysis. Cary: SAS Institute, Inc., p. 113.

Diebold, F. X. (1998). *Elements of Forecasting.* Cincinnati, Ohio: Southwest College Publishing, Inc., pp. 108–110.

Enders, W. (1995). *Applied Econometric Time Series.* New York: John Wiley and Sons, Inc., p. 112.

Frances, P. H. (1991). "Seasonality, non-stationarity, and the forecasting of monthly time series." *International Journal of Forecasting,* Vol. 7, pp. 199–208.

Granger, C. W. J., and Newbold, P. (1986). *Forecasting Economic Time Series,* 2nd ed. San Diego: Academic Press, pp. 43, 101–111.

Granger, C. W. J. (1989). *Forecasting in Business and Economics.* San Diego: Academic Press, pp. 93-108.

Makridakis, S., Wheelwright, S. C., and McGhee, S. (1983). *Forecasting: Methods and Applications.* New York: John Wiley and Sons, pp. 384, 625–627. Data are used with permission of John Wiley and Sons.

McCleary, R., and Hay, R. with Meidinger, E., and McDowell, D. (1980). *Applied Time Series Analysis.* Newberry Park: Sage, p. 80, 83, 312–318. Data are used with permission of Richard McCleary.

Meyer, K. V. (September 23–24, 1998). SAS econometric time series technical consultant. Cary, NC: SAS Institute, Inc. Personal communication about Dickey-Fuller seasonal root tests in SAS® PROC ARIMA.

Meyer, K. V. (July 19, 1999). SAS econometric time series technical consultant. Cary, NC:

SAS Institute, Inc. Personal communication about Dickey-Fuller seasonal root tests in SAS® PROC ARIMA.

Mills, T. C. (1990). *Time Series Techniques for Economists.* Cambridge: Cambridge University Press, pp. 171–172.

Neimira, M. P., and Klein, P. A. (1994). *Forecasting Financial and Economic Cycles.* New York: John Wiley and Sons, pp. 9–11.

Reilly, D. P. (June 27, 1999). Founder and Senior Vice-President of Automatic Forecasting Systems. Personal communication about modeling seasonality in AUTOBOX®.

SPSS, Inc. (1994). *SPSS Trends 6.1.* Chicago, IL: SPSS Inc., pp. 264–269.

Vandaele, W. (1983). *Applied Time Series and Box–Jenkins Models.* Orlando: Academic Press, p. 170.

Wei, W. (1990). *Time Series Analysis: Univariate and Multivariate Methods.* Boston: Addison-Wesley., pp. 162, 165, 167.

Chapter 6

Estimation and Diagnosis

6.1. Introduction
6.2. Estimation

6.3. Diagnosis of the Model
References

6.1. INTRODUCTION

The second and third stages of the Box–Jenkins model-building protocol are those of estimation and diagnosis. In Chapters 3 through 5, the reader has been introduced to preliminary considerations and the identification process of nonseasonal as well as seasonal time series models. This chapter explains the succeeding stages of estimation and diagnosis. In the estimation stage, three principal algorithms for estimating the identified model parameters are explained. In the diagnosis stage, the omnibus fit of the model and the significance of its estimated component parameters are assessed by various tests and protocols. Chapter 7 addresses the subsequent stages of metadiagnosis and forecasting. During metadiagnosis concurrent model evaluation and during forecasting predictive model evaluation is undertaken. After such model assessment, the optimal model is found as part of the Box–Jenkins model building strategy.

6.2. ESTIMATION

After identification of the model components, the parameters are estimated. Three principal algorithms used by popular statistical packages to estimate model parameters are unconditional least squares, conditional least squares, and maximum likelihood estimation. SAS permits the use of any of these techniques while SPSS employs maximum likelihood for model estimation allowing the researcher to choose between conditional and unconditional least squares when forecasting. Therefore, each of these algorithms is explained in this chapter. The first estimation technique discussed is that of conditional least squares, on which unconditional sums of squares, the next technique to be explained, is based.

6.2.1. CONDITIONAL LEAST SQUARES

This algorithm is based on minimization of residual variance. A function is estimated that has an error term. In this case, this function is a version of the ARIMA model. Consider first an integrated moving average (IMA) model. The model is already rendered stationary, so w_t, implying that Y_t was integrated and required differencing, is employed as a differenced indicator of the series variable. After first differencing the mean is zero. The model is

$$w_t = \varphi_1 w_{t-1} + e_t - \theta_1 e_{t-1}. \tag{6.1}$$

Table 6.1 displays the recursive conditional least squares estimation of this model. It contains IBM closing stock prices, extending from June 29, 1959, at time $t = 0$, to July 10, 1959, at time $t = 11$ (Box et al., 1994). Time $t =$

Table 6.1
Recursive Calculation of Least Squares

Time t	IBM Stock Price Y_t	w_t	e_t	θ_1 .53	e_{t-1}	$\theta_1 e_{t-1}$	a_t	$\theta_1 a_{t+1}$
-1	447.10	0.00	0.00	.53	0.00	0.00	0.00	0.00
0	445	-2.10	-2.10	.53	0.00	0.00	0.00	2.10
1	448	3.00	1.89	.53	-2.10	-1.11	3.97	0.97
2	450	2.00	3.00	.53	1.89	1.00	1.83	-0.17
3	447	-3.00	-1.41	.53	3.00	1.59	-0.33	2.67
4	451	4.00	3.25	.53	-1.41	-0.75	5.04	1.04
5	453	2.00	3.72	.53	-3.25	1.72	1.96	-0.04
6	454	1.00	2.97	.53	-3.72	1.97	-0.08	1.08
7	454	0.00	1.58	.53	-2.97	1.58	-2.03	-2.03
8	459	5.00	5.84	.53	-1.58	0.84	-3.83	-8.83
9	440	-19.00	-15.91	.53	5.84	3.09	-16.66	2.34
10	446	6.00	-2.43	.53	-15.91	-8.43	4.41	-1.59
11	443	-3.00	-4.29	.53	-2.43	-1.29	-3.00	0.00

-1 is the previous day to which the value of price is backcast. The closing stock price values are the observations found in the column Y_t. The values of the first differences of those Y_t values are found in the column labeled, w_t. This series is a first differenced, first-order moving average process of the type just described. Starting values are needed for e_{t-1}, w_{t-1}, and a_{t+1}, where a_{t+1} is the error generated when back forecasting (backcasting) the starting values. In higher order ARIMA models, starting values would be needed for the $w_{t-1}, \ldots, w_{t-p}$, and $e_{t-1}, \ldots, e_{t-q}$, and $a_{t+1}, \ldots a_{t+q}$. If the model had seasonal terms, then starting values would be needed for those as well. In Table 6.1 the starting values for the *unobserved* e_{t-1} and a_{t+1} are at first set to zero. The model is reexpressed as a function of its error term:

$$e_t = \theta_1 e_{t-1} + w_t - \phi_1 w_{t-1}$$
$$(1 - \theta_1 L)e_t = (1 - \phi_1 L)w_t \tag{6.2}$$

$$= \frac{(1 - \phi_1 L)w_t}{(1 - \theta_1 L)}.$$

In the first cycle, estimation of the starting values is computed by back-forecasting. In the second cycle, with the new starting values, the error terms are estimated through a process of forward recursion. Then the sum of the squared errors is computed as a criterion to be minimized with a nonlinear least squares algorithm (McCleary *et al.*, 1980):

$$S(w_t, \varphi_1, \theta_1) = \sum_{t=1}^{n} e_t^2 = \sum_{t=1}^{n} \left(\frac{(1 - \varphi_1 L)w_t}{(1 - \theta_1 L)} \right)^2. \tag{6.3}$$

To explain conditional least squares, we consider the simpler ARIMA(0,0,1) model

$$w_t = e_t - \theta_1 e_{t-1} = (1 - \theta_1 L)e_t$$
so that $$\tag{6.4}$$
$$e_t = \frac{w_t}{1 - \theta_1 L} = \sum_{i=0}^{\infty} \theta_1^i w_{t-i}.$$

In general, the model is estimated by minimizing the objective criterion of the sum of squared errors. For each value of θ_1 tried, an error term [Eqs. (6.2 through 6.4)] and its sum of squares are computed:

$$\sum_{t=1}^{\infty} e_t^2 = S(\theta) = \sum_{t=1}^{\infty} \left(\sum_{i=0}^{\infty} \theta_1^i w_{t-i} \right)^2. \tag{6.5}$$

The value of θ yielding the minimum sum of squares is chosen as the final estimate.

We return to Table 6.1 to elaborate on the estimation process, which entails backcasting the starting values at $t = 0$ and estimating the model at $t > 0$. Because

$$w_t = (1 - \theta_1 L)e_t \text{ for model estimation and}$$
$$w_t = (1 - \theta_1 F)a_t \text{ for backcasting}$$
where
$$F = \text{the lead operator,}$$

(6.6)

the equation error, $e_t = w_t + \theta_1 e_{t-1}$ can be reexpressed for back-forecasting as $a_t = w_t + \theta_1 a_{t+1}$. In column 2 the data for Y_t and its difference w_t are given. The e_{t-1} and a_{t+1} are given starting values of 0 for a particular selected starting value of θ_1.

The first cycle begins with backward estimation. The purpose is the backcast the starting value of e_{t-1} at $t = 0$. Therefore, the $\theta_1 a_{t+1}$, beginning at time period $t = 11$ is given a value of zero. With backward recursion, $t - 1$ is now $t + 1$. Hence, from $a_t = w_t + \theta_1 a_{t+1}$, $a_t = -3 + .53*0$, $a_t = -3$. At time $t = 10$, $\theta_1 a_{t+1} = .53*(-3) = -1.59$ and because $a_t = 4.41$, $w_t = 6.00$. As this process proceeds backward to time $t = 1$, a_t, a_{t+1}, and $\theta_1 a_{t+1}$ can be calculated. In this way, the starting value of $\theta_1 a_{t+1} = 2.10$ for time $t = 0$ can be backcast.

With the newly backcast starting values, the second cycle begins. At time $t = 0$, $w_0 = -\theta_1 a_{t+1}$. From $e_t = w_t + \theta_1 e_{t-1}$, the calculation of the values of e_t can proceed by forward recursion. Then the e_t, e_t^2 are stored for the selected value of θ_1. The sums of those squared errors are computed and stored for that value of θ_1, in both Table 6.1 and 6.2.

The value of θ_1 is incrementally changed and the process is reiterated. In Table 6.2, the sums of squared errors associated with particular values

Table 6.2

Estimation of Moving Average Parameter θ_1

Iteration	θ_1	SS error
1	0.52	361.88
2	0.54	361.60
3	0.56	361.50
4	0.58	361.55
5	0.60	361.73
6	0.62	362.00
7	0.64	362.35
8	0.66	362.72
9	0.68	363.10
10	0.70	363.43

of θ_1 are recorded. These sums of squared errors for each value of the parameter θ_1 are plotted as a function and stored in the computer (Fig. 6.1). This function guides the estimation of the best parameter value. In the estimation process, the new value of the parameter θ_1 is based on the movement from a previous value along a downward slope in the sum of squared errors function. Further changes in θ_1 eventually cease to reduce the sum of squared errors beyond some criterion (tolerated error) of convergence. If the sum of squared error function is deemed to attain a minimum value, then the process has converged upon the value of the parameter estimate. In short, the movement of the parameter θ_1 along its parameter space finally achieves a minimization of this sum of squared errors (fails to reduce it beyond some criterion). Convergence is attained and the iterations cease. The process has iterated to a solution.

6.2.2. UNCONDITIONAL LEAST SQUARES

The algorithm of unconditional least squares is almost identical to that of conditional least squares. The difference between them is in the computation of the starting values. In conditional least squares, the estimation is conditional on starting values of unobserved errors being set to zero, but in conditional least squares, the backcast values that may be closer to the real ones are used. In unconditional least squares, the starting values are simply set to zero (Ege *et al.*, 1993). If the series is sufficiently long, conditional and unconditional least squares estimation processes will yield very similar estimates.

6.2.2.1. Estimation of Autoregressive Parameters

The same estimation process can be extended to include AR, ARI, and ARIMA models. A moving average model may be expressed as an infinite order autoregressive model.

$$w_t = e_t(1 - \theta_1 L)$$

$$\frac{w_t}{(1 - \theta_1 L)} = e_t \qquad (6.7)$$

$$w_t(1 + \theta_1 L + \theta_1^2 L^2 + \theta_1^3 L^3 ...) = e_t$$
$$w_t = -\theta_1 w_{t-1} - \theta_1^2 w_{t-2} - \theta_1^3 w_{t-3} - \cdots - \theta_1^i w_{t-i} + e_t$$

and because $\pi_t = -\theta_i^t$

$$w_t = \pi_1 w_{t-1} + \pi_2 w_{t-2} + \pi_3 w_{t-3} + \cdots + \pi_i w_{t-i} + e_t. \qquad (6.8)$$

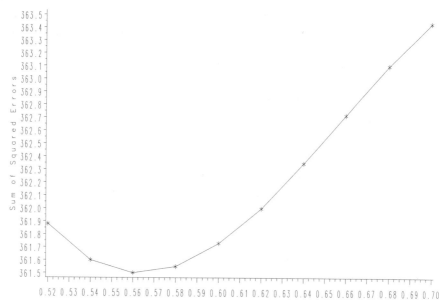

Figure 6.1 Parameter Estimation of θ_1.

Thus, this model can be reparameterized in terms of π parameters. The autoregressive process can be represented by a weighted sum of present and past values of the white noise process (Box *et al.*, 1994; Ege *et al.*, 1993). The sum of squared errors function can be inferred from this formulation, and that function is used to estimate the values of the parameters, π_i:

$$w_t = e_t + \sum_{i=1}^{\infty} \pi_i L^i$$

$$e_t = w_t - \sum_{i=1}^{\infty} \pi_i L^i. \tag{6.9}$$

For least squares estimation,

$$\sum_{i=1}^{\infty} \hat{e}_t^2 = \sum_{i=1}^{n} \left(w_t - \sum_{i=1}^{\infty} \hat{\pi}_i L^i \right)^2.$$

Of course, the absolute values of the π_i weights must have values less than 1 for the process to be stable and to be able to converge. This process iterates until the value of the parameter being estimated minimizes the sum of squared errors. At that point, the process has iterated to a solution.

Under conditions of stationarity and stability, the moving average and the autoregressive processes are interchangeable. An infinite or finite-order autoregressive process can be converted to a moving average process and estimated. Owing to the stationarity of the moving average parameter, the significant autoregressive parameters will taper off to nonsignificance after a few lags (Box and Jenkins, 1976). Conversely, these may be converted to a moving average model without much difficulty, and such models can be estimated in the manner described. These models are trimmed to those with preliminary identification and estimated. Partial autocorrelations can be estimated by fitting successive autoregressive parameters and computing the value of the last parameter at each stage of estimation. The least squares error criterion and signifi- cance tests of the parameters will determine the proper order of the process. Alternatively, the autoregressive process can be estimated by the Yule–Walker equations expounded in Chapter 4, Section 4.7. Bivariate correlations can be substituted for the theoretical autocorrelations and the partial autoregressive coefficients can be computed and used as starting values for the least squares estimation process:

For an AR1 process, $\hat{\varphi}_{11} = r_1$.

For an AR2 process, $\hat{\varphi}_{21} = \dfrac{r_1(1 - r_2)}{1 - r_1^2}$

$$(6.10)$$

and

$$\hat{\varphi}_{22} = \frac{r_2 - r_1^2}{1 - r_1^2},$$

where φ_{ii} is the partial autoregressive parameter, r_i is the bivariate correla- tion coefficient. With the Yule–Walker equations, the starting values of the autoregressive parameters and all of the autoregressive parameters can be estimated.

6.2.2.2. Estimation of ARIMA Model Parameters

In the event that the model is an ARIMA(1,1,1), it can be appropriately transformed to permit estimation. Just as one can convert a moving average model into an autoregressive model and vice versa, so one can convert a mixed model to a higher order autoregressive model, which may be sequentially estimated. Given an ARIMA(1,1,1) model with $w_t = Y_t - Y_{t-1}$,

$$(1 - \varphi_1 L)w_t = (1 - \theta_1 L)e_t.$$

Therefore,

$$e_t = \frac{(1 - \varphi_1 L)}{(1 - \theta_1 L)}\, w_t.$$

Thus, for unweighted least squares:

$$\sum_{t=1}^{n} e_t^2 = \sum_{t=1}^{n} (w_t - C_t V_t^{-1}(w_1, \ldots, w_{t-1})'), \tag{6.11}$$

where

C = covariance matrix of w_t and $(w_1, \ldots, w_{t-1})$ and

V = variance of $(w_1, \ldots, w_{t-1})$ if $\varphi_1 \neq \theta_1$,

$$= n^{-1}\frac{(1 - \varphi_1\theta_1)}{(\varphi_1 - \theta_1)^2}\begin{pmatrix} ((1 - \phi_1^2)(1 \cdots \varphi_1\theta_1)) & ((1 - \varphi_1^2)(1 - \theta_1^2)) \\ ((1 - \phi_1^2)(1 - \theta_1^2)) & ((1 - \theta_1^2)(1 - \phi_1\theta_1)) \end{pmatrix}$$

In this way, the mixed model can be converted to an autoregressive process, with attenuating coefficients, and it can be iteratively solved by least squares to minimize the error variance for the ARIMA parameters (Ege *et al.*, 1993).

6.2.3. MAXIMUM LIKELIHOOD ESTIMATION

Another numerical method used for parameter estimation of a nonlinear system of equations is maximum likelihood estimation. The Levenberg–Marquardt algorithm transforms a nonlinear model into a linear form for maximum likelihood estimation. This algorithm attempts to optimize the estimation process by combining an objective log-likelihood function, a conditional least squares estimation of starting values, a modified Gauss–Newton method of iterative linearization estimation, a steepest descent directional supervisor, and a step-size governor to enhance efficiency, with a convergence test to determine when to cease iteration. The integrated algorithm provides for efficient and reasonably fast maximum likelihood estimation of nonlinear models.

Maximum likelihood estimation usually begins with a likelihood function to minimize or maximize. A likelihood function is a probability formula. When observations are independent of one another, the probability of the multiple successive occurrences is the product of their individual probabilities. For example, in two coin tosses, the probability of a head in one toss is ½. The probability of two heads in two tosses is ½ × ½ = ¼. In a time series, white noise appears as random shocks and the individual shock is e_t. If the condition of ergodicity exists, each shock is independently and normally distributed, and therefore possess a normal probability distribution $[N \sim (0,\sigma^2)]$. The probability density function of a shock therefore is given by McCleary et al. (1980):

$$p(e_t) = \frac{e^{-e_t^2/2\sigma_e^2}}{\sigma_e\sqrt{2\pi}}. \tag{6.12}$$

The multiplicative constant $1/\sqrt{(2\pi)}$ can be dropped and the probability of the product of multiple shocks can be expressed (Box et al., 1994) as

$$p(a_1,a_2,\ldots,a_n) \propto \sigma_e^{-n} \exp\left[-\left(\sum_{t=1}^{n} \frac{e_1^2}{2\sigma_e^2}\right)\right]. \tag{6.13}$$

Taking the natural log of that function, the analyst can obtain the natural log of the likelihood function, conditional on the choice of the parameters:

$$LL(\varphi,\theta,\sigma_e) = -n\ln(\sigma_e) - \frac{\Sigma e_t^2.(\varphi,\theta)}{2\sigma_e^2}. \tag{6.14}$$

The first term on the right-hand side of Eq. (6.14) will be negative whenever the σ is positive and zero whenever σ is equal to 1. The second term on the right-hand side of this equation will always be negative. The e_t^2 can be conceptualized as $(y_t - \mu)^2$. When these scores are mean deviations, the second term on the right hand side appears to be the negative $\Sigma z^2/2$. In other words, the -2 log likelihood of the right-hand side of the equation is distributed as a χ^2 distribution. Therefore, the maximum (log) likelihood will occur when the sum of squared error term is at a minimum. When this -2 log likelihood is calculated for the null model—that is, the model with only the constant in it—and then subtracted from that of the full model—that is, the model with all of the parameters in it—the difference of these two log likelihoods is distributed as a χ^2 with p degrees of freedom, where p is equal to the number of parameters in the model. This is the amount of reduction of sum of squared error that is attributable to the inclusion of parameters in the model. This subtraction of log likelihoods is a likelihood ratio χ^2 with p degrees of freedom. If the likelihood ratio χ^2 is statistically significant, then the parameters included in the model significantly minimize the sum of squared errors, maximize the likelihood, and thereby improve parameter estimation.

The algorithm basically works after the fashion of a guided grid search. The grid search moves along values of the parameter being estimated. First, starting values of the parameter identified by the model are obtained from conditional least squares estimates, a series mean, or values preset by the researcher. A sum of squared errors, referred to as the old sum of squared errors, is computed for these starting parameter values. Vectors of slopes, of the sum of squared errors for each parameter being estimated, are computed. The vector with the steepest slope (largest derivative) is chosen for direction and step size guidance. With this selection of direction and step size, an incremental value is added to the starting value of the parameter to form a new value of the parameter being estimated. At this point, a new sum of squared errors is calculated and compared to the old sum of squared errors. The process is repeated as long as each new sum of squared errors is substantially smaller than the old sum of squared errors. At the beginning of the next cycle, the new starting value is now the old value of the parameter. Iterations continue until the reduction of the error sum of squares fails to exceed some limit, called the criterion of convergence.

An unmodified grid search without the aid of the steepest descent innovation has serious deficiencies. If, by chance, the starting value is close to this minimum value of the error sum of squares, then convergence takes place quickly. If, however, the starting value is far away, the process takes longer before convergence is reached. If the convergence criterion is quite small, the process may not converge for a long time, if ever. Therefore, a grid search method by itself leaves much to be desired in an estimation algorithm. To render it more useful, a mechanism of steepest descent is incorporated.

The steepest descent algorithm can be clarified with some elaboration. To facilitate efficient convergence of a grid search process that by itself might meander randomly, the Levenberg–Marquardt algorithm controls both the direction and the size of the step at each iteration. At the end of each iteration, it computes the derivatives of $S(\theta)$ in several directions and follows the direction of the steepest derivative of that function with respect to the parameter in question. For control of the step size, the algorithm assesses the speed of convergence. The farther away from the minimum of the lack of fit function, the larger the step size is made. Conversely, the nearer that minimum, the shorter the step size. The size of the step is basically controlled by the steepness of the descent toward the minimum. By multiplying the first derivative of the lack of fit function with respect to the parameter, the algorithm provides for control of the speed of movement toward convergence (Draper and Smith, 1981).

The modified Gauss–Newton method of iterative linearization applies a Taylor series linear approximation of the functional relationship between the log-likelihood function taken from the identified model and the parameter estimates of each of the parameters that derives from the nonlinear ARIMA model. Let us consider the nature of the derivative of our postulated model. Suppose a hypothesized IMA model is of the form

$$w_\mu = f(\xi_\mu, \theta) + \varepsilon_\mu \tag{6.15}$$
$$= \varepsilon_t - \theta_1 \varepsilon_{t-1}, \quad \text{where } \varepsilon \sim N(0, \sigma^2).$$

By definition,

$$\frac{\partial f(\xi, \theta_0)}{\partial \theta_i} = \frac{f(\xi, \theta_i) - f(\xi, \theta_0)}{\theta_i - \theta_0}.$$

Therefore, $\tag{6.16}$

$$f(\xi, \theta_i) = f(\xi, \theta_0) + \frac{\partial f(\xi, \theta_0)}{\partial \theta_i}(\theta_i - \theta_0).$$

Yet this holds for a particular point on the θ_i axis. The nonlinear ARIMA model is reformulated as a function of the error and approximated by a Taylor series linear approximation. If a higher order—for example, with derivatives taken to the ith power—Taylor series approximation were applied, with i successive derivatives taken, the factor by which these derivatives would be multiplied would be $1/i!$. Therefore, the general formula for the approximation includes division by $I!$, as shown in Greene (1997) and derived in Thomas (1983):

$$f(\xi, \theta_i) = f(\xi, \theta_0) + \sum_{i=1}^{p} \frac{1}{i!}\left[\frac{\partial^i f(\xi, \theta_0)}{\partial \theta_i^i}\right](\theta_i - \theta_0)^i + \varepsilon_t. \tag{6.17}$$

The equation may be expressed as a function of the likelihood or sum of squared errors. The sum of squared errors is

$$S(\theta) = \sum_{i=1}^{n}\left[w_t - f(\xi, \theta_0) - \sum_{i=1}^{p}\frac{1}{i!}\left(\frac{\partial f(\xi, \theta_0)}{\partial f(\theta_i)^i}\right)(\theta_i - \theta_0)^i\right]^2 \tag{6.18}$$

$$= \sum_{i=1}^{n} \varepsilon_t^2.$$

The left-hand side of Eq. (6.17) may be considered a new function to be minimized. Alternatively, its opposite, the log likelihood, may be the function to be maximized. In either case, the root of that function to be found may be set to zero, so that the approximation of the function can be solved. The equation can be reexpressed as a function of the parameter θ to show how the iteration process works:

$$\Delta_1 = \theta_1 - \theta_0 \tag{6.19}$$
$$\theta_1 = \theta_0 + \Delta_1.$$

The Δ approaches the criterion of convergence as θ_1 approaches its parameter estimate. Unless the criterion of convergence is properly set, this approach may oscillate excessively back and forth between positive and negative values of step size before converging on the final parameter estimate,

$$\theta_i = \theta_0 - \frac{f(\xi, \theta_0)}{\left(\dfrac{\partial f(\xi, \theta_0)}{\partial \theta_0}\right)}, \tag{6.20}$$

where θ_0 is the old value of parameter θ and θ_i is the new value of parameter θ.

The Levenberg–Marquardt algorithm clearly controls convergence of the estimation. If the slope of the lack of fit function is positive, then the value of the parameter is reduced. If the slope is negative, the value of the parameter is increased. Eventually, the value of the parameter θ approaches the point where the lack of fit function is minimized and the likelihood is maximized. This is the point where $f(\xi, \theta) = 0$, the root of the equation. At this time, the change in θ converges to a solution and iterations cease.

Other forms of maximum likelihood algorithms exist. Brockwell and Davis (1991) suggest that ARMA models are often estimated with algorithm based on the principle

$$\theta_1 = \theta_0 - d\frac{\partial S(\theta)}{\partial \theta}, \tag{6.21}$$

where d is some coefficient of step size. Here, the sign of the derivative of sum of squared errors with respect to the parameter θ will control the direction of the change in value of the parameter. Positive slopes indicating a growth in the sum of squared errors will decrease the value of the parameter, while negative derivatives indicating declines in the sum of squared errors result in an increase in the parameter value of Fig. 6.1. If the log likelihood replaces the sum of squared errors then the sign in the equation becomes a positive rather than a negative. This proceeds until convergence is attained. Another modification of this algorithm is the Newton–Raphson algorithm where d is replaced by the inverse of the negative of the Hessian matrix:

$$\theta_i = \theta_0 + \left(\frac{-\partial^2 S(\theta)}{\partial \theta_0^2}\right)^{-1}\frac{\partial S(\theta)}{\partial \theta_0}. \tag{6.22}$$

If the slope is positive in Fig. 6.1, the new parameter θ becomes less positive, and if the slope is negative, the new parameter θ becomes more positive. In this way, it eventually converges on the point where the slope is horizontal and the derivative is 0. The minus times the inverse of the second derivative matrix is the information matrix. The elements of the principal diagonal of this matrix are the asymptotic variances of the θ. The more peaked the slope, the more the information. The more the information, the larger the step size and the greater the change in value of the parameter. Conversely, the smaller the step size, the less the information, the flatter the curve, and the less the change in the parameter value. This control over the step size renders the convergence process much more efficient than a mere grid search (Long and Trivedi, 1993).

If θ_1 is the parameter that minimizes the $S(\theta)$, one can express the upper equation to show how the functional relationship may change in connection with the previous value of the parameter, θ_1. To do so, the algorithm moves the parameter θ_1 from its starting value, and the likelihood (or sum of squared errors) as well as the derivative of the likelihood function with respect to the parameter in several directions are computed. In order to expedite convergence, the procedure chooses the derivative with the steepest slope. The direction in which the movement of the parameter will proceed is the opposite of the sign of the slope. If the sign of the derivative is positive, then the value of the parameter will decrease. If the sign of the derivative is negative, then the value of the parameter will increase. As long as the partial derivative is nonzero, there is a tendency for the parameter θ to move in the next step in a direction to reduce the sum of squared errors. From the graph of this function in Fig. 6.1, we see that it is possible to iterate through the parameter values of θ until this lack of fit function arrives at a minimum. At this point the derivative of the function approaches zero. The reduction of the sum of squared errors ceases to improve beyond a limit of convergence, so that further iteration ceases.

In other words, the value of the sum of squared error function after each shift of the parameter θ_1 is calculated and recorded. The change in the value of the parameter is a function of the slope of the sum of squared error function. If the slope is negative, then the shift in the value of the parameter θ will be in a positive direction. If the slope is positive, then the shift in the value of the parameter will be in the negative direction. When the function attains a minimum, provided that the criterion of convergence is reached, the iterations cease (Draper and Smith, 1981; Eliason, 1993; Long and Trivedi, 1993; Wei, 1990).

McCleary *et al.* (1980) give four criteria of convergence: The Levenberg–Marquardt algorithm converges when any one of these criteria has been met. First, when the percentage reduction of $S(\theta)$ goes below a set limit, the iteration process will end. Second, when the percentage change in the value of the parameter (θ_1) goes below a specified level, the iteration process will stop. Third, when the number of iterations reaches a maximum limit, adjustable by the researcher, which has been set as the default limit for the program, the iterations will terminate. Fourth, when the last iteration reached a minimal ratio of change from the initial sum of squared errors or log-likelihood to the last one, the program will complete its iteration process.

In sum, this maximum likelihood estimation follows an iterative process. It begins with starting values of the model parameters to be estimated.

These starting values can be supplied by the user or by his selection of options within the program. SAS utilizes conditional least squares to obtain the starting values. SPSS gives the researcher the choice of user predefined or automatically set starting values. The Levenberg–Marquardt algorithm computes the sum of squared errors. Then it computes a Taylor series linear approximation of the model, from which a vector of correction factors is derived. The new value of the parameter is then corrected by the appropriate element of this vector. A new sum of squared errors is computed. If the new sum of squared errors is less than the old one, then the adjustment is made to the approximation. If the new sum of squared errors is greater than the old one, then the test for convergence is applied. That is, the change in the sum of squared errors is tested to see whether it is below the level of convergence—whether the new sum of squared errors is almost identical to the old one. If it is, then the iterations cease. If not, then the process cycles through another iteration (Draper and Smith, 1981; Eliason, 1993).

There may be potential problems with efficient maximum likelihood estimation, and the researcher should be aware of them. Sometimes, preliminary moving average estimates do not converge, in which case other initial starting values may be tried. This problem may stem from multicollinearity which flattens error surfaces along the parameter space. This flattening of the valleys makes it difficult for the algorithm to find a minimum of the sum of squared errors or a maximum of the log likelihood. When the parameter values are close to the bounds of stability or stationarity, these flat surfaces may also be found, and there may be a need to increase the number of iterations permitted for the algorithm to converge. If some of the parameters are not identifiable, it may be possible to trim them from the model. Sometimes convergence takes place on a local rather than a global minimum. The analyst may randomly try different starting values from the range of possible values to assure himself that convergence always takes place on the same optimal value of the parameter. If the parameter estimates remain the same, then the reliability of this solution would suggest that the solution is indeed the optimal one.

6.2.4. COMPUTER APPLICATIONS

Of the two statistical packages compared here, SAS allows the user to choose freely from three different algorithms for estimation: conditional least squares, unconditional least squares, and maximum likelihood estimation. SPSS uses only maximum likelihood estimation, but allows the user to choose between conditional and unconditional least squares for forecasting.

Although both SPSS and SAS employ a version of the Marquardt algorithm as described in Kohn and Ansley (1986), and Morf *et al.* (1974), SPSS now uses Melard's fast maximum likelihood algorithm (1984). If the analyst employs the maximum likelihood estimation in both models, the results are identical to the thousandths decimal place. Beyond that, differences begin to appear.

Although there is some controversy over which algorithm yields the best results under what circumstances, conditional least squares generally performs better with smaller data sets than maximum likelihood estimation. For very large data sets, conditional least squares is faster than maximum likelihood estimation, but maximum likelihood is believed to be more accurate (Vandaele, 1983; Brocklebank and Dickey, 1986). Because maximum likelihood estimation entails asymptotic estimation, it should be used only with larger data sets. Standard errors tend to be smaller and differences in iterated sum of squared errors are easier to detect with larger sample sizes. With smaller data sets, these differences are harder to detect and iterative maximum likelihood estimation can meander myopically about, doing more damage than good. In general, if the data set is small, it is advisable to avoid maximum likelihood estimation. But if parameter estimates are close to the bounds of stationarity or stability or if seasonal multiplicative models are estimated, either conditional least squares or unconditional least squares might yield better results. The algorithms may produce different results, and the user is advised to try several to get a sense of the possible variation. Conditional least squares attempts to obtain more accurate starting values, while unconditional least squares might use either the series mean or midpoint of neighbors. Both SPSS and SAS employ missing data replacement algorithms to replace missing values in the data set if the user has not already done so. Using interpolative procedures based on the work of Jones (1980) and Kohn and Ansley (1986), SAS and SPSS automatically replace the missing values from predictions from an infinite memory process of the previously nonmissing data, and these artificial values are updated at each stage of iteration. In the intervention or transfer function models covered later, which involve other input variables, the user must supply the missing values for those input variables.

Computer syntax specifying the type of model estimation to be invoked is available in the SAS PROC ARIMA procedure and shown in Program syntax example 6.1. In the ESTIMATE subcommand, the user may select the kind of estimation. The user may specify the order of the autoregression with a P=X, where X is the numeric order of the autoregression. The user may specify the order of the moving average with Q=Y, where Y is the numeric order of the moving average. If a multiplicative model is being estimated, then a P=(1)(12) or a Q=(1)(12) may be in order. If there is to be no mean term in the model, an option called NOINT (alternatively, the specification NOCONSTANT may be used) is added to the subcommand. PRINTALL and PLOT are usually advisable if one wishes to diagnose the estimation process. PRINTALL gives the iteration history and diagnostics; PLOT provides the ACF, IACF, and PACF of the model residuals.

Specification of the algorithm comes with the METHOD option. With any algorithm, the options available are CLS, ULS, and ML. These signify conditional least squares, unconditional least squares, and maximum likelihood, respectively. If the user fails to specify the algorithm of choice, CLS is used in default. The user may specify his starting values with the AR=, MA=, MU= options. In the SAS program syntax Example 6.1, the initial values of the CLS model for the regular and seasonal moving average parameters are 0.3 and 0.4, respectively. The initial value for the mean is 0.1. METHOD=ML is used to obtain maximum likelihood estimation, but maximum likelihood estimation uses starting values from conditional least squares. Therefore, if the user provides starting values for maximum likelihood estimation, the program begins conditional least squares estimation with those starting values and then uses the conditional least squares estimates as the starting values for the maximum likelihood estimation. With maximum likelihood, it may be advisable to limit the number of iterations with the MAXIT option. MAXIT=41 is used in the example. The CONVERGE= .0001 specifies the convergence criterion. To request unconditional least squares estimation, the user merely specifies METHOD=ULS. If the user does not specify which type of estimation is preferred, SAS invokes conditional least squares by default.

Program Syntax Example 6.1

```
PROC ARIMA DATA=SASSTOCK;
 IDENTIFY VAR = Yt(1,1) nlag=20;
 ESTIMATE Q=(1)(12) MU=.1 MA=.3,.4 PRINTALL PLOT NOINT
 METHOD=ML MAXIT=41 CONVERGE=.0001;
```

SPSS users may also wish to specify their ARIMA estimation syntax. They must remember that the first line of an SPSS command must begin in column 1 while continuations of the command must be indented at least one space, and delimited with a forward slash. The user does not have control over the choice of algorithm at the estimation stage: Melard's fast maximum likelihood algorithm is automatically invoked for parameter estimation. Only when he begins his forecasting does he currently have a choice of either conditional least squares or unconditional least squares as a forecast option. He can therefore include a /FORECAST CLS or a

/FORECAST EXACT option statement at the end of the procedural command syntax in Example 6.2 to obtain, respectively, either CLS or ULS forecast estimation. Although subcommands follow the forward slashes, the termination of the SPSS command is designated by a period.

Users can set the starting values of the parameters with the AR=, MA=, SAR=, SMA=, REG=, or CON= subcommands. The user can control the criterion of convergence as a percentage change in the parameter value with the /PAREPS=.001 subcommand. He can also control the criterion of convergence as a percentage of sum of squared errors with the /SSQPCT = .001 subcommand. Also, he can control the number of iterations with the MXITER=10 subcommand.

Program Syntax Example 6.2

```
ARIMA SPSSTOCK / Model=(0,1,1)(0,1,1) Constant/
   AR=0 / MA=.1/ SAR=0/ SMA=.1/CON=0.5/
   MXITER=41 /SSQPCT = .0001/FORECAST CLS.
```

In both the SAS and SPSS examples, the maximum number of iterations was set to 41 and the convergence criterion was set to .0001 of the sum of squared errors. Both use the maximum likelihood estimation algorithm. SPSS uses starting values of the parameters as .1 for the moving average and the seasonal moving average estimates, along with a starting value of .5 for the constant. If the algorithm does not converge, SAS permits the use of two other algorithms. With SPSS, the user may try to increase the MXITER option limit and/or change the SSQPCT criterion. Either or both of these adjustments might facilitate convergence.

These three algorithms are the principal estimation techniques employed by SAS and SPSS. Each algorithm has its own advantages and disadvantages. Abraham and Ledolter (1983) maintain that unconditional least squares works well when the parameters are not close to the bounds of invertibility. Brocklebank and Dickey (1994) find that conditional least squares is much faster than either unconditional least squares or maximum likelihood estimation on large data sets. Granger and Newbold (1986) note that unconditional least squares and conditional least squares are satisfactory for larger sample sizes, when their results approximate those of maximum likelihood estimation. Maximum likelihood estimation is based on large sample asymptotic estimators, which are asymptotically normally distributed, for which reason it is advisable to have long series of at least 50 equally spaced observations before applying it. Granger and Newbold (1986) claim that maximum likelihood estimation gives satisfactory results even with more limited-size samples. Unfortunately, maximum likelihood estimation is vulnerable to local minima. Therefore, it may be necessary to try randomly selected starting values within the permissible range of parameter values to be sure that the convergence uniformly takes place on the same parameter estimate. The maximum likelihood method, as an iterative procedure, consumes more computer time and resources than others. Often, the researcher would be well advised to try several of these methods. If convergent validity holds, then different algorithms using the same starting values and missing data replacement procedure should yield identical results. If they do not yield essentially the same results, then it is important to ascertain which of the models explains and fits better, a subject that will be examined in the next chapter on metadiagnosis and forecasting.

6.3. DIAGNOSIS OF THE MODEL

After estimation of the model, the Box–Jenkins model building strategy entails a diagnosis of the adequacy of the model. More specifically, it is necessary to ascertain in what way the model is adequate and in what way it is inadequate. This stage of the modeling strategy involves several steps (Kendall and Ord, 1990). Perhaps the first order of business is to assess the omnibus fit of the model. This entails being sure that the model converged upon a minimum sum of squared errors. The sum of squared residuals should be quite small so that the R^2 of the model would be quite large. Note can be made of the information criteria for benchmark or baseline reference. The second-stage individual parameter evaluations will be made in accordance with their reduction of the value of the information criteria to a minimum.

Evaluation of the individual parameter estimates should be the second order of business. Review of the parameters estimates may reveal adequacies or inadequacies of the model. Their significance, magnitude, intercorrelation, number, proximity to the boundaries of stationarity or invertibility, and estimation algorithm have implications for their retention or exclusion from the model. Stable and parsimonious models are preferred.

Parameter estimation should be attempted by different algorithms to see if they yield identical results. Different tests producing identical results on the same data provides concurrent validation of the estimation techniques employed. A kind of convergent validation can be inferred from this multimethod approach. The model exhibits reliability, stability, and relative robustness to variations in the estimations when this takes place. If the results from the various estimations differ substantially, that is evidence of what Leamer (1983) referred to as a fragile model. The magnitudes of the parameter estimates should be reasonable. The parameter estimates should lie well within the bounds of stationarity and invertibility. If necessary, their polynomials should be formulated and their roots should be tested for reality or complexity. If the parameter estimates are close to these bounds, then unit root tests might be in order. For example, if the absolute value of the first-order autoregressive parameter is close to 1, then the model may be nonstationary and differencing might be in order. Parameter estimates near the bounds of stability or invertibility might result in wild fluctuations at the initial stages of iteration, which might suggest misspecification of the model. In moving average models, the parameters should be within the bounds of invertibility. If the model is a second-order moving average model, the sum of θ_1 and θ_2 should be less than 1. θ_2 minus θ_1 should be less than 1. The absolute value of θ_2 should be less than 1 as well. If the model is a second-order autoregressive model, the φ coefficients should similarly lie within the bounds of stationarity.

Not only should the parameter estimates be of reasonable magnitude, they should be clearly statistically significant as well. Their t-ratios should be greater than 1.96. If the parameters are not significant, they should be trimmed from the model. If they are significant, they should remain within the model. Sometimes, parameters may be close to significance and should remain within the model anyway for reasons of theory testing and theory building. These statistical significances are merely estimates of the real significance and may vary somewhat from the real ones in the data-generating process.

The estimation process should have successfully converged upon the estimates. If the parameter estimates are too close to the bounds of stationarity or stability, the estimation process for the model may not have converged. If the model did not converge, there may be several reasons for it. The parameter estimates might be so intercorrelated that collinearity between them flattens the response surface of the sum of squared errors. The grid search on so flat a response surface might meander without convergence. Either increasing the maximum number of iterations permitted or loosening the criterion of convergence might facilitate iteration to a solution.

Collinearity between the parameter estimates should be examined. Usually, the statistical package includes a correlation matrix of parameters. Evidence of collinearity can be found in this matrix, from which it can be inferred that the response surface of the parameter space may level off or flatten out. When the parameter estimates are highly intercorrelated, one option is to reduce the number of intercorrelated items in the model.

Model diagnosis entails residual analysis as well. If the model is properly specified and the model parameters account for all of the systematic variance, then the residuals should resemble white noise. Residual analysis is performed with the autocorrelation and partial autocorrelation function. These correlograms can be examined with reference to modified Portmanteau tests of their associated significance levels. It should be remembered that the Portmanteau statistic might inflate the autocorrelation under conditions of short series or short lag times, for which reason the modified Ljung–Box statistic is used to provide better significance tests. White noise residuals do not have significant p values. These white noise p values of the residuals should not be less than 0.05. Graphically, white noise residuals have associated spikes that do not extend beyond the confidence interval limits. The ACF and PACF plots reveal these limits as dotted lines spreading out from the midpoint of the plot. When spikes protrude beyond the limits of two standard errors on each side of the central vertical axis of no autocorrelation, then the autocorrelation or partial autocorrelation of the residuals have significant spikes with p values less than 0.05. Indication of significant ACF or PACF residual spikes is empirical evidence of lack of fit.

The pattern of lack of fit will suggest the reparameterization of the model. Slowly attenuating autocorrelation functions suggest further differencing. Sharp and pronounced alternating spikes in the correlogram may suggest that overdifferencing has been invoked and that a lower order of differencing is in order. Seasonal spiking of slowly attenuating autocorrelation functions suggest that seasonal differencing may be in order.

Combinations of ACF and PACF patterns indicate whether the additional terms should be moving average or autoregressive. Gradual attenuation of the ACF with a few spikes and sudden decline in PACF magnitude suggest that autoregressive parameters should be added, whereas gradual attenuation of the PACF and a few finite spikes of the ACF with sudden decline of their magnitude suggest moving average terms should be used. If these patterns have seasonal spikes in the same direction, with no spikes in the opposite direction, then a purely seasonal model may be indicated. There may be alternating seasonal spikes indicating negative seasonal parameters. If there is seasonal spiking with an occasional spike in the opposite direction, a multiplicative seasonal model may be in order. The type of seasonal parameters would depend on the pattern of spikes characterizing the ACF and the PACF. Once these have been properly identified and estimated, the ACF and PACF of the residuals should appear as white noise.

The model needs to be tested by underfitting and overfitting. If the model is optimal, neither underfitting (dropping of questionable parameters) nor overfitting (including extra parameters) should yield a lower sum of squared errors. Diagnoses of these models come from the use of the R^2 statistic.

Based on the minimum sum of squared residuals, the R^2 of the model is found with the following formula:

$$R^2 = 1 - \frac{\sum\limits_{t=1}^{T} SS^2_{error}}{\sum\limits_{t=1}^{T} SS_{total}}. \tag{6.23}$$

These are not the only useful indicators of goodness of fit. The R^2 may be adjusted for degrees of freedom where the additional parameters tend to inflate this statistic:

$$\text{Adjusted } R^2 = \left(1 - \frac{k}{T}\right) r^2, \tag{6.24}$$

where k = number of parameters, T = number of observations. If the parameters are all accounted for in the model, then the residuals should consist purely of white noise or unsystematic random variation.

The model should be reasonable and parsimonious. It should be as elegant as possible. It must account for as much of the systematic variance as possible, leaving white noise residuals. The parameters of the model should be estimated with convergence of the model upon a minimum sum of squared residuals. The parameter estimates should not be highly intercorrelated and they should be significant. The diagnostic tests discussed in this section permit the assessment of these properties of statistical adequacy.

Diagnostic testing also requires assessment of the methodological adequacy of the model. If the research, sampling, and data collection were conducted properly, there should be no impairment of the internal or external validity of the data. Cook and Campbell (1979) point out that the researcher must methodologically guard against threats to the internal validity of a time series analysis. Short time series may deprive the analyst of adequate statistical power to estimate and find real significance. Not much has been written on the subject of power analysis for ARIMA models, but most writers pay homage to the caveat that the series needs to be long enough to possess enough power to detect and reject a false null hypothesis. For seasonal models or even longer cyclical models, there is a greater need for longer series. The series needs to include several seasons and cycles if these are to be detected and properly identified. For impact assessment models, the segment of the series before the intervention and the segment of the series after the intervention have to be long enough to be properly modeled. For all of these reasons, short time series threaten the validity and utility of time series models.

Cook and Campbell (1979) warn about flaws in research design or happenstance that jeopardize the statistical conclusion validity of interrupted time series quasi-experiments. They mention the lack of an equivalent control group, nonequivalent dependent variables being modeled, uncontrolled and/or unscheduled removal of treatment from the series, uncontrolled and/or unscheduled contamination of control and experimental groups by migration of subjects between groups, and inadequate archival recording of experimental processes as possible contaminants of the purity of the research process. For impact assessment studies, which will be explained in much greater detail in Chapters 8 and 9, the analyst has to be sure that his series is long enough for him to detect and model gradual, ramp, or delayed impacts. Instrumentation should have been reliably calibrated to ensure that data collection procedures had not been altered over time. If the subjects, exhibiting the trait being observed, measured, recorded, modeled, and analyzed, are changing over time, then sample selection bias threatens the internal validity of the sampling and measuring process. If too much of the data generating process is not being detected, then the data might not be a valid indicator of what is really happening.

One example of this problem is revealed in the development of the AIDS epidemic. In the early years, it was not clear what AIDS was. In the early 1980s, it was thought of as Kaposi sarcoma and PCP peumonia. By 1984 other opportunistic infections were included in the definition. By 1987, the scope of these infections had widened considerably. By 1992, the T cell count was also included as a criterion. Then the T cell count standard changed to broaden the definition further. Because of the frequency of this redefinition, it was difficult to find a series of 30 or more observations under the same definition. In 1991, the U.S. Centers for Disease Control and Prevention (CDC) suspected that it was obtaining data on about 85% of the actual AIDS cases in the United States. The CDC got its data from the state health departments which got their data from the hospitals in each state. One researcher, John Stockwell, in personal communication, expressed suspicions this level was much less than the real number of cases, many of which were being treated at a private and local level without being reported to the hospitals. Although the CDC publicly distributed its data on the number of reported AIDS cases and deaths per month, this suspicion in addition to the frequent changes in the definition of AIDS made it difficult to confidently model the growth of the epidemic in the domestic United States with ARIMA models.

Historical impacts on univariate time series should have been precluded by isolation of the phenomenon under study as much as possible. Univariate series being analyzed ought not to have been significantly or substantially affected by other events over time. Univariate time series analysis are studies of the history of a process. External influences might effect a shift in level or variance of the series being observed. Nonetheless, measuring the data with reactive tests may sensitize, fatigue, or mute the respondent and impair the validity of the data. Surveys conducted without regard to these facts may produce specious data. Any or all of these problems may impair the internal validity of the study.

External validity needs to be protected and assessed as well. The sampling should have been performed so as to avoid biasing the results. It helps to have had a control group and an experimental group. Care must be taken to see that people from one group do not migrate to another group during the experiment, which is what was reported to have happened during the early AZT clinical trials. Local history and selection may interact when members of the control group learn that they are not being given anything other than a placebo. They may drop out of the study to get into the control group. Attrition then takes place for reasons other than death from AIDS. Without a double-blind experiment, this kind of interaction can complicate matters. The researcher needs to study the research methodology employed to know whether and how much to trust the data. Problems such as these may affect the data collection and measurement. Once it has been determined that the model is adequate, the question of optimality of the model arises.

REFERENCES

Abraham, B., and Ledolter, J. (1983). *Statistical Methods in Forecasting.* New York: John Wiley and Sons, Inc., p. 253.

Brocklebank, J., and Dickey, D. (1986). *SAS System for Forecasting Time Series* Cary: SAS Institute, Inc., pp.77–83.

Brocklebank, J., and Dickey, D. (1994). *Forecasting Techniques using SAS/ETS Software: Course Notes.* Cary SAS Institute, Inc., p. 90.

Brockwell, P. J., and Davis, R. J. (1991). *Time Series: Theory and Methods.* 2nd ed. New York: Springer-Verlag. pp. 256–273.

Box, G. E. P., and Jenkins, G. (1976). *Time Series Analysis: Forecasting and Control.* 2nd ed. Oakland: Holden Day, p. 216.

Box, G. E. P., Jenkins, G., and Reinsel, G. (1994). *Time Series Analysis: Forecasting and Control.* 3rd ed. Englewood Cliffs, NJ: Prentice Hall, pp. 10, 46–47. Series D' data are used with permission of the author, p. 543.

Cook, T. D., and Campbell, D. T. (1979). *Quasi-Experimentation: Design and Analysis Issues for Field Settings.* Boston: Houghton Mifflin Co., pp. 207–293.

Draper, N., and Smith, H. (1981). *Applied Regression Analysis,* 2nd ed. New York: John Wiley and Sons, Inc., pp. 458–462, 530.

Ege, G., Erdman, D. J., Killam, R. B., Kim, M., Lin, C. C., Little, M. R., Narter, M. A., and Park, H. J. (1993). *SAS/ETS User's Guide, Version 6,* 2nd ed. Cary, NC: SAS Publications, Inc., pp. 130–134, 140–142.

Eliason, S. R. (1993). *Maximum Likelihood Estimation. Logic and Practice.* Newberry Park, CA: Sage Publications, pp. 12, 42–44.

Granger, C. W. J., and Newbold, P. (1986). *Forecasting Economic Series,* 2nd ed. San Diego: Academic Press, p. 93.

Greene, W. (1997). *Econometric Analysis,* 3rd ed. Englewood Cliffs, NJ: Prentice Hall, pp. 49–55, 113–115, 129–138, 197–203.

Jones, R. H. (1980). "Maximum Likelihood Fitting of ARMA Models to Time Series with Missing Observations." *Technometrics 22,* pp. 389–395.

Kendall, M., and Ord, K. (1990). *Time Series Analysis,* 3rd ed. New York: Oxford University Press, p. 110.

Kohn, R., and Ansley, C. F. (1986). "Estimation, Prediction, and Interpolation for ARIMA Models with Missing Observations." *Journal of the America Statistical Association,* **81,** pp. 751–764.

Leamer, Ed. (1983). "Let's Take the Con out of Econometrics." *American Economic Review,* **73,** pp. 31–43.

Long, J. S., and Trivedi, P. K. (1993). "Some Specification Tests for the Linear Regression Model," in Bollen, K. A., and Long, J. S., Eds., *Testing Structural Equation Models.* Newberry Park, CA: Sage Publications, pp. 76–78.

Makridakis, S., Wheelwright, S. C., and McGee, V. E. (1983). *Forecasting: Methods and Applications,* 2nd ed. New York: John Wiley and Sons, Inc., p. 891.

McCleary, R., and Hay, R., with Meidinger, E., and McDowell, D. (1980). *Applied Time Series Analysis for the Social Sciences.* Beverly Hills: Sage Publications, pp. 208, 210, 213–218, 280, 298.

Morf, M., Sidhu, G. S., and Kailath, T. (1974). "Some New Algorithms for Recursive Estimation of Constant Linear Discrete Time Systems." *I.E.E.E. Transactions on Automatic Control.* AC-19, 315–323.

Thomas, G. B. (1953). *Calculus and Analytical Geometry.* Addison-Wesley, pp. 573–579.

Vandaele, W. (1983). *Applied Time Series and Box–Jenkins Models.* Orlando: Academic Press, pp. 114–117.

Wei, W. S. (1990). *Time Series Analysis: Univariate and Multivariate Methods.* New York: Addison-Wesley, pp. 136–149.

Chapter 7

Metadiagnosis and Forecasting

7.1. INTRODUCTION

After the ARIMA models are assessed for adequacy, the analyst undertakes a metadiagnosis of the different ARIMA models. In this stage of the analysis, the researcher compares and contrasts competing models to determine which is the best explanatory model. On the one hand, the analyst may use concurrent tests of the information set, model fit, stability, and explanatory power, and parsimony. He may review the model for aspects of parameter size, number, scope, significance, and stability. As part of this evaluation, he should assess the model for its forecasting ability, stability, and robustness; or he may assess it for predictive precision, validity, and reliability. Because each model is an imperfect representation of the data-generating process, the analyst should compare and contrast the models for their fit, precision, scope, validity, and reliability in order to choose one that is optimal for his purposes.

This chapter presents metadiagnosis as a process and concentrates on the criteria that the analyst uses in this endeavor. The chapter is divided into concurrent and predictive perspectives. The researcher should compare

215

different algorithms whether he is model building or forecasting. Comparison of the results of the different algorithms permits assessment of the convergent validity of the model. If different algorithms yield virtually identical results, this outcome would be empirical evidence of convergent validity as defined by Campbell and Fiske. If the fit of the model provided by one algorithm is significantly better than that provided by another, then one model may have more validity than the other. The previous chapter discussed some of the techniques involved in diagnosing models. This chapter will discuss the tools and techniques designed to metadiagnose—that is, compare and contrast—models.

Models may be compared by the size, quality, and cost of data collection and cleaning (Granger, 1989). When large data sets are required, the cost of acquiring the information may be high. When the data have to be reviewed and cleaned of errors, the cost of the cleaning is higher for larger data sets. Minimum size requirements for different time series models will be addressed in greater detail in the final chapter.

Models are often compared according to standards of concurrent omnibus fitting statistics. Among them are those measuring goodness of it. There are also complementary lack of fit statistics. One family of goodness of fit criteria is the model R^2 and its variants. Complementing that family is another of lack of fit: the sum of squared errors, the residual variance, or the residual standard error. The number of parameters to be estimated is a measure of parsimony of the model. Because fit tends to improve with the addition of parameters modeled, several information criteria may be employed as standards of goodness of fit adjusted for the number of parameters to be estimated. Models may be compared according to the speed of estimation or model nonconvergence. The analyst can use these measures for concurrent metadiagnosis before he begins the forecasting.

Models can be compared in the longitudinal perspective as well: according to their predictive validity, precision of forecasting, or magnitude of forecast error. By posing critical questions, the researcher may compare the stability of the models: Regardless of starting values, does the model always converge on the same parameter estimates? Are these parameter estimates always statistically significant? When the model is estimated, how stable are the magnitudes and significance of the parameter estimates to other changes? Models can be compared according to fulfillment of their assumptions. If they violate assumptions, which models violate which assumptions? Can the model tolerate minor violations of those assumptions? Is it robust to more serious violations? He can also compare models according to their robustness in face of violation of elegance. This chapter will address the metadiagnosis of different models and their ability to forecast.

A substantial portion of metadiagnosis involves forecasting comparisons. Practitioners of social science, policy planning, and engineering find that metadiagnosis is central to their objectives. Comparative analysis of forecasts provide for fine tuning of the forecast. Forecast comparison provides the basis for statistical process adjustment and control. The forecast, in addition to the feedback, provides the basis for feedfoward to predict where the process will be given specific amounts of correction. In this way, the forecast helps determine the amount of adjustment for statistical process control. Therefore, forecasting comparisons are important objectives of the scientist, the policy planner, and the engineer.

The chapter discusses the forecasting process and its characteristic profiles. Forecasting allows assessment of predictive validity. With the help of a forecast function, the analyst makes a point forecast that he hopes is not biased. The time span over which the forecast extends is called the forecast horizon. On either side of the point forecast, confidence limits are constructed. The confidence limits defines the boundaries of the forecast interval. The forecasting error over the forecast horizon can be measured by the minimum mean square forecast error or the mean absolute percentage of forecast error. Both measures are useful criteria of metadiagnosis; together they form the basis of forecast profiles of different processes. The chapter also examines the forecast profiles characteristic of white noise, integrated processes, basic autoregressive, moving average processes, and ARMA processes. These profiles provide a basis for forecast and model evaluation. Therefore, the latter part of the chapter on metadiagnosis is devoted to the discussion of the theory and application of forecasting.

7.2. METADIAGNOSIS

The principal question is how can one compare and contrast several competing models to determine which is the better model. The better model will usually fit the data well. The general model goodness of fit needs to be evaluated. Commonly used measures of goodness/lack of fit include the mean error, the mean percent error, the mean absolute error, and the mean absolute percentage error. Applied to forecasts, these measures are:

$$\text{Mean prediction error} = \frac{1}{T}\sum_{t=0}^{T}(y_t - \hat{y}_t)$$

$$\text{Mean percent prediction error} = \frac{100}{T}\sum_{t=0}^{T}\frac{(y_t - \hat{y}_t)}{y_t}$$

$$\text{Mean absolute error} = \frac{1}{T}\sum_{t=0}^{T}|y_t - \hat{y}_t|$$ (7.1)

$$\text{Mean absolute percent prediction error} = \frac{100}{T}\sum_{t=0}^{T}\left|\frac{(y_t - \hat{y}_t)}{y_t}\right|,$$

where T is the total number of temporal periods (number of observations), y_t is the actual value and $\hat{y}_t$ is the forecast value at time t. These are average measures of percent and absolute error that can be used as indicators of forecast accuracy. Although there is no single absolute level above which the model is unacceptable, the smaller the measure of error, the better the model fits the data.

There are several measures of omnibus fit based on the sum of squares. First, there is the total sum of squares.

$$\text{Total sum of squares (SST)} = \sum_{t=0}^{T}(y_t - \bar{y})^2,$$ (7.2)

where $\bar{y}$ is the series mean.

Second, there is the sum of squared errors (referred to by SPSS as the adjusted sum of squares). SAS refers to this measure as the SSE. The smaller the sum of squared errors, for a given number of degrees of freedom, the better the model fit:

Sum of squared errors (SSE) (adjusted sum of squares)

$$= \sum_{t=0}^{T}(y_t - \hat{y}_t)^2$$ (7.3)

where $\hat{y}$ is the predicted value.

The mean square error or error variance, sometimes referred to as sigma squared, σ^2, is frequently used as a measure of lack of fit. This criterion serves as a good basis of comparison of different models:

$$\text{Mean square error (MSE)} = \frac{\sum_{t=0}^{T}(y_t - \hat{y}_t)^2}{T - k} = \frac{SSE}{T - k},$$ (7.4)

where T is sample size, and k is the number of parameters to be estimated. By simply taking the square root of the error variance, the analyst obtains another common criterion of lack of fit, the root mean square error (referred to as sigma, σ):

$$\text{Root mean square error (RMSE)} = \sqrt{\text{MSE}} = \sqrt{\frac{\text{SSE}}{T - k}}. \qquad (7.5)$$

From these measures, several measures of the proportion of variance explained may be constructed, including the R^2 and the adjusted R^2. Although the R^2 does not adjust for the number of variables in the model, the adjusted version attempts to compensate for inflation due to the number of predictor variables in the model:

$$R^2 = 1 - \frac{\text{SSE}}{\text{SST}}$$

$$\text{Adjusted } R^2 = 1 - \left[\left(\frac{T - 1}{T - k} \right) (1 - R^2) \right], \qquad (7.6)$$

where k is the number of parameters. The LaGrange multiplier is merely TR^2, where T, as has been noted in Eq. (7.1), is the total number of data points in the series. The adjusted R^2 and the R^2 of Amemiya use slightly different adjustments to compensate for the number of parameters being estimated. Both measures are included in the SAS forecasting output, and both attempt to provide a sense of overall fit per number of parameters being estimated. The better the specification of the model, the higher these criteria will be. The R^2 statistics typically range from a minimal value of 0 to a maximum value of 1. Adjusted R^2 of overparameterized models with poor fits can actually be less than 1. The closer these R^2 values are to 1, the greater the proportion of explained variation and the better the ability of the model to forecast:

$$\text{Amemiya's adjusted } R^2 = 1 - \left[\left(\frac{T + k}{T - k} \right) (1 - R^2) \right] \qquad (7.7)$$

where k is the number of parameters.

Another measure of fit is Harvey's random walk R^2, which takes the R^2 of the model and compares it to the R^2 of a random walk:

$$\text{Harvey's random walk } R^2 = \left(1 - \left(\frac{T - 1}{T} \right) \right) \frac{\text{SSE}}{\text{RWSSE}}, \qquad (7.8)$$

where

$$\text{RWSSE} = \sum_{t=2}^{T} (y_t - y_{t-1} - \mu)^2$$

$$\mu = \frac{1}{T-1} \sum_{t=2}^{T} (y_t - y_{t-1}).$$

A version of this Amemiya's adjusted R^2 is Amemiya's prediction criterion (APC). This is a degree of freedom corrected version of the sum of squared errors. The formula for the APC is

$$\text{Amemiya's prediction criterion} = \left(\frac{1}{T}\right) \text{SST} \left(\frac{T+k}{n-k}\right)(1 - R^2). \quad (7.9)$$

When the likelihood function of the model is calculated, the log of that number is usually a negative number. When one subtracts the log likelihood of the model with its parameters from the minus log likelihood of the model with only the intercept, the researcher obtains a number that when multiplied by -2 provides the likelihood ratio χ^2 of the model. This χ^2 is distributed as a χ^2 with the number of degrees of freedom equal to the number of parameters in the model. The higher this likelihood ratio χ^2, the more the additional parameters improve the fit of the model. Akaike's information criterion (AIC) and the Schwarz Bayesian criterion (SBC) are measures of this logged fit that attempt to adjust for added parameters in the model. These information criteria are designed to deal with the fit of the nonlinear models and to account for the number of the parameters in the model as well. They consist of the natural log of the MSE plus a penalty for the number of parameters being estimated:

$$\text{Akaike's information criterion} = T \ln(\text{MSE}) + 2k$$
$$\text{Schwarz Bayesian Information criterion} = T \ln(\text{MSE}) + k \ln(T), \quad (7.10)$$

where

$$\text{Mean square error} = \frac{1}{T-k} (\text{SSE})$$

$T = $ number of observations
$k = $ number of parameters.

Insofar as they deal with both the fit and the parsimony of the model, these information criteria provide a measure of efficient and parsimonious prediction. The lower value of an information criterion indicates the better model.

Parsimony of the model may be determined by the number of parameters (k) in the model. $K = p + q + P + Q + 1$ if there is a constant in the model. It is equal to $p + q + P + Q$ if there is no constant in the model. For this reason, the number of parameters estimated is often the basis of the degrees of freedom for the model, by which the sum of squares is divided to provide a measure of variance.

For the purposes of metadiagnosis, these statistical measures of goodness of fit are available and are often used to compare and contrast different aspects of the models (SAS, 1995). Whether the measures assess the goodness of fit, magnitude of error or the effectiveness, efficiency, or parsimoniousness of prediction, the comparative measures enable the analyst to assess competing model forecasts within the concurrent evaluation sample as well as into the forecast horizon of the future.

By a judicious application of these concurrent criteria, the analyst can derive a sense of which competing model is best. Occasionally, the analyst will find that one model is superior according to some criteria, while another model is superior according to other criteria. The forecaster must decide which of the competing criteria might render the different models more or less advantageous. For example, some models require more information than others. Some models provide a better fit but the number of parameters in them render them more complicated. Other models provide more parsimonious explanations but do not fit as well. The question arises as to which borderline parameters should be kept in the model or which algorithm should be used for estimation. Exponentially weighted smoothing models may not handle seasonality and trend as well as others. X-11 or X-12 models, which decompose a series into its component parts, involve complicated processes. Box–Jenkins models can combine some of the best features of both of these models, insofar as they can model cycles, stochastic trends, seasonality, and innovations, while providing an elegant, comprehensive, explanatory formulation of the process.

7.2.1. Statistical Program Output of Metadiagnostic Criteria

The statistical package printout includes an array of metadiagnostic indicators. Whether the analyst is using SAS or SPSS, the principal comparative measures of fit are included in the output of the programs. In SAS, the standard ARIMA procedure listing contains the standard deviation and sample size of the series. It also includes the iteration history of the sum of squared errors (SSE), the stopping values of these iterations in terms of SSE, the latest R^2 of SSE, the number of iterations, the error variance, the standard error, the number of residuals, the *AIC,* and the *SBC.* In SPSS, the standard ARIMA output listing includes the number of iterations, the adjusted sum of squared errors, the residual variance, the standard error of the residuals, the number of residuals, the *AIC* and the *SBC.* These comparative tests of the models are part of the standard output of the statistical programs under consideration here. We can obtain comparative tests not included in the standard output by applying auxiliary analysis to the forecasting procedures in the two packages. We now turn our attention to the theory and programming of this analysis.

7.3. FORECASTING WITH BOX–JENKINS MODELS

7.3.1. FORECASTING OBJECTIVES

The importance of forecasting is well understood. The philosopher Kierkegaard was reported to have observed, "Life has to be lived forward but can only be understood backward" and "Those who forget the past are condemned to repeat its mistakes." Moreover, C. F. Kettering is reported in 1949 to have said that "We should all be concerned about the future because we all have to spend the rest of our lives there" (2020 Vision, 1999). Forecasters, during the twentieth century, have developed better short-term as well as long-term forecasting capability. Consequently, forecasting has become increasingly useful and important in formulating educated estimates of things to come. As previously noted, strategists, policy makers, business executives, project managers, investors, and foremen resort to forecasting for help in strategic planning, investment, policy planning, resource procurement, scheduling, inventory maintenance, quality assurance, and resource mobilization in the short run. Nonetheless, the strategists and planners are aware that the basic and ultimate purpose of forecasting is to predict in the near term what will happen in order to avoid substantial cost or loss. The cost of poor prediction may be the loss of soldiers in war, jobs in an economy, job performance approval of public officials in politics, control in a production process, or profits in business. By having an informed and educated opinion of future probabilities, the planner can mobilize and deploy the necessary resources to facilitate or secure achievement of the objectives at hand and thereby reduce the substantial cost of miscalculation (Chatfield, 1975).

The forecaster has a number of methodological objectives as well. He needs to know how the forecast is to be used (Chatfield, 1975). He needs to assess the reasonableness of his model specification, to test the fit of his model, to quasi-predictively validate his model, and to compare forecasts of different models with respect to forecast accuracy and forecast error variance.

To accomplish these objectives, the forecaster collects sample data and divides the sample into two subsamples. The forecasting model is developed on the basis of the first subsample. The temporal period spanning this portion of the series is called the historical, estimation, or initialization period. The competing models are formulated on the basis of this period.

The second subsample is sometimes referred to as the holdout, evaluation, or validation subsample (Makridakis *et al.*, 1983). Alternative parameterizations are tested and refined on the basis of the evaluation or validation period (McCleary *et al.*, 1980).

The models are compared according to a variety of criteria. Evaluation of forecast error with reference to the holdout subsample, permits preliminary predictive validation of the model. The cost of acquiring the required information for the method employed also provides a standard of comparison. The simplicity or parsimony of the model is also important. Albert Einstein is said to have commented that things should be as explained as simply as possible but not more simply. Following the assessment of model fit and predictive validation, the forecaster may use the best model to forecast over the forecast horizon. The cost and ease of computation are important criteria. The forecast may be evaluated on the basis of its sophistication, which depends on the components of the forecast profile. The point and interval forecasts are important and often essential components of the forecast profile. Occasionally, the whole probability density distribution of the forecast is included to construct the forecast profile (Diebold, 1998). On the basis of such components, the forecaster may hazard a probability forecast. For example, he might say that it is almost certain that he will not win the lottery or that it is highly unlikely that the horse will win the race. These are probability forecasts for one point in time. The definition and duration of the forecast horizon are other criteria of comparison. The beginning, duration, and endpoint of the forecast horizon are bases upon which the forecast may be compared. The forecaster generally assumes that the circumstances surrounding the forecast are constant. This assumption fails as the forecast horizon is extended further into the future. Because the length of the forecast horizon may vary, the stability of the forecast over a particular horizon may be an issue. The value of the forecast depends on how well it holds up under changing vicissitudes (Makridakis *et al.*, 1983; Makridakis, 1984). Stable forecasts are clearly more reliable than unstable ones.

Different approaches may be used for forecasting. There are forecasts from exponential smoothing, which basically reduce to the moving average models that have already been covered. There are also the X-11 (or X-12) forecasts, which predict fairly well over a 12-month period or so. The more comprehensive Box–Jenkins methods of forecasting are generally very good for short-term forecasts. Regression analysis can be used with moving average models or series with deterministic trends, and autoregression models may also serve to predict over the longer run. What is more, there are methods of combining forecasts to improve reliability and to reduce forecast error. This exposition of the Box–Jenkins approach to forecasting includes a discussion of basic concepts—including the nature of the forecast function, forecast error, forecast variance, forecast profiles, review of measures of fit, and forecast assessment. The programming and interpretation of computer output follows, while the final section of the chapter compares the relative advantages and disadvantages of these forecasting methods and addresses the theory and programming of combining forecasts.

7.3.2. BASIC METHODOLOGY OF FORECASTING

Forecasting involves basic definitions and assumptions. A decision needs to be made at current time t and the optimal decision depends on expected future value of a random variable, y_{t+h}, the value being predicted or forecast. The number of time points forecast into the future forecast horizon is called the lead time, h. The value of the random variable for such a forecast is the value of y_{t+h}. A forecaster would like to obtain a prediction as close as possible to the actual value of the variable in question at the concurrent or future temporal point of interest. As a rule, the more accurate the prediction, the less the cost of miscalculation. As the forecaster develops his model on the basis of the historical or estimation sample, he makes the first assumption that his model is a stable definition of the underlying data-generation process. He conducts these tests on the validation period series. To do so, he extrapolates over the validation period and compares his predicted values to the actual values of the series. When he builds his model, he wishes to minimize the difference between his forecasts and the observed values of the process under examination during the validation period. This difference is known as the forecast error, e_t. One criterion for measuring the precision of prediction is the sum of squared forecast errors. A more commonly used criterion is the mean square forecast error (MSFE). This MSFE is the average difference between the true value and the predicted value,

$$\text{Mean square forecast error: MSFE}_t(y_{t+h}) = \frac{1}{T}\sqrt{\left(\sum_{i=1}^{T}\hat{y}_{t+h} - y_{t+h}\right)^2}, \quad (7.11)$$

where h = number of periods into the future horizon one wishes to forecast. This would be divided by T, but since $T = 1$, it is invisible. It would also be summed, but for one case, the sum is 1, so that is also invisible. It will, however, be shown that this can be estimated by the conditional expectation of $\hat{y}_{t+h}$:

$$\hat{y}_t(h) = E(y_{t+h} \mid y_t, y_{t-1}, \ldots, y_1), \quad (7.12)$$

where h is the forecast horizon lead time. This is an expectation that is conditional on the information up to and including time t. Having found that he is able to predict well over his validation period, the forecaster makes a second assumption, that his model is stable over the forecast horizon as well. The implication is that the *ceteris paribus* assumption holds. That is, all other potentially important influential factors remain essentially the same. It is this constant condition that permits the model to remain stable. On that basis, he proceeds to extrapolate values $(y_{t+1}, y_{t+2}, \ldots, y_{t+h-1})$ into the future to estimate y_{t+h}. As t extends the time of prediction, h, called the forecast horizon, the forecast emerges. These forecasts based only on the data up to the beginning of the forecast horizon are called unconditional forecasts or *ex ante* forecasts (Armstrong, 1999).

There are, of course, some caveats to these assumptions. There have to be enough data points for a forecast. The important data have to be collected. The most recent data obtained should be collected. These data have to be valid and cleaned. It may be noted that the *ceteris paribus* assumption may not correspond to reality. The forecaster needs to be especially knowledgeable within his field of prediction. He needs to know what external factors significantly impinge upon it. In many arenas, only a forecaster with comprehensive historical knowledge and solid situational understanding of the processes at work will be able to understand whether potentially influential factors remain the same or begin to significantly change. Only then can the expert forecaster know whether these assumptions hold or whether, because of their impact, important turning points ensue. Forecasts based on information drawn from the situation, over which the forecast horizon extends, are called ex post forecasts (Armstrong, 1999).

7.3.3. THE FORECAST FUNCTION

When the predicted values of the identified, estimated, and diagnosed model are plotted as a function of time, they represent a relationship referred to as the forecast function. This model may be an autoregressive model based on the previous lags of the dependent variable. It may be a moving average model based on the previous errors of the dependent variable. It is an additive, nonseasonal autoregressive, integrated, moving average model. Alternatively, it may be a more complex seasonal, multiplicative form of an ARIMA model. Whatever the ARIMA model, there are three basic parameterizations of the forecast function (Box and Jenkins, 1976). One of them is the actual difference equation model formulated, such that $Y_{t+h} = \varphi_1 Y_{t+h-1} + \varphi_2 Y_{t+h-2} + \cdots + \varphi_{p+d} Y_{t+h-p-d} - \theta_1 e_{t+h-1} - \theta_2 e_{t+h-2} - \cdots - \theta_q e_{t+h-q} + e_{t+q}$. Another expression is a weighted sum of random shocks with ψ weights, such that $Y_{t+h} = \sum_{l=0}^{\infty} \psi_l e_{t+h-l}$ where $\psi_0 = 1$ and $L\psi(L) = \theta(L)/\varphi(L)$. The other form is an autoregressive sum of previous values with π weights, such that $Y_{t+h} = \sum_{l=0}^{\infty} \pi_l Y_{t+h-l} + e_{t+1}$. Any of these formulations of the forecast function may be used to obtain point forecasts. The easiest way to understand the forecasting process is first to consider the difference equation form of the basic model.

7.3.3.1. An AR(1) Forecast Function

The difference equation parameterization of the basic model is simply the equation identified, estimated, and diagnosed. The analyst takes the equation, collects its terms, and expresses it as a regression equation. Then he merely extends the time subscript of the model one time period into the future, whereupon the analyst has the formula for the one-step-ahead forecast. For example, the process can be shown in the following equation:

$$(1 - \varphi_1)y_t = C + e_t$$
$$y_t = C + \varphi_1 y_{t-1} + e_t \qquad (7.13)$$
$$\hat{y}_{t+1} = C + \varphi_1 y_t + e_{t+1},$$

where C is a constant. If y is centered and C is set to zero, the computation for an autoregressive forecasting process can be illustrated with the aid of Table 7.1. A question arises as to initial values. For purposes of this example, the initial values at time $t = 0$ of the actual series data, Y_T, is set to 1 and

Table 7.1

Simulated AR(1) Model with 1-Step-Ahead Forecast Horizon

Time	$\varphi_1 = .6$		AR(1) Model	Forecast	Forecast error
t	$\varphi_1 Y_{t-1}$	e_t	$Y_t = \varphi_1 Y_{t-1} + e_t$	$\hat{Y}_{t+1} = \varphi_1 Y_t$	$Y_{t+1} - \hat{Y}_{t+1}$
0			1.000	0.000	
1	0.600	−0.230	0.370	0.222	0.329
2	0.222	0.329	0.551	0.331	0.120
3	0.331	0.120	0.451	0.270	0.140
4	0.270	0.140	0.410	0.246	−0.330
5	0.246	−0.330	−0.084	−0.050	0.348
6	−0.050	0.348	0.298	0.179	0.298
7	0.179	0.298	0.477	0.286	0.770
8	0.286	0.770	1.056	0.634	0.758
9	0.634	0.758	1.392	0.835	0.746
10	0.835	0.746	1.581	0.949	0.222
11	0.949	0.222	1.171	0.702	0.799
'2	0.702	0.799	1.501	0.901	−0.901

its forecast, $\hat{Y}_{t+1}$, is set to 0. For the first time period, $t = 1$, there is no forecast. The value of the forecast at this time period therefore is set to 0.0. The autoregressive parameter, ϕ_1, is set to 0.600. From this value of ϕ_1 times the starting value of Y_t, the value of 0.600 is obtained. When this amount is added to the random shock of -0.230, the response value of Y_t at time $t = 1$ becomes 0.370. The forecast ($\hat{Y}_{t+1}$) for the next time period, $t = 2$, is 0.600 times the 0.370, which equals 0.222. The forecast error is computed by subtracting the forecast at this time period from the actual value at the next time period, 0.551 minus 0.222, yielding 0.329. This process is repeated at $t = 2$, in order to obtain the new value of $Y_t = 0.551$ and a forecast equal to 0.331. The forecast error, the difference between this value and the value of the Y_t at that point in time, is 0.120. The process is iterated until the end of the series:

$$e_{t+h} = Y_{t+h} - \hat{Y}_{t+h}$$
$$= Y_{t+h} - f_{t+h} \tag{7.14}$$

where

$$\hat{Y}_{t+h} = f_{t+h} = \text{forecast at } t + h.$$

The forecast errors can be squared and added, and their average taken. Finally, the criterion value of the minimum mean square forecast error is used to compare the relative error of prediction of the different φ_1 models. The value of ϕ_1 that yields the minimum lack of fit and best prediction is the one estimated for the model. Similarly, this same minimum mean square forecast error criterion may be used to compare the fit of models with different parameters for metadiagnosis.

Because the AR(1) process can be reparameterized as

$$y_t = C + \varphi_1 y_{t-1} + e_t$$
$$(1 - \varphi_1 L)y_t = C + e_t, \tag{7.15}$$

and therefore

$$\hat{y}_t = E(y_t) = \frac{C}{(1 - \varphi_1 L)},$$

it can be expressed as an infinite moving average process. It may also be expressed as the accumulation or sum of an infinite order series of random shocks plus an autogressive component:

$$\hat{Y}_{t+h} - \sum_{j=0}^{\infty} \varphi_j^{h-1} e_{t+h-j} + \varphi^h Y_t. \tag{7.16}$$

These shocks can be interpreted as one-step-ahead forecast errors. At $h = 1$, the forecast error is not correlated. After time $h = 1$, these forecast errors are generally correlated.

The expectation of the forecast error of an unbiased forecast at time t is 0. The unconditional expectation of $E(y_t) = \mu$ and the relationship between the constant and the mean is

$$\lim_{h \to \infty} \hat{y}_t = E(y_t) = C \sum_{h=0}^{\infty} \varphi_1^h = \frac{C}{(1 - \varphi_1 L)} = \mu_y. \qquad (7.17)$$

If the series is centered, the unconditional expectation of the AR(1) is 0, whatever the lead time h. Although the point forecast may deviate from expectation, the unbiased forecast on the average will be equal to zero (Granger and Newbold, 1986). McCleary *et al.* (1980) demonstrate how stepping the AR(1) equation ahead one time period, and then taking the expectation, the analyst is left with

$$Y_{t+1} = E(e_{t+1} + \varphi_1 Y_t) = \varphi_1 Y_t. \qquad (7.18)$$

From the point of view of the forecast origin, t, the expected value of the one-step-ahead shock, e_{t+1}, is zero. With a centered (zero mean) series, the successive conditional forecasts for the first-order AR process are

$$E(Y_{t+1}) = \varphi_1 Y_t$$
$$E(Y_{t+2}) = \varphi_1^2 Y_t$$
$$\vdots \qquad\qquad (7.19)$$
$$E(Y_{t+h}) = \varphi_1^h Y_t.$$

The first-order autoregressive process has a characteristic forecast function. At the commencement of the forecast, there is an initial spike equal to the value of the autoregressive parameter times the value of the series. The direction of that spike depends upon the value of the autoregressive parameter. After the spike, there is an incremental increase in the value of the series along the forecast horizon. That increase exponentially converges to the value at the beginning of the forecast.

7.3.3.2. An IMA(1,1) Forecast Function

The general form of an integrated moving average (IMA) model is represented by the differenced response being equal to the current error minus a fraction of the previous error. This moving average process may be expressed as $Y_t - Y_{t-1} = e_t - \theta_t e_{t-1}$. Simulated data for such a model are found in Table 7.2. The differenced response is represented by w_t. The disturbance or shock at the current time t is represented as e_t and the previous disturbance or shock to the process is represented as e_{t-1}. The first-order moving average parameter, representing the component of stochastic trend, is θ_1.

Starting values are needed for w_t and e_{t-1}. These values may be obtained from a preexisting series, from the mean of the current series by presetting

Table 7.2

ARIMA(0,1,1) or IMA(1,1) Model

$$w_t = Y_t - Y_{t-1} = e_t - \theta_1 e_{t-1}$$

Time t	Response Y_t	Differenced response $w_t = Y_t - Y_{t-1}$	Current shock e_t	Moving average parameter θ_1	Previous shock e_{t-1}	Effect of previous shock $-\theta_1 e_{t-1}$
−1	[445.00]	0.00	0.00		0.00	0.00
0	445.00	0.00	0.00	0.20	0.00	0.00
1	439.50	−5.50	−5.50	0.20	0.00	0.00
2	435.60	−3.90	−5.00	0.20	−5.50	1.10
3	434.59	−1.01	−2.01	0.20	−5.00	1.00
4	434.01	−0.59	0.99	0.20	−2.01	0.40
5	436.20	2.20	2.00	0.20	−0.99	0.20
6	431.80	−4.40	−4.00	0.20	2.00	−0.40
7	431.20	−0.60	−1.40	0.20	−4.00	0.80
8	433.66	−2.45	2.17	0.20	−1.40	0.28
9	433.07	−0.59	−0.15	0.20	2.17	−0.43
10	432.04	−1.03	−1.06	0.20	−0.15	0.03
$h = 1$	432.29	0.25	0.04	0.20	−1.06	0.21
$h = 2$	432.29	0.00	0.00			

them to zero, or by the back-forecasting previously described. Differencing often has the effect of centering the differenced series around a zero mean. Therefore, the mean is set to 0. The choice of starting values may depend on the known series length, and/or how much of the series has unfolded at the point of consideration. In this example, the starting value for the response variable, Y_t, at time $t = 0$, is set to 445.0, and the starting values of e_t and e_{t-1} at this point are also set to 0. A previous starting value may be backforecast with conditional least squares.

The shock or disturbance at time t drives this process, along with a portion of the shock from the previous point in time, characterized by $-\theta_1 e_{t-1}$. By time $t = 1$, θ_1, which is 0.20 in this iteration of estimation, times the previous error (which is zero) yields 0.00, and this product is subtracted from the current error, e_t, of −5.50 to yield a response of −5.50. The actual decline of 5.50 is observed. This decline of the differenced value, w_t, means that response has dropped from 445 at $t = 0$ to 439.50 at $t = 1$. This process is iterated until the forecast horizon is reached. The magnitude of θ_1 is estimated by least squares. For pedagogical purposes, this value is assumed for this pass to be 0.20.

When the first-order moving average process extends into the forecast horizon indicated by time $h > 0$, the forecast value is predicated upon the expected value $E(e_{t+h}) = 0$. A one-step-ahead forecast may be computed from $-\theta_1 e_t$. Clearly, the expected value of the future shock at $h = 1$ continues to be zero. The one-step-ahead forecast at $h = 1$, $\hat{\omega}_{t+1} = e_{t+1} - \theta_1 e_t$, from this first-order moving average process allows a jump, based on the expectation of the future shock, $E(e_{t+1}) = 0$ and a portion of the current shock, $-\theta_1 e_t$. The effect of this first-order jolt is to jar the one-step-ahead forecast before the eventual forecast function stabilizes around the series mean, which in this case is zero. The two-step-ahead forecast at $h = 2$ has $\hat{\omega}_{t+2} = E(e_{t+2}) = 0$. The stabilization of the forecast of the moving average process around zero (or the series mean, if it is not zero) is reflected in the blank cells of Table 7.2 for $h = 2$. If this process were a third order process, there would be two jumps before the forecast function would be stabilized. If the undifferenced process were a qth-order moving average, there would be $q - d$ jumps before the function was stabilized, the number of jumps is the order of the moving average minus the order of differencing.

Farnum and Stanton (1989) summarize this procedure. The analyst must take note of the optimal model he has estimated. If he wishes to forecast h steps ahead, then he needs to replace t with $t + h$ as the subscript of each component of the model. Y_t will become $\hat{Y}_{t+h}$ for $h = 0,1,2$, etc. In the event that past values were being modeled as Y_{t-h}, then the h may assume the appropriate values for $0,1,2$, etc. Past errors may be represented by previous errors as required by the lags in the equation, so that e_{t-h} becomes e_{t-h} for $h = 0,1,2,3$, etc., while future errors e_{t+h} are given values equal to their expectation as they are set to 0 for all h. Stationary series, which require no initial differencing, have forecast functions that converge to their expected value, the series mean.

7.3.3.3. An ARIMA (1,1,1) Forecast Function

The forecast of the ARIMA (p,d,q) may be conceived as a linear combination of its random shock components and may be useful in determining the forecast variance. If we collect the autoregressive and moving average terms of this model, we can parameterize the ARIMA model as a series of weighted shocks. For example, a basic, centered ARIMA (p,d,q) model may generally be represented as

$$Y_t = \frac{\theta(L)}{(L)\varphi(L)}$$

$$= \frac{(1 - \theta_1 L - \theta_1^2 L^2 - \ldots)e_t}{(1 - L)^d(1 - \varphi_1 L - \varphi_1^2 L - \varphi_1^2 L^2 - \cdots - \varphi_1^{l-1}L^{l-1})}.$$

(7.20)

Expressed as a series of weighted shocks, this equation is

$$Y_t = \sum_{j=0}^{\infty} \psi_j e_{t-j}. \tag{7.21}$$

If one assumes that the series, Y_t, has already been differenced and centered as W_t, then

$$
\begin{aligned}
W_t &= \frac{(1 - \theta_1 L)e_t}{1 - \varphi_1 L} \\
&= (1 - \varphi_1 L)^{-1}(1 - \theta_1 L)e_t \\
&= (1 + \varphi_1 L + \varphi_1^2 L^2 + \cdots + \varphi_1^L L^L)(1 - \theta_1 L)e_t \\
&= (1 + \varphi_1 L + \varphi_1^2 L^2 + \cdots + \varphi_1^L L^L \\
&\quad - \theta_1 L - \theta_1 \varphi_1 L^2 - \theta_1 w_1^2 L^3 - \cdots) \\
&= (1 + (\varphi_1 - \theta_1)L + (\varphi_1^2 - \theta_1 \varphi_1)L^2 + (\varphi_1^3 - \theta_1 \varphi_1^2)L^3 \\
&\quad + (\varphi_1^4 - \theta_1 \varphi_1^3)L^4 + \cdots)e_t.
\end{aligned}
\tag{7.22}
$$

After the fourth or fifth ψ weight, the attenuation is so substantial that the remainder is often negligible. In other words,

$$
\begin{aligned}
\psi_1 &= \varphi_1 - \theta_1 \\
\psi_2 &= \varphi_1^2 - \theta_1 \varphi_1 \\
\psi_3 &= \varphi_1^3 - \theta_1 \varphi_1^2 \\
\psi_4 &= \varphi_1^4 - \theta_1 \varphi_1^3 \\
&\quad . \qquad . \qquad . \\
\psi_p &= \varphi_1^p - \theta_1 \varphi_1^{p-1}.
\end{aligned}
\tag{7.23}
$$

In general, the series can be divided into two components: those expected observations within the forecast horizon h, and those actual current and past observations $t, t - 1, t - 2$, etc. Pindyck and Rubenfeld (1991) show that the forecast function with h leads into the forecast horizon may be defined as a function of weighted shocks within the horizon and another summation of them during the evaluation period:

$$
\begin{aligned}
\hat{W}_{t+h} &= \psi_0 e_{t+h} + \psi_1 e_{t+h-1} + \cdots + \psi_h e_t + \psi_{h+1} e_{t-1} + \cdots \\
\hat{W}_{t+h} &= \psi_0 e_{t+h} + \psi_1 e_{t+h-1} + \cdots + \psi_h e_t + \psi_{h+1} e_{t-1} + \sum_{j=0}^{\infty} \psi_{h+j} e_{t-j}.
\end{aligned}
\tag{7.24}
$$

Because the φ and ψ parameters are estimated optimally to minimize the sum of squared errors or the mean square forecast error, by maximum likelihood, conditional, or unconditional least squares, the ψ weights may be estimated. With the holdout sample, it is easy to distinguish the estimated from the real ψ weight. Both real and estimated weights are necessary for obtaining a forecast error and a forecast interval.

7.3.4. THE FORECAST ERROR

The forecast error for lead time h is defined as the difference between the actual forecast and its conditional expectation, consisting of optimal estimates, which can be expressed as a linear combination of ψ weights:

$$e_t(h) = Y_{t+h} - \hat{Y}_t(h)$$
$$= \psi_0 e_{t+h} + \psi_1 e_{t+h-1} + \cdots + \psi_{h-1} e_{t+1}. \tag{7.25}$$

The same holds for a differenced and centered series, W_t; the forecast error over the forecast horizon h is the difference between the forecast and its conditional expectation:

$$e_t(h) = W_{t+h} - \hat{W}_t(h)$$
$$= \psi_0 e_{t+h} + \psi_1 e_{t+h-1} + \cdots + \psi_{h-1} e_{t+1}. \tag{7.26}$$

The forecast error denoted by $\hat{e}_{t+1}$ is the difference between the value of w_{t+1} and its forecast. For example, the expression of the forecast error of differenced series and uncentered series w_t that represents a first-order autoregressive process is:

$$\hat{e}_{t+1} = w_{t+1} - \hat{w}_{t+1}$$
$$= \mu + \varphi_1 w_t + e_{t+1} - (\mu + \varphi_1 w_t) \tag{7.27}$$
$$= e_{t+1}.$$

Measuring the forecast error permits formulation of the cost of such error. If this cost can be derived, it can be formulated as a function of the forecast error. With this function, the researcher can estimate the cost of forecast inaccuracy. Usually, the forecast with the smallest error is the one-step-ahead forecast. If the series is integrated or autoregressive, then the forecast error increases as prediction is projected into the forecast horizon. If the researcher is able to assess the increasing cost of inaccurate prediction, he may also be able to assess how far ahead it is affordable to forecast. Once he has properly formulated the forecast error, he can formulate the forecast error variance.

7.3.5. FORECAST ERROR VARIANCE

The one-step-ahead forecast variance at time $= t + 1$ time is

$$\text{Var}(\hat{e}_{t+1}) = \text{Var}(e_{t+1}) = \sigma_e^2, \tag{7.28}$$

and in terms of expectations, the variance of the forecast error is

$$
\begin{aligned}
\mathrm{Var}(e_{t+h}) &= E(e_{t+h})^2 \\
&= E(W_{t+h} - \hat{W}_{t+h})^2 \\
&= (\psi_0^2 + \psi_1^2 + \cdots + \psi_{h-1}^2)\sigma_e^2 \quad\quad (7.29) \\
&= \left(1 + \sum_{i=1}^{h-1} \psi_i^2 \right)\sigma_e^2.
\end{aligned}
$$

The optimum values for the ψ weights are found by minimizing the forecast variance or mean square forecast error. The optimum forecast is based on the conditional expectation of y_{t+h}, which derives from the expected value of the errors in the forecast horizon equalling 0, and the expected value for the past residuals, which are simply those from the estimated equation (Pindyck and Rubenfeld, 1991).

7.3.6. FORECAST CONFIDENCE INTERVALS

An estimate of the forecast error variance is needed in order to compute the confidence intervals of a forecast. This estimate is based on the sum of squared errors obtained after the final estimates of the parameters are made and can be found in Eq. (7.28). The denominator is the number of degrees of freedom for the error term. The asymptotic standard error is found by taking the square root of the forecast error variance. This may be used with a t value to form the forecast confidence interval:

$$
\text{Forecast interval of } \hat{y}_{t+h} = \hat{y}_{t+h} \pm t_{(1-\alpha/2,df)} \left(\sqrt{1 + \sum_{i=1}^{h-1} \psi_i^2} \right) \sigma_e^2, \quad (7.30)
$$

where

$t_{(1-\alpha/2,df)}$
$= (1 - \alpha/2)$th percentile of a t distribution with df degrees of freedom.

The forecasts may be updated with the assistance of the ψ weights. As the lead time gets larger, the interval generally becomes larger, although the exact pattern depends on the ψ weights. This forecast interval formula is analogous to the formula for the confidence interval of a mean.

The autoregressive parameterization of the forecasts is based on an infinite autoregressive function of previous observations. Owing to invertibility, a stationary moving average model can be parameterized as an infinite autoregressive model. The weights in this linear combination are called π

weights. A series parameterized in terms of these weights may be expressed as follows: Because

$$\frac{\varphi(L)}{\theta(L)} = 1 - \sum_{i=1}^{\infty} \pi_i L^i,$$

(7.31)

$$y_t = e_t + \sum_{i=1}^{\infty} \pi_t y_{t-i}.$$

The forecast function comprises an autoregressive sum of weighted past values of the series, and an autoregressive sum of estimated autoregressive response values as shown in Equation 7.32 (Ege *et al.*, 1993):

$$\hat{y}_{t+h} = \sum_{i=1}^{h-1} \hat{\pi}_t \hat{y}_{t+h-i} + \sum_{i=h}^{\infty} \hat{\pi}_i y_{t+h-i}.$$

(7.32)

On this basis, the one-step-ahead forecast may be formulated as follows:

$$\hat{y}_{t+1} = \pi_1 y_t + \pi_2 y_{t-1} + \pi_3 y_{t-2} + \ldots.$$

(7.33)

Similarly, the second-step-ahead forecast is

$$\hat{y}_{t+2} = \pi_1 \hat{y}_{t+1} + \pi_2 y_t + \pi_3 y_{t-1} + \cdots$$

(7.34)

The autoregressive parameterization of the function into forecast horizon components and actual components may be estimated by updating one step at a time so as to minimize the mean forecast error. Suppose α is the smoothing parameter. The larger the α, the more the recent observations are weighted in the calculation of the sum of squared errors or minimization of the mean square forecast error with the holdout sample. Then the updating formula for the forecast may be derived from the formula for simple exponential smoothing found in Chapter 2:

$$\hat{y}_{t+h} = \alpha y_t + \alpha(1 - \alpha) y_{t-1} + \alpha(1 - \alpha)^2 y_{t-2} + \ldots.$$

(7.35)

The smaller the smoothing parameter, α, the more the history of the series counts in determining the point forecast value. By using this updating approach explained in Chapter 2 in the section on exponential smoothing, the forecast can be extrapolated as well. For the reason that this procedure involves one step at a time and the forecast function predicted by this procedure is based on the most recent past estimates, the forecast function computed in this way is often referred to as the eventual forecast function.

7.3.7. FORECAST PROFILES FOR BASIC PROCESSES

Different forecast functions possess different error structures. The forecast functions therefore possess different forecast error variances. The forecast intervals clearly depend on the type of forecast error and its variance. Hence, different processes have different forecast profiles. It is helpful to consider the basic forecast profiles at this point, so the analyst can recognize them in conjunction with the models he is estimating. He may check these profiles against the type of process that produces them to be sure they match. More detailed examination of complicated profiles can be found in Box and Jenkins (1976), Box *et al.* (1994), Pindyck and Rubenfeld (1991), and Diebold (1998). For this reason, we examine common forecast profiles for white noise, integrated, AR, MA, and ARMA processes.

7.3.7.1. Forecast Profile for a White Noise Process

McCleary and Hay (1980) explain the forecast for the white noise process as the sum of the shocks from the point of origin of the forecast horizon. At that point, $y_t = e_t$. The ψ weights ($\psi_1 = \psi_2 = \cdots = \psi_h = 0$) are all equal to zero. The expectations of the shocks are all equal to zero. The product of these components remains zero. For the reason that all of the ψ weights are constant, the error variances at each point in the forecast horizon are equal to σ_e^2. Therefore, $\text{Var}(1) = \sigma_e^2$, $\text{Var}(2) = \sigma_e^2 \ldots$ $\text{Var}(h) = \sigma_e^2$. In other words, the forecast profile for the white noise process is constant. There is no deviation from the series mean and the forecast profile is one of parallel lines extrapolated to the end of the forecast horizon. An example of a white noise forecast profile is simulated over time. The forecast profile for this process begins at time period of 75. As the forecast horizon extends 24 periods into the future, this white noise process forecast function converges on the series mean of zero, since the series has been centered, and the constant forecast error variance provides for a parallel line forecast profile for the duration of the forecast horizon, as can be seen in Fig. 7.1.

7.3.7.2. The ARIMA(0,1,0) or I(1) Forecast Profile

The annual U.S. national unemployment series taken from the 1995 *Economic Report of the President* is an integrated series (see C7pgm1.sas listed on the Academic Press World Wide Web site for this textbook: http://www.academicpress.com/sbe/authors). Unemployment is equal to its previous year's tally in thousands plus some random shock. McCleary *et al.* (1980) show how an I(1) series is white noise after differencing and is formulated as $Y_t = Y_{t-1} + e_t$. The forecast is simply the updated equation. Since the expectation of the error equals zero, $E(Y_{t+1}) = Y_t$. Also, $E(Y_{t+2}) = E(Y_t + e_{t+1} + e_{t+2}) = Y_t$. The same holds for the forecasts $Y_{t+3} \ldots Y_{t+h}$. The expectation of the series is the latest series value plus the value of

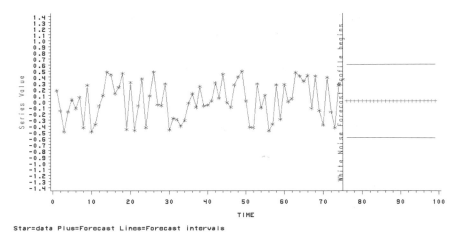

Star=data Plus=Forecast Lines=Forecast intervals

Figure 7.1 White noise forecast profile.

each shock in the forecast horizon. Because the expectation of each shock is zero, $E(Y_{t+h}) = E(Y_t + e_{t+h} + e_{t+h-1} + \ldots + e_t + e_{t-1} + \ldots) = Y_t$. As a result, the forecast value of the series at each temporal point in the forecast horizon is the series value at the origin of the horizon. This forecast function provides for a constant point forecast after differencing, leaving white noise residuals.

The forecast error variance is an integrated process. When the difference between the value of the forecast function of one time point and the previous one is taken, the error variance equals σ_e^2. For each period ahead in the forecast horizon, another σ_e^2 is added. This result renders the forecast error variance of an I(1) process cumulative. After h periods into the forecast horizon, the forecast variance is $h\sigma_e^2$. Consequently, the forecast error variance accumulates and the forecast interval spreads as the forecast function proceeds along the forecast horizon. An example of widening forecast interval from an integrated process are the IBM closing stock prices in the beginning of 1962 shown in Fig. 7.2.

7.3.7.3. Forecast Profile for the AR Process

AR(p) forecast profiles are recognizable by spreading forecast intervals as well. The spread exponentially declines until it levels out. The rate of declining spread is a function of the magnitude of the autoregressive parameters. In the AR(1) forecast profile, the addition of variance at each time point in the forecast horizon declines in order of the square of the

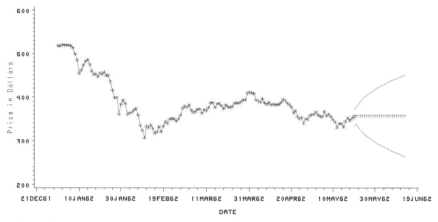

Figure 7.2 Forecast profile for ARIMA(0,1,0) model: IBM 1962 closing stock prices.

first autoregressive parameter estimate. In the AR(1) model, the forecast interval and function establish this pattern. Because an AR(1) model is characterized by the equation with C designating the constant

$$y_t = C + \varphi_1 y_{t-1} + e_t, \tag{7.36}$$

a forecast one, two, and h steps ahead will be:

$$\begin{aligned}
\hat{y}_{t+1} &= \varphi_1 y_t + C + e_{t+1} \\
\hat{y}_{t+2} &= \varphi_1 \hat{y}_{t+1} + C + e_{t+1} \\
&= \varphi_1^2 y_t + C(\varphi_1 + 1) \\
\hat{y}_{t+h} &= \varphi_1^h y_t + C(\varphi_1^{h-1} + \varphi_1^{h-2} + \cdots + \varphi_1 + 1).
\end{aligned} \tag{7.37}$$

As h increases along the forecast horizon, its temporal order of magnitude increases. When the parameter ϕ, which is less than unity, is taken to the h power, the exponentiated result diminishes in magnitude. As h becomes large, this exponentiated increment to the forecast function converges to zero, if the process has been centered. Otherwise, as h becomes large for a stationary series, this limit converges to the mean of the series:

$$\lim_{h \to \infty} \hat{y}_{t+h} = \frac{C}{(1 - \varphi_1)} = \mu_y. \tag{7.38}$$

Indeed, for AR forecasts, in general, a stationary series converges to its mean (Pindyck and Rubenfeld, 1991; Griffiths *et al.*, 1993).

7.3.7.3.1. The Forecast Interval for the AR(1) Process

By subtracting the expected value h steps ahead on the forecast horizon from the forecast function at that point in time, one can obtain the forecast error for the AR(1) process:

$$
\begin{aligned}
e_{t+h} &= y_{t+h} - \hat{y}_{t+h} \\
&= \varphi_1 y_{t+h-1} + C + e_{t+h} - \hat{y}_{t+h} \\
&= \varphi_1^2 y_{t+h-2} + C(\varphi_1 + 1) + e_{t+h} + \varphi_1 e_{t+h-1} - \hat{y}_{t+h} \\
&= \varphi_1^h y_t + C(\varphi_1^{h-1} + \cdots + \varphi_1 + 1) \\
&\quad + e_{t+h} + \varphi_1 e_{t+h-1} + \cdots + \varphi_1^{h-1} e_{t+1} - \hat{y}_{t+h}.
\end{aligned}
\tag{7.39}
$$

When one substitutes Eq. (7.37) for the estimated component on the far right-hand side of Eq. (7.39), the following formula for e_{t+h} is obtained (Pindyck and Rubenfeld, 1991; Gilchrist, 1976):

$$
e_{t+h} = e_{t+h} + \varphi_1 e_{t+h-1} + \cdots + \varphi_1^{h-1} e_{t+1}.
\tag{7.40}
$$

From the sequential squaring of the error components in Eq. (7.41), this forecast error is computed by the expanding forecast interval of the AR(1) process:

$$
\begin{aligned}
E(e_1^2) &= \sigma_e^2 \\
E(e_2^2) &= (\sigma_e^2 + \varphi_1^2 \sigma_e^2)^2 = (1 + \varphi_1^2)\sigma_e^2 \\
E(e_3^2) &= (\sigma_2^2 + \varphi_1^2 \sigma_e^2 + \varphi_1^4 \sigma_e^2)^2 \\
&= (1 + \varphi_1^2 + \varphi_1^4)\sigma_e^2 \\
E(e_h^2) &= (1 + \varphi_1^2 + \varphi_1^4 + \cdots + \varphi_1^{(2h-2)})\sigma_e^2.
\end{aligned}
\tag{7.41}
$$

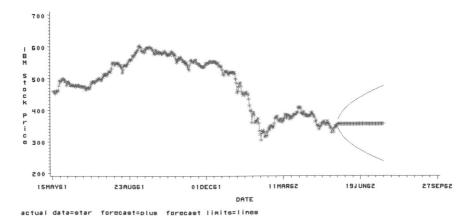

actual data=star   forecast=plus   forecast limits=lines

Figure 7.3 Forecast profile of AR(1) model: series B IBM stock prices (Box *et al.*, 1994).

The process yields exhibits a characteristic AR(1) forecast profile. The forecast error variance spreads with an exponential decline of the incremental spread as the forecast horizon extends. The computation of the forecast interval was shown in Eq. (7.30), and the shape of the forecast profile for this AR(1) process is displayed in Fig. (7.3).

7.3.7.3.2. The Forecast Interval for the AR(2) Process

Second and higher order autoregressive processes, such as $Y_t = C + \varphi_1 Y_{t-1} - \varphi_2 Y_{t-2} + e_t$, may be characterized by oscillation if the second-order autoregressive parameter, φ_2, is not of the same sign as the first-order autoregressive coefficient. This discrepancy in signs creates an apparent undulation in the forecast that is short lived and dissipates rapidly along the forecast horizon as the higher order leads are characterized by convergence to their mean limits. This kind of forecast profile is presented in Fig. 7.4.

7.3.7.4. Forecast Profile for the MA Process

The point forecast of an MA(1) process is derived from the formula from the MA model. The expectation of e_{t+1} through e_{t+h} equals 0. Therefore, as

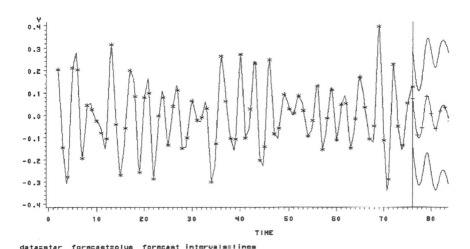

Figure 7.4 Forecast profile of AR(2) process.

the forecast function is extended, the forecast function rapidly converges to the series mean:

$$y_t = e_t - \theta_1 e_{t-1} + \mu$$
$$\hat{y}_{t+1} = E(y_t) + \mu - \theta_1 e_{t-1}$$
$$\vdots$$
$$\hat{y}_{t+h} = E(\mu + e_{t+h} - \theta_1 e_{t+h-1}) = \mu_y. \tag{7.42}$$

A stationary MA(1) process is a short, one-period memory process. For one or two periods ahead, the best forecast is the mean of the series (Pindyck and Rubenfeld, 1991). The forecast variance of such an MA series around that mean can also be computed as

$$
\begin{aligned}
E(e_{t+h}^2) &= E(y_{t+h} - \hat{y}_{t+h})^2 \\
&= E(e_{t+h} - \theta_1 e_{t+h-1})^2 \\
&= E(e_{t+h}^2 - 2\theta_1 e_{t+h-1} e_{t+h} + \theta_1^2 e_{t+h-1}^2) \\
&= (1 + \theta_1^2)\sigma_e^2.
\end{aligned}
\tag{7.43}
$$

7.3.7.4.1. The Forecast Interval of an MA(1) Process

The forecast error variance is formulated in Eq. (7.43). Owing to the short, one-period memory of this process, the value of the error variance remains the same after the initial shock. Although the forecast interval might expand during the first lead period, it would remain constant into the forecast horizon beyond that first lead. The interval is merely the point forecast plus or minus the 1.96 times the asymptotic standard error [the square root of the forecast error variance in Eq. (7.43)]. It is this forecast interval that comprises the forecast profile for the MA(1) process of the Democratic proportion of major party seats in the U.S. House of Representatives (Maisel, 1994; U.S. House of Representatives, 1998). The MA(1) model that explains the percentage of Democratic seats at this time is the proportion of Democratic seats in House$_t$ = 54.6 + (1 + 0.49L) e_t, and the forecast profile from this MA(1) is displayed in Fig. 7.5. After the first lead period, the forecast interval of the likelihood of percentage of a Democratic Congress remains constant, unless the Republicans exhibit obviously poor political judgement in coping with the political, economic, and social welfare issues that challenge them.

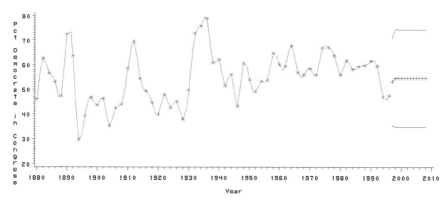

Figure 7.5 Forecast profile of MA(1) process: percentage of Democrats in U.S. House of Representatives. Data courtesy of Prof. Richard Maisel, NYU Graduate Sociology Department. Updated data available from the U.S. House of Representatives Web site (http:// clerkweb.house.gov/histrecs/hostory/elections/political/divisions.htm). Downloaded 10/25/98.

7.3.7.4.2. The Forecast Profile of an MA(2) Process

The forecast profile for the MA(2) model is very much the same. A simulation produces an MA(2) process $Y_t = (1 + 0.71L + 0.21L^2)e_t$. A forecast profile for the MA(2) model is produced. The point forecast is jostled only for the first two leads, suggestive of the shocks at lags 1 and 2 in an MA(2) process, and then it levels off at the series mean. We can see that the forecast profile is shown to get jarred during the first two leads, and then levels off on an interval parallel to that of the estimated forecast of the series mean. The number of shocks or disturbances prior to the leveling off of the series is equal to the order of the MA process. If there were six of these shocks or if the order of the MA did not end until lag 6, then there would be six disturbances in the forecast profile before the MA forecast profile leveled out. Figure 7.6 depicts this simulated process for the MA(2) series.

7.3.7.4.3. The Forecast Profile for the ARMA Process

The simple ARMA process has been theoretically developed in the discussions of Eq. (7.20) through (7.23). From the derivation and elaboration of these equations, it can be seen that the forecast interval of an ARMA(1,1) process is one whose limits over the forecast horizon may be expanding at a diminishing level:

$$Y_t = C + \varphi_1 Y_{t-1} + e_t - \theta_1 e_{t-1}. \tag{7.44}$$

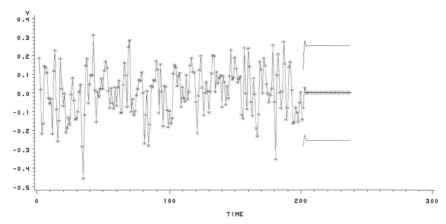

Figure 7.6 Forecast profile for MA(2) process. Simulation of $y(t) = (1 + 0.7L - 0.21L^2)e(t)$ process.

The existence of the autoregressive component provides for the division by $1 - \varphi_1$. The conditional expected value of the one, two, and h, step ahead forecasts are as follows:

$$
\begin{aligned}
y_{t+1} &= E(C + \varphi_1 y_t + e_{t+1} - \theta_1 e_t) = C + \varphi_1 y_t - \theta_1 \hat{e}_t \\
y_{t+2} &= E(C + \varphi_1 y_{t+1} + e_{t+2} - \theta_1 e_{t+1}) = C + \varphi_1 \hat{y}_{t+1} \\
&= \varphi_1^2 y_t + C(\varphi_1 + 1) - \varphi_1 \theta_1 \hat{e}_t
\end{aligned}
\tag{7.45}
$$

$$
\hat{y}_{t+h} = \varphi_1^h y_t + C(\varphi_1^{h-1} + \varphi_1^{h-2} + \cdots + \varphi_1 + 1) - \varphi_1^{h-1} \theta_1 \hat{e}_t.
$$

From this equation, one can infer that as the forecast horizon extends farther into the future, so that h becomes large, the increment added to the ARMA(1,1) forecast interval rapidly becomes negligible and reaches a limiting value:

$$
\lim_{h \to \infty} \hat{y}_{t+h} = \frac{C}{(1 - \varphi_1)} \Rightarrow C \Rightarrow \mu_y.
\tag{7.46}
$$

Figure 7.7 reveals how the ARMA(1,1) forecast profile undulates with exponentially diminishing amplitude and proceeds to converge upon the mean, based on Eq. (7.37), with a stationary series.

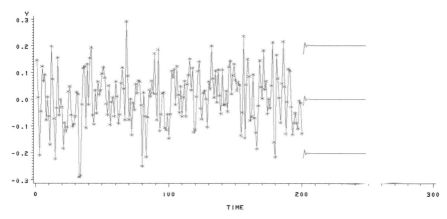

Figure 7.7   Forecast profile for ARMA(1,1) process.

7.3.7.4.4. Forecast Interval for the IMA Model   The more complex IMA(1,1) model exhibits a different forecast profile. The implicit integration in this model produces a gradually expanding forecast profile, as can be seen in Fig. 7.8. The farther into the forecast horizon the forecast is extended, the larger the forecast error. An example of this kind of process is the natural log of U.S. coffee consumption, the forecast profile for which is displayed in Fig. 7.9. This is a model that can be easily reparameterized as an ARIMA(1,1,0) model. The expanding forecast interval is suggestive of this fact. To be sure, when the model is reparameterized as such, the

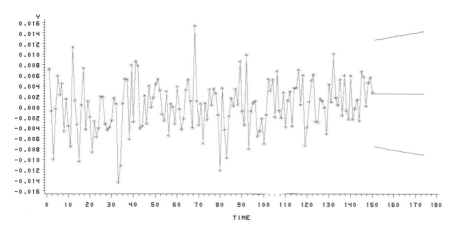

Figure 7.8   Forecast profile for IMA(1,1) model. Simulated IMA process.

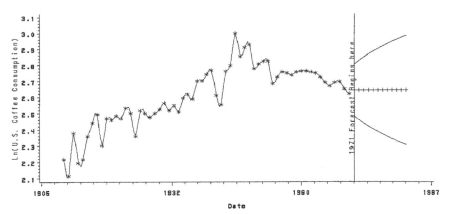

Figure 7.9   Forecast profile for ARIMA(0,1,1) model: U.S. coffee consumption, 1910 through 1970. Source: Rob Hyndman Time Series Data Library (http://www/maths.monash.edu.au/~hyndman/)—extracted 1997.

SBC is almost as low as that of the IMA(1,1) model. In the event that the model has multiple MA parameters, there will be shifts in the point forecast and forecast intervals. These apparent shifts will appear to be deviations from the smooth patterns the forecast points and intervals are following.

7.4. CHARACTERISTICS OF THE OPTIMAL FORECAST

It is useful to understand the nature of an optimal forecast. Granger and Newbold (1986) note that the information set of a finite sample longer than the memory of the series, given a known model, is necessary. Using a least squares estimation, the weight in the weighted sum of squared errors has to be chosen so as to minimize the sum of squared errors in finding the solution. What is more, there should be invertibility of the forecast. This way, an infinite autoregressive process could be expressed as a finite moving average process. An h step ahead optimal forecast can be expressed as an MA($h - 1$) model. The forecast error for the one-step-ahead forecast should be white noise. If the forecast error were otherwise, it might be improved by the incorporation of that other function and therefore would not be optimal. Meanwhile, the forecast error variance often increases along the forecast horizon. From Eqs. (7.29) and (7.30), we can see that the forecast variance increases with the value of h. From the forecast profiles, however, we can see that for stationary models, the forecast interval exhibits an initial expansion due to the autoregressive component shown in Figs. 7.3 and 7.4, and then begins to exhibit an asymptotic leveling off of that interval around the mean value, shown in Figs. 7.5 and 7.6. This may take place with diminishing levels, as can be seen in the forecast profile for the ARMA(1,1) process in Fig. 7.7 (Granger, 1989).

7.5. BASIC COMBINATION OF FORECASTS

There may be a need to enhance the reliability, robustness and accuracy of separate forecasts by combining forecasts from different statistical models. After all, Niels Bohr, a Nobel laureate in physics, said, "Prediction is very difficult, especially of the future." The imperfection of a single forecast was underscored by John Maynard Keynes, who is reported to have remarked, "There are two kinds of forecasters: those who don't know, and those who don't know that they don't know." David Hendry (1999) has since stated, "The things that can hurt us are those things that we don't know that we don't know." Because different methods of forecasting possess different assumptions, procedures, advantages, and disadvantages, it is a common contention that enhanced reliability can be derived from combining forecasts (Winkler, 1983, 1989; Makridakis, 1989). Combining forecasts permits the analyst to guard against mistakes, varying circumstances, failing assumptions, possible cheating, and variations in accuracy (Armstrong, 1985). If an individual forecast is already optimal or it encompasses the other forecast prior to combination, combining that forecast with another will not improve the forecast accuracy. However, Bates and Granger (1969) found that by combining forecasts it is generally possible to improve robustness as well as enhance the forecast accuracy of the separate forecasts.

This section discusses some methods for combining forecasts, as well as some problems that may stem from applying these methods. For example, an analyst might obtain separate forecasts from exponential smoothing, X-11, ARIMA, or autoregression (see Chapter 10). These two or more forecasts could be combined by any of the methods explained here. The first method is that of averaging the separate forecasts; the second is the variance–covariance method proposed by Bates and Granger (1969); and the third method is the regression method advocated by Granger and Ramanathan in 1984. More advanced methods of combining forecasts involving regression are broached in this section, and those involving autoregression will be discussed in Chapters 10 and 11.

Consider the first two methods. One method is to take the simple arithmetic average of the separate forecasts. The second method, the variance–

covariance method, requires a weighted average of the forecasts, with the inverse of the error variances serving as combining weights. One method of combining forecasts is to take the smoothed function of two one-step-ahead forecasts. One forecast may be based on one approach or model while the other forecast may emerge from another approach or another model. Let F_{1t} and F_{2t} be two forecasts that have been mean centered, with forecast errors ef_1 and ef_2, respectively. The combined forecast, CF_t, can be formulated as a function of these two component forecasts, shown in Eq. (7.47), if one assumes that both forecasts have been mean-centered:

$$CF_t = \omega F_{1t} + (1 - \omega)F_{2t}. \tag{7.47}$$

Meanwhile, the forecast error for the combined forecast can be expressed as

$$e_{CF_t} = (CF_t - Y_{t+1}) = \omega e_{F_{1t}} + (1 - \omega)e_{F_{2t}}. \tag{7.48}$$

The forecast variance for the combined forecast can be computed by squaring the forecast error:

$$\sigma^2_{CF_t} = \omega^2\sigma^2_{eF_1t} + (1 - \omega)^2\sigma_{eF_2t} + 2\omega(1 - \omega)\sigma_{eF_1t}\sigma_{eF_2t}. \tag{7.49}$$

The objective in this case is to minimize the forecast error variance of the combined forecast. To do so, we take the first derivative of Eq. (7.49) with respect to ω. Then we set this equal to zero and solve for ω to determine the optimal smoothing weight, ω:

$$\omega = \frac{\sigma^2_{eF2t} - \sigma_{eF1t}\sigma_{eF2t}}{\sigma^2_{eF1t} + \sigma^2_{eF2t} - 2\sigma_{eF1t}\sigma_{eF2t}}. \tag{7.50}$$

If there is no correlation between the two series, the optimum smoothing weight, ω, has almost the same ratio, except that it lacks the cross-product terms in the numerator and denominator, as:

$$\omega = \frac{\sigma^2_{eF2t}}{\sigma^2_{eF1t} + \sigma^2_{eF2t}}. \tag{7.51}$$

When the optimal value for ω is plugged into the equation, the minimum error and error variance for the combined forecast are obtained. In this way, one-step-ahead simultaneous forecasts are combined to produce an optimal forecast. The weights therefore depend on the accuracy of the forecast (Armstrong, 1999a, 1999b). Trimming the means used for forecasts might improve the accuracy of these forecasts (Armstrong, 1999b).

Another version of the preceding combination method is known as the Kalman filter, mentioned in Box et al., (1994), which combines a current prediction with a previous forecast. This approach uses exponential smoothing to update the prediction. In this way, the moving average can be given the appropriate weights. Two independent estimates are combined to form a smoothed prediction. Because one of these is a prior information set, this approach has been referred to as Bayesian forecasting. It attempts to improve the accuracy of a forecast with current data by minimizing the errors of the combined forecast. Minimizing the total forecast error permits derivation of an optimal smoothing weight for the combination of independent forecasts.

Makridakis, Wheelwright, and McGee (1983) as well as Granger and Newbold (1986) explain the variance–covariance method as an exponentially smoothed updating formula. This combined one-step-ahead forecast consists of exponentially smoothing the recent data (Y_t) and the prior forecast (F_t) in accordance with

$$F_{t+1} = \omega Y_t + (1 - \omega)F_t. \tag{7.52}$$

The smoothing constant ω is estimated to improve the forecast. If we know the variances of Y_t and F_t are Insert σ_1^2 and σ_2^2, respectively, we can compute the overall variance as the weighted sum of these variances, shown in Eqs. (7.48) and (7.49). In order to obtain the minimum variance, the variance, σ^2, can be differentiated with respect to the smoothing constant, ω. The result is set to equal zero and differentiated to obtain the minimum slope. The equation is solved to obtain the optimum smoothing constant, ω, which is then substituted into Eq. (7.47). With a similar updating of the variance of Y_t, the optimum forecast, based on current data and previous information, may be obtained.

To ascertain the optimum updating variance, Eq. (7.50) can be substituted into Eq. (7.49). At the same time, recall that the correlation between the F_t and Y_t is

$$\rho = \frac{\sigma_{F_1 Y_t}}{\sigma_{F_1}\sigma_{Y_t}}. \tag{7.53}$$

When one applies this correlation to Eq. (7.49), after Eq. (7.50) has been substituted into it, the resulting formula–following Holden et al., (1990)— for the variance of the updated forecast becomes

$$\sigma_{F_{t+1}}^2 = \frac{\sigma_F^2 \sigma_Y^2 (1 - \rho_{FY})}{(\sigma_Y - \rho_{FY}\sigma_F)^2 + \sigma_F^2(1 - \rho_{FY})}. \tag{7.54}$$

Together with the first-order solution, this equation permits automatic updating of the forecast by updating the earlier forecast with the new data. In this way, a minimum forecast error is maintained.

Another method for combining forecasts is the regression method. Granger and Ramanathan (1984) are credited with modernizing the earlier regression parameterization of the earlier Bates and Granger (1969) approach. Bates and Granger advocated that the coefficients be constrained to sum to unity and that the intercept be constrained to equal zero, whereas Granger and Ramanathan (1984) suggested that less biased results are obtained without these constraints. From separate forecasts, F_1 and F_2, a combined forecast CF_t could be obtained with the following regression formula:

$$CF_t = \alpha + \beta_1 F_{1t} + \beta_2 F_{2t} + e_t. \tag{7.55}$$

In this case, the intercept is not constrained to equal zero and the regression coefficients are not constrained to equal unity. The result is usually a more accurate combined forecast with smaller error variance and therefore a smaller 95% forecast interval around the predicted scores.

Scholars have made noteworthy suggestions about appropriate applications of forecasting models. Granger and Newbold (1986) note that the simple exponential smoothing for updating is generally optimally applied when the model is an ARIMA$(0,k,k)$ configuration. Those processes with autoregression incorporated within them may be handled better by pure ARIMA forecasts. Brown et al. (1975) are also cited as developing a means of monitoring the performance of the forecast by a cumulative error chart with control limits. When the forecast exceeds the limits, this is an indication that something is awry. Trigg and Leach (1964, 1967) developed a method of monitoring the forecast error with a tracking signal made up of the ratio of the smoothed error to the mean absolute deviation. An approximation of confidence limits of plus or minus two standard errors is ascertainable so that the signal should remain within them. When the forecast error exceeded those limits, the value of the smoothing constant could be increased to give more weight to current rather than past observations. Moreover, the Holt or Winters version (explained in Chapter 2) of exponential smoothing might better be applied, for the reason that they accommodate trend and seasonality better than does simple exponential smoothing.

7.6. FORECAST EVALUATION

Models can be evaluated not only with respect to their optimality, but also according to their ability to produce a good forecast. Indeed, metadiagnosis entails comparative evaluation of models according to, among other things, their forecasting capability. The forecast should be relatively cheaper than the others, given the value of the forecast generated (Granger and Newbold, 1986). The forecast should have face validity. It should make intuitive sense (Armstrong, 1999). The forecast should be rational, in that the forecasts should be efficient and unbiased (Clements and Hendry, 1998). The better the model, the more accurate the forecast. In terms of absolute accuracy, we can evaluate the forecast by different measures presented in Chapter 2. The accuracy of a forecast is evaluated against real data in the evaluation or holdout sample, if not against another known, tested, and established criterion. One common measure of this accuracy is the forecast error variance or mean square forecast error. The mean absolute percentage error, mean absolute error, mean error, mean absolute percent error, maximum error, and minimum error are criteria less sensitive to outlier distortion that are often preferred. Measures of error used should not be too sensitive, and there should be provisions for checking their measurements (Armstrong, 1999).

Forecast accuracy can depend on the forecast horizon. Some forecasts are more stable than others. How far into the future this horizon extends and where it ends must be known. In general, the farther into the future the forecast horizon, the more difficult it is to forecast. Some methods are better at short-run forecasting while other methods are better at long-run forecasting. The more stable the forecast, the more reliable the forecast. Armstrong (1999) notes that there should be consistency over the forecast horizon. Although this topic is broached here, the last chapter delves into the subject in greater detail.

The dispersion of the forecast interval is a measure of forecast accuracy. The width of the 95% confidence interval at a particular point of interest on the forecast horizon is another standard by which forecast accuracy can be measured, as is the shape of the forecast profile due to its probability distribution. The mean square forecast error (MSFE) is commonly used to assess the accuracy of the forecast, but because the MSFE is sensitive to outlier distortion, the mean absolute percentage error (MAPE) may be preferred.

There are other common measures of fit such as R^2, adjusted R^2, and Amemiya's adjusted R^2. Although these measures may be used for model evaluation, Armstrong suggests that R^2 and adjusted R^2 not be used for forecast evaluation (1999). There are also minimum information criteria— for example, the AIC and the SBC—that are used to assess parsimonious fit. These criteria are joint functions of the minimum forecast error and some form of penalty for the number of free parameters (degrees of freedom) in the model. The AIC tends to produce more overparameterized models, whereas the SBC applies a stronger penalty for the number of terms in the model than the AIC. The lower the value of the AIC or the SBC, the better the fit of the model.

Forecasts can be evaluated according to the information set required and acquired. The more data and the more recent the data, the better. The less biased and more valid the data, the better. Often, validity and reliability can be enhanced by using alternative sources for data collection. The data must be checked for and cleaned of input errors. The more recent data should be weighted more heavily (Armstrong, 1999). Some exponential smoothing methods do not require as much data as ARIMA models require in order to generate a good forecast.

Forecasts can be evaluated by the cost of the forecast, when the cost and function of forecast error are known. The cost of acquiring data may figure into this calculation. The more data required, the more costly the forecast may be (Granger, 1989).

Forecasts can also be evaluated in terms of their complexity or parsimony. The lesser the parameter redundancy and parameter uncertainty, the better the model used for forecasting. Simpler forecasts are preferred to complex forecasts, given the same level of accuracy (Diebold, 1998).

Relative forecast ability might be assessed in terms of the comparative abilities of different approaches. One criterion of relative forecast ability is that of forecast efficiency. Forecast efficiency of a model involves comparing the mean square forecast error of the model to some baseline model. The forecast error variance of the model under consideration may be derived from a baseline comparison. That baseline used is often the naive forecast, a forecast formed by assuming that there is no change in the value of the latest observation. Theil developed a U statistic (Makridakis *et al.*, 1983), which compares forecasts. It takes the square root of a ratio:

$$\text{Theil's } U = \sqrt{\dfrac{\dfrac{\sum\limits_{t=1}^{T} (\text{FPE}_{t+1} - \text{APE}_{t+1})^2}{(t-1)}}{\dfrac{\sum\limits_{t+1}^{T} (\text{APE}_{t+1})^2}{t-1}}},$$

where

$$F_{t+1} = \text{forecasted value}$$
$$X_{t+1} = \text{actual value}$$

$$\text{FPE} = \frac{F_{t+1} - X_t}{X_t}$$

$$\text{APE} = \frac{X_{t+1} - X_t}{S_t}.$$

The numerator of the ratio consists of the mean square error between the forecast and the naive baseline (average). The denominator of the ratio consists of the mean square of the average percentage error (baseline average). When there is no difference between the model forecast and the naive forecast, U equals zero. When the forecast error variance exceeds the mean square average baseline error variance, then the value of the U rises. The choice of the naive or baseline model is open to debate. Different scholars employ different models as their baseline. If alternative models have been systematically eliminated, the final model should be more accurate and more general. The more general the model, the more theory or parameters it encompasses. The more the model encompasses the variance inherent in the series being forecasted, the better fit the model has. Therefore, the more encompassing the model, the more useful the forecast.

Although the final chapter contains a more complete comparison of forecast abilities, this chapter takes note of major advantages and disadvantages of various models already presented. Models smoothed with single exponential weighting may not require long series. Some of the simpler smoothing methods—such as simple, double, or Holt's method—may not handle seasonality as well as others. The Winter's method can forecast series with both trend and seasonal turning points. Nonetheless, the simpler methods are cheaper and easier to compute, and often provide better forecasts than more complex methods—such as X-11 decomposition or Box–Jenkins methods (Hibon and Makridakis, 1999). Although X-11 decomposition models work well for basic signal extraction of regular long wave cyclical turning points, they sometimes may not comprehensively explain the underlying data generation process well. Box–Jenkins models generally handle both moving average and autoregressive problems along with some seasonal variation well and generally offer fairly good predictive validity over the short run. Sometimes, some regression or autoregression models, which will be covered later, have more explanatory power and predict better than others over the long run. Granger and Newbold say that Box–Jenkins methods give better series forecasts than other methods in the near term. How models can be combined to improve prediction and the advantages and methodology of combining models will be discussed in more detail in the final chapter.

7.7. STATISTICAL PACKAGE FORECAST SYNTAX

7.7.1. INTRODUCTION

At this time both SAS and SPSS have the ability to forecast. After the model is developed, it is used to generate the forecast values of the data set. The user determines how many periods into the future horizon he forecasts by setting the appropriate parameters. The residuals are easily produced. From those residuals, the standard errors of the predicted values are estimated. The upper and lower 95% confidence limits are also generated. With SAS the user has the capability of generating its forecasts with any of three types of estimation, whereas with SPSS the user may chose one of two types of estimation.

The default estimation for SAS forecasts is conditional least squares, whereas with it is unconditional least squares. Although both packages allow forecasts generated with unconditional least squares or conditional least squares, only SAS permits maximum likelihood to be used for the estimation of the forecasts. The forecast subcommands of the two packages generate these variables, which may then be used to construct the graphical forecast profile plot.

7.7.2. SAS SYNTAX

The SAS programming syntax for forecasting the natural log of U.S. coffee consumption is given in program c7pgm2.sas. This program syntax constructs a data set called COFFEE and sets up a time variable that begins with a value (set by the RETAIN statement) of 1910 and is a counter (constructed with the TIME + 1 statement) that increments by a value of one for each observation. The date variable is constructed with the INTNX function. The format is specified by the FORMAT statement. The data follow the CARDS statement.

```
/* c7pgm2.sas or c7fig9.sas */
title ' Forecast Profile for ARIMA(0,1,1) Model';
title2 'US Coffee Consumption 1910 through 1970';
title3 'Rob Hyndman Time Series Library: coffee.dat';

data coffee; RETAIN TIME(1909);
  input cofcon;
time + 1;
lcofcon=log(cofcon);
DATE = INTNX('year','01JAN1910'd,_n_-1);
format Date YEAR.;
cards;
 9.2
 8.3
 . . .
14.2
13.8
```

```
proc print;
run;
SYMBOL1 V=STAR C=BROWN I=SPLINE;
PROC GPLOT;
 PLOT COFCON * date/haxis=axis1;
title 'US Coffee Consumption 1910-1970';
RUN;
proc gplot;
 plot lcofcon*date;
title 'Log of US Coffee Consumption 1910-1970';
run;


/* A proc print is used to check the data to be sure the
 program is reading the data correctly.
The GPLOT plots the natural log of coffee consumption against
 the date so the analyst may view the data */

proc arima;
 i var=lcofcon(1) nlag=20;
 e q=1 printall plot noint;
 f lead=12 id=date interval=year printall out=fore;
run;
```

The FORECAST subcommand which begins with the letter f within the PROC ARIMA, sets up a 12-step-ahead forecast with the LEAD subcommand. The ID variable is set to equal DATE so the DATE variable will be used for identification of the observation and output with the data set. Because the LEAD is set to 12 periods and the interval is set to 'YEAR', a 12-step-ahead forecast will extend into the future 12 years. The PRINTALL statement will printout the data and the variables created by the FORECAST subprocedure. The output data set is constructed with the OUT = subcommand, and the output data set is called 'FORE'. FORECAST, UCL95, and LCL95 variables are generated, which along with the natural log of the U.S. coffee consumption make up the component variables of the forecast plot.

```
data new;
 merge fore coffee; by date;
if _n_ < 62 then forecast=.;
if _n_ < 62 then l95=.;
if _n_ < 62 then u95=.;
proc print;
run;
```

DATA NEW sets up a data set called NEW. The MERGE command matches
the data sets called 'COFFEE' and 'FORE' according to the key variable,
called DATE, into the NEW data set. Then the FORECAST variable and its
confidence limit variables, L95 and U95, are stripped of values before the
forecast begins. To clarify graphical presentation of the forecast profile
plot, values of the forecast components are set to missing (period) before
the forecast begins at observation 62. The data are printed to confirm
proper trimming.

```
symbol1 i=spline c=green v=star;
symbol2 i=spline c=blue v=plus;
symbol3 i=spline c=red;
symbol4 i=spline c=red;
axis1 label=('Date') ;
proc gplot data=new;
 plot (lcofcon forecast l95 u95)*date/overlay
 haxis=axis1;
title 'Forecast profile for ARIMA(0,1,1) model';
title2 'US Coffee Consumption 1910 through 1970';
title3 'Source: Rob Hyndman Time Series Data Library-extracted 1997';
title4 'World Wide Web URL: http://www.maths.monash.edu.au/~hyndman/';
run;
```

Then the forecast profile plot is constructed. The lines are defined by
their respective SYMBOL subcommands. The points are joined with the I=
JOIN subcommand. The type of data value is indicated by the V= option.
Actual data are represented by a STAR and the forecast is represented by
a PLUS sign. The forecast proceeds horizontally ahead, unless the natural
logged series is centered, in which case it lifts upward. Although the line
color is not shown here, the user may select the C= option to specify the
color of the line. In this syntax, green lines are chosen to represent actual
data. Blue and red lines are selected to respectively designate, the forecast
and its confidence intervals. The AXIS1 subcommand defines the label for
the horizontal axis. The data come from DATA NEW and an OVERLAY plot
is generated against the date. The forecast profile plot generated by SAS
may be found in Fig. 7.8.

7.7.3. SPSS Syntax

The SPSS syntax, contained in C7pgm3.sps, for generating the forecast is
different. The forecasts extend from 1970 through 1982 as can be seen in the
third line of the following SPSS program syntax. The variable used for the
ARIMA model is the natural log of U.S. coffee consumption from 1910

through 1970. The forecasts extend through 1982. The model from which the forecasts are generated is an ARIMA(0,1,1) without a constant model. The final line of command syntax shows that the user chooses conditional least squares with which to estimate the forecast function and interval.

```
* ARIMA.
TSET PRINT=DEFAULT CIN=95 NEWVAR=ALL .
PREDICT THRU YEAR 1982 .
ARIMA lcofcon
 /MODEL=( 0 1 1 )NOCONSTANT
 /MXITER 10
 /PAREPS .001
 /SSQPCT .001
 /FORECAST EXACT .
Title 'Forecast Profile for Coffee Consumption'.
Subtitle 'of IMA(1,1) Process'.
*Sequence Charts .
TSPLOT VARIABLES=lcofcon fit_1 lcl_1 ucl_1
 /ID= year_
 /NOLOG
 /MARK YEAR 1970 .
```

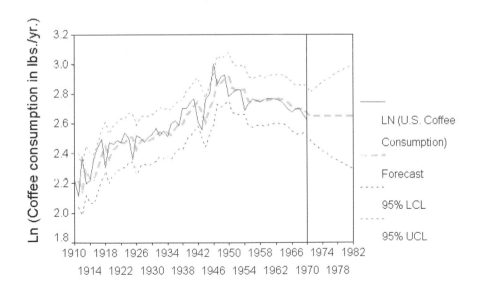

Figure 7.10   ln(U.S. coffee consumption), 1910 through 1970.

This procedure generates a forecast variable called FIT_1. It generates upper and lower confidence limit variables, called, UCL_1 and LCL_1, respectively. Along with the data, these newly generated variables comprise the component variables of the forecast profile plot. The forecast profile plot generated by SPSS TSplot command can be found in Fig. 7.10. The final chapter of this book will further examine the forecast accuracy of various models.

7.8. REGRESSION COMBINATION OF FORECASTS

It is also possible to improve the accuracy of forecasts by combining forecasts with regression analysis (Granger and Ramanathan, 1984). When one forecast does not encompass (subsume) the other, the researcher can use regression analysis to form the combining weights of the combined point forecast. The regression model of the combined forecast is a function of an intercept and two component forecasts. If one forecast encompassed the other, the regression coefficient of the encompassed forecast would be nonsignificant and the combined forecast would merely be a function of one of the two component forecasts. In such a case, the combination is of little use. If neither of two component forecast encompasses the other, then the combination is a function of the intercept and two significant regression coefficients, each of which is multiplied by the component forecast. Although the example here uses only two component forecasts, a forecast combination may entail the regression combination of more than two component forecasts (Granger, 1989).

The methodology for combining forecasts merits serious consideration. The overall sample is divided into an historical and an evaluation subset. Different models are developed on the basis of the historical subsample. In this example, one of the models is a Box–Jenkins ARIMA model and the other is a polynomial regression model. Forecasts from each of these models are generated and graphed. The forecasts span the evaluation period. The data sets from the respective forecasts and the evaluation sample are merged. Then the forecast from each separate model is compared with the actual data. The residual, its variance, and the mean absolute percentage error are calculated within the evaluation sample to assess the accuracy of the separate forecasts. In this way, the evaluation of the separate forecasts is performed on the basis of the data in the evaluation period. The forecast profiles are then graphed for visual comparison.

A combining regression is performed on the basis of the data within the evaluation period. The dependent variable in this combining regression is that of the real data in the evaluation sample: $Actual_{t+h} = a + b_1 F_{1,t+h} + b_2 F_{2,t+h} + e_{t+h}$. The predictor variables are the forecasts from the different models mentioned. The regression model output supplies an intercept and regression coefficients, which are sometimes called the combining weights. The predicted scores from the regression model are saved, because they are the combined forecast of the real data. The forecast interval around the forecast is formed from the upper and lower 95% confidence intervals around those predicted individual scores. These intervals are generally smaller than the forecast intervals around either of the $F_{1,t+h}$ and $F_{2,t+h}$ forecasts from the earlier models.

We can evaluate the combined forecast by subtracting the combined forecast from the actual data in the evaluation period. This subtraction yields the residual. The residual variance and mean absolute percentage error are computed. We can use these criteria to compare the accuracy of the combined forecast to that of each of its component forecasts.

For a programming example of this combining process, the U.S. defense and space equipment index reported by the U.S. Federal Reserve Board on its World Wide Web data base (U.S. Federal Reserve Board, 1999) is used. The index is a measure of the gross product value of defense and space equipment produced in the United States. The reference year for the index is 1992 and the gross product value for that year is 86.44. The data were selected because they are central to a contemporary American public policy controversy revolving around defense spending. Some pundits suggest that the demise of the Cold War, the fall of the Berlin Wall in 1989, and the dissolution of the Soviet Union in 1991 have led to a reduction of U.S. defense spending. These pundits hope that funds from this peace dividend could be marshaled to save social security and/or medicare from eventual depletion. Some hoped that it could increase funding of biomedical research and national health insurance. Other scholars and commentators contend that the decline in defense spending has seriously reduced the ability and readiness of the United States to provide for national defense, international security, foreign aid, and domestic disaster relief in a changing world. The index chosen shows the rate of decline in production of defense and space equipment from January 1988 through March 1999. These data were subset into an historical sample extending from January 1988 through December 1994, and an evaluation sample, extending from January 1995 through March 1999, shown in Fig. 7.11. A caveat should be noted, however, that these models were developed on the basis of an historical sample that does not span the time in which NATO allied countries attacked the Milosevic regime in Serbia for the latter's policies of mass murder of ethnic Albanian men, rape of ethnic Albanian women, destruction of their identification records, and the forced expatriation of surviving women and children from the province of Kosovo. Therefore, the forecasts here are almost extensions of what might have happened had the Milosevic regime submitted to NATO's demands that the Serbian military and paramilitary forces leave Kosovo, that those forced out be allowed to return, and that the inhabitants

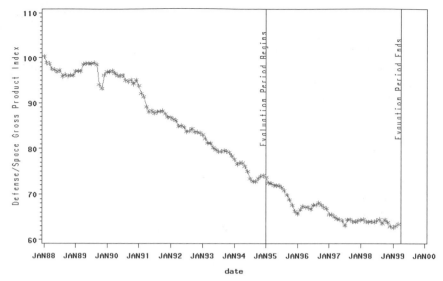

Figure 7.11 U.S. defense and space gross product value index. Series B52008 and T52008. 1992 value weight: 84.677. Source: Federal Reserve Board Statistical Release G.17 (http://www.bog.frb.fed.us/releases/g17/ipdisk/gvp.sa). Retrieval date: May 7, 1999.

reside under some autonomy guarded by a NATO-run peacekeeping operation. The revelations of Serbian commitment of crimes against humanity and NATO's military response demonstrate that the farther into the future the forecast horizon is extended, the more likely it is that unforeseen exogenous events may render earlier forecasts inappropriate or moot.

Two models were developed on the basis of the historical sample in program C7pgm4.sas. The first of the two forecasting models is that of an IMA model. The first forecast is generated and shown in Fig. 7.12.

The integrated moving average model, built from and fitted to these data, is

$$(1 - L)(\text{Defense/Space gross product value} - 0.316)$$
$$= (1 + 0.302L - 0.205L^{18})e_t$$

and generates the first forecast shown in Fig. 7.12. The second model is a polynomial regression model, with $R^2 = 0.976$, that has significant linear, quadratic, and cubic time period predictors, each of which has a $p < 0.01$, and is formulated as follows:

$$\text{Defense/SpaceGrossProductValue}_t = 97.035 + 0.197\text{Time}$$
$$- 0.011\text{Time}^2 + 0.00007\text{Time}^3 + e_t.$$

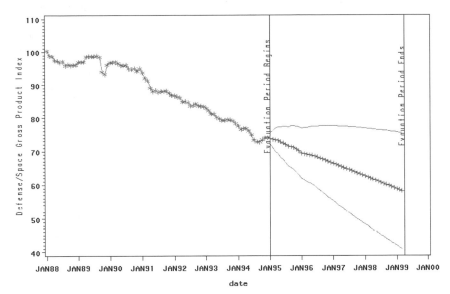

Figure 7.12 U.S. defense and space gross product value index. Source: Federal Reserve Board Statistical Release G.17 Series B52008 and T52008. 1992 value weight: 84.677. Model 1: IMA forecast.

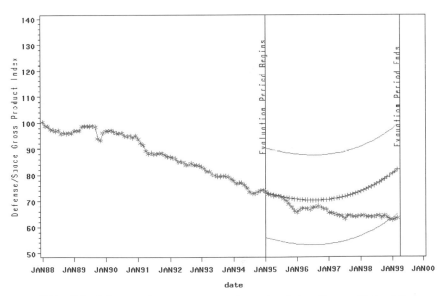

Figure 7.13 U.S. defense and space gross product value index. Source: Federal Reserve Board Statistical Release G.17 Series B52008 and T52008. 1992 value weight: 84.677. Model 2: Estimation of polynomial regression forecast from data in the estimation sample.

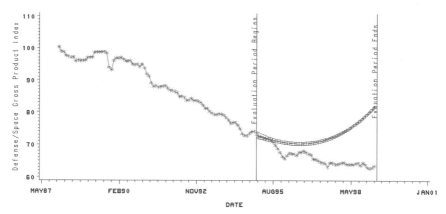

Figure 7.14   U.S. defense and space gross product value index. Source: Federal Reserve Board Statistical Release G.17 Series B52008 and T52008. 1992 value weight: 84.677. Model 2: Polynomial regression forecast.

The forecast profile of the cubic regression is depicted in Fig. 7.13. Although the fit is good in the historical period, the regression forecast in the evaluation period (shown in Fig. 7.14) seriously mispredicts as the forecast is extended further into the horizon of the evaluation sample.

The accuracy of these separate forecasts can be improved by regression analysis. An ordinary least squares (OLS) regression analysis of the real data in the evaluation period on the forecasts of the two models yields predicted scores that are the combined forecast. In this program, C7pgm4.sas, two models are developed on the basis of the historical sample. Forecasts from each of these models are projected over the time horizon of the evaluation sample. The actual data are then regressed on the two forecasts. The combined forecast regression, with an R^2 of 0.932 and all highly significant parameters ($p < 0.001$), estimates the equation used to project the point forecast over the time horizon of the evaluation sample:

$$\text{Combined Forecast}_t = 63.839 + 1.157F_{1t} - .782F_{2t} + e_t$$

The predicted scores from this regression model are in fact the combined forecast. The combining regression generates predicted these mean scores along with the 95% confidence limits around the mean within the evaluation sample.

The improved predictive forecast profile of this method is shown in Fig. 7.15. Graphical comparison of the combined forecast with the component forecasts reveals how the accuracy of the combined forecast is improved. For example, the predicted values cleave more closely to the real data than

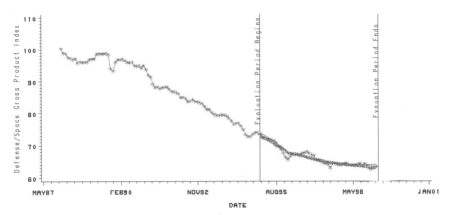

Figure 7.15 U.S. defense and space gross product value index. Source: Federal Reserve Board Statistical Release G.17 Series B52008 and T52008. 1992 value weight: 84.677. Model 3: Regression combination forecast within evaluation sample.

in the polynomial regression model. The confidence intervals are smaller than those in either component model.

We can perform a quick and partial evaluation of the combination of forecasts by comparing the models according to some criteria of measurement error. We could use the mean square error or the mean absolute percentage error for this comparison. Although the mean square error or error variance is commonly used, it is sensitive to outliers. Because the mean absolute percentage error is less vulnerable to the influence of outliers, the mean absolute percentage error is often preferred as a criterion of comparison. These measures can be computed for the historical sample as well as for the validation sample. In this example, these statistics are computed over the evaluation sample and displayed in Table 7.3. From Table 7.3, we can see that both component forecasts have a larger error variance and a larger mean absolute percentage error than the regression combination of forecasts. Moreover, the regression combination of forecasts

Table 7.3

Forecast Evaluation

Type of model	Mean square forecast error	Mean absolute percentage error
ARIMA forecast	5.319	2.794
Polynomial regression forecast	5.982	42.217
Regression combined forecast	0.607	0.008

reduces the dependency of the forecast on the statistical method employed, thereby yielding a more reliable and robust prediction.

Of course, it is possible to forecast beyond the time horizon of the evaluation sample as well. If the researcher has reason to perform only a short-run forecast beyond the evaluation sample, he might use the combined forecast produced from the combining regression as a series from which ARIMA forecasts and confidence intervals are extended beyond the evaluation sample. If he has reason to believe that there is more of a long-range trend in the series, he may use the original models to generate longer component forecasts that extend beyond the evaluation sample. With the combining regression developed in the evaluation sample, he may extend the combined forecast, based on the extended component forecasts, beyond the evaluation sample. On the basis of the extended combined forecast as the series under consideration, he may develop an ARIMA model that generates a forecast and its confidence limits that extend beyond the evaluation sample.

Before combining the forecasts, the analyst should note that there are matters of estimation, functional form, and power to be considered. Whether OLS regression analysis should be used for combining forecasts depends on the type of series that are being combined. If the series are stationary, the regression coefficients may be relatively small and nonsignificant. Under those circumstances, it may be preferable to combine forecasts with a simple or weighted average of the separate forecasts. The series used for forecasting in this example is that of an integrated moving average series, and this series is amenable to combining by OLS regression analysis. If the series under examination possesses autocorrelated errors, then OLS regression will be inefficient and some form of autoregression, discussed in Chapter 10, would be preferred for combining the separate forecasts. Alternatively, if the series under analysis exhibits deterministic trend and ARMA errors, then a detrending regression analysis followed by an ARIMA modeling of the errors could be used as the basis of forecast combination (Diebold, 1998).

Although linear combinations have been employed in this example, the functional form of the combining regression analysis need not be linear. It can contain squared and cubic terms. It may contain interaction terms. It can contain interactions between main effects and polynomial terms. It can also be intrinsically nonlinear. Different models can be tried until the R^2 of the combining regression is high enough or the SBC is low enough to render the combining model worthwhile.

Moreover, it is essential that the evaluation period be long enough that there are enough observations to confer sufficient statistical power on the combining regression model. If the evaluation period is not long enough, the regression coefficients of the component forecasts may not be significant when they otherwise would be. Of course, the higher the R^2 of the combining regression, the fewer the number of observations needed with a constant number of component forecasts. Clearly, these matters merit consideration before the analyst proceeds to combine forecasts with regression analysis.

REFERENCES

Armstrong, J. S. (1985). *Long-Range Forecasting: From Crystal Ball to Computer.* New York: Wiley, p. 444.

Armstrong, J. S. (1999a). "Forecasting Standards Checklist." *Principles of Forecasting: A Handbook for Researchers and Practitioners.* Dordrecht, The Netherlands: Kluwer Academic Publishers, World Wide Web URL: *http://www-marketing.wharton.upenn.edu/forecast/glossary.html.* Retrieved July 15, 1999.

Armstrong, J. S. (1999b, June). "Combining Forecasts." *Principles of Forecasting: A Handbook for Researchers and Practitioners.* Dordrecht, The Netherlands: Kluwer Academic Publishers (in press). Chapter (in press) presented at 19th International Symposium on Forecasting, Washington, D.C.

Bates, J. M., and Granger, C. W. J. (1969). "The Combination of Forecasts," *Operation Research Quarterly* **20,** pp. 451–486. Cited in Holden, K., Peel, D. A., and Thompson, J. L. (1990). *Economic Forecasting: An Introduction.* New York: Cambridge University Press, pp. 85–88.

Box, G. E. P., and Jenkins, G. M. (1976). *Time Series Analysis: Forecasting and Control.* Oakland, CA: Holden Day, pp. 129–132.

Box, G. E., Jenkins, G., and Reinsel, (1994). *Time Series Analysis: Forecasting and Control,* 3rd ed., Englewood Cliffs, NJ: Prentice Hall, pp. 164–169.

Brown, R. L., Durbin, J., and Evans, J. M. (1975). "Techniques for Testing the Constancy of Regression Relationships over Time (with discussion)." *Journal of the Royal Statistical Society B* **37,** 149–192. Cited in Harvey, A. (1991). *The Econometric Analysis of Time Series,* 2nd ed. Cambridge, MA.: MIT Press, p. 155.

C. F. Kettering (1949). 2020 Vision Seeing the Future (1999). World Wide Web URL: http://www.courant.comhttp://www.courant.com/news/special/future/. Downloaded 12 May 1999.

Chatfield, C. (1975). *The Analysis of Time Series: Theory and Practice.* London: Chapman Hall, pp. 98–107.

Clements, M., and Hendry, D. (1998). *Forecasting Economic Time Series.* New York: Cambridge University Press, p. 56.

Diebold, F. X. (1988). "Serial Correlation and the Combination of Forecasts." *Journal of Business and Economic Statistics,* **8**(1), 105–111.

Diebold, F. X. (1998). *Elements of Forecasting.* Cincinnati, Ohio: Southwestern College Publishing, pp. 34, 42–44, 110.

Ege, G., Erdman, D. J., Killam, B., Kim, M., Lin.,C. C., Little, M., Narter, M. A., & Park, H. J. (1993). *SAS/ETS® User's Guide.* Version 6, 2nd ed. Cary, NC: SAS Institute, Inc. pp. 99–183.

Farnum, N. R., and Stanton, L. W. (1989). *Quantitative Forecasting Methods.* Boston: PWS-Kent Publishing, pp. 484–486.

Gilchrist, W. (1976). Statistical Forecasting. New York: Wiley, pp. 105–106.

Granger, C. W. J. (1989). Forecasting in Business and Economics. San Diego: Academic Press, pp. 47–89, 105, 183–196.

Granger, C. W. J., and Newbold, P. (1986). Forecasting Economic Time Series, 2nd ed. San Diego: Academic Press, pp. 120–149.

Granger, C. W. J., and Ramanathan, R. (1984). "Improved Methods of Combining Forecasts." Journal of Forecasting 3, 197–204.

Griffiths, W. E., Hill, R. C., and Judge, G. G. (1993). Learning and Practicing Econometrics. New York: John Wiley and Sons, Inc., Chapter 20.

Hendry, D. F. (1999, June). "Forecast Failure in Economics: Its Sources and Implications." 19th International Symposium on Forecasting, Washington, D.C. Keynote address.

Hibon, M., and Makridakis, S. (1999, June). "M3-Competition." Paper presented at the 19th International Symposium on Forecasting, Washington, D.C.

Holden, K., Peel, D. A., and Thompson, J. L. (1990). Economic Forecasting: An Introduction. New York: Cambridge University Press, pp. 85–107.

Maisel, R. (1994). NYU Graduate Department of Sociology used and provided historical records of the Political Divisions in the U.S. House of Representatives pertaining to the Democratic percentage of the vote. New York.

Makridakis, S., Wheelwright, S. C., and McGee, V. E. (1983). Forecasting: Methods and Applications, 2nd ed. New York: Wiley, pp. 50–52, 65–67.

Makridakis, S. (ed.). (1984). "Forecasting: State of the Art." In The Forecasting Accuracy of Major Time Series Methods. New York: Wiley, pp. 1–3.

Makridakis, S. (1989). "Why Combining Works." International Journal of Forecasting 5, 601–603.

McCleary, R., and Hay, R. (1980). Applied Time Series Analysis for the Social Sciences. Beverly Hills, CA: Sage Publications, Chapter 4, pp. 205–215.

Pindyck, R. S., and Rubenfeld, D. L. (1991). Econometric Models and Economic Forecasts, 3rd ed. New York: McGraw-Hill, pp. 516–530.

SAS Institute, Inc. (1995). SAS/ETS® Software: Time Series Forecasting System. Version 6, 1st ed. Cary, NC: SAS Institute, Inc., pp. 244–245.

SPSS, Inc. (1994). SPSS Trends 6.1. Chicago, Ill: SPSS, Inc., pp. 14–15.

Trigg, D. W. (1964). "Monitoring a Forecast System." Operation Research Quarterly 15, 271–274. Cited in Farnum, N. R., and Stanton, L. W. (1989). Quantitative Forecasting Methods. Boston: PWS-Kent Publishing, pp. 528–545.

Trigg, D. W., and Leach, A. G. (1967). "Exponential Smoothing with an Adaptive Response Rate." Operation Research Quarterly 18, 53–59. Cited in Farnum, N. R., and Stanton, L. W. (1989). Quantitative Forecasting Methods. Boston: PWS-Kent Publishing, pp. 528–545, and Makridakis, S., Wheelwright, S. C., and McGee, V. E. (1983). Forecasting: Methods and Applications. 2nd ed. New York: Wiley, pp. 91, 114, 117.

U.S. Federal Reserve Board Release G.17 (1999). Series B52008 and T52008 have a 1992Gross Product Value reference point of 86.44. World Wide Web URL: http://www.bog.frb.fed.us/releases/g17/ipdisk/gvp.sa Downloaded 7 May, 1999.

U.S. House of Representatives (1998). Political Divisions of the U.S. Senate and House of Representatives, 34th to 105th Congresses, 1855 to Present. World Wide Web URL: http://clerkweb.house.gov/histrecs/history/elections/political/divisions.htm, downloaded 25 October, 1998.

Winkler, R. L. (1983). "The Effects of Combining Forecasts and the Improvement of the Overall Forecasting Process." Journal of Forecasting 2(3), 293–294.

Winkler, R. L. (1989). "Combining Forecasts: A Philosophical Basis and Some Current Issues." Journal of Forecasting 5, 605–609.

Chapter 8

Intervention Analysis

8.1. INTRODUCTION: EVENT INTERVENTIONS AND THEIR IMPACTS

The study of impact analysis shifts the reader from examination of the univariate history of a series to the examination of multiple time-dependent series. With impact analysis, the researcher assesses the response in a series to a discrete event or intervention input (Makridakis and Wheelright, 1987). These events or interventions are often unusual or singular. The intervention input may be a scandal, war, embargo, strike, or price change (Pack, 1987). The response series may be a popularity rating, a gross domestic product, industrial productivity index, or a level of sales. Gallup Poll approval ratings of how well the incumbent President is handling his job are tallied several times a month and provide ample examples of public appraisal of Presidential response to various events (The Gallup Poll, 1997). Gallup responses are coded as approve, disapprove, or no opinion. The percent approving the President's handling of his job is often used as a measure of public approval of his decisions, directives, actions, and policies.

The kind of impact that these interventions have on the public approval series may be statistically modeled. Some presidential responses to challenges have sudden and temporary impacts. For example, instantaneous but short-lived heightening of public approval followed President Ronald Reagan's April 1986 retaliatory bombing of Libya for a terrorist attack on a German disco patronized by American servicemen. In that terrorist attack, two American servicemen were killed. The American public immediately favored this kind of counterterrorist reprisal. The public felt that the response was appropriate given the nature of the problem, although news of unintended collateral damage may have attenuated residual approval. An immediate climb in President George Bush's presidential approval ratings followed the rallying of an alliance of nations and the waging of the Gulf War against Saddam Hussein after Iraq invaded the state of Kuwait. Bush's favorable ratings impact lasted longer than that from Reagan's retaliatory bombing, although Bush's popularity waned as the economy faltered. Other impacts are just as abrupt but more permanent in duration. The political fallout from the Watergate scandal resulted in an unprecedented decline in presidential approval ratings. President Richard Nixon's stature and popularity declined rapidly and failed to recover. Military strategists may study the impact of various wars on U.S. gross national product or the economic warfare waged by OPEC with its oil embargo and production cutback directed against Israel and her allies after the Yom Kippur War of 1973. An ecologist might study the decline in atmospheric pollutants following the passage of legislation requiring the installation of emission control devices on automobiles in an urban area, such as Los Angeles. A criminologist may wish to examine the effects of gun-control laws on the number of armed robbery arrests in a given area. A psychopharmacologist may wish to study the response time degradation on the perceptual speed of his patients that comes from their taking particular kinds of drugs, such as chlorpromazine. Another policy analyst studying the effect of seat-belt legislation on the number of automobile accident fatalities will be able to assess the impact on highway public safety. Impact analysis is useful for identifying and modeling the effect of events on a process under examination.

Some impacts are mere impulse responses in the series under examination. Not all impacts stem from exogenous events. Sometimes an outlying observation may represent an error in the value of an item of data input. If this outlier remains part of the series, it represents a mean deviation that could seriously bias autocorrelation and partial autocorrelation functions. If it is discovered later, the gross error in the data can be modeled as an observational outlier. If an outlier has a substantial impact that lasts throughout the series, then its presence may corrupt the ACF and PACF, thereby undermining specification of the model (Chang et al., 1988). Once an observation has been determined to be in error, it may be deemed missing. Missing values can be estimated by linear interpolation. We can then replace the missing value with the estimated value to prevent corruption of the modeling process. Outliers have basically two types of impacts, which will also be discussed in this chapter. Impact analysis permits the analyst to model the impact of an event or outlier, thereby describing and controlling for its effects.

8.2. ASSUMPTIONS OF THE EVENT INTERVENTION (IMPACT) MODEL

Impact analysis is predicated on particular assumptions. The system in which the input event and the impact response take place is assumed to be closed. Apart from the noise of the series itself, the only exogenous impact on the series is presumed to be that of the event or intervention. All other things are presumed to remain the same or to remain external to the system. Series are best analyzed when they are fairly stable and the intervention event alone precipitates the impact. If too many significant or important events affect the response series at about the same time, it may be difficult to partial out the effect of a particular intervention. Therefore, the system under observation should be one where the effects of a particular event under examination can be easily distinguished from others. This requirement often renders analysis of presidential approval a difficult one, because the presidential agenda includes many important decisions in a short span of time. Another assumption is that the temporal delimitations of the input event or phenomenon are presumed to be known. The time of onset, the duration, and the time of termination of the input event have to be identifiable. Because the presence or absence of an event is a deterministic rather than a stochastic phenomenon, the impact-generating events can be modeled by indicator variables, such as pulse or step indicators. Moreover, the ARIMA model (sometimes called the noise model), that describes the series before the intervention, has to be stable. The character of the noise model is assumed not to change after the beginning of the input event; this process is supposed to continue as before even after the inception of the intervention. The only apparent changes are assumed to stem from the impact of the event or intervention. Another assumption is that there are enough observations in the series before and after the onset of the event for the researcher to separately model the preintervention series and the postintervention series by whichever parameter estimation process he is using. For example, a researcher wishing to show the influence of war on annual gross domestic product would need to have yearly data dating back to the 1920s to have enough non-wartime data to model a comparison. From the emergence of the Cold War in 1947 until 1989, the

United States had been on some kind of wartime footing. During most of this time, at least one geographical part of the world has been entangled in a bloody and costly conflict of one kind or another. Committed to collective international, regional, and national security, the United States has been obliged to militarily support Third World allies against unfriendly forces in various parts of the world. Consequently, the United States has engaged in substantial tactical and strategic military research and development, along with weapons production and distribution, to confer upon itself and its allies security against military threat. In each of these areas, the United States has sought to maintain comfortable leads in the surface, bomber, naval, and missile, as well as computer, communication, command, control, and intelligence (C^4I), modernization races. Owing to this resource commitment and mobilization, the analyst would have to reach back into the 1920s for GDP levels during times of complete peace. This means that the intervention analysis of war on annual GDP would require a much longer series, the first segment of which would have to extend over portions of the 1920s, than an ordinary ARIMA model would need.

8.3. IMPACT ANALYSIS THEORY

The impact response model is formulated as a regression function. With the dependent variable representing the response series, the regression model contains independent variables consisting of an ARIMA noise model and an intervention function. More specifically, the response variable, Y_t, is a function of the preintervention ARIMA noise model plus the input function of the deterministic intervention indicator for each of the interventions being modeled:

$$Y_t = N_t + \sum_t \Sigma f(I_t), \tag{8.1}$$

where

$$N_t = \text{ARIMA preintervention model}$$
$$f(I_t) = \text{intervention function at time } t.$$

8.3.1. INTERVENTION INDICATORS

The $f(I_t)$ is a function of a deterministic (dummy) intervention indicator. The summation of these functions suggests that there can be more than one intervention and that all of the intervention indicators are ultimately included. The intervention indicator is an exogenous variable whose discrete coding represents the presence or absence of an input event. If the intervention function is a step function, then the value of $f(I_t)$ is 0 until the event begins at time T. At the onset of the event, the intervention function $f(I_t)$ is equal to 1. The intervention remains at 1 for the duration of the presence of the event, as can be seen in Eq. (8.2):

$$f(I_t) = S(t) \text{ when } S(t) = \begin{cases} 0 & \text{when } t < T \\ 1 & \text{when } t \geq T \end{cases} \tag{8.2}$$

$S(t)$ = step function.

If $f(I_t)$ is a pulse function, then a different condition obtains. Prior to the event, the intervention indicator is coded as zero. At the instance of onset, the intervention function is coded as one. It remains one for the duration of the presence of the event, which in the case of a conventional pulse is only one time period, or in the case of an extended pulse, the time period spanned by the duration of the event. A pulse function is shown in Eq. (8.3):

$$f(I_t) = P(t) \text{ when } P(t) = \begin{cases} 0 & \text{when } t \neq T \\ 1 & \text{when } t = T \end{cases}. \tag{8.3}$$

The step and conventional pulse functions are input variables. They are interrelated. Actually, the pulse function is merely a transformed step function; the pulse function is a differenced step function:

$$P(t) = (1 - L)S(t). \tag{8.4}$$

The coding for the indicator variables representing these input functions is given in Table 8.1. The coding of the intervention indicator can be used

Table 8.1

Indicator Coding for Pulse and Step Function Inputs

Pulse function		Step function	
Time (t)	$P(t)$	Time (t)	$S(t)$
1	0	1	0
2	0	2	0
3	0	3	0
4	0	4	0
5	0	5	0
6	1	6	1
7	0	7	1
8	0	8	1
9	0	9	1
10	0	10	1

to specify the presence or absence of a particular input phenomenon, such as one of those mentioned. In Table 8.1, note that the onset of the intervention input begins in time period 6 ($t = T$) and remains for only one period in the case of the pulse, but remains for the duration of the time periods in the case of the step function. The pulse function can also be used to code the transient presence of an observational outlier (McDowell *et al.*, 1990).

8.3.2. THE INTERVENTION (IMPULSE RESPONSE) FUNCTION

Changes in the level or shape of a series at the time of the impact of an input indicator are presumed to be responses to the intervention. For this reason, it is important to appreciate the general structure of an intervention function over time. One function will be considered at a time. The structure of the intervention function determines the shape of the impact over time on the series under consideration. The dependent series responds in a particular form because it is dependent on the intervention input. The response function is characterized according to whether it is basically one of a step or a pulse. A step function is generally formulated as

$$f(I_t) = S(t) = \frac{\omega(L)I_{t-b}}{1 - \delta_1 L} \tag{8.5}$$

where

$$\omega(L) = \omega_0 - \omega_1 L - \omega_2 L^2 - \cdots - \omega_s L^b,$$

whereas a simple intervention pulse function is formulated as

$$f(I_t) = P(t) = \frac{\omega_i(L)I_{t-b}(1 - L)}{1 - \delta_1 L}, \tag{8.6}$$

where

$$\omega_s(L) = \omega_0 - \omega_1 L - \omega_2 L^2 - \cdots - \omega_s L^s.$$

To define and explain the operation of the impulse response function $f(I_t)$, the components of the numerator of a simple step function, called a zero-order function, will be addressed first:

$$f(I_t) = S(t) = \omega_0 I_{t-b}. \tag{8.7}$$

8.3.3. THE SIMPLE STEP FUNCTION: ABRUPT ONSET, PERMANENT DURATION

Suppose the input event is properly represented by a dummy variable. A step function represents permanent change in a response. When the denominator of the step function in Eq. (8.5) reduces to unity, what remains is the simple step function in Eq. (8.7). The components of this step function include the regression coefficient of the intervention, ω_0, representing some gain or loss; the intervention indicator variable, I_{t-b}, which some call the change agent (McDowell *et al.*, 1980), coded as a 0 or 1 to indicate the absence or presence of the intervention; and the time delay involved for the impact to take effect, the subscript b. Suppose for the sake of simplicity, that the time delay is nonexistent—that is, $b = 0$. We can observe the nature of the step function by the change in level of the series at the time of onset of the event. A positive ω_0 would indicate a rise in the level of the series at the time of impact of the intervention. A negative ω_0 would indicate a drop in the level of the series at the time of impact of the intervention. The magnitude of the slope, ω_0, would indicate the size of rise or drop in the level. Therefore, one could compute, respectively, the magnitude of this regression effect and its variance by

$$\hat{\omega} = \frac{\sum_T Y_t I_t}{\sum_T Y_t^2}$$

$$(8.8)$$

$$\text{Var}(\hat{\omega}) = \frac{\sigma_e^2}{\sum_T Y_t^2}.$$

A distinction can be drawn between the time of incidence of the intervention and the time of impact on the response variable, Y_t. The b index in the $t - b$ subscript of the intervention indicator value represents the number of periods of delay between the instance of the intervention and the time of its impact on the response. If $b = 2$, there would be two periods of delay between the intervention and the time at which its impact is observed in the response variable Y_t. If the $b = 4$, then there would be a lapse of four time periods before the impact of the intervention was experienced by the response variable. If $b = 0$, the response to an input step function is abrupt and permanent.

To illustrate this relationship between the response and the intervention, a simplifying assumption that the noise model contributes nothing to the response Y_t is made. Assume for the sake of simplification that the value of the delay parameter $b = 0$ for this case. In this case, the value of $I_{t-b} = I_t$. The value of the response Y_t becomes dependent solely on the functional relationship between Y_t and I_t, in the case that $f(I_t)$ is a step intervention. Prior to the instance of the intervention, the value of t is $t < T$ and the value of I_t is 0, as can be seen in Fig. 8.1. At time $t = T$, the value of I_t becomes 1. At time $t > T$, the value of I_t remains 1.0. Let the regression coefficient, ω_0, be equal to 0.5, and if there is no delay, then the change in

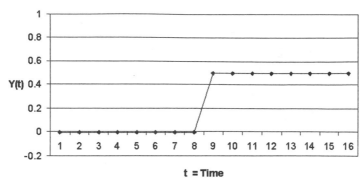

Figure 8.1   Zero-order step function.

the response Y_t can be seen to jump from 0.0 to 0.5 in one period of time. At the time of intervention, there will be an immediate increase in level of the response by a positive 0.5 value of Y_t. Of course, if the value of the regression coefficient were -1, then there would be a drop in the magnitude of the response by 0.5. The new level would be maintained for the duration of the series.

Suppose the value of the delay parameter were other than zero. If $b = 3$ and $\omega_0 = 0.50$, then there would be a lag of three time periods before the response function Y_t would be changed by a factor of the regression coefficient, multiplied by the value of I_t. Let the value of I_t suddenly jump from 0.0 to 1.0. The value of Y_t would abruptly be increased by a value of 0.50 three time periods later. If $b = 9$, then the delay before the change in Y_t would take place would be nine time periods. In this way, the delay parameter is subsumed as part of the step function. To the extent that the delay parameter is other than zero, the onset is less than instant. Nonetheless, when it kicks in, the onset is abrupt and permanent in this case.

8.3.4. First-Order Step Function: Gradual Onset, Permanent Duration

The first-order step function can be elaborated as a ratio that includes a numerator such as that already described, as well as a denominator that includes a single rate (decay) parameter (Makridakis *et al.*, 1983). The single rate parameter, δ_1, is part of the denominator of the first-order step response function. The order of the response function is equal to the number of rate parameters in the denominator:

$$S_t = f(I_t) = \frac{\omega_0 I_{t-b}}{1 - \delta_1 L}. \tag{8.9}$$

When the rate parameter $\delta_1 = 0$ or when the lag operator $L = 0$, the denominator reduces to unity and the formula reduces to that of Eq. (8.7). If the denominator reduces to unity, the onset of the impact is abrupt and instantaneous. If this rate parameter is greater than zero but less than unity and the input is a step function, there is a gradual increase in the level of the response until a permanent level is attained. The rate parameter controls the gradualness of the growth in the impact after onset. The lag operator, L, controls the lags at which this gradual increase is experienced. When there is one time lag in the function, the function is generally called a first-order impulse response function.

Consider the operation of the first-order response function to a step intervention. Suppose that the ARIMA noise model is pure white noise and is negligible. Assume also that the value of the series is zero prior to the intervention. Assume the value of the intervention indicator prior to the onset of intervention at time T is zero and also that the lag time before impact b equals zero as well. The general formula for this impact is

$$Y_t = \delta_1 Y_{t-1} + \omega_0 I_{t-b}. \tag{8.10}$$

At time $t < T$ this process yields

$$Y_t = \delta_1 \times 0 + \omega_0 \times 0 = 0. \tag{8.11}$$

When time $t = T$, the series jumps from zero to a level equal to the regression coefficient:

$$Y_t = \delta_1 \times 0 + \omega_0 \times 1 = \omega_0. \tag{8.12}$$

At time $t = T + 1$, the process adds an increment of the regression coefficient times the rate parameter to the equation:

$$Y_{T+1} = \delta_1 \times \omega_0 + \omega_0 \times 1 = \delta_1 \omega_0 + \omega_0. \tag{8.13}$$

At time $t = T + 2$, the process acts as though the rate parameter is an autoregression coefficient and adds a new level to the series. The added increment is equal to the previous series level times the rate coefficient:

$$\begin{aligned} Y_{T+2} &= \delta_1(\delta_1\omega_0 + \omega_0) + \omega_0 \times 1 \\ &= \delta_1^2\omega_0 + \delta_1\omega_0 + \omega_0. \end{aligned} \tag{8.14}$$

By induction, it can be seen that as long as the rate parameter, δ_1, is greater than 0 and less than 1, a smaller increment is added to the impact at each time period after the impact. The general function defining this process is

$$\begin{aligned} Y_{T+n} &= \delta_1(\delta_1^{n-1}\omega_0 + \delta_1^{n-2}\omega_0 + \cdots + \omega_0 \times 1) \\ &= \delta_1^n\omega_0 + \delta_1^{n-1}\omega_0 + \cdots + \omega_0. \end{aligned} \tag{8.15}$$

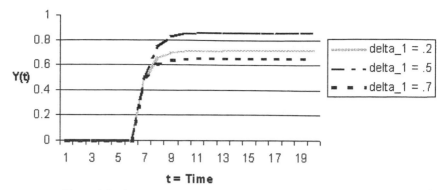

Figure 8.2   Step response functions for different decay parameters.

The shape of this impact is gradually increasing at a declining rate impact with permanent duration, which depends on the magnitude of the rate parameter, δ_1, as can be seen in Fig. 8.2.

In general, the shape of the first-order function may change as the magnitude of the rate parameter changes. The closer the value of the rate parameter, δ_1, is to zero, the more closely the response function approximates a sharp step upward from one time period to the following time period. The closer the value of the rate parameter, δ_1, is to unity, the more gradually and quickly the response approaches its upper limit. Three values of rate parameters are shown in Fig. 8.2 along with the corresponding response functions.

Let us consider how this process works. The assumption that the preintervention series is equal to zero is used for simplification. Also for simplification, assume that the delay parameter $b = 0$, the slope parameter $\omega_0 = 1$, and the rate parameter $\delta_1 = 0.5$. The impulse response function in this case equals

$$Y_t = \frac{\omega_0 I_t}{1 - L} = Y_{t-1} + \omega_0 I_t. \tag{8.16}$$

When time $t < T$, the intervention indicator equals 0 and the series resembles

$$Y_t = 0.5 \times 0 + \omega_0 \times 0 = 0. \tag{8.17}$$

At the time $t = T$,

$$Y_{t=T} = 0.5 \times 0 + \omega_0 \times 1 = \omega_0. \tag{8.18}$$

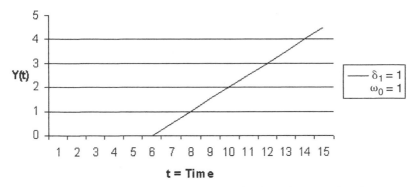

Figure 8.3 Step response function with delta_1 = 1 is a ramp function.

Then when $t = T + 1$,

$$Y_{t=T+1} = 0.5 \times \omega_0 + \omega_0 \times 1 = 1.5 \times \omega_0. \tag{8.19}$$

Then when $t = T + 2$,

$$Y_{t=T+2} = 0.5 \times 1.5\omega_0 + \omega_0 \times 1 = 1.75 \times \omega_0. \tag{8.20}$$

When $t = T + 3$,

$$Y_{t=T+3} = 0.5 \times 1.75\omega_0 + \omega_0 \times 1 = 1.875 \times \omega_0. \tag{8.21}$$

From these examples, the attenuation of slope as the series finds its new level is controlled by the rate parameter. When the rate parameter δ_1 is set to unity, this creates a constant linear trend that transforms the response into the ramp function shown in Fig. 8.3. Alternatively, if $\omega_0 < 0$ and $\delta_1 = 1$, then the ramp function is one that is declining instead of inclining. Figure 8.4 presents an example of a negative ramp function. Similarly, by allowing $\delta_1 < 1$, the ramp levels off as it decreases, as can be seen in Fig. 8.5.

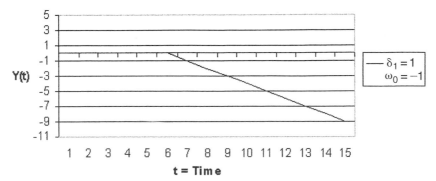

Figure 8.4 Negative ramp function with omega_0 < 0 and delta_1 = 1.

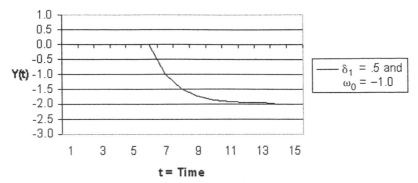

Figure 8.5 Scree response function with $\omega_0 = -1$ and $\delta_1 = 0.5$.

To recapitulate, a first-order step response, $Y_t = \omega_0 I_{t-b}/(1 - \delta_1)$, $= \omega_0(1 + \delta_1 L + \delta_1 L^2 + = \delta_1 L^n) I_{t-b}$, has a response series, a regression slope, a rate parameter, a delay parameter, and an intervention indicator. The Y_t is the response series. I_t is the intervention indicator, a dummy variable coded to indicate absence or presence of the intervention. The b in I_{t-b} is the time delay parameter, the number of time periods in the delay before the effect of the intervention is experienced. The regression level parameter is ω_0. Positive values of omega, ω_0, assure a positive level, whereas negative values of ω_0 provide for a negative level. The rate parameter in a first-order step function is δ_1. The rate parameter controls the decay or attenuation rate of the level of the response. A step function with $\delta_1 = 1$ has no leveling off of slope. With a rate parameter midway between one and zero, there is a gradual attenuation of slope. With $\delta_1 = 0$, there is an abrupt, vertical increase the size of ω_0, the regression coefficient, in slope of the series. If $\delta_1 < 0$, then there will be attenuating oscillation as long as $|\delta_1| < 1$. The latter condition is referred to as the lower boundary of system stability.

The analyst has to be wary of violations of the boundaries of stability. If a boundary of system stability is violated, the change does not converge to a limit. When $\delta_1 < 1$, the series oscillation, instead of attenuating, oscillates with increasing amplitude. When $\delta_1 > 1$, the series trends upward with increasing amplitude toward an infinite slope. Either violation produces an explosive or chaotic process that for practical purposes becomes unmanageable.

8.3.5. Abrupt Onset, Temporary Duration

When a time plot reveals a series level that suddenly shifts at the time of intervention but immediately returns to its previous level, the impact of the exogenous intervention can be modeled by a pulse response function. The pulse function has been formulated in Eq. (8.6). To understand the functional relationship between the pulsed input and the response, it is helpful to consider what is happening before, during, and after the intervention.

Some simplifying assumptions are made: that the value of the series prior to the time of intervention is zero, and that the delay time, b, is zero. At the time of intervention, $t = T$, the presence of the intervention, I_t, coded as one at this point in time, has an impact on the series. The magnitude of the sudden impact is measured by the regression coefficient, ω_0. In these terms, the pulsed impact can be represented by the elaborated formula in Eq. (8.22):

$$Y_t = \delta_1 Y_{t-1} + \omega_0 I_t (1 - L)$$
$$= \delta_1 Y_{t-1} + \omega_0 I_t - \omega_0 I_{t-1}. \tag{8.22}$$

When the step function is differenced, it becomes a pulse function. The elaboration of the pulse function can be seen in Eq. (8.22): The effect of this differencing is to subtract the lagged value of $\omega_0 I_{t-1}$, the last product on the right-hand side of the equation, from the $\omega_0 I_t$, the next-to-last product on the right-hand side.

This first-order pulse response function has a structure that endows it with a sudden peak and a more or less gradual return to its previous value. This response function can be represented by some delta parameter times the lagged response value plus some regression coefficient times the pulse function. Given the simplifying assumptions, when time $t < T$, the value of the intervention indicator, I_t, and its lagged value, I_{t-1}, are both zero and the value of the lagged series value, Y_{t-1}, is zero. Therefore, before the time of intervention, the value of the series, Y_t, is zero. In Table 8.2, the values of the parameters, shown at each point in time, facilitate understanding of the calculations. The coding of these parameters before the intervention may be seen in the first two rows of data in Table 8.2.

If we presume that the value of the ω_0 regression coefficient is 0.5 and that the value of the δ_1 parameter is 0.3, then the value of the series before the intervention can be computed from

$$Y_{t<T} = 0.3 \times 0 + 0.5 \times 0 - 0.5 \times 0 = 0. \tag{8.23}$$

When $t = T$, the time of the intervention, the value of the intervention indicator changes to unity. Its lagged value is still zero. But now the series value suddenly jumps to a peak of 0.5:

$$Y_{t=T} = 0.3 \times 0 + 0.5 \times 1 - 0.5 \times 0 = 0.5. \tag{8.24}$$

After the intervention time, the level of the series begins to diminish. At $t = T$, the lagged value of the intervention is still zero. But at $t = T + 1$,

Table 8.2

Pulse Response Function of Abrupt Onset, Temporary Duration:
$$Y_t = \delta_1 Y_{t-1} + \omega_0 I_t (1 - L)$$

Time	Y_t	δ_1	Y_{t-1}	ω_0	I_t	ω_0	I_{t-1}	$\omega_0 I_t(1 - L)$
$T - 2$	0.0000	0.3	0.0	0.5	0	0.5	0	0.00
$T - 1$	0.0000	0.3	0.0	0.5	0	0.5	0	0.00
T	0.5000	0.3	0.0	0.5	1	0.5	0	0.50
$T + 1$	0.1500	0.3	0.5	0.5	1	0.5	1	0.00
$T + 2$	0.0450	0.3	0.15	0.5	1	0.5	1	0.00
$T + 3$	0.0135	0.3	0.045	0.5	1	0.5	1	0.00

the lagged value of the series, Y_{t-1}, equals 0.5. The lagged value of the intervention indicator is now unity. However, the differencing takes effect. The differencing subtracts the regression effect at its previous value from that at its current value. What remains is the value of the rate parameter times the lagged value of the series. The net effect is that of a reduction in the value of the series, as can be seen in Eq. (8.25). The value of the series declines from 0.5 to 0.15:

$$Y_{t=T+1} = 0.3 \times 0.5 + 0.5 \times 1 - 0.5 \times 1 = 0.15. \tag{8.25}$$

Similarly, at the next time period, $t = T + 2$, the same process is at work. The pulse effect of subtracting the regressed lagged intervention indicator from its current value reduces this value further:

$$Y_{t=T+2} = 0.3 \times 0.15 + 0.5 \times 1 - 0.5 \times 1 = 0.045. \tag{8.26}$$

At this point in time, all that remains is the lagged value of the series of 0.15 multiplied by the rate parameter of 0.3 to yield 0.045, further diminishing the level of the series. After the passage of several periods of time, the level of the series declines to its previous value, a graph of which is shown in Fig. 8.6.

The sharpness of attenuation is controlled by the positive magnitude of the rate parameter. When the rate parameter $\delta_1 = 1$, there is no decay: The effect is that of a step function. When the value of the delta is less than but close to unity, there is slow decay. A value of 0.8 or 0.9 would mean very gradual attenuation of the level of the series as time passes. In contrast, a value of 0.1 or 0.2 yields a much steeper and rapid decay of the level of the series with the passage of time.

8.3.6. ABRUPT ONSET AND OSCILLATORY DECAY

When a pulse input function possesses a negative rate parameter, the series response function with a pulse input assumes a different shape,

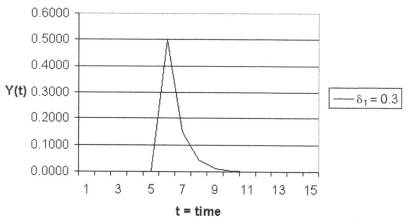

Figure 8.6 Pulse response function: abrupt onset, temporary duration; delta_1 = 0.3, omega_0 = 0.5.

$$Y_t = \omega_0/(1 + \delta_1)I_{t-b}(1 - L) = \delta_0(1 - \delta_1 L + \delta_1 L^2 - \cdots)I_{t-b}(1 - L).$$

Suppose there is a first-order decay process—with only a single small negative δ_1 rate parameter—then the response reaches a peak and then decays with oscillation. If the rate parameter has a value of -1, then unattenuated oscillation takes place. This is an example of a nonstationary process. The rate parameter may range from -1 to 0. The closer δ_1 is to -1, the more unattenuated the oscillation, whereas the closer the negative rate parameter is to zero, the more the decay in the oscillation. When the pulse input is used, the oscillation fluctuates around zero. When a step input is used with a negative pulse and a first-order decay, the oscillation fluctuates around the level of the first regression peak, ω_0. An example of a pulse input function with a first-order negative decay is given in Fig. 8.7.

8.3.7. GRADUATED ONSET AND GRADUAL DECAY

Sometimes the impulse response function appears to be one of gradual onset coupled with more or less gradual decay. The researcher can construct this compound response function by combining two step input functions. The gradual onset is produced by a second-order (two rate parameters in the denominator) step function, whereas the temporary duration can be produced by the subtraction of the third lag of the same function. Consider the following higher order response function equation:

$$Y_t = \frac{\omega_0 I_t - \omega_0 I_{t-3}}{(1 - \delta_1 L - \delta_2 L^2 - \delta_1 L^4 - \delta_2 L^5)} \tag{8.27}$$

$$= \delta_1 Y_{t-1} + \delta_2 Y_{t-2} + \delta_1 Y_{t-4} + \delta_1 Y_{t-5} + \omega_0 I_t - \omega_0 I_{t-3}.$$

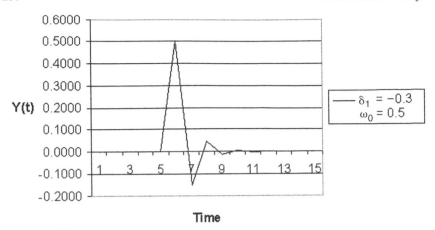

Figure 8.7 Pulse response function: abrupt onset, temporary duration; $\delta_1 = -0.3$, $\omega_0 = 0.5$.

When we expand this equation, with the same simplifying assumptions used before, over the time line before and after intervention, in Table 8.3, we display the calculated impact response in Fig. 8.8. If we eliminated from the denominator, the third and fourth delta parameters, we would have a simpler function. With two delta parameters, we would have a second-order response function. Second-order response functions can be used to introduce varying or undulating decay rates. The roots of the delta polynomial, $(1 - \delta_1 L - \delta_2 L^2)$, control the extent of dampening of the decay rate. In the second-order case, complex roots underdampen and yield undulation, real roots dampen and therefore attenuate decay, whereas real and equal roots critically dampen the decay rate (Box *et al.*, 1994).

We can also combine the same response functions with different parameters to generate compound response functions. By conjoining different response functions with various parameters, we can construct a variety of compound response functions. Readers interested in compound response functions may refer to Mills, 1990 and to Box *et al.*, 1994.

8.4. SIGNIFICANCE TESTS FOR IMPULSE RESPONSE FUNCTIONS

Most statistical packages employ standard T tests, in which the parameter is divided by its asymptotic standard error, to test for the statistical significance of the parameters. Both SAS and SPSS employ these tests to evaluate the parameter significance. Although statistical packages usually

Table 8.3

Data for Compound Response Function with Graduated Onset and Gradual Decline Response

$$(Y_t = \delta_1 Y_{t-1} + \delta_2 Y_{t-2} - \delta_1 Y_{t-4} - \delta_1 Y_{t-5} + \omega_0 I_t - \omega_0 I_{t-3})$$

Time	δ_1	δ_2	ω_0	I_t	Y_{t-1}	$\delta_1 Y_{t-1}$	$\delta_2 Y_{t-2}$	$\omega_0 I_t$	Y_{1t}	Y_{2t}	$Y_t = Y_{1t} - Y_{2t}$
1	0.5	0.3	3	0	0.00	0.00	0.00	0	0.00	0.00	0.00
2	0.5	0.3	3	0	0.00	0.00	0.00	0	0.00	0.00	0.00
3	0.5	0.3	3	0	0.00	0.00	0.00	0	0.00	0.00	0.00
4	0.5	0.3	3	0	0.00	0.00	0.00	0	0.00	0.00	0.00
5	0.5	0.3	3	0	0.00	0.00	0.00	0	0.00	0.00	0.00
6	0.5	0.3	3	1	0.00	0.00	0.00	3	3.00	0.00	3.00
7	0.5	0.3	3	1	3.00	1.50	0.00	3	4.50	0.00	4.50
8	0.5	0.3	3	1	4.50	2.25	0.00	3	5.25	0.00	5.25
9	0.5	0.3	3	1	5.25	2.63	0.90	3	6.53	3.00	3.53
10	0.5	0.3	3	1	6.53	3.26	1.35	3	7.61	4.50	3.11
11	0.5	0.3	3	1	7.61	3.81	1.58	3	8.38	5.25	3.13
12	0.5	0.3	3	1	8.38	4.19	1.96	3	9.15	6.53	2.62
13	0.5	0.3	3	1	9.15	4.57	2.28	3	9.86	7.61	2.25
14	0.5	0.3	3	1	9.86	4.93	2.51	3	10.44	8.38	2.06
15	0.5	0.3	3	1	10.44	5.22	2.74	3	10.97	9.15	1.82
16	0.5	0.3	3	1	10.97	5.48	2.96	3	11.44	9.86	1.58
17	0.5	0.3	3	1	11.44	5.72	3.13	3	11.85	10.44	1.41
18	0.5	0.3	3	1	11.85	5.93	3.29	3	12.22	10.97	1.25

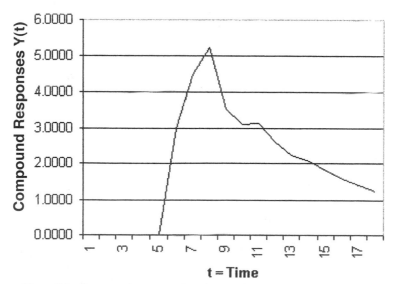

Figure 8.8 Compound response function: graduated onset, gradual decay.

employ such T tests for assessment of the statistical significance of the components for the response functions, T tests may not be the most appropriate for the analysis of impacts. Box and Tiao (1975, 1978) point out that the dynamic characteristics of the intervention and the serial correlation in the series bias such significance tests. Likelihood ratio tests have been suggested for use instead. Suppose a pulse response function represents an observational (additive) or innovational (with more lasting effect) outlier. Chang et al. (1988) and Box et al. (1994) suggest that the significance of the function may be found by the following formulas.

A significance test for an innovational outlier is

$$\lambda_{I,T} = \frac{\hat{\omega}_{I,t}}{\sigma_e}$$

and a significance test for an observational (additive) outlier is

$$\lambda_{O,T} = \frac{\tau \hat{\omega}_{O,t}}{\sigma_e}, \tag{8.28}$$

where

$$\hat{\omega} = \text{regression coefficient}$$

$$\sigma_e = \text{asymptotic standard error}$$

$$\tau = \sqrt{\sum_{i=0}^{n-T} \pi_1^2}$$

$$\pi_i = \frac{\theta_i(L)}{\varphi_i(L)}.$$

A more detailed theoretical derivation of the likelihood ratio tests may be found in Box and Tiao (1975). We can test the significance of other impulse response functions by such likelihood ratio tests. Yet likelihood ratio tests usually require large sample sizes for proper assessment, and sufficiently large sample sizes may not always be available.

8.5. MODELING STRATEGIES FOR IMPACT ANALYSIS

There are two basic modeling strategies used in intervention analysis. In the preferred and conventional approach, the preintervention series is modeled first, and the impact is modeled afterward. In the alternative approach, the modeling of the impact is undertaken on the whole series before the modeling of the residual noise model. We will address the former strategy first.

8.5.1. THE BOX–JENKINS–TIAO STRATEGY

With both strategies for intervention analysis, the analyst reviews the literature concerning the preintervention series, the intervention input, and the observed impact of the input on the post intervention series. He must examine the timing of the onset, duration, and termination of the input event under examination so he can distinguish real from spurious impacts. With the conventional strategy, the analyst then divides the sample into two segments, the preintervention series and the postintervention series.

With the conventional strategy, the analyst should be sure there are enough observations inthe preintervention model for separate ARIMA modeling. Then he graphs the preintervention series, examines it, and checks it for outliers. If there are outliers in the preintervention series and if there are enough observations prior to the incidence of the outliers, they may be identified, deemed missing values, and replaced by means of adjacent observations or by linear interpolations. After initial replacement of outliers, the analyst should recheck the series for outliers; if they exist, they should be replaced. This process should be reiterated until all outliers are smoothed out.

With an ARIMA model building protocol, the researcher transforms to stationarity, identifies, estimates, and diagnoses the ARIMA noise model for the preintervention series. Alternative noise models are estimated. After he finds the residuals of the preintervention models to be white noise, the researcher may compare models of the preintervention series. He metadiagnoses alternative models and selects the optimal noise model. An assumption is made that this noise model remains stable throughout the analysis and that any change in the process follows from the impact of the event or input.

A review of the source and nature of the intervention follows the modeling of the preintervention series. The researcher assesses the source of the exogenous input and determines whether the impact stems from the step or pulse input process. Then he codes the source of the input as a dummy variable to represent the presence or absence of the exogenous event. A graph of the series is plotted. The researcher must be careful to distinguish between a pulse, whether singular or seasonal, event input and an outlier. If there is a distinct outlier, he needs to deal with the outlier separately from the impact of the event. The timing of the input event is usually known and can often be distinguished from extraneous events with unknown timing. If the impact of the input event suddenly appears and suddenly disappears, it may be represented by a pulse function. If the source of the pulse is an observational outlier whose timing was not previously known, it may also be smoothed out or coded as a separate pulse. If there are plenty of observations prior to the appearance of an observational outlier, then the preoutlier ARIMA noise model up to the impact of that outlier is estimated. The detected observational outlier may be modeled as a pulse input or smoothed out. If the outlier has an innovational, that is, more or less lasting, effect on the series, it needs to be modeled as an impulse response function of a pulse effect plus noise. (Box *et al.*, 1994). The researcher should use likelihood ratio tests to detect the significance of a possible outlier or pulse function. If the outliers are ignored and left in the series, they may seriously bias the ACF and PACF of the series (Mills, 1990). Such biases may impair specification of the noise or transfer function model. For these reasons, detection, identification, and modeling of outliers may be important (Chang *et al.*, 1988; Mills, 1990) and may have to be based on a scholarly study of the situation. Even though the timing of outliers is generally unknown, they may be modeled to preclude biasing specification of the model.

Once the outliers have been modeled or replaced, the quantity of the other inputs should be addressed. There may be no impact, one impact, or multiple impacts. If there are multiple inputs, they may be separated from the primary input by enough time and data to allow them to be modeled as well. The analyst should test alternative explanatory inputs of phenomena. He needs to examine the nature of each of these alternative explanatory inputs, particularly their duration. If the event abruptly occurs and abruptly disappears, the event may be modeled as a pulse function. If the event suddenly occurs and remains for a short duration, it can be modeled as an extended pulse function. (Extended pulses are coded as unity as long as the event is present and zero at all other times). If the event appears and remains, it can be coded as a step function. The analyst can assess the fit of the model after formulation and inclusion of the input. In these ways, he can model the nature of the deterministic input.

There are several ways to assess the nature of each impact (Vandaele, 1983; Mills, 1990). First, the researcher should review the literature to gain an idea what kind of impact to expect. He can formulate a null hypothesis of no impact and a research hypothesis based on the literature and theory. In his first assessment of impact, he can test this hypothesis with observation of the postintervention response. Second, the researcher should check to see whether the noise model remains stable over the whole series. This entails an iterative process of checking to see whether the ARIMA noise model parameters remain significant both before and after the intervention impact, and whether their sign and magnitude remain stable as well. The ARIMA noise is modeled separately before and after the intervention. If the noise model parameter values remain stable before and after the intervention, the noise model appears to be stable. If the noise model is stable across the time span of the whole series, then an ARIMA model for the noise can be reliably formulated on the basis of the preintervention series. If the system is fairly isolated from other impacts, then the residuals after the onset of intervention should come from the impact of the intervention alone. The researcher should test the response series for isolation from other possible impacts by defining input variables for plausible alternative inputs and confirming they have no significant input by testing for significance and reviewing the residuals.

The researcher may begin testing his research hypothesis by modeling the impact. The first impact is indicated by the regression coefficient, ω_0, at time lag b. Subsequent impacts can be formulated as a ratio of the numerator to the denominator parameters: $\Sigma\omega_i(L)/(1 - \Sigma\delta_i(L))$. Depending upon whether the intervention indicator is a pulse or a step function, the impact should assume the shape indicated in the graph of the series at and after the impact of the intervention on the response series. The analyst tries to identify the impact model from the change in the series following the intervention. To do so, he focuses on several aspects of the impact. He considers the nature of onset and duration of the change in the postintervention series. He may focus on a change in mean level, a change in slope, or even a change in variance of the series. He notes whether this change at the onset of the intervention is sudden or gradual. He also examines the duration of the change for transience or permanence. He checks to see whether the postintervention process levels off, oscillates, or decays. From these aspects of the shape of the impact, he guides the construction of the impulse response function. Whether the onset is abrupt or gradual and whether the change is transitory or permanent determines the nature of the response. The change in shape of the output series will determine what kind of response function the analyst endeavors to model.

8.5.2. FULL SERIES MODELING STRATEGY

The analyst may have reason to try an alternative strategy for intervention analysis. When he graphically examines the series before, during, and after the event or intervention, he might encounter one of three situations that make an alternative modeling strategy preferable. First, the series may not be long enough to be segmented into pre- and postintervention segments. The researcher might decide that circumstances require modeling the impact first and the noise last. Second, the impact of the intervention might appear to have an overwhelming influence on the level or slope or variance of the series, making it reasonable to model the impact first and the residual noise later. Third, under other circumstances, if the salient shape of the impact, as seen from the time plot, is found to be transient, the whole series may be used as a basis of assessment of the nature of the impact (McCain and McCleary, 1979).

An analyst might try to model the intervention by reviewing the cross-correlation function (CCF) between the deterministic input indicator and the response series. The cross-correlation function is similar to the autocorrelation function except that it is computed as a correlation between the input variable and output series. The cross-correlation function is asymmetric: Significant positive spikes indicate that the input variable variations lead the corresponding variations in the output variable, and significant negative spikes indicate possible feedback from the output to the input variable. The delay in the response will be apparent in the cross-correlation. The cross-correlation function depends on the inverse filtering out (a process known as prewhitening, which will be discussed in detail in the next chapter) of the autocorrelation of the input series to preclude contamination of the cross-correlation between the input and output series. In the case of intervention models, the input is deterministic, coded as a dummy variable, and is not prewhitened. For these reasons, the shape of the impulse response weights may not be proportional to the cross-correlations. Moreover, negative spikes on an intervention that has not yet taken place make no sense and may well be ignored. Even though the cross-correlation function may be indicated in the SAS programming syntax, the shape of the postintervention pattern in the graphical time plot is really the theoretic and empirical basis for the identification of the impact (Brockelbank and Dickey, 1986; Box *et al.,* 1994; Woodfield, 1987; Woodward, 1998). Nonetheless, the lag time between incidence of the event and impact is easily identifiable from a cross-correlation plot.

The analyst should generally seek to model impacts from external interventions. He should not arbitrarily use pulse functions to control for random irregular residuals (Vandaele, 1983). To test the intervention hypothesis, he must examine the approximate T test statistics for the parameters hypothesized. If any parameter T values are less than 1.96 and nonsignificant, he needs to try to eliminate the parameters. If any parameters are significant, he may retain those parameters in the model. He needs to examine the residuals for white noise to see whether he has modeled all of the significant variation.

A step or pulse can be modeled easily, merely by including the input indicator coupled with the proper time delay parameter. If the size of the spikes in the time plot becomes pronounced at the second lag after impact, then the delay time should be two periods. If the spike is at the third lag, then the dead or delay time should be three time periods. A pulse response function would be represented by an instantaneous spike in the response series, whereas a step response function would be represented by an abrupt and permanent change in the response series.

If the change in the series appears delayed but sharp and temporary, then the pulse function may be coupled with a delay parameter as in Eq. (8.22) to model this process. When the response polynomials have been modeled so that there is a noise as well as a response function, the residuals should be diagnosed, checked, and refined so that the residuals are ultimately white noise. Ultimately, the researcher will have to metadiagnose the alternative models. The residuals will have to be checked for white noise. Different estimation techniques may have to be tried. The parsimony of the model will have to be compared with minimum information criteria, such as the AIC and SBC. The model with the smallest SBC will be deemed the optimal one.

In sum, impact analysis modeling strategy involves several steps. If there are enough observations in the pre- and postintervention data sets for separate modeling, the preintervention ARIMA noise model is undertaken first. The transformation and differencing of the series is first performed to effect stationarity. Unit-root tests—for example, the augmented Dickey–Fuller test—may be used to determine whether stationarity has been attained. An ARIMA noise model is developed with the assistance of the ACF and PACF. If an ARMA noise model is formulated, the corner method or EACF may be used to find the optimal order. The model is identified with the help of the ACF and PACF; then the model is estimated with conditional least squares, unconditional least squares, or maximum likelihood. Diagnosis of the model involves review of the residuals, checked against the portmanteau or modified portmanteau Q statistic, which indicate whether this ARIMA noise model is adequate. Metadiagnosis facilitates identifying the optimal ARIMA noise model.

Once the preintervention ARIMA noise model is formulated, the impact on the postintervention series may be modeled. If there the impact on the response series is very large in relation to other variation in the series, or if there are not enough observations in the overall series for separate modeling of the pre- and postintervention data sets, then the impact of the intervention on the series is modeled first. Alternatively, the postintervention series is reviewed and the impact is modeled. Identification of the transfer function is based on what is known and observed about the impact of the intervention. The time sequence plot shows the change from preintervention to postintervention. Modeling the impact involves observing the changes in mean level, slope, or variance of the series at particular time lags after intervention, which indicate the delay time for the impacts. The analyst will examine the onset and duration of the response. Whether the onset is abrupt or gradual and whether the duration is constant or temporary will determine the type of parameters tested. Sudden and constant changes are attributed to step functions. Sudden and instantaneous changes are modeled with pulse functions. Gradual and permanent increases may be modeled with step functions with first-order decay rates. Sudden and decaying responses are modeled with pulsed functions with first-order decays.

Gradual onset and gradually decaying responses may be modeled with compound functions. Estimation of the impact parameters should reflect what is known and observed about the impact of the intervention. The impact model should be as theory-driven as possible. Diagnosis of the impact model includes hypothesis testing about the impact parameters and entails trimming the impact model of nonsignificant effects as well as retaining theoretically and statistically significant effects. The likelihood ratio or T statistics will indicate which ARIMA and intervention parameters should be retained. When the residuals are white noise, then the adequacy of the model of the intervention model will be established. Alternative models should be compared according to their explanatory power, explanatory appeal, parsimony, AIC, or SBC. All other things being equal, the model with the lowest information criterion should be the optimal model. This model programming strategy holds for modeling the impact of events as well as outliers.

8.6. PROGRAMMING IMPACT ANALYSIS

Basic intervention analysis is possible with both SAS and SPSS. At the time of this writing, the SAS Econometric Time Series module has much more power and flexibility than does the SPSS Trends module for handling complicated impact analysis. Both SAS and SPSS permit the researcher, of course, to code either pulse or dummy input variables. Both SAS and SPSS permit the differencing of these variables. Both permit the inclusion of multiple discrete deterministic input variables into an ARIMA model. Therefore, both permit the modeling of simple step and pulse input functions as independent variable in a multiple time series model. For the simplest of intervention models, either SPSS or SAS does very well.

In the simplest of intervention models, SPSS syntax utilizes a dummy (step or pulse) intervention indicator, called X. The ARIMA procedure syntax merely models the ARIMA process on the data before the intervention. Then the whole data set is included. The intervention indicator is added to the ARIMA syntax by the "with" option. Remember, SPSS command syntax has to begin in the first column of the syntax window. For example:

```
ARIMA Response with X/
  Model=(0,1,1)(0,1,1)12 constant/
  /MXITER 10
  /PAREPS .001
  /SSQPCT .001
  /FORECAST EXACT
```

If the residuals are white noise, the model will have been fit, but reality is often not so kind to the analyst. It often presents much more challenging problems that require a more sophisticated modeling of the impulse response. If a more complex form of the intervention is desired, the SPSS coding would have to be approximated manually by the investigator or set up with the aid of some more sophisticated "compute" statements with which SPSS can construct new variables.

Of the two packages, SAS permits the automatic modeling of an impulse response function with a ratio of a numerator function of ω parameters to a denominator function of δ parameters. SPSS has developed two modules called DecisionTime and WhatIf? to allow automatic modeling. Because SAS permits custom design of the response function, SAS is used for parameter estimation of the identified response function parameters—including the ω_i of the numerator and δ_i values of the denominator—and is strongly preferred at present for pedagogical applications of intervention analysis.

Although both statistical packages have developed menu driven procedures that provide black-box automatic modeling of the impulse response function, these procedures have little pedagogical utility and are not covered here. The SAS Time Series Forecasting system and the SPSS Decision Time and What If modules endeavor to mechanically arrive at a model. The SAS ETS system provides for more flexible custom design of the impulse response function. For this reason, the design of the impulse response function with this package is explained and applied here.

Programming the impulse response function with SAS is simple. In the identification subcommand, the response variable is identified, differenced with parentheses around the order of differencing, and centered with the CENTER option; then the input variable X is cross-correlated with the CROSSCORR option. Centering is usually preferred with intervention analysis because this simplification facilitates focus on the deviations from the mean after intervention. Both RESPONSE and input series, X, conventionally receive the same differencing. An example of this subcommand in the ARIMA procedure is

```
PROC ARIMA;
IDENTIFY VAR=Response(1) CENTER CROSSCORR=X(1);
```

The ESTIMATE subcommand follows the IDENTIFY subcommand in the syntax sequence. The ARIMA noise model parameters are estimated, so that if the model is an ARMA(1,1) model, the first portion of the ESTIMATE subcommand would have the P=1 and Q=1 options noted. The PRINTALL and PLOT options follow. If maximum likelihood estimation is requested, then METHOD=ML MAXIT=40 would follow.

Then the INPUT option specifying the impulse response function is utilized within the same ESTIMATE subcommand. Suppose that the impulse response function is being modeled as

$$\text{Response}_t = \frac{(\omega_0 - \omega_1 L - \omega_2 L^2 - \omega_4 L^4)}{1 - \delta_1 L - \delta_2 L^2} X_{t-3}. \qquad (8.29)$$

The ESTIMATE subcommand right under the IDENTIFY subcommand within the same ARIMA procedure would be

```
ESTIMATE P=1 Q=1 PRINTALL PLOT METHOD=ML MAXIT=40
INPUT=3$(1 2 4)/(1 2) X;
```

The 3$ indicates the time delay between presence of the intervention and impact. There are three time periods of dead time or delay before impact in this case. The (1 2 4) numerator indicates the lags of omega parameters being estimated after the initial ω_0, while the denominator (1 2) terms indicate the lags of delta parameters being estimated. This ratio of polynomials is multiplied by the X intervention indicator. With this syntax, SAS can estimate the parameters of the impulse response function.

SAS offers flexibility and variety in its ability to model impact analysis. Most forms of impact may be modeled with this software. X may be either a step variable, so that X=S1, or a pulse variable, so that X=P1. (S1 and P1 are pedagogically used to indicate previous step and pulse constructions of the X variable, although proper specification of the input variable at this stage of the computer program is an X.) An abrupt and permanent impact may be modeled with an INPUT=(S1) option. An abrupt yet temporary form of impact may be modeled with the inclusion of a first-order rate parameter, δ_1, in the denominator with an INPUT=(/(1)P1) option. A gradual and permanent form may be obtained with the model INPUT = ((1)/(1)S1) specification. In each of the foregoing cases, the first-order rate parameter in the denominator indicated by (/(1)) has an estimate in the output less than 1.0 to prevent unattenuated oscillation. Another gradual (or graduated) yet permanent kind of impact may be programmed with INPUT = ((1 1 1 1 1)S1). An oscillatory and permanent type of impact can be obtained with ((1)/(-1)S1). In these ways, various types of impact can be identified with SAS (Leonard, 1998; Woodfield, 1987; Woodward, 1996–1998).

Once the ARIMA noise model is combined with this impact analysis, the parameter estimates are given along with their T tests. Then a residuals analysis is permitted with the ACF and PACF of the residuals, which, if the model fits, should appear to be white noise.

8.6.1. A EXAMPLE OF SPSS IMPACT ANALYSIS SYNTAX

SPSS ARIMA provides for including event or intervention input variables in a combination regression–ARIMA model, where the input variables are dummy variables. In the field of American electoral studies, the prominent controversies pertains to the proper classification of elections. In 1955, V. O. Key, Jr. proposed a "Theory of Critical Elections," according to which, there are critical elections in which the realignment of the voting patterns of the electorate is both abrupt and protracted. The realignment in the configuration of interest groups, pressure groups, political organizations, and controversies causes the dominant political party identification, affiliation, and support to shift from one party to the other. It alters the basis of political competition and controversies for years to come and serves as a basis for classification of periods in American electoral history (Niemi and Weisberg, 1976).

Various scholars seek to define periods in United States history. They endeavor to classify elections to determine the delineation of these periods. Campbell *et al.* (1960) in a classical study of *The American Voter* refined the classification of elections. They suggested that elections be classified into maintaining, deviating, and realigning types. In a maintaining election, the basic pattern of party loyalty is continued. In the deviating election, the minority party wins temporarily owing to a temporary defection of voters from the majority party. In the realigning election, the electorate is transformed into a new configuration of party loyalty.

Gerald Pomper (1972) developed a two dimensional typology for the classification of presidential elections. One dimension represents the continuity or change in the electorate, whereas the other dimension represents the victory or defeat of the majority party. Where the electoral cleavages do not change, the maintaining election is the one where the majority party wins and the deviating election is the one where the majority party is defeated. Where the electoral cleavages are transformed, the converting election is one where the majority party gains more electoral support to win and the realigning election is one where the majority party is defeated by a shift in characteristic voting patterns.

The proper classification of elections has implications for the periodization of the electoral history. If periods of electoral history are characterized by stable partisan attachments, then periods are delineated by critically realigning elections. Neimi and Wiesberg (1976) have suggested that one period of electoral history extends from the Civil War and covers the reconstruction era. This period ends around 1894 or 1896. The next period they call the populist or Bryan era, which begins somewhere between 1892 and 1896. This period extends through the progressive era in the early 1900s to the time of the Great Depression 1929. The next critical election was that of 1932, which ushered in the New Deal. The next period extended from the New Deal until the early 1990s (Heffernan, 1991). It can be argued that after the 1980 realignment, the next critical election took place in 1994. However, that matter may not be resolved till after the 2000 election when observers can see whether the current partisan attachments are maintained or revert to those of the pre-1994 era. Controversies over the proper periodization continue to this day.

There has been a debate over whether the critical election that ushered in the era of populism and Bryanism was that of 1894 or 1896. To define the nature of the competition characterizing the political milieu, it is necessary to know when the period began. Avery Leiserson (1958), Jensen (1970), Neimi and Weisberg (1976), and Gerald Pomper (1976), among others, suggest that 1896 witnessed a movement of voters from one party to the other. Burnham (1982) writes of the "system of 1896" in implying that the critical realignment stemmed from this time. Other scholars, such as Burnham (1970), Kleppner (1970), Heffernan (1991), and Maisel (1999) have suggested that the critical election was held in 1894. Whether the period began in 1894 or 1896 can be determined by ascertaining whether the election of 1894 was a deviating election with a temporary defection of voters or a critical election with a more or less stable realignment of them. This period extends to the time of the Great Depression in 1929/32. With the proper periodization, we are better able to understanding American political history.

The political debates during these campaigns manifest represent the political interests. Although Kleppner (1970) admits that the economic depression realigned the electorate for the election of 1894, he gives more emphasis to the social and religious interests during the 1896 campaign. The Panic and depression of 1893, like the Great Depression of 1929, threatened the livelihood of many people and thus realigned the interests and political loyalties of the political electorate. The Democratic party, the party of easy (free silver) money, pietism, and personal liberty, gained massive voter support from the impact of the depression. The Republican party of reform and hard (gold standard) money lost substantial support.

After the election of 1894 but before the election of 1896, partisan alignment shifted. Although the easy money Democrats advocated a progressive income tax, free silver, tariff reduction, more railroad and trust regulation, and opposition to the gold standard in the 1896 campaign, The Republicans advocated maintaining the gold standard, sound money, economic recovery, employment, and prosperity. The Republican position was helped by the improvement of the economic situation and discovery of Gold in South Africa, providing for easier money, and economic recovery (Boller, 1984). Kleppner maintains that the real political configuration was based on religious and not economic values. In 1896, Catholic and Lutheran voters defected from the Democrats to support McKinley. Jensen (1970) suggests that McKinley introduced pluralism to American politics. Burnham (1982) notes that there was a massive mobilization of new immigrants. Once the depression effects abated, those predispositions regained control

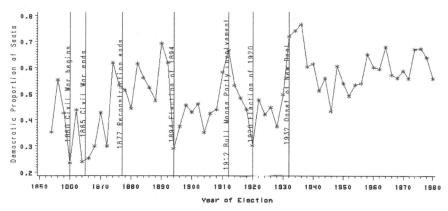

Figure 8.9 Graphical time plot of 1973 oil embargo/price rise: impact on U.S. petroleum product imports in millions of dollars. Data courtesy of U.S. Department of Commerce, Stat-USA Web site.

with a net movement toward a more broad-based, less pietistic, more prosperous McKinley Republicanism.

Whether the election of 1894 was a deviating or critical election can be tested by intervention analysis (Heffernan, 1991). The research question is whether the 1894 is a deviating or critically realigning election. Electoral party loyalty is measured by the (mean centered) percent of Democratic seats in the United States House of Representatives. The data come from the World Wide Web Site of the Clerk of the House of Representatives. The null hypothesis is that the 1894 election is a maintaining election and not a significant deviation from the status quo. Review of the time plot suggests that there is a deviation in the democratic percent of the seats in the House of Representatives, but the location of the observation remains slightly within the confidence limits of the individual forecasts. The two research hypotheses are that the 1894 is either a deviating (instantaneous pulse) or a critically realigning (extended pulse) election. If 1894 election can be better modeled as an instantaneous pulse, it can be construed as a deviating election. If that election can be better modeled as an extended pulse function, this election could be interpreted as a critically realigning election over the long run (Heffernan, 1991). The better fitting model should determine the interpretation.

An analyst seeking to determine whether the election of 1894 had no impact, an instantaneous impact, or a sustained impact on the percentage of seats held by members of the Democratic Party in the United States House of Representatives would first formulate a model of that Democratic percentage prior to the 1984 election. The researcher, using the centered variable named CDEMPROP, identifies, estimates, and diagnoses an ARIMA noise model prior to this controversial event. In this model, the series is mean-centered prior to analysis. Therefore, the model includes no constant. The ARIMA(1,0,0) of CDEMPROP with no constant is the preintervention model that leaves white noise residuals with these data and the SPSS syntax in C8PGM1.SPS:

```
* ARIMA.
TSET PRINT=DEFAULT CIN=95 NEWVAR=ALL .
PREDICT THRU END.
ARIMA cdemprop
 /MODEL=( 1 0 0 )NOCONSTANT
 /MXITER 10
 /PAREPS .001
 /SSQPCT .001
 /FORECAST EXACT .
ACF
 VARIABLES= err_1
 /NOLOG
 /MXAUTO 16
 /SERROR=IND
 /PACF.
```

To model the impact of intervention of the election of 1894, the series is expanded to include the pre- and postintervention (centered) election Democratic percentages of U.S. lower house Congressional seats, and the variable indicating the presence of the election of 1894 would be added to the first line of the SPSS ARIMA program. Several models are tested. In one model, the election of 1894 is defined as an instantaneous pulse, and in another model, the election of 1894 is defined as an extended pulse that lasts till 1914. In alternative tests of these models, the election of 1912 is controlled for by introducing a Bull Moose Party variable. This variable is defined as a pulse in 1912 to control for any voting changes induced by reconfiguration of interests, debates, and pressures stemming from that campaign. Alternative models use the election of 1896 instead of the election of 1894 as the event generating the impact. An 1896 pulse and an 1896 extending through 1930 are tested here. The better fitting model is the one selected as the basis for determining whether the election of 1894 was a deviating or critically realigning election.

A Bull Moose variable is introduced to control for the reconfiguration of party attachment during the election of 1912. The Democrats nominated a liberal governor, Woodrow Wilson. The regular Republicans nominated a prominent conservative, William Howard Taft. The progressive Republicans formed the Bull Moose Party and nominated Teddy Roosevelt. Boller

(1984) writes that the Progressive platform called for better factory working conditions, agricultural aid, women's suffrage, democratic election of Senators, a federal income tax, natural resource conservation, federal tariff and trust regulation, along with other proposals for popular election and social justice. Wilson, with the help of attorney Louis Brandeis, formulated a platform called "The New Freedom," which de-emphasized the regulation advocated by Roosevelt. The energetic and popular candidates, Roosevelt and Wilson, began to dominate the debate, as genial Taft slipped from salience. The relevant issue agenda included different versions of progressive or liberal positions. As the ideology of conservatism waned and the split within the Republican party undercut support for Roosevelt, Wilson won the election. The voter defection during this campaign is best controlled for by a Bull Moose election variable for the year of 1812.

Because there are other events or interventions that can confound the effect of the election of 1894, they are included in the model to control for their potentially confounding effects. The effect of the progressive realignment in 1912 is controlled for by inclusion of an instantaneous pulse dummy indicator for 1912 called BULLMOOS, whereas the effect of the New Deal realignment is controlled for by an extended pulse dummy indicator for 1932 through 1980 called NEWDEAL in the postintervention segment of the program. This program segment tests the impact of the election of 1894 while controlling for effects of these potentially confounding realignments.

Inclusion of the NEWDEAL variable is needed as a control for the 1932 critical realignment. The campaign of 1932 revolved around the issue of the Great Depression and the economic and financial actual and threatened devastation it wrought on America. Widespread asset depreciation, property loss, unemployment, poverty, and evictions created a climate of desperation. These economic ills were associated with Herbert Hoover. His attempts to launch public works did not engender recovery. Desperate, discontented veterans camped out in Washington in protest of their plight and in beseech of relief (Amendola, 1999). When Franklin D. Roosevelt promised farm aid, public development of hydroelectric power, a balanced budget, and regulation of large scale corporate power, he seemed to offer a way out of this predicament (Andries *et al.*, 1994–1998), there was a mobilization of previous nonvoters and a massive shift of voters to his party. This great realignment lasted until 1980 as can be seen from the preliminary graphical time plot in Figure 8.9.

In the following program segment, the election of 1894 is first tested as an extended pulse (E1894S) and later tested as an instantaneous pulse (E1894P). This test helps determine whether the election is a deviating or critically realigning election. The events or interventions that being tested or controlled for are included after the WITH option of the ARIMA command. The output of these models can be compared.

```
Title 'Test of the Critical Election hypothesis'.
* ARIMA.
TSET PRINT=DEFAULT CIN=95 NEWVAR=ALL .
PREDICT THRU END.
ARIMA cdemprop WITH e1894s bullmoos newdeal
 /MODEL=( 1 0 0 )NOCONSTANT
 /MXITER 10
 /PAREPS .001
 /SSQPCT .001
 /FORECAST EXACT .
*Diagnosis of the Critical Election Hypothesis'.
ACF
 VARIABLES= err_3
 /NOLOG
 /MXAUTO 16
 /SERROR=IND
 /PACF.
*Sequence Charts .
TSPLOT VARIABLES= demproph
 /ID= year
 /NOLOG
 /FORMAT NOFILL NOREFERENCE
 /MARK criteltn.
title 'Test of Deviating Election Hypothesis'.
* ARIMA.
TSET PRINT=DEFAULT CIN=95 NEWVAR=ALL .
PREDICT THRU END.
ARIMA cdemprop WITH e1894p bullmoos newdeal
 /MODEL=( 1 0 0 )NOCONSTANT
 /MXITER 10
 /PAREPS .001
 /SSQPCT .001
 /FORECAST EXACT .
ACF
 VARIABLES= err_4
 /NOLOG
 /MXAUTO 16
 /SERROR=IND
 /PACF.
```

After the specification of the extent of the forecast and the confidence intervals with the TSSET command, the ARIMA(1,0,0) model without a constant is specified. The several models are compared and it is found that both the 1894 electoral pulse and the 1894 extended pulse are significant. The diagnosis for white noise residuals follows. The more extended electoral variable is more significant. The hypothesis that the election of 1894 has an extended effect is confirmed by the better model. Detailed comparison of the models is shown in the next section.

8.6.2. AN EXAMPLE OF SAS IMPACT ANALYSIS SYNTAX

To evaluate the impact of the election of 1894, our strategy is to compare what happened with what would have happened, had there been no election of 1894. First, SAS program C8PGM1.SAS. produces a graphical time plot of the raw data (Fig. 8.9). The changes in the series attributable to this event in 1894 are modeled with SAS syntax. The procedure employed is explained in the preintervention phase, the forecasting phase, and the impact assessment phase. The preintervention ARIMA noise model is applied to the whole series and the residuals are modeled to represent the interventions. Within the impact assessment phase, there is an identification, estimation, diagnosis, and metadiagnosis phase. The residuals should be the whitest noise possible.

To program the preintervention phase in SAS, the data set, DATA PREINT, is constructed. The data are contained within the program. Two forms of the intervention variable that represent the election of 1894. These variables are E1894P and E1894S. These two variables represent hypothesized effects of the election of 1894. The E1894P represents the deviating election. This variable is a pure pulse dummy variable coded as one for the year of the election and zero otherwise. E1894S represents the lasting effect of a critical election; this variable is an extended pulse dummy variable that is coded one from 1894 through 1930 and zero otherwise. The strategy is to determine whether only the instantaneous pulse dummy is the best intervention variable, substantiating the hypothesis that the election of 1894 was a deviating election, or whether the extended pulse dummy, substantiating the hypothesis that the election of 1894 was a critical election, is significant as well. If the extended pulse dummy is significant, then that election can be justifiably described as a critical election.

Other indicator variables are constructed to represent pulses, phases, or level shifts. These event variables are employed to control for transient influences of potentially confounding effects. In the modeling that is performed here, indicator variables for the Election of 1912, for the election of 1920, and the New Deal phase of the analysis are constructed. These are coded as 1 when the event occurs and 0 otherwise. Accordingly, BULL-MOOS is coded as 1 during 1912 and 0 otherwise. E1920P is coded as 1 during 1920 and 0 otherwise. The NEWDEAL variable is coded as 0 prior to 1932 and 1 until 1980.

Therefore, the preintervention data predates the election of 1894. The year variable is incremented in steps of 2 years because the congressional elections take place every other year. A 'subsetting IF' statement extracts only the data prior to 1894 for analysis in DATA PREINT. A PRINT procedure permits checking of the data to be sure that it is being read correctly by the program. A title statement designates this portion of the analysis as the "Preintervention Series." If the title appears in a graphical plot and is too long, the font will be automatically reduced in size with a warning and that maximum permissible title size will depend on the linesize option setting. The preintervention series, consisting of only 44 observations, is estimated as an ARIMA(1,0,0) model. At this step the analyst takes a slight liberty with protocol. Because of the small preintervention sample size, there is low power and what might not pass for significance may be significant when more data values are added to the sample. The final preintervention series is modeled as an AR(1) accordingly and called the "Preintervention Noise Model."

```
Data preint;
 set congress;
year = year + 2;
if year < 1894;
proc print;
run;

symbol1 i=join c=green v=star;
symbol2 i=join c=blue v=diamond;
axis1 label=(a=90 'Democratic Proportion of Seats')
order=(.20 to .8 by .1); proc gplot data=preint;
 plot (Demproph) * year/overlay
 href=1860,1865,1877,1892 vaxis=axis1 annotate=anno1b;
title 'Preintervention Democratic Proportion of Congress';
run;

proc arima data=preint;
 i var=demproph center;
 e p=1 noint printall plot;
 f lead=18 id=year out=fore1;
title 'Preintervention Democratic Proportion of House Seats';
title2 'Test of AR1 model';
run;
```

More specifically, the preintervention ARIMA noise model is developed. In the SAS syntax of DATA PREINT, the data set CONGRESS is read into the data set PREINT with the SET command. All observations prior to the year of the 1894 election are included, while those following this date are excluded, with the 'subsetting IF' statement in the third line of the program.

The time series data are then printed for review with the procedure PRINT. A forecast from the intervention model is then compared to that of the preintervention series in a graphical plot.

The ARIMA procedure below the PRINT procedure models the preintervention series. The identify subcommand begins with an I, the estimate subcommand begins with an E, and the forecast subcommand begins with an F. The preintervention ARIMA noise model is almost the same one developed in the SPSS syntax above. It is an ARIMA(1,0,0) model without an intercept (constant) owing to centering of the series. The IDENTIFY subcommand, I VAR=DEMPROPH CENTER, generates the ACF, IACF, and the PACF of the centered series. The CENTER option centers the series by subtracting its mean. The ESTIMATE subcommand indicates that the series has an autoregressive parameter at lag 1. This estimation algorithm is that of conditional least squares. The printed estimation history and significance tests are requested along with the ACF and PACF plots of the residuals.

The input series is examined for stationarity. Unit root tests may be applied. Alternatively, the rate of attenuation of the ACF or PACF may be examined to determine whether differencing is necessary. If differencing of the input series is in order, it is invoked here. Once the proper order of differencing has been invoked, stationarity is attained.

Before the program can proceed to an analysis of the postintervention series, some other preprocessing is necessary. New variables are constructed out of the forecast and confidence limits. The temporal increment has to be adjusted so that each yearly increment is actually a biannual increment. Reassignment of year values accomplishes this objective. The reconstructed values are then saved for subsequent graphical presentation.

```
data prepa;
 set fore1;
year = year + 2;
proc sort; by year;
run;

data prep1;
 set prepa;

/* *************************************************** */
/* Renaming the variables from the first forecast */
/* profile and saving the renamed variables       */
/* *************************************************** */
dph = demproph;
fc1 = forecast;
```

```
1951 = 195;
u951 = u95;

/* ********************************************** */
/* Preparation for merging with other data sets  */
/* dropping the old variables                     */
/* and setting earlier values to missing for      */
/* values to be hidden in the graph               */
/* ********************************************** */

drop demproph forecast 195 u95;
if year < 1894 then fc1 = .;
if year < 1894 then 1951=.;
if year < 1894 then u951=.;

/* ********************************************** */
/* Because of biannual increment redefinition     */
/* of years is necessary to control forecast output */
/* ********************************************** */

if year = 1912 then year = 1928;
if year = 1911 then year = 1926;
if year = 1910 then year = 1924;
if year = 1908 then year = 1922;
if year = 1907 then year = 1920;
if year = 1906 then year = 1918;
if year = 1905 then year = 1916;
if year = 1904 then year = 1914;
if year = 1903 then year = 1912;
if year = 1902 then year = 1910;
if year = 1901 then year = 1908;
if year = 1900 then year = 1906;
if year = 1899 then year = 1904;
if year = 1898 then year = 1902;
if year = 1897 then year = 1900;
if year = 1896 then year = 1898;
if year = 1895 then year = 1896;
if year = 1894 then year = 1894;
proc sort; by year;
run;
proc print;
title3 'Review of Preintervention data';
run;
```

Preparatory to analyzing the shape of the impact after each event, an ASCII time plot is generated from a data set called DATA WHOLE. A time plot of the series is programmed with the PROC TIMEPLOT command.

```
data whole;
 set congress;


proc timeplot;
 plot demproph;
 id year;
title 'Time Plot of Democratic Proportion of House Seats';
run;
```

Using YEAR as an ID variable helps the researcher model the response to the input. Identifying the observations by YEAR facilitates counting the number of lags between one event and another. The inclusion of the value of the plotted variable provides for accurate assessment of pulses or level shifts. This permits accurate analysis of the time plot.

After we model the preintervention series with an ARIMA noise model, we preliminarily model the impact of the election input on the response series. First, we examine the SAS ASCII time plot output to ascertain the general shape of the response to the input. The time plot reflects the deep drop in Democratic proportion of House seats in 1894 and the Republican gains in each election thereafter until 1898. After some fluctuation, there is another drop in Democratic control as Teddy Roosevelt, with a "Square Deal" that incorporated fair labor and business regulation, swept the country in 1904 in a landslide victory. The Democrats recover in 1912 with the splitting of the Republican Party into the conservative Republicans and the progressive Bull Moose Party. When in the 1920 electoral campaign, Warren Harding advocated a return to normalcy from the hardships of the war, President Wilson had already suffered a stroke. Although Wilson hoped the election to become a referendum on the League of Nations, it became an evaluation of his health, which allowed the Republicans to take control of the Presidency and won widely throughout Congress. Only after 1920 did the Democrats begin to recover again.

```
                    Time Plot of Democratic Proportion of House Seats

       Year        Proportion of   min                                             max
        of          Democratic     0.2359550562                        0.7655172414
    Election      Seats in House    *------------------------------------------------*
       1876           0.53         |                             D                    |
       1878           0.51         |                          D                       |
       1880           0.44         |                   D                              |
       1882           0.62         |                                   D              |
       1884           0.56         |                              D                   |
       1886           0.52         |                          D                       |
       1888           0.47         |                     D                            |
       1890           0.69         |                                       D          |
       1892           0.62         |                                   D              |
       1894           0.29         |    D                                             |
       1896           0.38         |           D                                      |
       1898           0.46         |                  D                               |
       1900           0.43         |                D                                 |
       1902           0.46         |                  D                               |
       1904           0.35         |         D                                        |
       1906           0.42         |               D                                  |
       1908           0.44         |                 D                                |
       1910           0.58         |                              D                   |
       1912           0.67         |                                     D            |
       1914           0.53         |                          D                       |
       1916           0.48         |                   D                              |
       1918           0.44         |                 D                                |
       1920           0.30         |     D                                            |
       1922           0.48         |                    D                             |
       1924           0.42         |               D                                  |
```

According to our modeling strategy, we model the preintervention series first. At first glance, the ACF and the PACF of the series appear to yield white noise. Upon reflection, it can be observed that the sample size is low for the preintervention series and that apparent statistical nonsignificance can be due to sparse data. Therefore, this process is repeated with the full series included. The characteristic patterns of the ACF and PACF of the full series are those of an AR(1) ARIMA model. Inserting a low order AR(1) is consistent with the first part of the modeling procedure. Autocorrelation within the input series would render the statistical estimation inefficient. The standard errors become compressed. Therefore, the insertion of low order regular and, if necessary, seasonal AR components, is necessary to control for this effect on the significance tests. The ARIMA specification used as the basis of this analysis is therefore an ARIMA(1,0,0). Because the series has been centered, no constant is needed.

```
                        ARIMA Procedure
                Name of variable = DEMPROPH.
                Mean of working series =          0
                Standard deviation     =  0.12471
                Number of observations =         64
                        Autocorrelations
Lag Covariance Correlation -1 9 8 7 6 5 4 3 2 1 0 1 2 3 4 5 6 7 8 9 1
  0  0.015553    1.00000   |                      |********************|
  1  0.0085358   0.54883   |                   .  |***********          |
  2  0.0061667   0.39651   |                  .   |********             |
  3  0.0044287   0.28476   |                 .    |******.              |
  4  0.0029589   0.19025   |                 .    |****  .              |
  5  0.0010559   0.06789   |                 .    |*     .              |
  6 0.00041973   0.02699   |                 .    |*     .              |
  7 -0.0004303  -0.02766   |                 .   *|      .              |
  8  -0.000495  -0.03183   |                 .   *|      .              |
  9 -0.0004871  -0.03132   |                 .   *|      .              |
 10 0.00085925   0.05525   |                 .    |*     .              |
 11  0.0023297   0.14980   |                 .    |***   .              |
 12  0.0022281   0.14326   |                 .    |***   .              |
 13  0.0025995   0.16714   |                 .    |***   .              |
 14  0.0016638   0.10055   |                 .    |**    .              |
 15  0.0006645   0.04273   |                 .    |*     .              |
 16  0.0007348   0.04725   |                 .    |*     .              |
                        "." marks two standard errors

                   Partial Autocorrelations
    Lag Correlation -1 9 8 7 6 5 4 3 2 1 0 1 2 3 4 5 6 7 8 9 1
      1    0.54883   |                 .    |***********          |
      2    0.13637   |                 .    |***   .              |
      3    0.03204   |                 .    |*     .              |
      4   -0.01331   |                 .    |      .              |
      5   -0.09399   |                 .   **|      .              |
      6   -0.00550   |                 .    |      .              |
      7   -0.04147   |                 .    *|     .              |
      8    0.01430   |                 .    |      .              |
      9    0.01126   |                 .    |      .              |
     10    0.12169   |                 .    |**    .              |
     11    0.14358   |                 .    |***   .              |
     12   -0.00961   |                 .    |      .              |
     13    0.03747   |                 .    |*     .              |
     14   -0.09697   |                 .   **|      .              |
     15   -0.06626   |                 .    *|     .              |
     16    0.04610   |                 .    |*     .              |
```

The postintervention modeling is found in the **PROC** ARIMA DATA=
WHOLE; command. A few points of modeling strategy need to be noted.
After modeling the preintervention series, the SAS cross-correlation con-

nection is specified. Within IDENTIFY statement, a CROSSCORR option indicates the input variable name. More than one input variable may be designated. In this case, the option, CROSSCORR=E1894S designates one input variable, named E1894S. Because the input variable gets the same differencing as the response variable, if the response variable were first differenced, indicated by the (1), the same differencing in the computer syntax would appear after the response and input variables.

```
IDENTIFY VAR= VOTE(1) CENTER CROSSCORR = VOTE(1)
```

It is important to remember event intervention models utilize discrete dummy variable(s) as input(s) that do not require prewhitening (which is discussed in detail in Chapter 9). However, the cross-correlation function, invoked by the CROSSCORR subcommand, is helpful in displaying the lag in and shape of the impact.

The precise presentation of the dates and corresponding response values in this output permits precise specification of the lags in the INPUT statement. We model the delay time first. The delay time can be ascertained from the graphical or *ASCII* time plots, or from the delay in the cross-correlation function before positive spikes are observed. The delay time is the lag time of between the occurrence of an event and the time of observed impact of that event on the response. If we think that there was a delay of impact in the election of 1912, we can model this with the BULLMOOS variable. Suppose we observe that after 1912 there was a lapse of 4 (2 year) periods before an appreciable drop in Democratic control is observed. The modeling of this 8 year delay (when years are incremented by 2) is performed with the 4$ in the INPUT=(4$(...)BULLMOOS); subcommand.

The level shifts in the value of Democratic proportion of control over the House is modeled next. Suppose an inspection of spikes in the ASCII time plot after the event of the Bull Moose election of 1912 revealed that there are spikes at lags 3 and 5. The researcher could model the Bull Moose Party delayed effect as part of the ESTIMATE subcommand as INPUT= 3$(2)BULLMOOS;. Because a constant at the current time is assumed within the parenthesis, this statement estimates the first impact after a delay of 3 periods, and estimates another delayed effect occurs 2 periods thereafter. If, however, there were spikes at 0, 3, and 5 lags, the researcher could model these spikes with INPUT=(3,5)BULLMOOS in the ESTIMATE subcommand. In this way, the researcher can estimate the pulses that appear in the graph of the response function. These parameters are included as part of the INPUT option of the ESTIMATE subcommand.

Several event or intervention models are then tested in the program. To test whether the election of 1894 is a deviating election, the election of 1894 is modeled as an instantaneous pulse.

```
proc arima data=congress;
 i var=demproph center crosscorr=(e1894p bullmoos newdeal);
 e p=1 noint printall plot input=(e1894p bullmoos newdeal);
 f lead=6 id=year out=fore3;
title 'Test of Change in Democratic Proportion of Congress';
title2 'Test of AR1 and Instantaneous Pulse Function model';
run;
```

When the Bull Moose Party and New Deal variables are included as controls, the E1894P instantaneous pulse election variable is statistically significant. The T statistic exceeds an absolute value of 1.96, which disconfirms the null hypothesis of no effect of the election of 1894. At least the election of 1894 appears to be a deviating election.

Conditional Least Squares Estimation

Parameter	Estimate	Approx. Std Error	T Ratio	Lag	Variable	Shift
AR1,1	0.39314	0.12312	3.19	1	DEMPROPH	0
NUM1	-0.19148	0.08786	-2.18	0	E1894P	0
NUM2	0.14272	0.08700	1.64	0	BULLMOOS	0
NUM3	0.14774	0.03662	4.03	0	NEWDEAL	0

Before concluding that this test of the deviating election hypothesis is adequate, it is necessary to review the residuals from this model to determine whether they appear to be white noise. This residual review can be performed by examination of the ACF and PACF or the modified portmanteau tests.

Autocorrelation Check of Residuals

To Lag	Chi Square	DF	Prob	Autocorrelations					
6	7.77	5	0.169	-0.077	0.183	0.010	0.156	-0.215	0.028
12	12.00	11	0.354	-0.208	0.004	-0.084	-0.005	-0.076	0.013
18	15.75	17	0.541	-0.021	-0.017	-0.205	0.022	-0.014	-0.013
24	19.60	23	0.666	-0.117	0.094	0.043	0.039	-0.031	0.110

Because the ARIMA model coupled with these intervention variables controls for all systematic variation in the system, the residuals are white noise. If the ARIMA model did not control for all of the systematic variation, the modified Portmanteau tests for the residuals would be significant. A review of the CCF, ACF, and/or PACF of the residuals would permit diagnosis of spikes that remained unmodeled. Revision of the model would follow. In this case, the model accounts for the systematic variation. Although the election of 1894 is a deviating election, the stability of the realignment is evanescent; it is really just a temporary defection of the voters.

If the partisan realignment is more stable, then what appears to be a deviating election can actually be a critical election. If there is evidence that the partisan realignment is maintained for a few years, then the election under consideration can be construed as a critical election. To be sure, if the realignment is maintained until the next critical election of 1932, then the election of 1894 is a critical election. A review of the time plot in Fig. 8.9 suggests that this effect might hold for several elections. Visual inspection can lead to subjective conclusions. More objective standards are preferred. We test the hypothesis that the election of 1894 is a critical election with an extended pulse Election of 1894 variable, named E1894S. Several models to test this hypothesis are constructed. One model is the simple extended pulse model. Another model is the first-order attenuated pulse. A third model controls not only for the Bull Moose Party deviation of 1912 but also for the Harding takeover from Wilson four years later. If, with any of these types of controls, the extended pulse is found to be significant, then evidence exists to support the hypothesis that the election of 1894 was also a critical election, followed by protracted partisan realignment among the voters. Each of these models is tested. The simple extended pulse (temporary step function) can be tested first with the following syntax.

```
proc arima data=congress;
  i var=demproph center crosscorr=(e1894s bullmoos newdeal);
  e p=1 noint printall plot input=(e1894s bullmoos newdeal);
  f lead=18 id=year out=fore;
title 'Test of Change in Democratic Proportion of House';
title2 'AR1,Step Function, with Bull Moose model';
```

In the output from this model, the effect of the New Deal is clearly significant. Neither the effect of the Bull Moose Progressive Movement nor that of the election of 1894 extended pulse is statistically significant.

Conditional Least Squares Estimation

Parameter	Estimate	Approx. Std Error	T Ratio	Lag	Variable	Shift
AR1,1	0.37543	0.12347	3.04	1	DEMPROPH	0
NUM1	-0.05190	0.04507	-1.15	0	E1894S	0
NUM2	0.15627	0.09064	1.73	0	BULLMOOS	0
NUM3	0.12799	0.04226	3.03	0	NEWDEAL	0

A review of the modified portmanteau tests shows that this model accounts for the systematic variation in the system.

```
                  Autocorrelation Check of Residuals
  To    Chi                      Autocorrelations
 Lag   Square DF    Prob
   6     5.47  5  0.361 -0.069  0.136  0.093  0.155 -0.148  0.020
  12     7.04 11  0.796 -0.106 -0.028 -0.056 -0.072 -0.003 -0.020
  18    10.20 17  0.895  0.068 -0.012 -0.136 -0.046  0.091 -0.048
  24    16.51 23  0.833 -0.179  0.119 -0.054  0.021 -0.078  0.097
```

The residuals are white noise also. If the number of observations were substantially larger, the researcher could be sure that this evidence disconfirmed the hypothesis of the critical election of 1894. Because this relative sparseness could limit the power of the tests, one more model is considered. A review of the graphical time plot reveals that a realignment can be more temporary than an extended pulse function coupled with the effects of the Bull Moose Party activity of 1912, the effect of the Wilson stroke prior to the election of 1920, and the New Deal realignment would indicate. A pulse dummy is included to represent the effect of the 1920 election campaign. A model is therefore tested where there is attenuation of the 1894 realignment when the effects of the progressive Bull Moose Party, the election of 1920, and the New Deal are taken into account. The model that represents this specification is

$$\text{DEMPROPH}_t - \mu =$$
$$\{(\omega_0 - \omega_1 L)/(1-\delta_1 L)\}\text{E1894P}_t + \text{BULLMOOS}_t + \text{E1920P}_t + \text{NEWDEAL}_t + e_t/1-\phi_1 L)$$

and the program that estimates it is

```
data cong3:
 set congress;
proc arima;
 i var=demproph center crosscorr=(e1894s bullmoos e1920p newdeal);
 e p=1 noint printall plot input=((1)/(1)c1894s bullmoos e1920p newdcal);
 f lead=6 id=year out=fore3;
title 'Test of Change in Democratic Proportion of Congress';
title2 'Test of AR1, Extended Pulse Function, Bull Moose,1920 model';
title3 'Optimization by Modeling Bull Moose Party';
run;
```

This model is estimated by conditional least squares estimation. In it, there are controls for the progressive Bull Moose Party activity of 1912, the election of 1920 downfall of Wilson, and the New Deal realignment. The noise model AR parameter is significant. Controlling for its effects precludes inefficient and incorrect significance tests. When the effects of these events and noise are controlled for, all of the component parameters of the election of 1894 extended and attenuated pulse are statistically significant. These results suggest that when the other effects are controlled for the election of 1894 had temporary rather than permanent critical partisan realigning effects.

Conditional Least Squares Estimation

		Approx.				
Parameter	Estimate	Std Error	T Ratio	Lag	Variable	Shift
AR1,1	0.40273	0.12620	3.19	1	DEMPROPH	0
NUM1	-0.19889	0.06951	-2.86	0	E18948	0
NUM1,1	-0.19503	0.06551	-2.98	1	E18948	0
DEN1,1	0.64018	0.22961	2.79	1	E18948	0
NUM2	0.13711	0.08328	1.65	0	BULLMOOS	0
NUM3	-0.15775	0.08283	-1.90	0	E1920P	0
NUM4	0.09409	0.04351	2.16	0	NEWDEAL	0

The combined effects of the 1894 election and three other events appear to account for the overall partisan realignment. A review of the residuals indicates that all of the significant systematic variation is taken into account by this model.

Autocorrelation Check of Residuals

To	Chi			Autocorrelations					
Lag	Square	DF	Prob						
6	8.53	5	0.129	-0.098	0.201	0.060	0.190	-0.179	0.050
12	11.88	11	0.373	-0.092	0.021	-0.136	0.078	-0.101	-0.016
18	19.44	17	0.304	-0.142	0.094	-0.204	-0.136	0.005	-0.011
24	21.64	23	0.542	-0.020	0.087	-0.102	0.052	-0.042	-0.000

In retrospect, the researcher may chose to compare the models to see which fit the data best. By relying on the SBC or error variance for each model, the researcher can decide which of the models provides the best fit. We observe that three models fit. We visually compare the ACF and PACF of each of the models. We review the models for their SBC. The deviating election model, with an instantaneous 1894 electoral pulse, has an SBC of -110.77. The critical election model, with an extended pulse from 1894 through 1928, has an SBC of -107.24. The attenuated realignment model has an SBC of -105.42. According to this standard of fit and parsimony, the election of 1894 was primarily a deviating election rather than one of critical realignment. If we allow for attenuation of the critical realignment effect, controlled for by the election of 1920 as well, we can interpret the 1894 election to have declining realignment effects. If our series were much longer, we could compare models over varying time spans to see which model parameters are more stable and reliable. We could also evaluate the forecasts of the respective models. We could generate forecasts 6 periods in advance and compare the MSFE or MAPE of each of the models. Either we could evaluate the fit, parsimony, parameter constancy, or the forecast accuracy of the models.

After we have conducted our hypothesis testing, drawn our conclusions, and evaluated our models, we can select the optimal equation. On the basis of the SBC, we chose the model with the lowest SBC. That happens to be the model with the 1894 electoral pulse whose output for the response variable is:

```
Model for variable DEMPROPH
Data have been centered by subtracting the value 0.5084619457.
No mean term in this model.

Autoregressive Factors
Factor 1: 1 - 0.39314 B**(1)

Input Number 1 is E1894P.
Overall Regression Factor = -0.19148

Input Number 2 is BULLMOOS.
Overall Regression Factor = 0.142719

Input Number 3 is NEWDEAL.
Overall Regression Factor = 0.147738
```

This equation can be formulated as

$$(Demproph_t - 0.0508) = (-0.191E1894P_t + 0.143BULLMOOS_t$$

$$+ 0.148NEWDEAL_t) + \left(\frac{e_t}{(1 - 0.393L)}\right),$$

where

$$Demproph_t = \text{Democratic Proportion of Seats in}$$

$$\text{House of Representatives at time } t \qquad (8.30a)$$

$$E1894P_t = \text{Instantaneous effect of election of 1894}$$

$$BULLMOOS_t = \text{Instantaneous effect of Bull Moose Party}$$

$$NEWDEAL_t = \text{Extended effect of New Deal.}$$

We reassess our finding. How does history help disconfirm the hypothesis that the election of 1894 is a critical election? The test of the critical election disconfirms a categorical assertion that this election critically realigns partisan dominance. When a simple extended pulse for the election of 1894 is shown to be statistically insignificant, this evidence disconfirms an unqualified assertion that the election in question was a critical one. Is the partisan shift of the deviating election of 1894 combined with other events stable enough to be deemed critical realignment?

Graphical evidence reveals that the low level of Democratic control of House seats was not maintained from 1894 through 1930. During the two elections following 1894, there was continued Democratic recovery in the House of Representatives from the low point of 1894. Then there was some fluctuation before the Democrats resumed their Congressional recovery. By 1908 the progressive movement split the ranks of the Republican opposition to allow for a huge Democratic recovery of seats. Clearly, the realignment of 1894 was short-lived. It took the illness of Wilson to allow Democratic control to fall to the level of the 1894 alignment, after which Democratic recovery resumed. These two electoral events were able to engender new deviations that cast doubt on the thesis of a critical realignment in 1894. Although there appears to be a slight and insignificant difference in levels of Democratic dominance between the post Civil War era and the pre-World War period after 1894, instability of Democratic dominance casts doubt on the thesis that the 1894 election was a critical election. The fact that there is a stable level shift in Democratic dominance after 1932 is not at issue here.

If the impact of the election of 1894 is qualified by the intervention of other electoral events, then this slightly attenuated step upward in Democratic control following the 1894 electoral event shows that the decline in Democratic proportion of House seats was neither permanent nor stable. If the researcher examines the parameters, he observes that the drop in Democratic control was reversed but the reversal was attenuated as time passes. This is the impact that the first-order extended pulse function models. The other three indicator variables account for major shifts in the Democratic control that destroyed the stability of a 1894 realignment. The Bull Moose Party indicator variable represents the dividing of the opposition and the resurgence of Democratic control in 1912. Although the 1920 election indicator was not statistically significant in earlier models, it is almost significant and therefore retained in this model. The 1920 election pulse indicator represents a loss of control due to Wilson's stroke the previous year and Harding's takeover of the Presidency. The positive impact of the New Deal in 1932 represents a critical and stable realignment that lasts until 1980. This New Deal indicator is another extended pulse that spans this period of time.

These other indicators account for the sharp increase and decline in the Democratic control. When they are entered, the researcher finds that they support the hypothesis of a short-lived duration of the 1894 partisan shift in control. When he examines the nature of this realignment, he sees that this evidence supports the thesis of an initial (deviating) but not lasting (and hence not critical) election.

Our modeling of this impact yields estimated parameters.

```
Model for variable DEMPROPH

Data have been centered by subtracting the value 0.5084619457.

No mean term in this model.

Autoregressive Factors
Factor 1: 1 - 0.40273 B**(1)

Input Number 1 is E1894S.
The Numerator Factors are
Factor 1: -0.1989 + 0.19503 B**(1)

The Denominator Factors are
Factor 1: 1 - 0.64018 B**(1)
Input Number 2 is BULLMOOS.
Overall Regression Factor = 0.137112

Input Number 3 is E1920P.
Overall Regression Factor = -0.15775

Input Number 4 is NEWDEAL.
Overall Regression Factor = 0.09409.
```

This output can be formulated as

$$Demproph_t - 0.508 = \frac{(-0.199 + 0.195L)}{(1 - 0.64L)} E1894S_t + 0.137 BullMoos_t$$

$$-0.158 E1920P_t + 0.09 NewDeal_t$$

$$+ \frac{e_t}{(1 - 0.403L)}, \tag{8.30b}$$

where

$$Demproph_t = \text{Democratic Proportion of House of}$$
$$\text{Representatives seats at time } t$$
$$BullMoos_t = \text{1912 electoral pulse}$$
$$E1920P_t = \text{1920 electoral pulse}$$
$$NewDeal_t = \text{New Deal realignment extended pulse}$$

This model of attenuated pulse (with a step function $E1894S_t$ multiplied by the rational polynomial) of partisan realignment represents a decomposition of total impact into relative impacts from each of these event interventions. From the signs of the regression parameter estimates, the analyst can tell whether the impact on the Democratic control is enhanced or reduced. From the magnitude of the regression coefficients, the analyst can assess their relative contribution. From the nature of the impulse response function of the 1894 electoral event, the analyst can glean a sense of the attenuation of the partisan realignment the election of 1894. Although the more statistically parsimonious model is that of the deviating election, this model theoretically explains the nature of the impacts of these other events and thereby facilitates our understanding of American electoral history.

A scenario or "what if" analysis can sometimes provide a new and helpful perspective in the examination of hypotheses. To test the effects of the 1894 election and the following modeled events, we examine the preintervention pattern. In so doing, we ask the question, "What would have happened if there were no deviating election of 1894 or subsequent influential events?" We model the preintervention series. Then, we can generate a forecast profile until the next critical realigning election in 1932 from the ARIMA(1,0,0) preintervention model. The preintervention forecast and its intervals depict a scenario of what would have happened had the election of 1894 or the subsequent modeled events not impacted the series. The forecast and its limits are renamed and saved. Fluctuations of Democratic proportion of seats in the House of Representatives beyond these forecast limits would indicate statistically significant and substantial impacts on the preintervention system. Due to sparse preintervention data, the confidence interval around the forecast is wide and the width of the interval easily encompasses the variation of Democratic proportion of House seats (Fig. 8.10). With insufficient data, previously deemed significant events of the 1894 election and the Bull Moose Party participation do not appear to be significant after all. A full-series modeling strategy instead of the conventional two-step strategy can obviate this problem. Even with enough data, the election of 1894 appears to be a demarcation point and contributing factor but not the event that defines the partisan alignment until the onset of the New Deal. When we have sparse data, the scenario

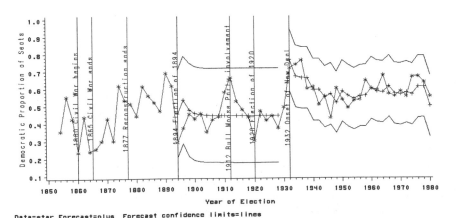

Data=star Forecast=plus Forecast confidence limits=lines

Figure 8.10 A forecast profile from the pre-1894 Democratic proportion of House seats series provides a ceteris paribus baseline set of expectations for the no event scenario analysis.

analysis is not sufficient to use as a basis for our conclusions; we may need to the full intervention model before drawing conclusions.

If we extend the scenario analysis to include electoral events beyond 1932, a forecast profile is generated over the New Deal period. The saved forecast profiles from the preintervention model and from the model up to 1932 are superimposed on the actual data to reveal what would have taken place without other event interventions. In this case, the election of 1932 reveals a Democratic proportion of seats in the House that exceeds the earlier forecast limits. From this significant change, it is clear that the election of 1932 marks a level shift or sea change in the configuration of political affiliation and partisan support. In this way, scenario analysis can be useful in assessing the nature of regime or level shifts in a series. The SAS program syntax for the graphical scenario analysis is

```
/* Preparation of a Scenario Analysis Profile Plot */

data prep2;
 set fore2;
  if year < 1932 then forecast = .;
  if year < 1932 then λ95=.;
  if year < 1932 then u95=.;
proc sort; by year;
proc print;
title3 'Complete data for fore2';
 run;
data all;
 merge prep1 prep2; by year;
  if dph < 1896 then dph = .;
proc print;
title 'Merging Prep1 and Prep2';
 run;
symbol1 i=join c=green v=star;
symbol2 i=join c=blue v=plus;
symbol3 i=join c=red;
symbol4 i=join c=red;
symbol5 i=join c=green v=star;
symbol6 i=join c=blue v=plus;
symbol7 i=join c=red;
symbol8 i=join c=red;
axis1 label=(a=90 'Democratic Proportion of Seats');
proc gplot data=all;
  plot (Dph fc1 l951 u951 demproph forecast l95 u95) * year/overlay
  href=1860,1865,1877,1894,1912,1932 vaxis=axis1
  annotate=anno2;
title justify=L 'Figure 8.10 Democratic Proportion of House of
                 Representatives Seats';
title2 'Forecast represents Null Hypothesis of No Change';
title3 'After Elections of 1894 and 1932';
run;
```

To determine the nature of the election of 1894, we have tested the different hypothetical models by intervention and scenario analysis. Although the deviating election model seems to be the statistically optimal one, the attenuated realignment model has theoretical explanatory power. The subsequent scenario analysis, although plagued by limited sample size, showed that the big shifts in Democratic control were within the confidence limits of the pre-1894 election situation. It can be argued that as the number of observations in our series grows larger, the confidence intervals may become more compressed and impacts deemed insignificant might become significant. Hence, a full-series intervention model would be needed to test our hypotheses. In the Chapter 12, the proper sample size needed to perform this analysis with confidence will be addressed. Without an understanding of what sample size is needed, we have to run the risk of the perils of low statistical power—namely, a tendency to a Type II error—to glean the information we need. For our failure to control all relevant factors with the proper sample size, our theoretical assessment must be deemed tentative in nature.

8.6.3. EXAMPLE: THE IMPACT OF WATERGATE ON NIXON PRESIDENTIAL APPROVAL RATINGS

8.6.3.1. Political Background of the Scandal

One of the greatest political scandals in the history of the United States presidency to date was that of the Watergate affair. To understand American political history, it is helpful to examine the impact of the Watergate scandal on President Richard M. Nixon's public approval and political support. What happened, and what kind of public impact did it have? The question posed by the Gallup Poll is "Do you approve or disapprove of the way the President is handling his job" The answer categories from which respondents choose one are: Approve, do not approve, and no opinion. The monthly average of the percentage of respondents approving presidential job performance is used to gauge public approval. In general, the Gallup Organization in Princeton, New Jersey, conducts these polls usually twice a month and publishes the results for general review (Gallup Opinion Index, 1969–1974). Even though the number of observations over time (in this case, after the incidence of the conviction of the Watergate burglars) may not be as many as we would ordinarily wish, it may be worthwhile to apply this form of impact analysis to enhance our understanding of the impact of Watergate on the public approval of the presidential job performance at the time.

Because revelations of the Watergate scandal implicated President Nixon in serious crimes against the state, a student of American history or political science needs to examine both the background and the illegalities of Watergate. The *sub rosa* activity of the Nixon Administration made for a time of tumult before the break-in at the Watergate Democratic National Committee Headquarters. Journalistic investigative reporting and the Senate Watergate Committee hearings exposed of a multitude of operations of political espionage, dirty tricks, and sabotage of the democratic process, assuring the American public that a national nightmare, of which there were only rumors before, had indeed been brought to life by the Nixonians. This revelation led to a precipitous decline in legitimacy, trust in government, political support, and approval of the Presidential performance.

The roots of Watergate stemmed from the dark side of the Nixon administration. Although Nixon originally approved the Huston plan for covert illegal surveillance of political opponents on January 23, 1970, J. Edgar Hoover, Director of the FBI, allegedly objected and the plan was reportedly quashed. In his fascinating transcription of the new Nixon tapes, Professor Stanley I. Kutler reveals that on June 17, 1971, Nixon verbally reendorsed the plan for officially sanctioned, surreptitious, illegal political activity directed at designated dissidents and other imagined Nixon political enemies during the Vietnam War (Kutler, 1997). With these shadowy activities, Nixon began his fall from grace (Kutler, p. 454). Nixon believed that there was a conspiracy out to get him, and he tried to enlist other officials in the belief that this cabal had to be sought out and destroyed (Kutler, pp. 8, 9, 10, 14–16). He wanted to break into the Brookings Institution, the Rand Corporation, and the Council of Foreign Relations to steal national security information, which he would selectively leak to the press in order to tarnish the historical image of the Democratic Party (Kutler, pp. 6, 11, 17, 24). One of his aides, John Ehrlichman, expressed the desire to break into the National Archives to steal such information as well (Kutler, p. 30). Specifically, Nixon wanted disparaging information about plans for and implementation of operational attempts on the life of Fidel Castro, the Bay of Pigs invasion fiasco, the Cuban Missile Crisis, and the origins of the Vietnam War. Nixon had his aide, Charles Colson, hire E. Howard Hunt, a former CIA political officer and old school chum, for political espionage and sabotage. Hunt had told Colson early on that if the truth had been known, Kennedy would have been destroyed (Russo, 1998). Early in June of 1971, Nixon had Erlichman tell Director of Central Intelligence Richard Helms that he had to have the file on the "the Cuban Project." It was the considered and well-founded opinion of Helms, who had been director of covert operations and heavily involved with anti-Castro activities, and Lawrence Houston, the CIA General Counsel, that Nixon wanted to use these documents for partisan political purposes. Helms reluctantly released only three files, after 3 years of stalling. The reluctance to divulge the dirt and refusal to block the FBI investigation into the Watergate espionage got Helms transferred from his position of Director of Central Intelligence to that of U.S. Ambassador to Iran (Russo, pp. 421–423, 580n).

Nixon really wanted to reinaugurate McCarthyism (of the Joseph, not the Eugene, stripe) (Kutler, pp. 11, 18). He wanted a congressional committee to investigate internal security (Kutler, 11, 18). When Daniel Ellsberg leaked the Pentagon Papers to the *New York Times* and *Washington Post,* a secret White House unit was formed, ostensibly to plug national security leaks, but actually to surreptitiously combat a presumed conspiracy believed to consist of antiwar activists and political opponents of Richard Nixon (Kutler, pp. 13–19). Egil Krogh placed Howard "Eduardo" Hunt and G. Gordon Liddy, chief of security for the Committee to Re-Elect the President (CREEP), in charge of a "special investigations unit," colloquially called "the plumbers." This unit broke into the office of Dr. Lewis Fielding, psychiatrist of Daniel Ellsberg, in order to gather compromising information on him. Nixon ordered the members of this unit to collect information on political opponents for purposes of character assassination in the press (Kutler, pp. 34–36).

The Nixon administration planned and carried out espionage against their political opposition. Nixon and his aides considered using the Secret Service to spy on opposing political candidates (Kutler, p. 40). Nixon wanted the IRS to harass his opposing political candidates, Senators Muskie and Humphrey (Kutler, pp. 28–30). They had Howard Hunt and another White House investigator, Tony Ulasewicz, dig for poltical dirt in the private life of Senator Ted Kennedy (Kutler, p. 29). Under the direction of Hunt and Liddy, they formulated a plan code-named GEMSTONE for surreptitious entry, electronic eavesdropping, agents in place, and prostitute-escorted entrapment, among other things. In the economical version that was authorized by head of CREEP and Attorney General of the United States John Mitchell, the Watergate burglars targeted the Democratic National Committee Headquarters at the Watergate Hotel for surreptitious espionage and surveillance. In late May 1972, they planted the bugs. Malfunctions developed. On June 17, 1972—*the anniversary of Nixon's verbal reendorsement of the infamous Huston scheme for coordinated illegal surreptitious entries, thefts of political documents, recruitment of campus informants, and an array of measures by which the political opposition would be neutralized*—when they reentered the Watergate Offices of the Democrats to replace the defective bugs, their break-in was detected by the security guard and they were apprehended by DC police shortly thereafter.

How was this information to be used? The collected information along with covert action was used to sabotage the political campaigns of opponents. In fact, Nixon had E. Howard Hunt, a White House consultant, forge cables to falsely implicate President John Kennedy in the murder of

President Diem of South Vietnam. One field operative, Donald Segretti, run by another White House aide, Dwight Chapin, let mice loose at a campaign rally of one of the opposing candidates and over-ordered pizzas for another candidate's campaign workers. Nixon operatives forged a letter that tarnished Democratic candidate for President Edmund Muskie that had him appear to Canadians as "Canucks," a term considered by more cultured individuals to be a vulgar ethnic slur (Kutler, p. 454). Nixon operatives spread false rumors of homosexuality about another opposing candidate, Senator Henry Jackson (Kutler, p. 454). Nixon's people, under his direction, management, and financial support, were insecure of their capability to win in the arena of the intellect, political discourse, and public persuasion. In sabotaging and disrupting the campaigns of their political opponents, Nixon's covert operatives engaged in wholesale sabotage of the electoral process.

From June 17, 1972, when the Watergate political espionage team was captured, a cover story in fact was fabricated by President Richard M. Nixon and his cronies (Kutler, pp. 3,55). The tactics and strategy of public impression management diverted Nixon's attention from other critical issues and resulted in the first resignation of an American president. The intervention analysis reveals the impact of Watergate on President Richard Nixon's public approval ratings. Nixon's public approval during some of the roots of the Watergate scandal is depicted in Fig. 8.11.

Who broke into the Watergate? The perpetrators of the Watergate break-in on June 17, 1972, were arrested as they attempted to break in, steal documents, and repair eavesdropping equipment that they had already installed in the Democratic National Committee offices at the Watergate office complex in Washington, DC. The penetration group consisted of

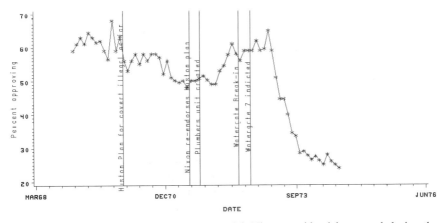

Figure 8.11 The roots of the Watergate scandal: Nixon presidential approval during the origins.

Bernard Barker (a former assistant to Bay of Pigs political officer Howard Hunt), James McCord (a former agency security officer in charge of electronic surveillance), Eugenio Rolando Martinez (a "connected" boat captain), Virgilio Gonzales (a virtuoso of locks and picks), and Frank Sturgis (a swashbuckling former "soldier of fortune" who was with Castro when he took over Cuba in 1959). Three other members of this cadre, Reinaldo Pico, Felipe DeDiego, and Alfred Baldwin, their lookout at the Hotel across the street, escaped prosecution. Overall, they were being directly supervised by Hunt and Liddy. While the White House pretended to be ignorant of these mischievous activities, Hunt's White House phone number was found in the address book belonging to one of those arrested. Even when Liddy, Hunt, and the "Cubans" were indicted, the matter was not publicly identified as a White House caper and Nixon's popularity did not yet suffer.

The plan to cover up involvement of Nixon and his White House aides entailed the crimes of obstruction of justice and perjury. For the apprehended team to remain silent and serve out their jail terms without revealing the sources of their activities, Hunt sought financial support for the men and their families from the White House through White House Counsel John Dean. From the viewpoint of the "Cubans," and Hunt who felt responsible for them, this support was necessary for their families until they were released from prison and got new jobs. At one point, Nixon sought to get the CIA to obstruct the FBI inquiry into the Watergate break in by having General Walters speak to people in the FBI and tell them to back off. One of Nixon's aides, Charles Colson, suggested the possibility of framing the CIA for what happened (Kutler, p. 61).

To what extent was fear of unraveling the cover based on fact? The White House cover story was that they wanted to protect the Bay of Pigs secrets harbored by those arrested, as if the fiasco had not been known to Castro far enough in advance. In fact, Castro had advance knowledge of the impending Bay of Bigs invasion. Security for the invasion was deplorable. The plans for the "secret" invasion had leaked into the Cuban community in Miami weeks before the invasion, C.I.A. official Lyman Kirkpatrick tried to get the operation aborted, but he was told that it was too late and was overruled (Kirkpatrick, 1968; Russo, 1997). Cuban representatives to the UN protested that the CIA was preparing to invade, and Castro prepared the trap for the invasion.

Hunt served as a CIA political officer for the Bay of Pigs invasion. Cuban exiles involved in the Bay of Pigs invasion were promised that the U.S. military would provide air cover for the invasion, but President Kennedy blocked the provision of promised air cover. Castro's forces fought, decimated, and captured the surviving members of landing parties. The Kennedy brothers felt humiliated by their defeat. Afterward, Hunt maintained a liaison between Bobby Kennedy, then Attorney General, who was directing covert operations against Cuba, and Sergio Arcacha Smith's Cuban Revolutionary Council (CRC), which was one of the groups implementing his marching orders and which was reportedly the group that in the summer of 1961 passed on reports of Russian missiles being deployed in Cuba. Of course, Russian Colonel Oleg Penkovsky, before his capture by the K.G.B., had already advised his M.I.6 and C.I.A. contacts of Russian plans to deploy these missiles and these deployments were later confirmed by C.I.A. U-2 surveillance flights. Hunt and Barker also served as case officers of Cuban exiles involved in contingency planning of a reinvasion of Cuba, purportedly to be an autonomous enterprise to be launched from Central America. Nixon suggested to Ehrlichman that he pursuade the CIA that the FBI investigation would unravel the cover of a series of covert anti-Cuban operations with which Hunt had been associated (Russo, 1998). The cover-up was revealed to Judge John Sirica in a letter from James McCord, the team surveillance expert, and part of it was reported to the Senate Watergate Committee by John Dean, Counsel to the White House. From Dean's viewpoint, the provision of this hush money could be considered extortion and obstruction of justice. Nixon's taped orders approving the procurement and disbursement of the hush funds are what at the time appeared to implicate him in the cover-up.

What was the source of the "hush money" and how was it distributed? The resources that made the cover-up possible were indirectly accessible. Nixon came up with the idea and John Mitchell, Attorney General, seems to have been the one who made the arrangements. The hush money was obtained in cash from Thomas Pappas in exchange for securing an Ambassadorship for Henry Tasca in Greece (Kutler, pp. 217–218), although Frank Sturgis used to tell acquaintances that the money came from fugitive Robert Vesco. After the conviction of the Hunt, Liddy, and the other Watergate burglars, Nixon's popularity began to plummet. It became necessary to involve others White House officials in the disbursement of the hush funds. Nixon's personal attorney Herb Kalmbach directed Tony Ulasewicz, a former New York City policeman who served the Nixon White House as a private investigator, to skulk around town depositing little brown bags filled with $100 bills at prearranged drop spots for surreptitious retrieval. When Helen Hunt, Howard's wife, died in a plane crash with thousands of dollars of hush money, part of it went undelivered. More had to be obtained and John Dean got worried that there might not be an end to these demands. John Dean, turning state's evidence, soon revealed that the cover-up extended to the Oval Office and that Nixon had been warned there was a cancer on the Presidency that had to be excised. The cover-up did not, however, stop there.

In addition, on June 19, 1972, when the FBI drilled a White House safe to retrieve contents in their investigation of the Watergate affair, L. Patrick Gray, Acting Director of the FBI, found incriminating material and was convinced by White House staffers to destroy this material. In so doing, Gray himself became implicated in the cover-up.

Meanwhile, the diligent investigative reporting of Bob Woodward and Carl Bernstein of the *Washington Post* penetrated much of the cover, exposing enough nefarious activities to give rise to and help the investigation. In their odyssey, they claim to have been guided by leads from a highly placed, well-positioned official with connections to the national security establishment, whom they code-named "Deep Throat." To this day, Woodward, Bernstein, and Ben Bradlee, their editor, have kept the identity of this reliable source a secret.

Phase one of the Watergate scandal included the origins of the scandal and the exposure of one of many covert operations. The origins of the political skullduggery reside in the Huston plan for surreptitious entry, illegal surveillance, and covert disruption of the political enemies and its verbal reendorsement by Nixon on June 17, 1991. The story of the Watergate burglars and their foremen culminated in their conviction on January 30, 1973, and this was indeed a watershed event. With the disclosure of this cover-up, conviction of the Watergate Five plus Hunt and Liddy, and the implication of Nixon's involvement, Nixon fell from respect and his public approval began to dive. This revelation exposed one of a collection of covert activities that threatened the Constitutional structure of fundamental personal and political freedom.

The second phase began when James McCord revealed the existence of an organized White House cover-up and perjuries in a letter to Judge John Sirica on March 23, 1973. H. R. "Bob" Haldeman, John Ehrlichman, and John Dean were fired on April 30, 1973. By the end of June 1973, John Dean had implicated President Nixon in the cover-up in his testimony before the Senate Watergate Committee. As evidence emerged that the White House had been heavily involved in political skullduggery, the cover-up and its attendant obstruction of justice began to really unravel. For Nixon, the beginning of the end had come (Fig. 8.12). For years trust in government had suffered from one governmental disaster after another. The Bay of Pigs fiasco, the official cover-up of Lee Harvey Oswald's involvements in the interest of preventing another war (Russo, 1998), and the official propaganda about the origins of the Vietnam War, and the success of the U.S. military in waging it (Kutler, p. 37) all compounded public cynicism about government.

The third phase of the Nixon's fall from public grace was characterized by the exposures of high crimes following from the legislative and judicial quest for evidence. On May 17, 1973, the Senate Watergate Committee began televised hearings. For a short while there was a wait-and-see attitude

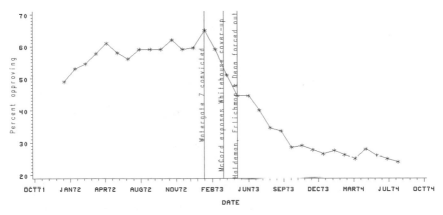

Figure 8.12 Impact analysis of Watergate scandal on Nixon presidential approval rating. The opening of the floodgates: McCord exposes and Dean implicates.

on the part of the public. John Dean began admitting to prosecutors that Nixon and he had discussed the cover-up at least 35 times and testified before the Senate on June 30 that Nixon was clearly implicated in approving the hush money. Alexander Butterfield in July disclosed the existence of the White House taping system, whereupon the Senate and the Special Prosecutor began to subpoena the tapes. Nixon demurred on the basis of "executive privilege" (Fig. 8.13).

The White House was under siege. More and more White House aides

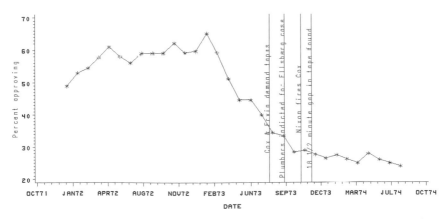

Figure 8.13 Impact analysis on Watergate scandal on Nixon presidential approval rating. The first cascade: The demand for evidence; the courts and legislature besiege the Nixon White House.

got into trouble. This scandal saw the indictment, arrest, conviction, and imprisonment of more White House officials than ever before in American history. Evidence traced from the revelations of Dean and others brought to light more evidence of wrongdoing. The White House plumbers were indicted in early September 1973, and it was disclosed that Nixon had inspired the break-in of the office of Dr. Lewis Fielding, the psychiatrist of Daniel Ellsberg, the Pentagon researcher who had released the Pentagon Papers to the press. The plumbers, who included most of the Watergate burglars plus some others, broke into the office of Dr. Fielding to check for damaging information on Ellsberg following his release of the secret history of American involvement in the Vietnam War. On one level, the plumbers were convicted for these scandalous legal offenses. On another level, a fascinating story underlay the secret history of American involvement and prosecution of the Vietnam War. The drama gained new excitement when Nixon ordered the Special Prosecutor, Archibald Cox, fired in October of 1973. Aggravating suspicion of a cover-up, an 18½ minute gap was found in the recording on one of the tapes. Chief of Staff Alexander Haig attributed it to "sinister forces." After this cascade of evidence, a new phase began (Fig. 8.14).

The last phase was that of Presidential impeachment. On May 17, 1974, the Senate began impeachment hearings. On July 23, the House of Representatives voted articles of impeachment. Those articles accused Nixon of failing to take care that the laws were faithfully executed, abuse of power, obstruction of justice, and sabotage of the democratic process. They were not compartmentalized personal peccadillos. The articles charged that high crimes and misdemeanors against the state—warranting removal of Nixon

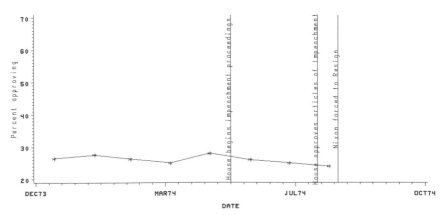

Figure 8.14 Impact analysis of Watergate scandal on Nixon presidential approval rating. The final deluge: Impeachment, taped evidence, and forced resignation.

from office—had been committed. Then on August 5, a tape revealing Nixon authorizing the hush money payments to the burglars was discovered. This was the proverbial "the smoking gun," the first piece of conclusive evidence of obstruction of justice. As more and more evidence was uncovered, this collection provided the evidentiary basis for the ultimate disgrace and downfall of an American president. The release of the tapes provided for the deluge of evidence that the investigations needed.

More concern about obstruction of justice with destruction of evidence grew. Nixon was implicated in the authorization of the support and maintenance of the cover-up, and hence was clearly guilty of obstruction of justice. The House Judiciary Committee voted articles of impeachment, and Nixon was forced to resign or face almost certain conviction. After he weighed the odds, 68 months after taking office, Nixon resigned on August 8, 1974, and left town retaining his pension and Secret Service protection. Little did the skulkers in the Watergate know that their efforts to re-elect the President would boomerang as they did. The flood of evidence had overwhelmed fortress White House. After Vice-President Gerald Ford became president, he pardoned Nixon, which may have cost Ford any chance of reelection.

In sum, the results of the analysis show that after the conviction of Hunt, Liddy, and the other Watergate burglars on January 30, 1973, Nixon's popularity began to plummet. In this analysis, the scandal begins as of February 1, 1973, since that was the date by which the conviction had been reported in the press. Others might begin it at June 17, 1972, or after the indictment of the Watergate burglars. Each of these approaches would result in a different impact model. Even a different algorithm for missing data replacement (when Presidential trial heats rather than job approval polls were conducted) might change the finely tuned specification of the model. McCord accelerated the political unraveling by revealing the White House cover-up. After Dean implicated Nixon and Butterfield told of the tapes, the coverup came undone as the clamor rose for release of the taped evidence. It was the tapes that provided the evidence of culpability. Nixon was ultimately forced from office in disgrace. Much can be learned from the public reaction to these events. From the graphical analysis, it can be seen that the conviction of the Watergate burglars severely damaged presidential job approval and precipitated the removal of a president from office. We now turn our attention to whether this analysis can show what damaged the President most.

8.6.3.1.2. *Programming the Watergate Impact Model*

This intervention model is programmed with SAS. The data are input into an Excel file and then converted to a SAS data set called NIXAPP5.SD2 with a conversion program called DBMSCOPY®. The program below sets up a LIBNAME for the directory and reads in the data.into a data set called NIXON. The approval rating is called APPROV. It consists of the average percentage of respondents to the Gallup Poll approving the job performance of the president. If a poll began in one month and ended in the next, it was considered to be within the month of its termination. If there were no other polls during that month, the value of the poll beginning but not ending in that month was used. When Gallup conducted presidential trial heats instead of these polls, the means of the adjacent ratings were imputed. These data were culled from publications of the Gallup Organization (1969–1974). A date variable is constructed with the INTNX function, which defines and names the month and year of each observation from the point of inception. Missing values are truncated so that the data set ends at the time of Nixon's resignation. A scandal variable is constructed from the dates formed. The scandal variable is a step function dating from the time of conviction of the Watergate burglars on January 30, 1973. The scandal variable is coded 1 from February 1, 1973 when the news was disseminated through August 8, 1974, when Nixon was forced to resign. Several models were programmed and compared. I selected the model that fit best according to the Schwartz criterion. Ultimately, the choice of the best model is a question of art and judgment in the trade-off between explanatory power, parsimony, and the whitest noise. The best general model to define the impact this scandal had on Nixon's presidential approval is formulated as

$$(1 - L)(Y_t - \mu) = (\omega_0)(I - L)I_{t-3} + \frac{(1 - \theta_{13}L^{13})e_t}{(1 - \phi_1 L - \phi_3 L^3)} \qquad (8.31)$$

or

$$(1 - L)(\text{Approval}_t - \mu) = \omega_0 \text{Scandal}_{t-3}(1 - L)$$
$$+ \frac{(1 - \theta_{13}L^{13})e_t}{(1 - \varphi_1 L - \phi_3 L^3)} . \qquad (8.32)$$

When the parameters are estimated by the program the model becomes

$$(1 - L)(\text{Approval}_t - .522) = -9.736 \text{Scandal}_{t-3}(1 - L)$$
$$+ \frac{(1 - .356L^{13})e_t}{(1 - .351L - .365L^{13})} . \qquad (8.33)$$

In order to explain the rationale behind the SAS programming, we first list the command log

```
/* ************************************************** **/
/* SAS  LOG of Program of Watergate Scandal C8PGM2.SAS */
/* Blank lines were deleted from the log file to conserve space */
/* ************************************************** * ** */

NOTE: Copyright (c) 1989-1996 by SAS Institute Inc., Cary, NC, USA.
NOTE: SAS (r) Proprietary Software Release 6.12 TS020
 Licensed to NEW YORK UNIVERSITY, Site 0011830001.

1 options ls=80 ps=55;
2 title 'Impact Analysis of Watergate Scandal';
3 title2 'on Nixon Presidential Approval';
4 title3 'Percent of Respondents approving of Way President';
5 title4 'is Handling his Job: Source Gallup Poll Monthly';
6 title5 'January 1969 thru Aug 1974';
7 libname inp 'e:statssas';

NOTE: Libref INP was successfully assigned as follows:
 Engine: V612
 Physical Name: e:statssas

8 data nixon;
9  set inp.nixapp5;
10  time + 1;
11  date = intnx('month','01jan1969'd,_n_-1);
12  scandal=0;
13 if date > '30jan73'd & date < '01Sep74'd then scandal=1;
14 label scandal='Scandal from Convictn of WGate 7 2 end';
15 if _N_ < 69;
16 format date monyy5.;

NOTE: The data set WORK.NIXON has 68 observations and 11 variables.
NOTE: The DATA statement used 0.69 seconds.

17 proc print;
18 title6 'Nixon Era';
19 run;

NOTE: The PROCEDURE PRINT used 0.42 seconds.

23 data anno1;
24  input date date7. text $ 9-50;
25  function='label'; angle = 90 ; xsys='2'; ysys='1';
26  x=date; y=45; position='B';
27 cards;

NOTE: The data set WORK.ANNO1 has 5 observations and 9 variables.
NOTE: The DATA statement used 0.22 seconds.
```

```
35 /* generates the x-axis value above the reference line */
36 data anno2;
37  input date date7. text $ 9-50;
38  function='label'; angle = 90 ; xsys='2'; ysys='1';
39  x=date; y=50; position='B';
40 cards;
```

NOTE: The data set WORK.ANNO2 has 4 observations and 9 variables.
NOTE: The DATA statement used 0.25 seconds.

```
47 /* generates the x-axis value above the reference line */
48 data anno3;
49  input date date7. text $ 9-50;
50  function='label'; angle = 90 ; xsys='2'; ysys='1';
51  x=date; y=50; position='B';
52 cards;
```

NOTE: The data set WORK.ANNO3 has 8 observations and 9 variables.
NOTE: The DATA statement used 0.34 seconds.

```
63 /* generates the x-axis value above the reference line */
64 data anno4;
65  input date date7. text $ 9-50;
66  function='label'; angle = 90 ; xsys='2'; ysys='1';
67  x=date; y=50; position='B';
68 cards;
```

NOTE: The data set WORK.ANNO4 has 3 observations and 9 variables.
NOTE: The DATA statement used 0.19 seconds.

```
74 /* generates the x-axis value above the reference line */
75 data anno5;
76  input date date7. text $ 9-50;
77  function='label'; angle = 90 ; xsys='2'; ysys='1';
78  x=date; y=50; position='B';
79 cards;
```

NOTE: The data set WORK.ANNO5 has 3 observations and 9 variables.
NOTE: The DATA statement used 0.17 seconds.

```
86 axis1 order=(20 to 70 by 10) label=(a=90'Percent approving');
87 symbol1 i=join c=blue v=star;
88 proc gplot data=nixon;
89  plot approval * date /
90  href='23jan70'd '17jun71'd '03sep71'd,'17Jun72'd,'15sep72'd
91  vaxis=axis1 annotate=anno1;
92 title justify=L 'Figure 8.11 The Roots of the Watergate Scandal';
93 title2 'Nixon Presidential Approval during';
94 title3 'The Origins';
95 run;
```

```
101 axis1 order=(20 to 70 by 10) label=(a=90 'Percent approving');
102 symbol1 i=join c=blue v=star;
```

NOTE: The PROCEDURE GPLOT used 1.95 seconds.

```
103 proc gplot data=nixon;
104 plot approval * date /
105   href= '30jan73'd,'23Mar73'd,'30apr73'd '04sep73'd
106   vaxis=axis1 annotate=anno2;
107   where date > '01Dec71'd;
108 title justify=L 'Figure 8.12 Impact Analysis of Watergate Scandal';
109 title2 'on Nixon Presidential Approval Rating';
110 title3 ' The Opening of the floodgates';
111 title4 'McCord Exposes & Dean Implicates';
112 run;
```

```
114 axis1 order=(20 to 70 by 10) label=(a=90 'Percent approving');
115 symbol1 i=join c=blue v=star;
```

NOTE: The PROCEDURE GPLOT used 9.49 seconds.

```
116 proc gplot data=nixon;
117 plot approval * date /
118   href= '17May73'd,'03jun73'd,'25jun73'd, '16jul73'd,'23jul73'd,
119   '04sep73'd,'20Oct73'd,'21nov73'd
120   vaxis=axis1 annotate=anno3;
121   where date > '01jan73'd & date < '01Mar74'd;
122 title justify=L 'Figure 8.13 Impact Analysis of Watergate Scandal';
123 title2 'on Nixon Presidential Approval Rating';
124 title3 ' The First Cascade: The Demand for Evidence';
125 title4 'The Courts & Legislature Beseige Nixon White House';
126 run;
```

```
128 axis1 order=(20 to 70 by 10) label=(a=90 'Percent approving'),
129 symbol1 i=join v=star;
```

NOTE: The PROCEDURE GPLOT used 2.17 seconds.

```
130 proc gplot data=nixon;
131 plot approval * date /
132   href= '17may74'd,'23jul74'd,'08aug74'd
133   vaxis=axis1 annotate=anno4;
134   where date > '01Dec73'd;
135 title justify=L 'Figure 8.14 Impact Analysis of Watergate Scandal';
136 title2 'on Nixon Presidential Approval Rating';
137 title3 'The Final Deluge: ';
138 title4 'Impeachment, Taped Evidence, & Forced Resignation';
139 run;
```

```
142 /* ********************************* */
143 /* Pre-Watergate Nixon Era            */
144 /* Model A is chosen for parsimony    */
145 /* Model F is chosen as best fitting  */
146 /* ********************************* */
```

NOTE: The PROCEDURE GPLOT used 2.47 seconds.

```
148 data prewater;
149 set nixon;
150 if date < '01feb73'd;
```

NOTE: The data set WORK.PREWATER has 49 observations and 11 variables.
NOTE: The DATA statement used 0.26 seconds.

```
151 proc print;
152 title3 'Pre-watergate Nixon Era';
153 run;
```

NOTE: The PROCEDURE PRINT used 0.01 seconds.

```
155 /* does approval need differencing */
156 proc arima data=prewater;
157  i var=approval center stationarity=(adf=(0,1,2,3,4,5,6)) nlag=20;
158  run;
```

NOTE: The PROCEDURE ARIMA used 0.2 seconds.

```
160 proc arima data=prewater;
161  identify var=approval center nlag=25;
162  e p=(1) noint printall plot;
163 title3 'Pre-Watergate Model A AR1 No Seasonal AR';
164 /* residuals not wn */
165 run;
```

NOTE: The PROCEDURE ARIMA used 0.01 seconds.

```
168 proc arima data=prewater;
169  identify var=approval center nlag=25;
170  e p=(1,3) noint printall plot;
171 title3 'Pre-Watergate Model B AR (1,3)';
172 /* parsimonious & resids are wn */
173 run;
```

NOTE: The PROCEDURE ARIMA used 0.08 seconds.

```
175 proc arima data=prewater;
176  identify var=approval center nlag=25;
177  e p=(1) q=(3) noint printall plot;
178 title3 'Pre-Watergate Model C ARMA(1,3) component';
```

```
179 /* Residuals are not wn */
180 run;
```

NOTE: The PROCEDURE ARIMA used 0.02 seconds.

```
182 proc arima data=prewater;
183  identify var=approval center nlag=25;
184  e p=(1,3) q=(4) noint printall plot;
185 title3 'Pre-Watergate Model D: AR2 MA 1 components';
186 /* MA1 term is ns */
187 run;
```

NOTE: The PROCEDURE ARIMA used 0.14 seconds.

```
189 proc arima data=prewater;
190  identify var=approval center nlag=25;
191  e p=(1,6) q-(3,13) noint printall plot;
192 title3 'Pre-Watergate Model E ARMA(1,6-3,13) components';
193 run;
```

NOTE: The PROCEDURE ARIMA used 0.02 seconds.

```
195 proc arima data=prewater;
196  identify var=approval center nlag=25;
197  e p=(1,3,6) noint printall plot;
198 title3 'Pre-Watergate Model F 3 AR components';
199 run;
```

NOTE: The PROCEDURE ARIMA used 0.13 seconds.

```
201 proc arima data=prewater;
202  i var=approval center nlag=25;
203  e p=(1,3) q-(13) noint printall plot;
204 title3 'Pre-Watergate Model G AR2 MA(13)';
205 run;
```

NOTE: The PROCEDURE ARIMA used 0.02 seconds.

```
208 proc arima data=prewater;
209  i var=approval(1) center nlag=25;
210  e p=(1) noint printall plot;
211 title3 'Pre-Watergate Model H diff AR1';
212 run;
```

NOTE: The PROCEDURE ARIMA used 0.02 seconds.

```
214 proc arima data=prewater;
215  i var=approval(1) center nlag=25;
216  e p=(1,3) noint printall plot;
217 title3 'Pre-Watergate Model I diff ar(1,3)';
218 run;
```

NOTE: The PROCEDURE ARIMA used 0.02 seconds.

```
221 proc arima data=prewater;
222   i var=approval(1) center nlag=25;
223   e p=(1,3) q=(13) noint printall plot;
224 title3 'Pre-Watergate Model J diff ar(1,3)MA(13)';
225 run;

227 /* ********************************** */
228 /* Model J is selected for best SBC    */
229 /* Residuals are wn                    */
230 /* All terms are significant           */
231 /* No substantial collinearity         */
232 /* ********************************** */
233
235 /* The Complete Nixon Era */
```

NOTE: The PROCEDURE ARIMA used 0.02 seconds.

```
239 data water;
240   set nixon;
241   time + 1;
```

NOTE: The data set WORK.WATER has 68 observations and 11 variables.
NOTE: The DATA statement used 0.27 seconds.

```
242 proc print;
```

NOTE: The PROCEDURE PRINT used 0.01 seconds.

```
243 proc timeplot;
244   id time date;
245   plot approval;
246 title3 'Nixon Era';
247 run;
```

NOTE: The PROCEDURE TIMEPLOT used 0.08 seconds.

```
248 proc arima data=water;
249   i var=approval(1) center crosscorr=(scandal(1)) nlag=20;
250   e p=(1,3) q=(13) input=(scandal) noint printall plot;
251 title justify=L 'Impact Analysis of Watergate Scandal';
252 title2 justify=L 'testing for scandal ';
253 run;
```

NOTE: The PROCEDURE ARIMA used 0.04 seconds.

```
255 proc arima data=water;
256   i var=approval(1) center crosscorr=(scandal(1)) nlag=20;
257   e p=(1,3) q=(13) input=(3$scandal) noint printall plot;
258   f id=date lead=24 out=nixres;
259 title justify=L 'Impact Analysis of Watergate Scandal';
```

```
260 title2 justify=L 'Parsimonious Model';
261 run;
```

NOTE: The data set WORK.NIXRES has 71 observations and 7 variables.
NOTE: The PROCEDURE ARIMA used 0.19 seconds.

```
263 data forec;
264 set nixres;
265  if date < '30jan73'd then forecast = .;
266  if date < '30jan73'd then l95 = .;
267  if date < '30jan73'd then u95 = .;
```

NOTE: The data set WORK.FOREC has 71 observations and 7 variables.
NOTE: The DATA statement used 0.17 seconds.

```
268 proc print data=forec;
269 run;
```

NOTE: The PROCEDURE PRINT used 0.01 seconds.

```
270 axis1 order=(20 to 70 by 10) label=(a=90'Percent approving');
271 symbol1 i=join v=star c=blue;
272 symbol2 i=join v=plus c=green;
273 symbol3 i=join c=red;
274 symbol4 i=join c=red;
275 proc gplot;
276  plot (approval forecast l95 u95) * date/overlay
277  vaxis=axis1 href='30jan73'd '23mar73'd '23jun73'd annotate=anno5;
278 title justify=L 'Figure 8.17 Impact Analysis of Watergate Scandal';
279 title2 'on Percent of public approving Presidential job performance';
280 title3 'Data from The Gallup Organization web site';
281 title4 'World Wide Web URL: http://www.gallup.com/';
282 footnote justify=L 'data=star, forecast=plus, 95% confidenceintervals=
    lines';
283 run;
```

The complete SAS program constructing and annotating the preceding graphs can be found in the SAS program 8.2. After a review of the graph, the ARIMA preintervention model is identified. The monthly Gallup Poll presidential job approval percentage appears to be an autoregressive process. Owing to detection of MA-type spikes in the residuals of the differenced, AR identification, several models are tested. Model J, which has the lowest SBC, is selected as the best fitting model. With inspection of the ACF and PACF, we can hypothesize that the ARIMA noise model may be a differenced ARMA model with autoregressive lags at 1 and 3, and one moving average component at lag 13. Lines 221 through 225 of the SAS program log above reveal the syntax for programming the preintervention model.

```
Data prewater;
        set nixon;
        if date < '01Feb73'd;
        proc print;
        title3 'Pre-watergate Nixon Era';
        run;

Proc arima data=prewater;
        identify var=approv(1)center nlag=25;
        e p=(1,3) q=(13) noint printall plot;
title3 'Pre-Watergate Model J diff ar(1,3) MA(13)';
        run;
```

After this preintervention model is identified, the model is estimated using conditional least squares. We find the *T* ratios of the AR components to be significant.

Conditional Least Squares Estimation

		Approx.		
Parameter	Estimate	Std Error	T Ratio	Lag
MA1,1	0.47998	0.16842	3.03	13
AR1,1	-0.49209	0.12335	-3.99	1
AR1,2	0.36113	0.12023	3.00	3

The subsequent residuals are diagnosed as white noise.

Impact Analysis of Watergate Scandal
on Nixon Presidential Approval Rating
Pre-watergate Nixon Era
ARIMA Procedure

Autocorrelation Check of Residuals

To Lag	Chi Square	DF	Prob	Autocorrelations					
6	6.39	3	0.094	0.016	-0.077	-0.138	-0.248	-0.108	0.134
12	12.60	9	0.181	0.190	0.194	-0.027	-0.147	-0.007	-0.080
18	17.58	15	0.286	0.009	0.016	-0.005	0.065	-0.232	-0.075
24	25.84	21	0.213	-0.032	0.180	0.109	0.038	-0.041	-0.201

The ARIMA noise model is identified, estimated, diagnosed, and metadiagnosed with the SBC and is presumed to be stable throughout the whole study. It is this model that we can carry over into the model of the complete series. All changes to the series found in the time plot are attributed to the intervention.

After the noise model is diagnosed and resolved, the intervention component needs to be modeled. There are several steps to this process. The first step is to augment the data set by the whole Nixon era. The augmentation is accomplished by construction of a new data set named DATA WATER. All of the NIXON data set is brought down and subsumed within DATA WATER with the SET command. The next step is to set up the identification procedure. The preintervention model is identified, differenced, and centered to effect stationarity and simplification. Identification now entails construction of cross-correlation syntax. The same differencing that is applied to the response series APPROV is applied to the SCANDAL intervention variable in the cross-correlation syntax. Remember that there is no real prewhitening of the input variable with an intervention analysis, even though the cross-correlation syntax is being employed (Box *et al.*, 1994; Brocklebank and Dickey, 1986; Woodfield, 1987; Woodward, 1998).

The next step is to carry over the preintervention model. This is accomplished with the estimation subcommand. The estimation of the preintervention series is carried over into this syntax with the E P=(1.3) Q= (13) NOINT PRINTALL PLOT portion of the ESTIMATE subcommand. The last step is to include and define the input response function to the SCANDAL intervention variable, which is done with the CROSSCORR subcommand within the IDENTIFY subcommand and the INPUT subcommand within the ESTIMATE subcommand. The programming syntax for the full series can be found in lines 235 through 261.

```
235 /* The Complete Nixon Era */

239 data water;
240   set nixon;
241   time + 1;
242 proc print;

243 proc timeplot;
244   id time date;
245   plot approval;
246 title3 'Nixon Era';
247 run;

248 proc arima data=water;
249   i var=approval(1) center crosscorr=(scandal(1)) nlag=20;
250   e p=(1,3) q=(13) input=(scandal) noint printall plot;
251 title justify=L 'Impact Analysis of Watergate Scandal';
252 title2 justify=L 'testing for scandal ';
253 run;
```

```
255 proc arima data=water;
256  i var=approval(1) center crosscorr=(scandal(1)) nlag=20;
257  e p=(1,3) q=(13) input=(3$scandal) noint printall plot;
258  f id=date lead=24 out=nixres;
259 title justify=L 'Impact Analysis of Watergate Scandal';
260 title2 justify=L 'Parsimonious Model';
261 run;
```

In order to model the input response function, we turn our attention to the time plot, the five lines of programming for which may be found under the PROC PRINT statement of the above program. Both the TIME counter, which counts the observations and facilitates lag estimation, and the DATE variable are used as ID variables by which to identify each of the monthly approval ratings. Although the graphical plots before this nicely show the overall picture, they do not carefully identify each observation with its corresponding date. For intervention modeling, the time plots in SAS are very useful. Careful examination of the time plot in Fig. 8.15 at the time of intervention and impact suggests the kind of model to be programmed.

Figure 8.15 reveals that the time following the conviction of the Watergate burglars on January 30, 1973 coincided with the decline in Presidential approval. In March 1973, about two months later, James McCord, the surveillance expert of the Watergate penetration team, disclosed to Judge John Sirica a felonious conspiracy to cover up criminal involvement of high officials, and presidential job approval began to slide downhill. From the appearance of these sudden shocks to the approval rating, we can observe that news of the Watergate break-in aroused suspicions among the *cognoscenti*. However, not until Hunt and Liddy and the other Watergate Five were convicted did Nixon's job approval ratings begin to slump. Only when

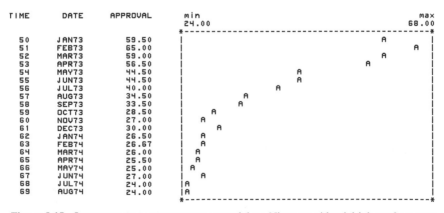

Figure 8.15 Impact analysis of Watergate scandal on Nixon presidential job performance.

Nixon let his chief aides Haldeman and Ehrlichman go was there a pause in the fall of the job approval ratings.

A little more than two months later, on June 25, 1973, John Dean openly implicated Nixon in the cover-up, whereupon presidential job approval resumed its decline. Although Nixon had really been involved in the cover-up from the beginning, the full extent of this involvement would not become evident until the new Nixon tapes were released. Nixon and his aides had tried to pin the Watergate escapade on the CIA and then to use the CIA to block the FBI investigation into their partisan political espionage. By falsely claiming it would compromise too much about the Bay of Pigs covert activity, Nixon tried to get the FBI to back off of their exposure of the political espionage activities of the Committee to Re-Elect the President, affectionately known as the "CREEP." When the burglars were nabbed, Nixon launched the plot to obtain the hush money by trading on an "ambassadorship" in return for the cash. When Alexander Butterfield testified to the existence of the White House taping system that might contain evidence of these activities, in July 1973, Special Prosecutor Archibald Cox and Senator Sam Ervin, chairman of the Senate Watergate Committee, subpoenaed the tapes. Nixon, of course, resisted full disclosure and his approval fell further. In September, Egil Krogh's "plumbers" were indicted for the surreptitious entry and burglary of the office of Dr. Lewis Fielding, the psychiatrist of Daniel Ellsberg, and Nixon's public persona suffered and his job approval slid further. On October 30, Nixon fired Archibald Cox and an 18½ minute gap was discovered on a tape made shortly after the Watergate break-in. Nixon's job approval slid yet further. By the time the scandal had reached its conclusion, 21 Nixon aides had been tried and convicted of crimes. Nixon was gradually and painfully forced from grace, public respect, and political office.

Using these changes as indicators, we examine the cross-correlation function and observe a single negative pulse at lag 3. A pulse function is constructed from a differenced scandal dummy variable. The impact is lagged by 3 months. Just before the incidence of this lag, McCord exposed the White House cover-up and perpetrator perjury, and at about the time of this lag, Haldeman, Ehrlichman, and Dean were fired by Nixon, suggesting that high White House Aides may have been involved. By July 1973, Butterfield had revealed the existence of the White House tapes, and Nixon had resisted turning over subpoenaed tapes. In September of that year the White House plumbers were indicted for the Fielding break-in. The nature of the series is one where the regression coefficients have the following structure. A differenced scandal input would create such a pulse. The impact of these events is represented by the differenced scandal lagged by 3 months.

$$\omega_0 \text{Scandal}_{t-3}(1 - L) \tag{8.34}$$

At this point, it should be remembered that the same differencing that was applied to the response variable is applied to the intervention input.

The regression weights are estimated to find the values for their coefficients. In this model, conditional least squares estimation is employed.

Conditional Least Squares Estimation

Parameter	Estimate	Approx Std Error	t Value	Pr > |t|	Lag	Variable	Shift
MA1,1	0.35594	0.13300	2.68	0.0096	13	APPROVAL	0
AR1,1	-0.35149	0.11097	-3.17	0.0024	1	APPROVAL	0
AR1,2	0.36478	0.11365	3.21	0.0021	3	APPROVAL	0
NUM1	-9.73637	2.61901	-3.72	0.0004	0	scandal	3

Variance Estimate	8.896762
Std Error Estimate	2.982744
AIC	325.3777
SBC	334.0132
Number of Residuals	64

* AIC and SBC do not include log determinant.

All significant coefficients owing to the model reduce the presidential approval following scandal-connected events. The fit statistics—including the standard error estimate, SBC, and AIC—are presented. The correlation matrix in the output reveals that the largest intercorrelation among the parameters was -0.176, from which we can infer that multicollinearity is not problematic. Then the autocorrelation check of the residuals is presented. If these are white noise residuals, then the systematic variation has been accounted for by the model parameters and the model fits. Although a model with fewer AR terms or no MA term may be parsimonious, the residuals are more white noise from this chosen model. When only white noise residuals remain, as indicated by the insignificant modified Q Tests, these model parameters seem to account for all significant impact (Fig. 8.16).

When we review the graphs of these sequence of events, we perceive events that drive down the approval ratings at lags 3, 6, and 8 after the disclosure of the cover-up. What events are associated with these drops in approval? Following the burglary conviction, James McCord's exposure of the cover-up, and the firing of top White House Aides Haldeman, Ehrlichman, and Dean were accompanied by a precipitous decline in presidential approval. When the televised Senate Watergate hearings convened, there was a short let-up while both sides built their cases. For two months, the Senate Watergate hearings were televised. But once John Dean, former White House counsel, implicated Nixon in the cover-up and obstruction

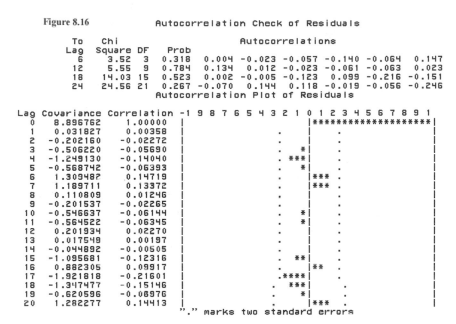

Figure 8.16 Impact analysis of Watergate scandal: best-fitting model.

of justice, approval of the presidential performance started its downhill slide again. The legislature and the courts demanded the tapes and the president demurred, citing "executive privilege"; there was a slow-down of decline. However, once the plumbers were indicted the rate of decline increased again. The mathematical equation describing this phenomenon is based on the following output.

```
                Model for variable APPROVAL
                Data have been centered by subtracting the value -0.522388006.
                No Mean term in this model.
Period(s) of differencing = 1.

                Autoregressive Factors
Factor 1: 1 + 0.35149 B**(1) - 0.36478 B**(3)

Moving Average Factors
Factor 1: 1 - 0.35594 B**(13)

Input Number 1 is Scandal with a shift of 3.
Period(s) of Differencing 1
Overall Regression Factor -9.73637
```

From this output, we can reconstruct the formula for Nixon's presidential approval ratings before and during the Watergate scandal as

$$(1 - L)(\text{Approval}_t - 0.522)$$

$$= (-9.736)\text{Scandal}_{t-3}(1 - L) + \frac{(1 - 0.356L^{13})e_t}{(1 + 0.351L - 0.365L^3)}. \quad (8.35)$$

The series for Nixon's presidential approval rating can be mathematically explained as a nonstationary series $I(1)$. After differencing, this presidential approval series is typically an autoregressive series with components at lags 1 and 3 prior to the Watergate scandal. The onset of that scandal is modeled as a differenced step (pulsed) input. The impact of the scandal is associated with a plummeting value of Nixon's approval rating. The approval was plummeting after the conviction of Hunt, Liddy, and the burglars. About three time periods (months) afterward, McCord revealed the existence of a conspiracy to cover up the full extent of involvement. Nixon's implication came with John Dean testimony in June of 1973. By July, Nixon refused to turn over the subpoenaed tape recordings, and his approval fell further. Eventually, Nixon was forced to turn over the tapes, which contained evidence that he had been a principal to obstruction of justice, abuse of power, sabotage of the democratic process, and failing to take care that the laws were faithfully executed, whereupon his legitimacy crumbled and he was forced to resign lest he be certainly convicted of high crimes and misdemeanors. The end result, a general public disillusion and dissatisfaction with his presidency, was reported by more than three-quarters of the people polled. In the end, the later exposure of Nixon's darker deeds overwhelmed him and his place in history. The model and the forecast profile generated by it is presented in Fig. 8.17.

In the analysis of political scandals, there is a caveat. This formula for the decline in presidential approval rating for President Richard M. Nixon's administration highlights a dimension of Nixon's popularity among the people, but does not necessarily hold for that of any other president. Charisma was a coat that did not fit Nixon well. His expressions, mannerisms, and style were stilted and awkward. The circumstances of the Nixon administration were rather unique and, assuming the nation's leaders learn from the errors of their predecessors, there is no reason to suspect that this tragedy for the country is destined to recur in the near future. Nixon had promised to extricate the United States from an unpopular military quagmire in Vietnam. Contrary to promises, he actually appeared to extend the war into Cambodia in 1970. His military activities against North Vietnam engendered protest at home while the economy was hobbled by balance-of-payments problems in 1971, by leaving the gold standard later that year, and then by the OPEC oil embargo and production cutbacks in the 1970s.

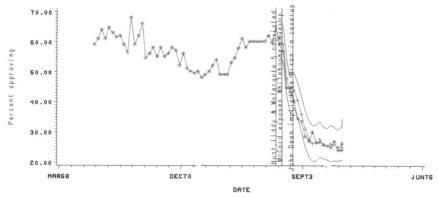

Figure 8.17 Impact analysis of Watergate scandal on percent of public approving presidential job performance. Data from the Gallup Organization Web site (http://www.gallup.com).

Governmental credibility over the Vietnam war had crumbled by summer of 1971; trust in government suffered, as political discontent and unrest percolated. Not all political scandals have the capability to undermine the foundations of a presidency. For the country to learn from its history, its historians and political scientists must carefully study it and accurately report it. More will be said about the limitations of this kind of analysis at the end of the chapter.

Other presidential scandals need not resemble Watergate. The key decision makers, their organizations, and the institutions involved as well as the configurations of power within them must be analyzed amid the political context and historical background. A complete analysis of presidential scandals might have to candidly and completely examine the Teapot Dome scandal, the surprise of the Japanese attack on Pearl Harbor, the Bay of Pigs fiasco, the Gulf of Tonkin incidents, the Watergate scandal, the Iran–Contra scandal, and the Clinton–Lewinsky scandal. Only from an honest analysis of a more or less complete set of political scandals could the commonalities and their impacts be deduced. Although these findings would be of great interest to serious students of political science, they might not find much government sponsorship outside of the intelligence agencies.

To illustrate how presidential scandals differ in their impact, consider the Clinton–Lewinsky scandal. Monica Lewinsky was a young White House intern fresh out of college with whom President William Jefferson Clinton had a brief inappropriate relationship. The Republican Congress during the tenure of President Clinton gave the appearance of an inquiry in search of a scandal. Time and again, unfounded accusations gave rise to inquiries. Time and again, none of these inquiries yielded evidence of crimes linked to the White House. President Clinton, apart from this lapse of judgment, has been an extraordinary talented, politically adept, diplomatically adroit, and extremely intelligent president. In international affairs, he made decisions that put an end to genocide in Bosnia–Herzegovina, expanded NATO, mobilized NATO in attacking Serbia for its massacres of Kosovars, facilitated an end to "the Troubles" in Ireland, brokered two peace deals in the Mideast, and extended a hand of friendship to the Chinese people. He twice forced Saddam Hussein to promise to live up to UN agreements to allow UN weapons inspectors to continue their efforts to verify disarmament of weapons of mass destruction. When Hussein finally reneged on his UN agreements, President Clinton initiated air strikes against Iraq designed to reduce Hussein's ability to threaten his neighbors or the world. When Osama Bin Laden declared war against the United States and sponsored attacks against U.S. Embassies in Tanzania and Kenya, Clinton retaliated against Bin Laden's camp. He spoke before the United Nations, rallying countries to a war against terrorism before the terrorists obtained weapons of mass destruction. For his efforts in support of international security and peace, it has been reported that he was under serious consideration for the Nobel Peace Prize.

Domestically, the U.S. economy was prosperous and robust at the time of the scandal, even though other economies had suffered. The Asian economies had begun to cave in. The Russian economy was almost imploding. Instabilities were apparent in Latin American economies. Although the U.S. economy suffered from a drop in demand for exports from those economies, domestically things were going fairly well.

Many scholars—including, Edward Tufte—have noted that economic prosperity may determine the kind and extent of support that a president enjoys. Measures of such prosperity, such as the Conference Board index of consumer confidence or the University of Michigan index of consumer sentiment, have been used to indicate this kind of well-being. Nonetheless, this was not used as a factor in these analyses to model the impact of the Watergate or Lewinsky scandal, because of a desire to focus on the impact of the Watergate scandal in the former case and in the latter case for lack of sufficiently large sample size. Notwithstanding these constraints, the general economic well-being of the American citizen remains an important factor in public approval of the quality of political leadership.

While Congress and the political comedians fixated on the investigation of Clinton's indiscretion with a White House intern, the public became disenchanted with the lack of congressional focus on matters of real national interest. That a popular president had been seduced first by a young intern, hounded by a special prosecutor, ensnared in a perjury trap, deprived by the courts of legal support and real executive security, stripped of his privacy by having his personal peccadillos dumped in lurid detail onto the World

Wide Web by Congress, and divested of personal respectability by having his personal reputation besmirched before the world made the public wonder how insensitive and prurient the minds of the Republican Congressmen really were. Columnists began to complain that Congress was disgusting the country and leading it astray, and wondering what kind of grotesque country the United States had become. Images of sexual McCarthyism emerged and people began to sympathize with the persecuted rather than the prosecutors. The White House counterattacked that this was a donothing Congress willfully negligent of the needs of the people. The Republicans failed to gain all the seats they expected to in the election of 1998. The Republican Speaker of the House, Newt Gingrich of Georgia, whose relationship with a younger aide to the House Agricultural Committee would later become an issue during his divorce proceedings, resigned. The next Republican Speaker of the House, Robert Livingston of Louisiana, was discovered to have been guilty of sexual indiscretion and therefore forced into resignation. Clinton's popularity began to increase while that of the Republican-dominated Congress began to decrease. The public saw Clinton as basically an effective president who essentially should be forgiven for a real mistake. In a Gallup Poll in early November, 66% of national adults wanted Clinton not to be impeached and to remain in office.

Clinton's political opposition complained that he was just too slick and always one step ahead of them. When Clinton committed a personal and sexual peccadillo by having a liaison with a young, talkative White House intern, he gave her something to brag about. Rumors began to spread. This time they spread through the conservative spyvine back to the special prosecutor. After having been put on the stand and having publicly denied sexual involvement, he was forced to admit an inappropriate intimate relationship with the intern. Clinton, however, was blessed by the unsavory character of many of his most prominent political enemies and persecutors. Time and again, he was blocked by a Republican congressional majority from passing social reform legislation in the best interests of minorities and the needy in the country. In areas of campaign finance reform, tobacco legislation to protect the public health, health insurance, funding for more teachers and educational facilities, and pro-choice legislation favored by most women, the Republican Congressional majority protected the special interests and thwarted Clinton (Bentley, 1998). In the meantime, he built up a lot of faith, credit, and trust on the part of the people who believed that he was trying to do the right thing for the country. When he found himself caught by the Republicans and upbraided by friendly Democrats for this serious lapse of judgment, the more issue-oriented people in the country rallied around him rather than see him be politically lynched by his rabid Republican opposition. The majority of the mass public wanted the country spared another prolonged, offensive, insensitive, Republican investigation from which they could expect minimal yield. The public's sensibilities had been offended *ad nauseum* by these congressmen. They preferred that Clinton be censured and the Congress move on to attend to the pressing interests of the country. Others marveled at the kind of interpretation of the Constitution that would deprive a political system of stability and flaw it with vulnerability by allowing the whimsical irresponsibility of a young White House intern to bring down a Presidential administration.

Within this situation, Clinton's Gallup Poll presidential job approval ratings at first declined slightly, but then recovered (Fig. 8.18). While evaluations of Clinton's personal character suffered, his Gallup Poll monthly job approval average remained between 60 and 66%. Although the possibility that he had lied under oath threatened a charge of perjury, there was no credible evidence of suborning perjury or obstruction of justice. Much depended on whether these constituted "high crimes and misdemeanors" of the type the framers of the Constitution or the House of Representatives interpreted them to be. Although the scandal contributed to a decline in approval of the personal character of the President and threatened Clinton with impeachment, it was surprisingly accompanied by a general rise in Clinton's Gallup Poll presidential job approval as he masterfully dealt with situations and crises that challenged him. Although many members of the U.S. Senate expressed disapproval of President Clinton's misbehavior, the Senate ultimately acquitted him, on February 12, 1999, of crimes alleged in the articles of impeachment passed by the Republican-dominated House of Representatives, for lack of evidence, proof, or seriousness of the crimes. In other words, not all presidential scandals overturn a very popular president who clearly made a mistake.

8.7. APPLICATIONS OF IMPACT ANALYSIS

Impact analysis permits the study of input and output phenomena in the time domain. It has clear applications in the modeling of regime changes, impacts of external events (including policy changes), scenarios, contingencies, or even outliers in time series analysis. In contrast to cross-sectional research, impact analysis allows examination of the temporal sequence necessary for confirmation of sequential or causal relationships. It permits careful modeling of various forms of impact of one or more events on a response series. If a graphical analysis suggests a change in regime, indicated by a change in the level of a response series, and an objective test—e.g., Chow or Likelihood Ratio test—confirms such a structural change, then impact analysis might be in order. By forecasting from the preintervention series, intervention analysis permits comparison of the impact of an intervention with that preintervention forecast. In this way, the net difference between what might have happened, *ceteris paribus,* had there been no intervention and the impact of the intervention becomes clear. Intervention analysis enables the analyst to model a change in situation with the inclusion of an independent dummy variable. If the change in regime is gradual rather than sudden, the analyst can model that change in situation or regime by an impulse response function of an intervention variable. Various shapes of impact may be modeled by combining the components of gradual or sharp onset with those of sharp or gradual attenuation or oscillation of effect. These techniques are part and parcel of interrupted time series analysis (McCain and McCleary, 1979; McDowell *et al.,* 1980).

The accuracy of impact analysis is contingent upon the fulfillment of the assumptions mentioned earlier. The analyst can test alternative explanations for the observed impact on the series by including other event indicators in a multiple time series intervention analysis. The input indicator variable is deterministic, representing the presence or absence of an event. Multiple input functions may be modeled. The other deterministic indicators should be variables representing the plausible alternative explanations for the impact. If and when those impulse functions are shown to be nonsignificant, they are eliminated as explanatory variables, unless required for specification of other variables. If and when other deterministic inputs are significant, the input combinations will represent the combined driving forces of the response series. For example, if there are two significant step inputs, each with two possible values, the multiple input of the two indicators yields four combinations that can drive the response series. In option one, both inputs will have a value of zero, which may be deemed a reference point from which others may deviate. In option two, both inputs will have a value of unity. In option three, one input will have a value of unity and the other will have a value of zero, and in the final option, the input that in option three that had a value of unity will have a value of zero and the input that had a value of zero will have a value of unity. To represent k combinations of categories, it will be necessary to include k-1 dummy variables, regardless of whether all k-1 dummy variables are statistically significant. In this way, the final regression model can control for multiple complex and compound explanations, as well as interactions.

Assessment of interventions on a series is a valuable tool for policy analysis. If the system is relatively closed and there are only a few impacts on the series, then this kind of analysis is empirically very useful. It has the advantage of demonstrating temporal sequence, which is necessary for establishing a causal model. These models are flexible. These intervention models can entertain two or more separate interventions at different points in time. They may involve interactions between other inputs. Indeed, if there are two or more inputs, it may behoove the researcher to test whether there is a joint effect of these inputs, over and above their separate impacts (Ege, *et al.*, 1993). They may involve continuing interventions—for example, extended pulse functions or step functions. Although this is not sufficient to establish causality, the method provides a valuable statistical technique for detecting, testing, and modeling empirical evidence of causal relationships. When a conflict develops between statistical and theoretical fit, the model building should be theory-driven. Nonetheless, the temporal sequence is one component of causality in statistical models that cross-sectional research designs do not capture (Nagel, 1961; Campbell and Stanley, 1963; McDowell *et al.*, 1979). Application of intervention models in time series may elucidate these complex relationships.

Furthermore, impact analysis with its pulse function input may be used for modeling outliers, unusual data points that may be the product of coding errors. Such an error can produce what is called an observational or additive outlier, which can be modeled by a pulse intervention indicator. To model this outlier, the ARIMA series model [in this case, an ARIMA(1,0,1) process] is added to the pulse response function $(1 - L)I_{t-b}$ with intervention indicator I_t, time delay b, regression coefficient ω_1:

$$Y_t = \frac{(1 - \theta_1 L)}{(1 - \varphi_1 L)} e_t + \omega_1 (I - L)I_{t-b}. \tag{8.36}$$

This is a simple example with only one outlier. Actually, the series may have several outliers. The model for such a series would require an outlier response function for each outlier. In that case, there would be as many components beginning with ω_i on the right-hand side of the equation as there were outliers in the series. This kind of model is similar to the impact model of the Watergate scandal, which has multiple pulse inputs.

If the outlier has a persistent or permanent effect on the level and variance process of the series, it is called an innovational outlier. Innovational outliers are more complex than observational outliers. The formula defining such outliers was given by Mills (1990). With an innovation outlier the presence of the extraordinary shock effects a sustained general response through the noise model of the data-generating process:

$$Y_t = \frac{(1 - \theta_1 L)}{(1 - \varphi_1 L)} e_t + \frac{(1 - \theta_2 L)}{(1 - \delta_1 L)} [\omega_1 (I - L)I_{t-b}]. \tag{8.37}$$

where $\delta_1 = \varphi_1$.

The impulse response or transfer function is divided by the autoregressive parameters and multiplied by the moving average parameters before being added to the noise model. For a more detailed discussion of the analysis of innovational outliers, readers can consult Mills (1990) or Box *et al.* (1994). The procedure for modeling these outliers is the same as for modeling the standard impact analysis model. If an innovation outlier is estimated, the polynomial denominator of the transfer function will be the same as the autoregressive polynomial in the noise model of the series, in which case they are functionally equivalent. That is, δ_1 and ϕ_1 should be the same in sign, number, and magnitude.

Outlier detection requires an iterative process. First, the ARIMA model is estimated for the portion of the series under consideration (whether the preintervention or the postintervention series). The residuals and residual variance are obtained from this series. If interventions or innovational outliers occur near the end of the series, there will have to be enough observations that follow their occurrence for them to be properly detected, diagnosed, and modeled. Asymptotic standard errors are computed from the residual variance and standardized t statistics are computed for each innovational outlier. When these t statistics exceed a value of 3, the point is identified as an outlier with a significant impact (Box *et al.*, 1994). For observational outliers, the variance calculations are slightly different. These outliers can be smoothed out by assigning them the value of the mean of the residual. The process is reiterated until all outliers are identified, replaced, modeled, or removed.

8.8. ADVANTAGES OF INTERVENTION ANALYSIS

Time series research designs have very substantial advantages over other conventional research designs. A time series research design may be required to detect changes in level, slope, or regime of a process. Sometimes the impact of the intervention, treatment, or event is not applied instantaneously. A time series research design may be needed to detect a gradual, threshold, delayed or varying effect, which might go unobserved in a more conventional cross-sectional design. Cook and Campbell (1979) note that this kind of design is useful in detecting temporal change, such as the maturation of a trend prior to the intervention. If the researcher takes large, representative, and equivalent samples for control and experimental groups, he may be able to properly assess these effects (Campbell and Stanley, 1963). A principal advantage of a quasi-time-series experiment is that it focuses on the sequence of events, some of which may be input and others of which may be responses. Along with the covariation of input and response, this sequence of these events is necessary for the inference of causality. The modeling of the type of response reveals a sense of the structure of the impulse response to an input event. The shape of the response facilitates understanding of the nature of the effect, as it were. The advantage of multiple observations is that it is possible to detect and model various forms of impact that often escape detection and observation in more conventional cross-sectional designs.

8.9. LIMITATIONS OF INTERVENTION ANALYSIS

The researcher needs to understand the principal advantages and disadvantages of his design. Time series quasi-experiments may be afflicted with problems that threaten the internal and external validity of the analysis. Cook and Campbell (1979) note a number of threats to internal validity inherent in this kind of design. Problems with instrumentation may confound the design. The researcher must be sure that the series is properly defined conceptually and operationally before data collection. Proper administration of the data collection and maintenance of the records throughout the process is necessary. The time intervals must be made small enough to capture the process to be studied. If there is trend, cycle, or seasonality inherent in the series, then the instrumentation must be calibrated to units of temporal measurement appropriate to the capture, detection, and identification of these components. Calibration must be maintained. Without a large enough sample size for the preintervention and postintervention series, there will not be enough power to detect the differences of trend, cycle, seasonality, noise, or impact necessary for modeling an intervention analysis. Moreover, there must be sufficient protection against possible alternative historical impacts on the process to ensure internal validity. A concurrent control group for baseline comparison may be used to guard against such threats. A control group series isolates part of the series from impact stemming from the event, intervention, or treatment. The control of group series can then be concurrently compared to the impacted series. If the impact generates sample attrition, then selection bias may also creep into the study, which can be guarded against with the use of a control group series.

There may arise threats to internal validity that preclude confirmation of a causal relationship between two variables. Some threats to internal validity are not completely overcome by application of a time series quasi-experiment. Without concurrent isolation and establishment of a control or baseline series, it may be not be possible to conclusively demonstrate that the intervention alone generated the observed impact. Cook and Campbell (1979) mention several of these threats to internal validity. A control group helps keep maturation of the subjects within the time frame of the quasi-experiment from confounding the results. It can isolate subjects from test reaction bias. It may be necessary to shield against differential attrition of subjects from the groups owing to fatigue, demoralization, or other external pressures. Random assignment to the experimental and control groups is necessary to protect against differential selection of subjects into the groups and differential attrition of subjects. The use of equivalent control groups may also protect against interaction of selection and maturation biases, on the one hand, or selection and historical biases, on the other. Without the baseline series separated from the experimental series, there is no guarantee of protection against the confounding of learning, fatigue, or other carryover effects. With separate control and experimental groups, it is sometimes possible to employ unobtrusive measures that in and of themselves isolate the subjects to prevent compensatory equalization, imitation, or competition between or within the two groups. The unobtrusive measures may be necessary to preclude demoralization, which could bias the intervention effects, among those receiving the less favorable treatment. In many cases, the advantages far outweigh the disadvantages in the application of this kind of analysis. If the researcher guards against these contaminating problems so they do not plague the particular impact analysis, interrupted times series or intervention analysis may prove very valuable.

Another threat to internal validity is insufficient closure of the system under examination. There may be hidden factors at play that are not readily apparent to the analyst. The failure to model these factors is called specification error. Specification error can result in biased estimation. If the omitted variable is now in the error term, and if there is a positive correlation between the omitted and an included variable, then the error is now related to an included variable. This will increase the magnitude of the estimated coefficient of the included variable. If there is a negative correlation, then the estimation of the parameter is a reduction in the magnitude of the regression coefficient of the included variable. In either case, the parameter estimation of the included variable is biased. The significance tests can also be biased and spurious relationships may be mistaken for real ones. A conflict between closure, completeness, and consistency may be impossible to overcome, according to Gödel's Incompleteness Theorem (Kline, 1980).

Even though these models are occasionally called causal models, it is important to dispel the myth that causality, strictly speaking, is really being proven. For this reason, we need to consider the limitations of impact analysis in the demonstration of predictive causality. Although supporting evidence for a causal relationship may be developed by a time series quasi-experiment, the quasi-experiment does not, strictly speaking, prove causality. Although temporal sequence may be necessary for causality and may be shown by such a quasi-experiment, temporal sequence by itself is insufficient to prove causality. As David Hume has written, the habitual observation of a sequence of an event followed by another event is not a valid test of a causal relationship. When this physical proximity and temporal sequence appears to be invariable, the antecedent event is presumed to be a cause and the subsequent effect is presumed to be an effect. Just because a temporal sequence was repeatedly observed in the past does not imply that sequence will always take place in the future (Nagel, 1961).

If a sample is small and unrepresentative, the causal relationship might appear to hold. If the sample is expanded, it might be shown that under some circumstances the relationship does not hold. Since the material implication of logical induction is not empirically guaranteed, the early empiricists—Hume and Comte—insisted that established causality must be directly observed rather than inferred. Operationalism emerged as defining phenomena in terms of their measuring instruments and measurements, from which causality must be observed for its existence to be established (Cook and Campbell, 1979).

To infer a law from a single case would be to commit a universalistic fallacy. To believe that the Watergate scandal is characteristic of all political scandals is to commit that same fallacy. For a theory of the impact of scandals on the presidency, it is necessary to examine a representative number of scandals and their political, economic, and sociocultural environments. To be sure, the Watergate scandal undermined the presidency of Richard Nixon. His Gallup Poll presidential job approval ratings plummeted as his implication in the illegalities became more apparent. Politically, Nixon presided over an unpopular war and defeat in Southeast Asia, the failure to support Taiwan fully in its conflict with mainland China, and the development by the oil producing states of an oil weapon against Israel and its allies, including the United States. Trust in government was already a casualty of the Vietnam war and the Johnson administration. Perhaps in these respects, the Watergate scandal was sufficient to reveal extensive governmental corruption and to precipitate the impeachment and probable conviction of a president, which would force his resignation. The reader is warned not to commit the universalistic fallacy—overgeneralizing from a single case to a universalistic law—thereby concluding that political scandal guarantees the removal of a U.S. president.

If a president and his administration clearly work to foster what the people think are its best interests, enough faith and credit may be built up that people will support him even if he gets into trouble. During the administration of President Clinton, the exposure of an inappropriate liaison with a White House intern was not enough to fatally undermine President Clinton's Gallup Poll job approval ratings. President Clinton and his administration presided over a healthy and prosperous economy. He advocated campaign finance reform, tobacco legislation that would protect children from addiction and poisoning that comes from protracted smoking, and better school facilities that the Republican party opposed. He advocated protection of Social Security while Republicans pushed for tax cuts and scandalmongering. Although the disclosure of the inappropriate relationship embarrassed Clinton and the administration, the American mass public had more faith in him than it did in the opposition and consistently opposed his impeachment. Clinton's public approval ratings in general increased from the onset of the scandal right up to the congressional election, in which the Republicans failed to gain the customary number of seats. A graph of his public approval ratings from the onset of the crisis (Fig. 8.18) shows how resilient those ratings were in the face of investigation and exposure. The Conference Board index of consumer confidence is also graphed to show how one might be related to the other.

Essentialist philosophers added other criteria for establishing causality. In defining cause and effect as a necessary, sufficient, inevitable, and infallible functional relationship, they maintained that the cause refers to a constellation of variables that when taken together are both necessary and sufficient for an effect to occur. Whether precursors to events are necessary and/or sufficient for other effects to occur may require controlled experiments rather than naturalistic surveys. Controlled experiments involve random assignment of subjects to experimental and control groups as well as pre- and postintervention observations. Therefore, the impact analysis discussed here is not, strictly speaking, a controlled experiment. At best, it is a quasi-experiment riven with possible drawbacks. One drawback is the lack of differentiation between a control and an experimental group. Another problem is the lack of random assignment to control and experimental groups. Without these safeguards, impact analysis does not qualify as a controlled experiment. At best, it constitutes what Cook and Campbell (1979) call a time series quasi-experiment.

John Stuart Mill took the requirements of establishing causality one step

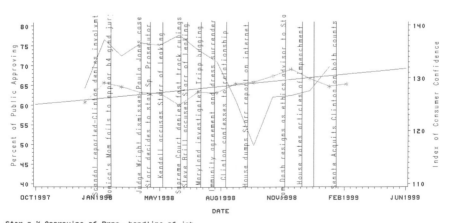

Star = % Approving of Pres. handling of job
Line = Index of consumer confidence
Straight Line = Trend line of Pres. job approval

Figure 8.18   Gallup Poll: Percent approving Clinton handling job.

further. He concurred with Hume that cause and effect must be related and that cause must precede effect in time. The mere temporal sequence of day and night does not mean that day causes night or that night causes day. He maintained that alternative explanations of the causal relationship need to be tested and eliminated, in which case the simple impact analysis model may not under all circumstances allow for the inclusion of enough independent variables to test all plausible alternative explanations. It is helpful to understand that under many circumstances alternative explanations can be tested with multiple input variables in an intervention analysis. This presumes that the inputs take place during the overall time span of the response series. Impact analysis allows a detailed assessment of the functional relationship of the impact of particular events or sets of events on a particular series, but it is necessary to examine in detail not just one such event, but a representative number of them before tendering generalizations about them.

REFERENCES

Ammendola, Guiseppe (1998). *From Creditor to Debtor: The U.S. Pursuit of Foreign Capital— The Case of the Repeal of the Withholding Tax.* New York: Garland Publishers; also personal communication, November 21, 1998.

Andries, L., *et al.* (1994–1998). *Encyclopaedia Britannica CD 98 Multimedia Edition,* version 98.0.09, New Deal articles.

Bechloss, M. R. (1997). *Taking Charge: The Johnson White House Tapes 1963–1964.* New York: Simon and Schuster, pp. 63–64. LBJ is worried about the CIA connections and fears that news that Castro or Krushchev was behind the JFK assassination might spawn widespread demand for retaliation that could lead to war. LBJ was the source of the official cover-up. Russo (1998) notes this as well.

Bentley, S. (1998) aptly suggested how much women appreciated his position against the anti-abortionists. (Personal communication: November 30, 1998.)

Boller, P. F., Jr. (1984). *Presidential Campaigns.* New York: Oxford University Press, pp. 171–172.

Box, G. E. P., and Tiao, G. C. (1975). "Intervention Analysis with Applications to Economic and Environmental Problems." *Journal of the American Statistical Association,* **70**(349), Theory and Methods Section, pp. 70–79.

Box, G. E. P., and Tiao, G. C. (1978). "Applications of Time Series Analysis," in *Contributions to Survey Sampling and Applied Statistics In Honor of H. D. Hartley,* New York: Academic Press, pp. 203–219.

Box, G. E. P., Jenkins, G. M., and Reinsel, G. C. (1994). *Time Series Analysis: Forecasting and Control.* 3rd ed. Englewood Cliffs, NJ: Prentice Hall, pp. 388, 462–479.

Brocklebank, J. C., and Dickey, D. A. (1986). *SAS System for Forecasting Time Series.* Cary, NC: SAS Institute, Inc., p.160.

Burnham, W. D. (1970). *Critical Elections and the Mainsprings of American Politics.* New York: Norton, p. 15.

Burnham, W. D. (1982). *The Current Crisis in American Politics.* New York: Oxford University Press, pp. 17, 71, 97, 174–176, 180.

Campbell, A., Converse, P., Miller, W., and Stokes, D. (1960) *The American Voter.* New York: Wiley.

Campbell, D. T., and Stanley, J. C. (1963). *Experimental and Quasi-Experimental Designs for Research.* Chicago: Rand-McNally, pp. 37–40, 55–56.

Chang, Ih, Tiao, G. C., and Chen, C. (1988). "Estimation of Time Series Parameters in the Presence of Outliers." *Technometrics* **30** (2), 193–204.

Cook, T. D., and Campbell, D. T. (1979). *Quasi-Experimentation: Design and Analysis Issues for Field Settings.* Boston: Houghton Mifflin, pp. 58, 261–293.

Ege, G., Erdman, D. J., Killam, B., Kim, M., Lin., C. C., Little, M., Narter, M. A., and Park, H. J. (1993). *SAS/ETS User's Guide. Version 6,* 2nd ed. Cary, NC: SAS Institute, Inc., pp. 122, 142–145, 177–182.

The Gallup Organization (1969–1974). "President Nixon's Popularity," *The Gallup Opinion Index.* Princeton, NJ: Gallup Organization. Dec. 1969, 1; Nov. 14–16, 1970, 2; Dec. 1970, 2; Jan. 1971, 1; Dec. 10–13, 1971, 3; Jan. 7–10, 1972, 5; June 23–26, 1972, 3; Nov.–Dec., 1972, 1; Feb 22–25; Mar. 1973, 2; Aug. 17–1, 1973, 4; Sept. 1973, 4; Dec. 1973, 2; Mar. 1–3, 1974, Apr. 12–15, 1974, 9; June 21–24, 1974, 1, 5; July 1974, 3. These data are used and reprinted with permission of The Gallup Poll, 47 Hulfish Street, Princeton, NJ 08542.

The Gallup Organization (1969–1974). "President Nixon's Popularity," *Gallup Opinion Monthly Opinion Index,* Report '92, Feb. 1973, pp. 2–3. These data are used and reprinted with permission of The Gallup Poll, 47 Hulfish Street, Princeton, NJ 08542.

The Gallup Organization (1997). The Gallup Poll, accessible on the World Wide Web at http://www.gallup.com. These data are used and reprinted with permission of The Gallup Poll, 47 Hulfish Street, Princeton, NJ, 08542.

Heffernan, R. J. (1991). *The Restorative Process in the American Polity: Party Competition for the House of Representatives and the lesser Chambers of the States.* New York: New York University. This well written, unpublished doctoral dissertation employs intervention analysis to determine whether the election of 1894 is a critical election.

Jensen, R. (1971). *The Winning of the Midwest.* Chicago: University of Chicago Press.

Kleppner, P. (1970). "The Political Revolution of the 1890s: A Behavioral Interpretation." Chapter 12, pp. 184–194.

Kline, Morris (1980). *Mathematics: The Loss of Certainty.* New York: Oxford University Press, p. 261.

Kirkpatrick, Lyman (1968). *The Real CIA.* New York: Macmillan, pp. 188–197.

Kutler, Stanley I. (1997). *Abuse of Power: The New Nixon Tapes.* New York: The Free Press, pp. 13–19, 59.

Leiserson, A. (1958) *Parties and Politics An Institutional and Behavioral Approach.* New York: Alfred A. Knopf, p. 166.

Leonard, M. (1998). SAS Institute, Inc. Cary, NC (Personal communications). Mike was helpful in advising me on aspects of SAS programming of impact analysis.

Maisel, R. (1999). Personal communication. September, 18, 1999. Maisel and Tuckel had discovered evidence of party machine mobilization of immigrant voters.

Makridakis, S., Wheelwright, S. C., and McGee, V. E. (1983). *Forecasting: Methods and Applications.* New York: Wiley, p. 485. Makridakis *et al.* take the position that the levels of differencing of the input variable define the order of a transfer function rather than the number of delta parameters, suggested by Box *et al.* (1994), pp. 383–392.

Makridakis, S., and Wheelwright, S. C. (1987). *The Handbook of Forecasting: A Manager's Guide,* 2nd ed. New York: John Wiley and Sons, p. 215.

McCain, L., and McCleary, R. (1979). "The Statistical Analysis of the Simple Interrupted Time Series Quasi-Experiment." In *Quasi-Experimentation: Design and Analysis Issues for Field Settings.* Cook, T. D., and Campbell, D. T., Eds.) Boston: Houghton Mifflin, pp. 233–294.

McCleary, R., and Hay, Jr. R. A. (1980). *Applied Time Series Analysis for the Behavioral Sciences.* Beverly Hills, CA: Sage, pp. 247–248. Data from this book are used and reprinted with permission of the author.

McDowell, D., McCleary, R., Meidinger, E. E., and Hay, R., Jr. (1980). *Interrupted Time Series Analysis.* Newberry Park, CA: Sage Publications, Inc., pp. 64–65.

Mills, T. C. (1990). *Time Series Techniques for Economists.* New York: Cambridge University Press, pp. 235–247.

Nagel, Ernst (1961). *The Structure of Science.* New York: Harcourt, Brace, & World, pp. 73–74.

Niemi, R. G., and Weisberg, H. F. (Eds.) (1976). *Controversies in American Voting Behavior.* San Francisco: W. H. Freeman and Co., pp. 359–360.

Pack, D. (1987). A Practical Overview of ARIMA Models for Time Series Forecasting. In *The Handbook of Forecasting: A Manager's Guide.* Makridakis, S., and Wheelwright, S. C., Eds.). New York: Wiley, p. 215.

Pomper, G. (1972). "Classification of Presidential Elections" in Silbey, J. H., and McSeveney, S.T. (Comps.) (1972). *Voters, Parties, and Elections; Quantitative Essays in the History of American Popular Voting Behavior.* Lexington, Mass: Xerox Publishing, p. 5.

Russo, G. (1998). *Live by the Sword: The Secret War against Cuba and the Death of JFK.* Baltimore, MD: Bancroft Press, pp. 16, 150, 600. Russo argues that Lee Harvey Oswald learned of secret anti-Castro activities of the U.S. government and sought to bring them to an end, which he succeeded in doing. Also see Bechloss (1997), pp. 63–64, 421–423.

SPSS, Inc. (1994). *SPSS Trends 6.1.* Chicago, IL: SPSS, Inc. pp. 137–151.

Tuckel, P. (1999). Personal communication. (September 21, 1999).

Vandaele, W. (1983). *Applied Time Series and Box–Jenkins Models.* Orlando, FL: Academic Press, pp. 333–348.

Wei, W. (1993). *Times Series Analysis: Univariate and Multivariate Methods.* New York: Addison Wesley, pp. 184–205.

Wood, D. (1992). ICPSR, University of Michigan. Personal communication (Summer, 1992).

Woodfield, T. J. (1987). "Time Series Intervention Analysis Using SAS ETS Software," *SUGI Proceedings* Cary, NC: SAS Institute, Inc. Courtesy of Donna Woodward, SAS Institute, Inc.

Woodward, D. (1996–1998). SAS Institute, Inc. Cary, NC (Personal communications). Donna was frequently helpful in advising me on various aspects of SAS programming of impact analysis.

Chapter 9

Transfer Function Models

9.1. DEFINITION OF A TRANSFER FUNCTION

A dynamic system may exist where an input series seems related to an output series. The relationship between the exogenous (sometimes called the forcing) series, X_t, and the endogenous response series, Y_t, is a functional one. The input series may be a pulse or step process like those functions examined in the previous chapter, or it can be a continuous process driving another series. Much as light can be interpreted as discrete photons or continuous waves, transfer functions can be interpreted as those having pulsed discrete inputs that approximate continuous inputs. That is to say, the input series under examination is periodically sampled although it may have values between the periodic sampling times.

In the case of a transfer function model, both the input and output series are time series, and the endogenous series is a function of the exogenous input series that is driving it. These transfer function models are generally formulated as $Y_t = v(L)X_t + n_t$. These models have two components. The $v(L)X_t$ is the transfer function component and n_t is the ARMA or ARIMA noise model component. The transfer function component consists of a response regressed on lagged autoregressive endogenous variables and lagged exogenous variables, whereas the noise model component is a time

353

series (ARMA or ARIMA) error model. Graphs of the $v(L)X_t$ relationships over time are generally referred to as transfer functions. The simplest case is a bivariate relationship between two time series, occasionally called a leading indicator, ARMAX (ARMA with a cross-correlation between input and output), or TFARIMA (transfer function ARIMA) model. The researcher uses these models to predict the Y_t response series from the leading indicator, $v(L)X_{t-b}$, which leads by b periods. Of course, a bivariate model, consisting of a pair of input and output series, can be extended to include multiple input series. Because there is more than one series involved in such a model, these models are sometimes referred to as multiple time series ARIMA or MARIMA models. We focus on two transfer function modeling strategies, and begin by addressing the Box–Jenkins modeling strategy for bivariate cases. When these separate inputs are added together to yield the output series, they constitute a linear transfer function. We will also consider another approach, called the Dynamic Regression or Linear Transfer Function Method (Pankratz, 1991), which is recommended in cases of multiple simultaneous inputs. This chapter thus continues our examination of the theory and programming of multiple time series analysis.

9.2. IMPORTANCE

Wherever and whenever time-dependent processes are examined, questions arise about the relation, transfer, and impact of one series on another over time. When the structure of that impact is important, transfer function models are important. Examples of these phenomena abound in economics, business, and engineering, among other fields. In economics, leading indicators or transfer function models are used in forecasting business cycles. A transfer function model can show how a change in net imports is affected by a change in the exchange rate. Another economic transfer function model reveals how personal disposable income drives real nondurable consumption in the United Kingdom (Mills, 1990). In business, this kind of relationship is that of advertising driving sales (Makridakis *et al.*, 1983). Another example of a transfer function is a combination of forecasts, where the driving series are the forecasts, with ARMA errors. Statistical and engineering process control are based on modeling the transfer functions between inputs and outputs and the construction of feedback monitoring and feedforward control loops in these systems (Box *et al.*, 1994). Such statistical process control systems are essential in remote-control or other kinds of servomechanisms. Although statistical and engineering process control are beyond the scope of this book, the transfer function models discussed in this chapter are fundamental components within many complex systems. Examples of bivariate and multiple input transfer function models will be used to illustrate the theoretical explanation and programming applications of these models.

9.3. THEORY OF THE TRANSFER FUNCTION MODEL

9.3.1. THE ASSUMPTIONS OF THE SINGLE-INPUT CASE

The transfer function model, consisting of a response series, Y_t, a single explanatory input series, X_t, and an impulse response function $v(L)$, is predicated on basic assumptions. The input series may be deterministic, as explained in the last chapter, or stochastic, as explained in this chapter. The input also includes a stochastic noise component, e_t, which may be autocorrelated. It is assumed that the discrete transfer function and the noise component are independent of one another. Moreover, it is presumed that this relationship is unidirectional with the direction of flow from the input to the output series. If the exogenous input series and the endogenous output series are stochastic, both variables are usually centered and differenced if necessary, to attain a condition of stationarity. They are usually, but not necessarily, deseasonalized to simplify modeling as well. Although previous X_t observations may influence concurrent or later Y_t observations, there can be no feedback from Y_t to X_t. In other words, the X_t in a transfer function must be exogenous, and regardless of whether it is discrete or continuous, the transfer function is assumed to be stable.

9.3.2. THE BASIC NATURE OF THE SINGLE-INPUT TRANSFER FUNCTION

The basic formulation of the transfer function is $Y_t = v(L)X_t + n_t$. The impulse response function is actually a lagged polynomial with impulse response weights, v_i. This lagged polynomial, with its entire set of v_i weights, may be formulated as $v(L) = v_0 + v_1 L + v_2 L^2 + \ldots$. The impulse response weights represent the change in the output series as a result of a unit change in the explanatory variable at the indexed time. Each of the impulse response weights may be interpreted as responses to a pulse input at a point in time. At time $i = t$, the magnitude of the output variable per unit change in the input variable is indicated by the magnitude of the coefficient, v_t. After one period of time has elapsed, the response of the endogenous variable is equal to the product of the coefficient times the value of the input variable at that time plus the same products at previous time periods.

That is, the output response is equal to the sum of the products of the impulse response weights times the value of the input variable, from the inception of the process through the current time period. If $v_i < 0$, the direction of the impulse response from the current time is opposite that of the value of the input variable. If $v_i > 0$, the direction of the response from the current time is the same as that of the value of the input variable. The transfer function can therefore be expanded:

$$Y_t = v(L)X_t + e_t$$
$$= v_0X_t + v_1X_{t-1} + v_2X_{t-2} + \cdots + v_LX_{t-L} + e_t. \quad (9.1)$$

In theory this transfer function may be of infinite order. As an infinite series, $v(L)$ converges as $|L| \le 1$. In other words, if the series is absolutely summable, then

$$\sum_{L=0}^{\infty} |v_L| < \infty. \quad (9.2)$$

When the discrete transfer function is absolutely summable, it converges and is considered to be stable. The transfer functions considered here are assumed to be stable (Box et al., 1994). In practice, the values may taper off after awhile, rendering them effectively finite. This total effect is the gain of the transfer function (Vandaele, 1991):

$$\sum_{L=0}^{\infty} v_{t-L} = \text{Gain}. \quad (9.3)$$

The output variable and the input variable(s) are assumed to have been transformed to stationarity. Mean-centering the input and output series also simplifies the modeling and is recommended in this kind of analysis.

9.3.2.1. A Discrete Transfer Function with Stochastic Input

To illustrate the dynamic meaning of these impulse response weights in a transfer function, attention is turned toward the dynamic transfer function process of a response series and a stochastic input series. Of primary interest here is the structure of the impulse response weights. Remember that the impulse response weight at each sampling period of time is deemed to be the response to the change in the input series from the previous to the current time period. The cumulative effect of those responses becomes the focus of attention now. Even though the input series might not be deterministic, the significant weighted effects are related to inputs at specific time periods, defined by the transfer function. The modeling process is explained as these weights are identified, estimated, diagnosed, metadiagnosed, and then possibly used for forecasting. For example, consider transfer function model in Eq. (9.4). The order of the function has been found to have a lag of 3. At time t, the V_t is estimated to be 0.1. At time $t - 1$, the $V_{t-1} = 0.6$. At time $t - 2$, the V_{t-2} coefficient is 0.3. And at time $t - 3$, the impulse response coefficient is -0.2. The linear transfer function model, minus the stochastic noise component, is therefore

$$Y_t = 0.1X_t + 0.6X_{t-1} + 0.3X_{t-2} - 0.2X_{t-3}, \qquad (9.4)$$

where

$$v_t = 0.1$$
$$v_{t-1} = 0.6$$
$$v_{t-2} = 0.3$$
$$v_{t-3} = -0.2.$$

Makridakis et al. (1983) graphically depict the process of transfer in a table similar to that of Table 9.1. A study of Table 9.1 facilitates understanding of the dynamic process. When time $= 1$, the value of Y_t (in the rightmost column) can be calculated from the product of the coefficient of X_t and the value of X_t for that time period. At time $t = 1$, the value of X_t can be found in the second column from the left, and the coefficient for X_t may be found in the equation at the head of the table. The product of the value of X_t (20) and the coefficient (0.1) is 2. At time $t = 2$, the impulse response Y_t, is a composite of inputs at the current and previous time. The input at the current time is the product of the value of X_t (30) and the coefficient (0.1) and has a value of 3. This value of 3 is found at the intersection of

Table 9.1

Transfer Function Process, $Y_t = 0.1X_t + 0.6X_{t-1} + 0.3X_{t-2} - 0.2X_{t-3}$

Time	Value of X_t	1	2	3	4	5	6	7	8	9	10	Value of Y_t
1	20	2										2
2	30	12	3									15
3	40	6	18	4								28
4	50	-4	9	24	5							34
5	60		-6	12	30	6						42
6	50			-8	15	36	5					48
7	40				-10	18	30	4				42
8	30					-12	15	24	3			30
9	20						-10	12	18	2		22
10	10							-8	9	12	1	14

time $= 2$ in the columns and time $= 2$ in the rows. The input at the previous time is represented by X_{t-1}. The coefficient of X_{t-1} is 0.6 times 20 (the value of X_t), which equals 12. This value is found at the intersection of time $= 2$ in the rows and time $= 1$ in the columns. The total value of Y_t at $t = 2$ is the sum of the values for Y_t at $t = 1$ and $t = 2$. That is, $3 + 12 = 15$, which is found in the second row all the way on the right. In this way, the impulse response for a particular time is computed down the table. The process produces the values found for Y_t in the rightmost column of the table from the inception to the end of this process.

9.3.2.2. The Structure of the Transfer Function

The structure of the transfer function can be defined by its constituent parameters. The impulse response weights are coefficients of a rational distributed lag model. The notion that impulse response weights are expressed as a ratio is inherent in the name of a rational distributed lag model. The impulse response weights v_t consist of a ratio of a set of s regression weights to a set of r decay rate weights, plus a lag level, b, associated with the input series, and may be expressed with parameters designated with r, s, and b subscripts, respectively. The order of the transfer function refers to the levels of (r, s, b), respectively.

The order of delay or dead time is represented by the value of b. This is the time delay between incidences of changes in input, X_t, and the apparent impact on response, Y_t. The structure of the response weights is also specified according to a set of lag weights, from time lag $= 0$ to time lag $= L$. The delay time b, sometimes referred to as dead time, determines the pause before the input begins to have an effect on the response variable: $(L)^b X_t = X_{t-b}$.

The order of the regression is also represented by the values of s, which designates the number of lags for unpatterned spikes in the transfer function. The number of unpatterned spikes is $s + 1$. Together, these components comprise the transfer function. The formula can be found in Eq. (9.5). The time delay is designated by the $t - b$ subscript of the input variable. The numerator of the ratio consists of $s + 1$ ω_s regression weights, from time $= 0$ to time $= s$. These coefficients, with the exception of the first, have negative signs.

The order of decay is designated by the value of r as well. This parameter represents the patterned changes in the slope of the function. The order of this parameter signifies the number of lags of autocorrelation in the transfer function. The denominator of the transfer function ratio consists of decay weights, δ_r from time $= 1$ to r. The magnitude of these weights controls the rate of attenuation in the slope. If there is more than one decay rate, the rate of attenuation may fluctuate. The transfer function formula is

$$\text{Transfer function } v(L) = v_0 X_t + v_1 X_{t-1} + v_2 X_{t-2} + \cdots + v_f X_{t-b}$$

$$= \frac{\omega(L)}{\delta(L)} (L)^b X_t = \frac{\omega(L)}{\delta(L)} X_{t-b} \tag{9.5}$$

$$= \frac{(\omega_0 - \omega_1 L - \omega_2 L^2 - \cdots - \omega_s L^s) X_{t-b}}{1 - \delta_1 L - \delta_2 L^2 - \cdots - \delta_r L^r}.$$

The levels of the parameters determine the structure of the transfer function. If we suppose that the b parameter is set to L^2 then $(L)^2 X_t = X_{t-2}$. There are $s + 1$ ω regression weights. The size of the s parameter indicates how many regression coefficients and at what lags these coefficients comprise the numerator. The order of regression (plus 1 for ω_0) designates the number of unpatterned spikes. If $s = 2$, then the numerator of the ratio is $\omega_0 - \omega_1 L$. The size of the r parameter determines the order of decay (rate of slope attenuation). The r parameter controls the pattern in the slope. If $r = 1$, then the transfer function would have a denominator equal to $(1 - \delta_1 L)$ and would be one of first-order decay. If $r = 2$, then the function would have a denominator equal to $(1 - \delta_1 L - \delta_2 L^2)$ and would be one of second-order decay. The structure of the transfer function model are characterized by these parameters as well as the patterns of impulse response associated with them.

9.3.2.3. A Discrete Transfer Function with Deterministic Input

With a discrete transfer function, each of the impulse response weights can be interpreted as a response to a pulse or a step input. In either case, $X_t = I_t$. Table 9.2 presents formulations of common transfer function response models. All the formulations in Table 9.2 have a delay time designated by the parameter b. The first three models are ones with decay rates of $\delta_r = 0$. The structural parameter representing the rates of decay, r, equals zero for these models. Models 1, 2, and 3 have ω_s regression coefficients, the order of which is $s = 0, 1,$ and 2. The number of significant regression coefficients $s + 1$ equals 1, 2, and 3, respectively, in these models. Models 4, 5, and 6 have δ_r decay rate parameters, the order r of which equals 1, 1, and 2, respectively. To illustrate the structure of the transfer functions, some discrete transfer function models and their impulse responses for different levels of r and s are illustrated in Table 9.2 (Box and Jenkins, 1976).

Although transfer function models may have pulse, step, or continuous inputs, the pulse and step inputs are employed to illustrate the characteristic patterns of these models. Figures 9.1 through 9.12 show the response patterns for these models.

Table 9.2

Basic Transfer Function Model Structures

Model	r	s	b	Model	Impulse response
1	0	0	b	$Y_t = \omega_0 X_{t-b}$	$j < b: v_j = 0$ $j = b: v_j = \omega_0$ $j > b: v_j = 0$
2	0	1	b	$Y_t = (\omega_0 - \omega_1 L)X_{t-b}$	$j < b: v_j = 0$ $j = b: v_j = \omega_0$ $J = b + 1: v_j = -\omega_1$ $J > b + 1: v_j = 0$
3	0	2	b	$Y_t = (\omega_0 - \omega_1 L - \omega_2 L^2)X_{t-b}$	$j < b: v_j = 0$ $j = b: v_j = \omega_0$ $j = b + 1: v_j = -\omega_1$ $j = b + 2: v_j = -\omega_2$ $j > b: v_j = 0$
4	1	0	b	$Y_t = \dfrac{\omega_0}{1 - \delta_1 L} X_{t-b}$	$j < b: v_j = 0$ $j = b: v_j = \omega_0$ $j + b: \delta_1 v_{j-1} = 0$
5	1	1	b	$Y_t = \dfrac{(\omega_0 - \omega_1 L)}{(1 - \delta_1 L)} X_{t-b}$	$j < b: v_j = 0$ $j = b: v_j = \omega_0$ $j = b + 1: v_j = \delta_1 \omega_0 - \omega_1$ $j > b + 1: v_j = \delta_1 v_{j-1}$
6	2	2	b	$Y_t = \dfrac{(\omega_0 - \omega_1 L - \omega_2 L^2)}{(1 - \delta_1 L - \delta_2 L^2)} X_{t-b}$	$j < b: v_j = 0$ $j = b: v_j = \omega_0$ $j = b + 1: v_j = \delta_1 \omega_0 - \omega_1$ $j = b + 2$ $v_j = (\delta_1^2 + \delta_2)\omega_0 - \delta_1 \omega_1 - \omega_2$ $j > b = 2: v_j = \delta_1 v_{j-1} + \delta_2 v_{j-2}$

Characteristic patterns of the responses to these discrete functions are displayed in a series of graphs for both step and pulse input functions. When these characteristic patterns are detected, the trained analyst has a clearer notion of what kind of impulse response function is at work. Model 1 exhibits characteristic patterns following pulse, $X_{t-b}(1 - L)$, and step, X_{t-b}, inputs. That is, when the input is designated as a pulse, the input is simply a first-differenced level shift. There are simple practical rules for determining these parameters, subject to some variation due to sampling. The delay or dead time is the number of time periods between intervention and impact. The decay rate is zero when there is no decay. When there is first-order exponential decay, then the decay rate is less than unity. If there is oscillatory or compound exponential decay, the order of decay is 2 or more. The number of unpatterned startup terms is usually $s + 1 - r$.

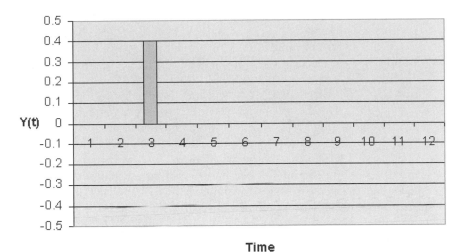

Figure 9.1 Model pulse response.

The discrete pulse response pattern for model 1, $Y_t = 0.4X_{t-3} (1 - L)$, with order of ($r = 0, s = 0, b = 3$) and a pulse input, is shown in the bar graph contained in Fig. 9.1 (a line graph might be more appropriate if the response appears to be continuous).

Figure 9.2 illustrates the discrete step response pattern for model 1, $Y_t = 0.4X_{t-3}$, with transfer function structural parameters ($r - 0, s = 0, b = 3$) and a step input. In contrast to the pulse response, the reader observes a clear step response.

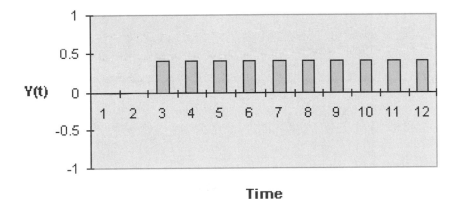

Figure 9.2 Model 1 response from step input.

The double pulse response pattern for a model 2, $Y_t = (0.4 + 0.3L)X_{t-2}$ $(1 - L)$, is characterized by $s + 1 = 2$ spikes, with $b = 2$, can be found in Fig. 9.3. In this case, the input does not occur until $t = 6$, but there is a two-period delay, so the first spike appears in period 8. The second spike, which follows, has a magnitude somewhat less than the first.

The model 2, $Y_t = (0.4 + 0.3L)X_{t-2}$, with step input and order parameters $(r = 0, s = 1, b = 2)$ exhibits a graduated step response pattern for $s + 1$ regression weights (Fig. 9.4). That is, there are two regression weights. Also, there are s augmentations of response before the peak of the response is attained at $s + 1$. The input appears at time $t = 6$, and then there is a two period lag before the impact becomes apparent at time period 8. For the step input, the pulse response weight of 0.4 kicks in at period 8. By the next period, the next impulse response of 0.3 is added to the first and the top of the step is reached. For this input, the response of 0.7 is continued during subsequent periods.

Model three, $Y_t = (0.4 + 0.3L + 0.2L^2)X_{t-6}(1 - L)$ with order $(r = 0, s = 2, b = 6)$, is an attenuated multiple pulse response to a pulse input with zero-order decay rate. It has three spikes in the response pattern that corresponds to each of the $s + 1$ regression coefficients in the model. Figure 9.5 displays the delayed response pattern for this pulse input. When there is pulse input, there is no other spike in the response pattern. The pulse takes place at lag 6 while the impact, owing to a delay of 2, appears two lags later, in period 8. The first regression weight is 0.4, the second is 0.3, and the last is 0.2. The pulses are shown in order of their appearance.

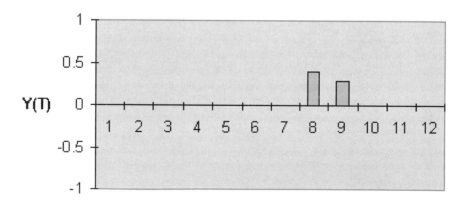

Figure 9.3 Model 2 response from pulse input.

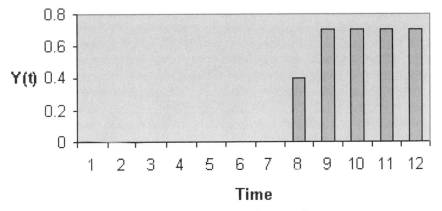

Figure 9.4 Model 2 response from step input.

Model 3 with step input, $Y_t = (0.4 + 0.3L + 0.2L^2) X_{t-2}$, exhibits a graduated onset of permanent impact shown in Fig. 9.6. The intervention begins in period 6, but owing to a delay of 2, the impact does not appear until time period 8. At that point, the response is equal to the first regression weight. One period later, the response is augmented by the next regression weight. Finally, the peak is reached with the augmentation of the last regression weight. When the spikes are considered all together, there are $s + 1 - r$ unpatterned spikes (0.4 and 0.3 and then the addition of 0.2 times the input), by which time the impact reaches a peak. Once the top of the step occurs, the impact remains constant.

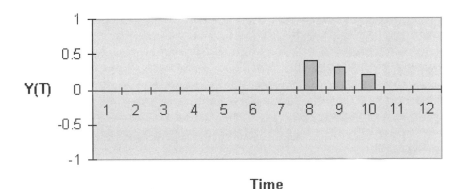

Figure 9.5 Model 3 response from pulse input.

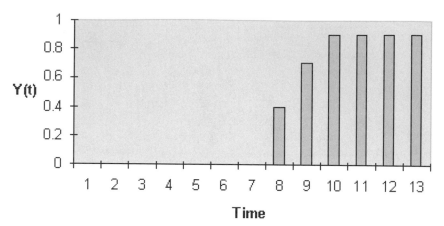

Figure 9.6 Model 3 response pattern from step input.

Model 4 with pulse input, $Y_t = [0.4/(1 - 0.5L)] X_{t-2} (1 - L)$, is the first transfer function in this series that contains an abrupt impact with a first-order decay (Fig. 9.7). The structural parameters for the pulse input of this function are ($r = 1$, $s = 0$, $b = 2$). There are $s + 1$ regression coefficients. In this case, there is one regression coefficient with the pulse input. The intervention occurs at lag 6, but there are two periods of delay before it is observed at time period 8. The single pulse has a magnitude equal to the regression weight of 0.4. In the next time period, the response consists of the autoregressive half of the previous response. In the next time period, that autoregressive response has a magnitude of only 0.2. In the next time

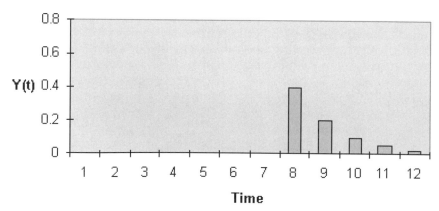

Figure 9.7 Model 4 response pattern from pulse input.

period, there is only half of the previous response remaining. Of the total spikes, there are $s + 1 - r$ unpatterned regression spikes. The response to such pulse input is one abrupt onset and exponential decay. The bounds of system stability require that the size of the decay parameter remains between plus and minus unity.

The model 4 response function with step input, $Y_t = [0.4 /(1 - 0.5L)] X_{t-2}$, is shown in Fig. 9.8. The response pattern is one of asymptotic growth or gradual onset and permanent duration. Both of these patterns are typical of first-order decay in the transfer function. The same delay of two periods is observed until impact at period 8. The increment to the impact is one-half of the earlier impact until it levels off.

Model 5, $Y_t = [(0.4 + 0.8L)/(1 - 0.5L)] X_{t-3}(1 - L)$, is distinguished from model 4 by three essential differences (Fig. 9.9). First, model 5, unlike model 4, has one extra regression coefficient in the numerator. This extra coefficient is lagged one period behind the first. With $s = 1$, the pattern may exhibit two distinguishing startup spikes, before decay takes effect. Second, this response function contains a first-order decay parameter, $0.5L$, in the denominator. Third, there is also a three-period delay so when the intervention takes place at time 3, the impact is not observed until period 6. For this pulse input, there are $s + 1$ unpatterned initial spikes, as well as a gradual attenuation of decreasing slope after the two initial spikes.

For the step input for model 5, $Y_t = [(0.4 + 0.8L)/(1 - 0.5L)] X_{t\ 3}$, shown in Fig. 9.10, there is gradual onset and permanent response duration after the $s + 1$ unpatterned startup spikes. The input takes place at time 3. The delay time for model 5 is three periods, before the input attains impact. At period 6, the first impact is observed. The magnitude is deter-

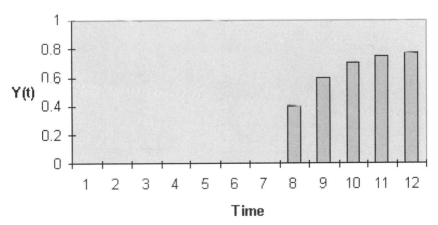

Figure 9.8 Model 4 response pattern from step input.

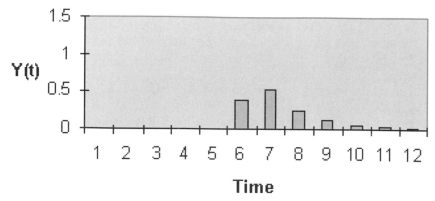

Figure 9.9 Model 5 response pattern from pulse input.

mined by the first regression coefficient, 0.4, at period 6. At the next time period, the magnitude of the response is determined by the decay rate parameter (0.5) as well as those of the first and second regression weights: 0.4 and 0.8. With the rate parameter at 0.5, this means that half of the last response ($0.5 \times 0.4 = 0.2$) is added to the new response, which is the sum of 0.4 and 0.8. The total accumulation for period 7 is 1.4. In short, the accumulation of response for the step input is half the value of the response at each previous time lag, after the startup spike.

The model 6 transfer function $Y_t = [(0.5 + 0.6L + 0.4L^2)/(1 - 0.5_1L -$

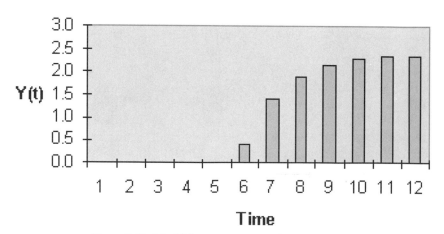

Figure 9.10 Model 5 response pattern from step input.

$.25L^2)]X_{t-3}$ $(1 - L)$, has structural parameters $(r = 2, s = 2, b = 3)$. This transfer function shown in Fig. 9.11, has $(s + 1 = 3)$ three regression weights, a second-order decay, and a time delay of 3. Because $s + 1 = 3$, there are three initial unpatterned spikes, before decay takes effect. Whereas the first-order decay follows a single rate of attenuation after the initial spikes, the second-order decay has a quadratic polynomial decay after the initial spikes. The response pattern depends on the roots of the characteristic equation of this denominator polynomial, $(1 - \delta_1 L - \delta_2 L^2)$. If the roots are real, the pattern exhibited may be more or less damped. If the roots are complex, the characteristic pattern will be one of sinusoidal oscillation. In other words, for this system to be stable, three conditions must hold: (a) $-1 < \delta_2 < 1$; (b) $\delta_1 + \delta_2 < 1$; (c) $\delta_2 - \delta_1 < 1$. The ω_s coefficients are 0.5, 0.6, and 0.4, respectively. In this model, the δ_r coefficients are 0.5 and 0.25, while the delay parameter remains 3. The pattern is one of both gradual onset and gradual (slightly quadratic) decline.

With a model 6 step input, $Y_t = [(0.5 + 0.6L + 0.4L^2)/(1 - 0.5L - 0.25L^2)]X_{t-3}$, the first few unpatterned spikes may be more or less distinctive, depending upon the similarity of the magnitude of the regression coefficients (Fig. 9.12). If the regression coefficients are similar, startup spikes may be indistinguishable. If the regression coefficients have significantly different magnitudes, then the initial spikes may appear to be noticeably unpatterned. The asymptotic growth after those spikes will be more or less quadratic, depending upon the relative magnitudes of the delta parameters

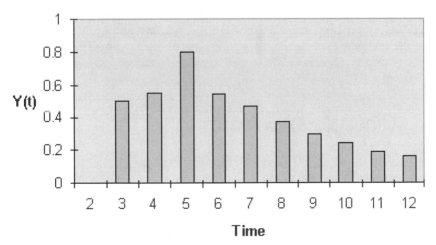

Figure 9.11 Model 6 response pattern from pulse input.

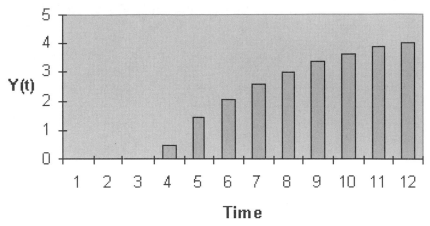

Figure 9.12   Model 6 response pattern from step input.

in the denominator (Fig. 9.12). In this instance, there is no perceptible fluctuation in the attenuation of growth. These characteristic patterns exhibit some features of the basic discrete transfer functions.

The transfer function may have either discrete deterministic input or stochastic continuous input. Although the last chapter addressed the basic nature of the response function to deterministic step and pulse input, where $X_t = I_t$, this chapter discusses the response function to input X_t, where it is a stochastic series. Whether the input is deterministic or stochastic, the functional relationship between the input and output series needs to be modeled. Modeling a transfer function includes identification, estimation, diagnosis, forecasting, metadiagnosis, and programming. For the bivariate case, the classical Box–Jenkins approach will be employed. For the multiple input case, the regression or linear transfer function approach will also be explained.

9.4. MODELING STRATEGIES

9.4.1. THE CONVENTIONAL BOX–JENKINS MODELING STRATEGY

9.4.1.1. Graphing and Preprocessing the Series

In this work, two modeling strategies will be discussed. The classical modeling strategy, particularly suited to bivariate cases, is presented by Box and Jenkins (1976). The alternative regression strategy, also known as the linear transfer function modeling strategy, which does not apply prewhitening, is presented in the case of the multiple input models, where this approach is most suitable. In this instance, the series is preprocessed. First, the two series need to be graphed or plotted. These plots should be examined for unusual patterns or outliers. The data are checked for errors, and any observational outliers are smoothed or modeled. Both series should be centered. Both the input and output series are transformed into stationarity, which may require a natural log, Box–Cox, power, or differencing transformation. Moreover, the series should be deseasonalized if possible. Deseasonalization, although not necessary, removes external sources of variation that could complicate the identification process (Makridakis *et al.*, 1983). Finally, the input series should be checked for exogeneity by a Granger causality test, described in the subsequent section on exogeneity.

9.4.1.2. Fitting an ARMA Model for the Input Series

After the preprocessing, an ARMA model is fit for the input series. In this case, it is recognized that autocorrelation within the input series may contaminate the cross-correlation between the input and output series. Box and Jenkins propose neutralizing this autocorrelation contamination with a prewhitening filter. This inverse filter, developed from the input series, is then applied to both input and output series.

9.4.1.3. Prewhitening

This filter is an inverse transformation, which turns the input series into white noise. If there is autocorrelation within the input series, there will be a need for prewhitening. If there is no autocorrelation within the input series, it is possible to do without the prewhitening (Liu and Hanssens, 1982). Once the prewhitening filter is applied to both the input and the output series, it removes the corrupting influence of the autocorrelation within the input series while maintaining the same functional relationship between the two series. Instead of solving for X_t, the equation is inverted to solve for e_t. After the prewhitening filter has been applied to both the output series and the input series, those series are said to have been prewhitened. Since the same factors are multiplied by the output and the input series, the functional relationship between them remains unchanged.

The prewhitening filter is formulated from the existing ARMA model. Suppose for a first-order ARMA model that the form of the ARMA model of the input series X_t

$$(1 - \varphi_1 L - \varphi_2 L^2 - \cdots - \varphi_p L^p)X_t = (1 - \theta_1 L - \theta_2 L^2 - \cdots \theta_q L^q)e_t \tag{9.6}$$

may be abbreviated by $\varphi_t(L)X_t = \theta_t(L)e_t$.

The inverse transformation converts the input series to white noise. Because

$$X_t = \frac{\theta_x(L)}{\varphi_x(L)} e_t,$$

$$e_t = \theta_x^{-1}(L)\varphi_x(L)X_t.$$ (9.7)

By applying this same filter to the output series, the output series is prewhitened and P_t is obtained:

$$P_t = \theta_x^{-1}(L)\varphi_x(L)Y_t.$$ (9.8)

The cross-correlation between two series subjected to the identical transformation remains the same. By transforming a set of nonorthogonal relations into a set of orthogonal relations, the prewhitening eliminates contamination of the cross-correlation by the autocorrelation of the input series. Because the relationship between the prewhitened output series and the prewhitened input series is now a dynamic function of white noise input, there is no autocorrelation to contaminate the cross-correlation function between P_t and $v(L)e_t$. The transformed output is now proportional to the impulse response function plus the transformed noise:

$$\theta_x^{-1}(L)\varphi_x(L)Y_t = \theta_x^{-1}(L)\varphi_x(L)X_t + \theta_x^{-1}(L)\varphi_x(L)e_t.$$

Because

$$\omega(L)(L^s)\varphi^{-1}(L) = v(L),$$ (9.9)

and

$$n_t = \theta_x^{-1}(L)\varphi_x(L)e_t,$$
$$P_t = v(L)e_t + n_t.$$

The cross-correlation should now accurately reflect the structure of the impulse response function (Box *et al.*, 1994). Therefore, with prewhitening, the pattern of cross-correlation should accurately reflect the impulse response weights, with some allowance for sampling error.

9.4.1.4. Direct Estimation of the Transfer Function Structure by Examination of the Cross-correlation Function

After the input and output series are prewhitened, direct estimation of the transfer function impulse response weights is made possible from examination of the cross-correlation function. The shape of the cross-correlation between those two prewhitened series reveals the pattern of (r, s, and b) parameters of the transfer function (Box and Jenkins, 1976).

In single-input MARIMA models, the cross-correlation *between the prewhitened input and prewhitened output series* reveals the structure of the transfer function model. Whereas in univariate ARIMA, the moving average or autocorrelation forms the basis of the *within-series* dynamics of the model, the cross-correlation between a prewhitened input series, e_t, and a prewhitened output series, P_t, is essentially a Pearson product moment correlation of the dynamics between the two series.

How the cross-correlation function (CCF) is computed and how it is interpreted are important. Like the Pearson Product Moment correlation, the cross-correlation is basically the covariance between the input and output series divided by the product of the standard deviation of one series times the standard deviation of the other series. Consider the numerator of the CCF, which is the covariance between the two series. The covariance of two variables is simply

$$Cov(X_i, Y_i) = \frac{1}{n} \sum_{i=1}^{n} (X_i - \overline{X})(Y_i - \overline{Y}), \tag{9.10}$$

where i = the observation, and n = the number of observations. If the process is stationary, then the autocovariance (j) = autocovariance $(-j)$ within the same process. When the cross-covariance is plotted against a time axis, it is called the cross-covariance function. Unlike the autocovariance or autocorrelation function, the cross-covariance and cross-correlation functions between two different processes are not symmetrical. Summing over time after differencing and using j as order of the cross-covariance is reformulated as the sum of the products of the mean deviations of each series at each point in time.

Cross-covariance$_{xy}(j)$

$$= \frac{1}{n} \sum_{t=1}^{n-j} (X_t - \overline{X})(Y_{t+j} - \overline{Y}) \text{ when } j => 0 \tag{9.11}$$

$$= \frac{1}{n} \sum_{t=1}^{n+j} (X_{t-j} - \overline{X})(Y_t - \overline{Y}) \text{ when } j =< 0,$$

where n = number of observations after subtracting the order of differencing, and j = order of cross correlation. Not only can the j subscript can assume a negative or positive value; when different lags (or leads) of the cross-covariance are used as the point of reference, different numbers of Y values are used for the computation after subtraction for differencing. As will be shown, the magnitude of the cross-covariance can also differ depending on the lag j under examination. The denominator of the cross-correlation (CCF) consists of the product of the sum of the standard deviations over time of one series times the sum of the standard deviation of the other series over time of the other series:

$$\text{Standard deviation } S_{xx} = \frac{1}{n-j} \sqrt{\sum_{t=1}^{n-j} (X_t - \overline{X})^2}$$

$$\text{Standard deviation } S_{yy} = \frac{1}{n-j} \sqrt{\sum_{t=1}^{n-j} (Y_t - \overline{Y})^2}$$

(9.12)

where n = number of observations, and j = lag of cross correlation. If a cross-correlation were used to test the leading indicator relationship of X_t to Y_{t+j}, the CCF with subscript $t + j$ equals the ratio of the cross-covariance to the product of the standard deviations of the two series:

$$\text{Cross-correlation}_{xy} \text{ (where } j \le 0) = \frac{\sum_{t=1}^{n+j} (X_{t-j} - \overline{X})(Y_t - \overline{Y})}{\sqrt{\sum_{t=1}^{n} (X_t - \overline{X})^2} \sqrt{\sum_{t=1}^{n} (Y_t - \overline{Y})^2}}.$$

(9.13)

$$\text{Cross-correlation}_{xy} \text{ (where } j \ge 0) = \frac{\sum_{t=1}^{n-j} (X_t - \overline{X})(Y_{t+j} - \overline{Y})}{\sqrt{\sum_{t=1}^{n} (X_t - \overline{X})^2} \sqrt{\sum_{t=1}^{n} (Y_t - \overline{Y})^2}}.$$

The interpretation of the CCF(j) indicates the transfer function direction between the series and delay between incidence and impact. Because this coefficient is asymmetric, after some delay time, b, if the CCF(j) > 0, then X_t is correlated after some delay b with Y_{t+j}. Prior to that time period the CCF(j) will not be significant. Afterward, if CCF($j + k$) > 0 then the input series and the response series will be related with $j + k$ lags difference. The shape of the CCF over time will resemble the response functions described earlier. If the impact is one of an autoregressive process, after b delay periods and j lags the cross-correlation parameter may be exponentiated to the jth power. Because it is assumed that the transfer proceeds from the input series to the output series, the positive side of the CCF(j) is used to define the nature of the impact.

Although one of the assumptions of the transfer function is that the relationship proceeds from X_t to Y_t, it is possible for the CCF($-j$) to indicate a reverse effect, feedback, or simultaneity. As Figs. 9.13 and 9.16 show, the CCF has the appearance of a Cartesian graph of positive and negative values against time. This asymmetry means that $r_{xy}(1) \ne r_{yx}(1)$ and that $r_{xy}(1) \ne -r_{xy}(1)$. If the CCF($-j$) > 0, then X_{t-b-j} is cross-correlated with Y_{t-b}. In other words, after the delay time, b, if CCF($-j$) > 0 then X_{t-b-j} leads Y_t. If this were the case, significant cross-correlation spikes would be observed on the negative side CCF and the assumption of no feedback would be violated. Hence, only the positive $CCF(j)$ is used for identification. If there is evidence of $CCF(-j) > 0$, then there is evidence of feedback and a more complicated dynamic simultaneous equation may be in order. For these reasons, the cross-correlation is useful in assessing the time delay, b, and the positive or negative direction of transfer between the input and the output series.

To understand how the CCF reflects the nature of the transfer function, it is important to understand how the cross-correlation function reflects the impulse response weights. The e_t is the uncorrelated white noise and ε_t is the transformed noise from the noise model n_t. In Eq. (9.9), one obtains a formula for the prewhitened series and that is redisplayed here for convenience:

$$
\begin{aligned}
P_t &= v(L)\theta_x^{-1}(L)\varphi_x(L)X_t + \theta_x^{-1}(L)\varphi_x(L)n_t \\
&= v(L)X_t e_t + \theta_x^{-1}(L)\varphi_x(L)n_t \\
&= v(L)X_t e_t + \varepsilon_t.
\end{aligned}
\tag{9.14}
$$

If we premultiply both sides by e_{t-j} and take the expectations, we obtain

$$
\begin{aligned}
E(e_{t-j}P_t) &= v_0 E(e_{t-j}e_t) + v_1 E(e_{t-j}e_{t-1}) + \cdots + v_j E(e_{t-j}e_{t-j}) \\
&+ E(e_{t-j}\varepsilon_t).
\end{aligned}
\tag{9.15}
$$

Because $E(e_t P_t) = 0$ for all $t \neq j$ and $E(P_{t}\varepsilon_t) = 0$ (since they are not correlated), this equation may be expressed as

$$
\text{Cross-covariance}_{P_t e_t}(j) = v_j \sigma_{e_t}^2.
$$

Therefore,

$$
\begin{aligned}
v_j &= \frac{\text{Cross-covariance}_{P_t e_t}(j)}{\sigma_{e_t}^2} \\
\\
&= \frac{\sigma_{P_t}\text{Cross-covariance}_{P_t e_t}(j)}{\sigma_e \sigma_e \sigma_{P_t}} \\
\\
&= \rho_{P_t e_t}(j)\frac{\sigma_{P_t}}{\sigma_{e_t}} \quad \text{for } j = 0,1,2,\dots .
\end{aligned}
\tag{9.16}
$$

The impulse response weights are therefore a function of the cross-correlations, ρ_j, at lag j in the cross-correlation function.

Significance of the cross-correlation function is given by the formula

$$
SE_{ccf} = \sqrt{\frac{1}{(T-j)}},
\tag{9.17}
$$

where T is the number of observations (n), and j is the number of lags (Wei, 1993). The printout of the cross-correlation function can be seen in the output in Figs. 9.13 and 9.16. In Fig. 9.13, the dotted lines represent the plus or minus 2 standard errors, or the 95% confidence limits of the cross-correlation coefficient. Significant correlations extend beyond those limits as they do in the ACF and PACF. From the significant sample cross-correlations, the impulse response weights may be estimated from the formula used in Eq. (9.16). If the CCF attenuates slowly, the relationship between the input and output series is not stationary. In this case, the researcher should consider further differencing of the input and output series to attain stationarity before proceeding with the analysis (Box et al., 1994). The transfer function structural parameters may be found either by matching the characteristic patterns of the positive CCF with those impulse response functions described earlier or by the corner method about to be explained.

Table 9.3 illustrates the computation of a cross-correlation between two series. The data used are segments of business cycle historical indicator data, courtesy of the U.S. Department of Commerce. The quarterly unit labor cost of all persons from the business sector data (rescaled to 1992 = 100) was selected as the input series, and the annual rate of corporate profits after taxes in billions of dollars (1992 = 100) was selected as the output series. Two cross-correlations are computed. The first is the cross-correlation at $j = 0$ lags. The second is the cross-correlation at $j = 1$ lag. From the formula for the cross-correlation (Eq. 9.13), it can be seen that there is one case lost each time the lag j increases. In the cross-correlation at lag 1, in Table 9.3, the Y_t series has been adjusted so that lag $j = 0$ is missing and lag $j = 1$ is moved up a row. Not only do the numbers of cases in their computation differ, the magnitudes of the cross-correlations differ according to the lag at which they are computed. Because the number of cases in them differ, their sample size and standard errors differ slightly as well.

9.4.1.4.1. Exogeneity

One of the assumptions of a transfer function model is that there is unidirectionality in the relationship between the input and output series. In other words, it is presumed that there is no feedback from the output to the input series. One test for exogeneity is the Granger causality test. When two time series are related, it is necessary to be sure that the X_t is exogenous with respect to Y_t. To test this form of exogeneity, the following autoregressive equations are estimated:

$$Y_t = \sum_{s=1}^{\infty} \beta_{1s} Y_{t-s} + \sum_{s-1}^{\infty} \gamma_{1s} X_{t-s} + \nu_1$$

$$X_t = \sum_{s=1}^{\infty} \delta_{2s} X_{t-s} + \sum_{s-1}^{\infty} \varepsilon_{2s} Y_{t-s} + \nu_2$$

(9.18)

Table 9.3

Cross-Correlation Computation of Unit Labor Cost and Corporate Profits (Annual Rate, $ Billions)[a]

Numerator Calculations

Year Qtr	X_t lbrcost (1992 = 100)	$(X_t - \bar{X})$	Y_t corprofit	Cross-correlation (0) $j = 0$ $(Y_t - \bar{Y})$	$(X_t - \bar{X}) \times (Y_t - \bar{Y})$	Cross-correlation (1) $j = 1$ $(Y_{t+1} - \bar{Y})$	$(X_t - \bar{X}) \times (Y_{t+1} - \bar{Y})$
1993-1	101.8	-0.8	280.8	-40.4	30.6	-30.4	23.0
1993-2	102.5	-0.1	290.8	-30.4	2.7	-29.0	2.5
1993-3	102.3	-0.2	292.2	-29.0	6.9	-6.6	1.6
1993-4	101.5	-1.1	314.6	-6.6	6.9	-30.5	32.2
1994-1	102.3	-0.2	290.7	-30.5	7.2	-2.8	0.7
1994-2	102.9	0.4	318.4	-2.8	-1.0	8.3	3.0
1994-3	102.8	0.3	329.5	8.3	2.4	19.4	5.5
1994-4	102.7	0.1	340.6	19.4	2.6	37.6	5.0
1995-1	103.1	0.6	358.8	37.6	21.9	32.5	19.0
1995-2	102.9	0.4	353.7	32.5	11.8	41.6	15.1
1995-3	103.2	0.7	362.8	41.6	27.2		15.1
				$\Sigma(y_t - \bar{y})$			$\Sigma =$ 107.5
Numerator Components =				119.1			

Denominator Calculations

Year Qtr	X_t lbrcost	$X_t - \bar{X}$	$(X_t - \bar{X})^2$	Y_t corprofit	$Y_t - \bar{Y}$	$(Y_t - \bar{Y})^2$
1993-1	101.8	-0.8	0.6	280.8	-40.4	1630.0
1993-2	102.5	-0.1	0.0	290.8	-30.4	922.5
1993-3	102.3	-0.2	0.1	292.2	-29.0	839.4
1993-4	101.5	-1.1	1.1	314.6	-6.6	43.2
1994-1	102.3	-0.2	0.1	290.7	-30.5	928.6
1994-2	102.9	0.4	0.1	318.4	-2.8	7.7
1994-3	102.8	0.3	0.1	329.5	8.3	69.3
1994-4	102.7	0.1	0.0	340.6	19.4	377.4
1995-1	103.1	0.6	0.3	358.8	37.6	1415.8
1995-2	102.9	0.4	0.1	353.7	32.5	1058.0
1995-3	103.2	0.7	0.4	362.8	41.6	1732.8
$\bar{X} =$ 102.5		$\Sigma(x_t - \bar{x})^2 =$ 2.9 Stdev = 1.7		$\bar{Y} =$ 162.8		$\Sigma(y_t - \bar{y})^2 =$ 9024.8 Stdev = 95.0
Cross-correlation (0) =	0.73			Cross-correlation (1) =		0.66

[a] 1992 = 100 for both series.

If feedback (Granger noncausality) obtains, the ε_{2s} parameter would have to be statistically significant. For exogeneity or unidirectional association to exist, there can be no feedback inherent in these linear projections. For there to be no feedback, the ε_{2s} parameter would have to be statistically nonsignificant.

Another test of exogeneity is the cross-correlation function. Two preconditions must hold. First, both the input and output series have to be identified properly, leaving white noise residuals after identification. Then both series have to be prewhitened by the appropriate inverse filter. If the cross-correlation function then exhibits significant negative spikes, it means that the direction of the relationship appears to be going from the postulated endogenous series to the postulated exogenous series. If there are both significant positive and negative spikes, then this is *prima facie* evidence of simultaneity or feedback. Feedback is a violation of the assumption of a unidirectional relationship from the exogenous to the endogenous series. When two series are not prewhitened, apparent feedback shown in Fig. 9.13 cross-correlations may result from the failure to trim out contaminating autocorrelation of the input series by introducing lower order AR terms in the model or failure to prewhiten.

Figure 9.13

One of the problems with labor costs leading the corporate profits is that if the analyst looks far enough into the past, he finds negative spikes in the series. Negative spikes signify feedback from the corporate profits to the unit labor costs. When profits become high enough, production may begin to expand and enjoy economies of scale, or automation and computerization may be implemented, any combination of which may reduce the unit labor costs. When the cross-correlation function is applied to this relationship, it can be seen that there are spikes in the negative as well as the positive part of the function. There are significant cross-correlations at time periods -5 and -7 as well as at 0 and 1. If any contaminating autocorrelation in the input series were removed by prewhitening or inclusion of AR terms, these negative spikes would suggest feedback in this relationship. Such feedback would violate the assumption of exogeneity of the input series and unidirectionality of the relationship between the input and output series. Because this apparent feedback violates a basic assumption of the transfer function model, this relationship is rejected as amenable to transfer function modeling and another example will be used. Therefore, the researcher should check for exogeneity as part of the preliminary consideration of the series, prior to modeling the transfer function.

9.4.1.4.2. *Linear Transfer Function Method of Identification of the Transfer Function Structural Parameters*

The transfer function r, s, and b coefficients can be identified directly from an inspection of the stationary, prewhitened cross-correlation function. Alternatively, a method called the linear transfer function method can be used. With this modeling strategy, we can render both series stationary, then add a lower order AR or ARSAR term (to partial out contamination of the within-series autocorrelation), after which the response series may be regressed on a distributed lag of the input series such that $Y_t = v_1 X_t + v_2 X_{t-1} + v_3 X_{t-3} + v_4 X_{t-4} + v_5 X_{t-5} + v_6 X_{t-6} + \ldots$. We standardize the v coefficients by dividing the absolute value of the maximum v weight into all of the weights. We plot the magnitude of the standardized v_i impulse response weights against the time lags of the input series. From a comparison of the actual pattern in the CCF or standardized impulse response weights with common theoretical transfer functions, we can derive the structure of the transfer function.

There are some practical guidelines (rather than exact rules) by which we can identify the structure of the transfer function. After the series is preprocessed in the ways described, these rules provide guidelines by which the general pattern of the transfer function can be identified. Fine-tuning the identification process may require some trial and error with a view toward testing the parameters for significance and minimizing residuals or the information criteria.

First, we identify the dead or delay time. The b parameter is simply the dead or delay time between input and apparent impact. The number of periods after reference time period, lag 0, before a significant positive spike appears on the cross-correlation function signifies the delay time. If the first significant spike is at the zero-reference point on the cross-correlation function, there is no delay time. If the first significant spike is at the first period after that point, the delay time is a lag of one period. If the first significant spike appears three lags after the point of input, there is a delay of three periods. The delay time is therefore easy to identify.

Second, we can identify the decay pattern. The decay parameter, r, represents the autoregressive decay in the process. The decay parameters indicate the portion of the weights that have a defined pattern. There is the case of the zero-order response function. If there are no decay parameters and the impulse response weights reach their permanent magnitude immediately, then $r = 0$ and the input is a step function. If $r = 0$ and the function is that of a pulse, then the input is a pulse function (a first-differenced step input). Whether step or pulse function, the onset of the response will be delayed by b time periods. There is also the case of the first-order response function. If there is only one decay parameter such that $r = 1$, there is usually exponential decay. If the decay parameter remains within its bounds of stability, exponential decay can characterize the slope. There is also the case of a second-order response function. If there are two decay parameters remaining within their bounds of stability, then the impulse response function could be a damped exponential or a dampened sine wave, depending on the roots of the polynomial $(1 - \delta_1 L - \delta_2 L^2)$. If the roots are real, the spikes would follow a pattern of uneven exponential attenuation, whereas if the roots are complex, response function would form a pattern of oscillation (Box et al., 1994; Wei, 1993). Assuming that the roots are real, common transfer functions can be defined with second- or lower-order response functions.

Pankratz (1991) notes that the pattern of decay is preceded by startup spikes. The number of these startup spikes generally corresponds to the order of the decay. In other words, there are usually r startup spikes before the decay begins. If there is first-order decay, there will usually be one startup spike. If there is a second-order decay, there will usually be two startup spikes before the decay commences. The number of startup spikes helps identify the order of decay (Pankratz, 1991).

Third, there are $s + 1$ unpatterned spikes generated by the ω_s regression weights. The one is added to account for the initial ω_0 weight. These weights need not follow a pattern; they can be completely unpatterned. After subtracting the r weights that exhibit a pattern, we find $s + 1$ unpatterned spikes in the model. In ideal situations, these patterns are easy to identify, but in real situations sampling variation may complicate the pattern by adding another source of variation. Although this variation can complicate distinguishing startup from unpatterned spikes, there are $s + 1$ unpatterned spikes found after delay time has passed. With proper application of these general rules, the cross-correlation function is used after prewhitening to identify the transfer function. Estimation of the transfer function parameters as well as the ARIMA noise model parameters follows (Box *et al.*, 1994).

9.4.1.4.3. Identification of Transfer Function Structure with the Corner Table

Another method proposed by Liu and Hanssens (1982) and expounded upon by Tsay (1985) involves the use of the corner method. Where additional assistance is required, the corner table is used to determine the structure of the transfer function. This method is recommended by Tsay (1985) where the autocorrelations do not taper off quickly; in other words, if the model is not stationary or contains unit or near unit roots, the corner table method can be used. Pankratz indicates that this method can handle the problem with autocorrelation in the input series. From the pattern inherent in the corner table, the r, s, and b parameters can be ascertained, even if the model has not been prewhitened. Before we examine this protocol, it may be helpful to examine the prewhitening, the cross-correlation function, and the corner table in detail.

The nature of the transfer functions can be identified from the structure of a corner table or C array. The corner table consists of determinants of matrices of standardized transfer function weights. This corner table is an $M + 1$ by M matrix made up of $c(f, m)$ elements. Each $c(f, m)$ element is a determinant of standardized impulse response weights. Standardization is performed by dividing the particular impulse response weight by the absolute value of the maximum impulse response weight. In each determinant, the standardized weights are designated by $\eta_{ij}(= v_{ij}/|v_{i,\max}|)$, where the subscript i is omitted from Fig. 9.19 for simplification. The c elements of the corner table have subscripts f and m. Subscript $f(f = 0, 1, \ldots, M)$ is the row number of the corner table and subscript $m(m = 1, 2, \ldots, M)$ is the column number of the corner table. The determinants $c(f, m)$ are constructed as follows:

$$c(f, m) = \begin{vmatrix} \eta_f & \eta_{f-1} & \cdots & \eta_{f-m+1} \\ \eta_{f+1} & \eta_f & \cdots & \eta_{f-m+2} \\ \cdot & \cdot & \cdots & \cdot \\ \cdot & \cdot & \cdots & \cdot \\ \eta_{f+m-1} & \eta_{f+m-2} & \cdots & \eta_f \end{vmatrix} \tag{9.19}$$

Table 9.4
The Corner Table (C-Array)

	m columns										
f	1	2	3	$\cdots$	$\cdots$	r	$r+1$	$r+2$	$\cdots$	$m = M + 1$	
0	0	0	0	0	0	0	0	0	0	0	
1	0	0	0	0	0	0	0	0	0	0	
·	·	·	·	·	·	·	·	·	·	·	b rows
·	·	·	·	·	·	·	·	·	·		
$b-1$	0	0	0	0	0	0	0	0	0	0	
b	x	x	x	x	x.	x	x	x	x	x	
·	x	x	x	x	x.	x	x	x	x	x	s rows
$b+s-1$	x	x	x	x	x	x	x	x	x	x	
$b+s$	x	x	x	x	x	x	0	0	0	0	
$b+s+1$	x	x	x	x	x	x	0	0	0	0	
·	x	x	x	x	x	x	·	·	·	·	
M	x	x	x	x	x	x	·	·	·	·	
	(r columns)						·	·	·	·	

For each explanatory variable, x_j, of the free-form distributed lag model, one can construct an element $c(f, m)$ that has a value of the determinant if $f > 0$, $m > 0$, and $\eta_j = 0$ if $j < 0$. Actually, the element $c(f, m)$ has a value of 0 or close to 0 due to random and/or sampling error if $j < 0$. When the corner table or C-array is constructed in this way, it contains a structure, shown in Table 9.4, from which the order of the transfer function may be derived. Within the corner table, we represent the values of the elements by zeros or x's. The cells with zeros represent relatively small weights, whereas the cells with x's represent relatively larger weights. From the patterns of zeros and x's we are able to derive the transfer function structure.

The pattern of the matrix of f rows by m columns reveals the order of the transfer function. The f rows are indexed from zero through M. The upper rows of the matrix will consist of zeros. There are b rows of zeros before we reach rows of x-marked cells. We find the delay time by counting the upper rows of zeros. Following the b rows of zeros (row 0 through row $b - 1$), there are s rows (extending from row b through row $b + s - 1$) of x-marked cells before we encounter a rectangular block of zeros in the lower right section of the table. This rectangular block of zeros begins in row $b + s$. There will be a distance of r columns from the first column of the table to the first column before the lower right block of zeros. In other words, the block of zeros therefore begins in column $r + 1$. From this characteristic pattern of the tabular matrix, we can identify the order of the transfer function (Lui and Hanssens, 1982; Mills, 1990; Pankratz, 1991; Lui *et al.,* 1992).

9.4.1.5. Estimation of the Transfer Function

We can estimate the transfer function by conditional least squares, unconditional least squares, or maximum likelihood. Maximum likelihood may require more data points than the other two. Sums of squared residuals are found and the iterations continue until those sums of squared residuals do not improve significantly.

9.4.1.6. Diagnosis of the Transfer Function Model

Diagnosis and metadiagnosis of the transfer function model takes place next. When the iterations converge, they yield estimates of the parameters. We test these parameters for significance against their standard errors. To test the model for adequacy, the parameters estimated should be significant. Moreover, the decay parameters should conform to the bounds of stability for transfer function models. If the model is one of first-order decay, then $|\delta_1| < 1$. This means that the δ_1 parameter estimate should not be too close to the value of 1.00. If the parameter is 0.96, then the model may be unstable and be in need of further differencing. If the model is one of second-order decay, the three conditions of system stability must hold: (a) $\delta_2 + \delta_1 < 1$; (b) $\delta_2 - \delta_1 < 1$; and (c) $|\delta_2| < 1$. None of the parameters should be nonsignificant. If the parameters are not significant, we prune them from the model. When the estimated parameters appear to be significant, and the nonsignificant ones are trimmed from the model, the model residuals should be white noise. The residuals can be diagnosed by their ACF and PACF along with use of the Box–Ljung Q test.

9.4.1.7. Metadiagnosis of the Transfer Function Model

Metadiagnosis entails comparative evaluation of alternative transfer function models. If there are spikes in the ACF and PACF of the residuals, new parameters that could account for those spikes are tested. If these parameters are significant, the alternative models are compared according to their residuals or their minimum information criteria, such as sums of squared residuals, the Akaike information criterion, or the Schwartz criterion. Metadiagnosis can also include the comparative evaluation of the forecasts generated by those models. The MSFE and MAPE are generally used for evaluation of the forecast against the validation sample, although the MAPE is often preferred.

9.4.1.8. Formulation of the Noise Model

If the Box–Jenkins approach is employed, the noise model of the input series is identified before prewhitening. Theoretically and ideally, the noise model is independent of the transfer function model. The residuals remaining after the identification, estimation, diagnosis, and metadiagnosis of the transfer function are reexamined. In particular, the noise model (the ARIMA model for the residuals from the transfer function) is reexamined.

9.4.1.8.1. Identification of Noise Model Parameters

The residuals from the transfer function are examined. All necessary differencing should have been performed. If the residuals exhibit ARMA characteristics, a particular procedure can be invoked to assist the proper identification of the ARMA order. The researcher can use the ACF, PACF, and the extended sample autocorrelation function (ESACF) to identify the proper ARMA order. The bounds of stationarity and invertibility should be considered to be sure that the parameters identified yield a stable model. If the parameters cleave closely to those bounds and stationarity becomes an issue, the parameters should be tested for nonseasonal and seasonal unit roots. From these considerations, he can identify the proper ARMA parameters. Those parameters can then be estimated and diagnosed.

9.4.1.8.2. Estimation of Noise Model Parameters

The estimation may be undertaken by the algorithms already discussed in the chapter on estimation. They are conditional least squares, unconditional least squares, or maximum likelihood. If the parameters are stable and the model converges, then further diagnosis is in order. Nonsignificant parameters are trimmed from the model.

9.4.1.8.3. Diagnosis of Noise Model

Diagnosis of the model includes a review of the model assumptions. Is the model congruent with those assumptions? Does the model make sense? The estimated parameters should not be too close to the bounds of stationarity and invertibility. If the parameters are not close to those bounds, then they will be stable. If the parameters are stable and account for all of the variation in the noise model, the ACF and PACF of their residuals should reveal white noise.

If there are any outliers apparent in the residuals, then the series should be checked for the outliers. Smoothing or modeling the outlier should be considered. For example, modeling an observational outlier can involve the use of another pulse function to be added to the model.

If the parameters are unstable, the residuals would not be white noise. There could be significant spikes in the ACF or PACF of the residuals. The estimated parameters might not be stable because the coefficient values might be too large or have the wrong signs. It is theoretically assumed that there should not be a multicollinearity problem. A check of the correlation matrix among the parameters would be in order. If the parameters are intercorrelated, they could be unstable. Changes in some parameters could change the values of other parameters, and the model could be difficult to fit. Another assumption is that the transfer function is uncorrelated with the ARIMA noise model. In fact, the noise model parameters could be correlated with the transfer function parameters, and changes in the transfer function might change the nature of the noise model. A cross-correlation function check is used between the noise and transfer function model to test the assumption of independence between the transfer function and noise model.

If these correlations between the parameters are substantial or high, the model may have difficulty converging to final estimates, in which case more differencing and remodeling may be necessary. When the iterations to parameter estimates converge, the estimates should be found to be significant and the ACF and PACF of the residuals should reveal white noise (Lui *et al.*, 1992).

Upon diagnosis of the ARMA noise model, we fine-tune the model. The nonsignificant parameters may be pruned from the model. If the residuals are not yet white noise, model reidentification should follow, with either new parameters that need to be added or old ones that need to be remodeled. Reformulation of the noise model would entail reformulation of a new prewhitening filter and a remodeling of the transfer function. Alternative models may be tested against one another with minimum information criteria and/or with residuals best resembling white noise.

9.4.1.8.4. *Metadiagnosis and Forecasting*

When alternative models are compared with one another to find the optimal model, they may also be compared for model explanation, fit, or forecast accuracy. If they are being evaluated for explanatory scope, they can be compared according to the amount of theory encompassed. If they are being evaluated for explanatory efficiency, they can be assessed by their adjusted R^2 or minimum information criteria. If they are being evaluated for model fit, the criteria by which they are compared can be minimum information criteria or the sum of squared residuals. If they were well estimated, their parameters estimates should have the right sign, a reasonable magnitude, and stability. When alternative models are used to generate forecasts and those forecasts are evaluated for accuracy, the forecasts can be compared with the mean square forecast error or the minimum absolute percentage error. We can forecast h leads into the forecast horizon, based on a model that includes both a transfer function and a noise component according to the following formula (Box *et al.*, 1994; Granger, 1999):

$$Y_{t+h} = \delta_1 Y_{t+h-1} + \cdots + \delta_{p+d+r} Y_{t+h-p-d-r}$$
$$+ \omega_0 X_{t+h-b} + \cdots + \omega_{p+d+s} X_{t+h-b-p-d-s}$$
$$+ e_{t+h} - \theta_1 e_{t+h-1} - \cdots - \theta_{q+r} e_{t+h-q-r},$$

where

t is the time period
h is the lead time period
p is the order of autoregression   (9.20)
d is the order of differencing
r is the order of decay
b is the delay
s is the order of regression
and
q is the order of moving average.

The forecast error variance and forecast interval limits are

$$\text{Var}(h) + \sigma_e^2 \sum_{j=0}^{h-1} \eta_j^2 + \sigma_e^2 \sum_{j=0}^{h-1} \psi_j^2, \qquad (9.21)$$

where

$$\eta_j = \text{error of } v_j$$
$$\hat{Y}_{t+h} = \pm 1.96[V(h)]^{1/2}.$$

From the definition of the ψ weights given earlier in the chapter on forecasting and these formulas, the analyst can compute the forecasts and forecast intervals (Fig. 9.13). Although some analysts use the MSFE as the conventional criterion of predictive validation of the model, other researchers prefer the MAPE, because it is not so vulnerable to outlier distortion. Those who prefer the minimum squared forecast error claim that, unlike MAPE, it is not as susceptible to distortion because of estimates being close to zero (Fildes et al., 1998). From the metadiagnosis, the analyst can select the optimal model and then plot the forecast. Before proceeding to the more complicated problems of multiple input models, an example of a single input transfer function modeling process is presented.

9.4.1.9. Programming a Single Input Transfer Function Model Using the Conventional Box–Jenkins Strategy

A single-input transfer function model can be constructed from the relationship between U.S. per capita personal disposable income (PDI) driving or influencing personal consumption expenditures (CE) from 1929 through 1994. The series data, measured in 1987 constant dollars, were obtained from the National Income and Product Accounts of the United States and the Survey of Current Business, July 1994 and March 1995, from the Bureau of Economic Analysis, U.S. Department of Commerce.

Once the data are gathered, the researcher should consider his strategy and choice of statistical package. To illustrate transfer function model building, SAS is chosen because it has excellent comprehensive transfer function modeling capability. SAS users can employ either the Box–Jenkins approach or the linear transfer function approach. If researchers adhere to the Box–Jenkins approach, they can automatically prewhiten the input and output series with the inverse filter formed from the noise model of an input series. The inverse prewhitening filter neutralizes autocorrelation in the input series that would bias the parameter estimates of the transfer function. Adherents of this approach identify the parameters of the transfer function model—including the delay, decay, and regression parameters of the transfer function—with the cross-correlation function. Some scholars have argued that prewhitening is necessary to remove the corrupting autocorrelation from the input series before modeling the transfer function. They suggest that without the prewhitening approach, the cross-correlations may not accurately reflect the impulse response weights (Brocklebank and Dickey, 1984; Box *et al.*, 1994; Woodward, 1997). Because SPSS does not automatically prewhiten the series, researchers who prefer the Box–Jenkins approach would prefer SAS.

SAS also permits the researcher to model the transfer function by the linear transfer function approach. Following the linear transfer function approach, the analyst includes lower order AR terms in the noise model of the input series to control for autocorrelation bias, and then the inputs are included. Other scholars maintain that this approach is sufficient to remove the corrupting autocorrelation from the input series (Liu and Hanssens, 1982; Tsay, 1985; Pankratz, 1993). The transfer function parameters are derived from inspection of the cross-correlation function or from the corner table. Researchers who prefer the linear transfer function approach to modeling can also use SPSS.

Although the SPSS ARIMA procedure at the time of this writing can handle discrete and continuous inputs, SPSS is currently developing a new time series analysis and forecasting module, called Decision Time, that allows for either single or multiple, deterministic and/or stochastic predictors. With the SPSS ARIMA procedure, the user must code the intervention himself, but with Decision Time the module permits automatic coding of the event or intervention. Neither SPSS module possesses automatic prewhitening capability, but the Decision Time module will permit the user to define the structure of the transfer function, with the exception of fixing the values of the parameters. Instead of permitting conventional Box–Jenkins modeling, both SPSS modules require the user to model with the linear transfer function method or a method that involves the regression of predictors with ARIMA modeling of the residuals. For this reason, the SPSS procedures are not discussed in the section on Box–Jenkins modeling strategy, and SAS is used to demonstrate the Box–Jenkins approach to transfer function modeling.

This section presents a step-by-step explanation of the program of the conventional transfer function modeling of the relationship between per capita personal disposable income and personal consumption expenditures. The complete file for this model is called C9PGM1.SAS. Some preliminary programming matters need to be addressed. An OPTIONS statement limits the number of columns to 80, with the LINESIZE (abbreviated LS) = 80 option. This options statement causes output to be formatted to 80-column width, a format that is easy to read on any computer monitor. There are four TITLE statements. These allow adequate description of the project, subproject, data source, and procedure for future reference.

The directory (folder) in which the data set is located must be defined, and this specification is done in the LIBREF (sometimes called LIBNAME) statement. The directory is abbreviated INP in the LIBNAME statement as follows: LIBNAME INP 'C:\STATS\SAS';. This means that data sets prefixed with an INP will be found in the specified directory C:\STATS\ SAS';. Each SAS program is divided into DATA steps and PROC steps. Then the DATA step must be given a name. In this program it is called NEW, by the DATA NEW; statement. Another command is issued to read in the data. The source directory and the source data set have to be identified with this command. From the LIBNAME statement the directory in which the data set is located is abbreviated INP. The data set PDI_CE.SD2 located in 'C:\STATS\SAS' is used. The command that imports this data set into the data set called NEW is SET INP.PDI_CE;. DATA NEW; has now gotten its data from PDI_CE.SD2 and redefined that data as its own. In this way, the directory and data set are defined and the data are imported.

The preprocessing of the data follows. The IF _N_ > 66 THEN DE-LETE; statement eliminates the irregular data set length in the two series. By deleting observations number 67 and higher in the series, this statement guarantees that both of the series have the same length. The DATE = INTNX function defines the date in years from 1929 onward. Each observation is dated by the year of observation. Extraneous variables DATE_ and YEAR_, which were created earlier, are dropped. The DATE variable created by the DATE function is then formatted according to a four-column YEAR designation. Figure 9.14 depicts the two series, with their stochastic trends, plotted against time in years.

The next paragraph defines the graph of the two series. Because we have reviewed SAS overlay graph program syntax before, we only cursorily review it now. From the graph review, we can observe that the two series exhibit similar patterns. The first AXIS statement defines the label and rotates it 90 degrees so it fits along the vertical axis. The SYMBOL statements sequentially define the joining of points, the color, and the shape of the

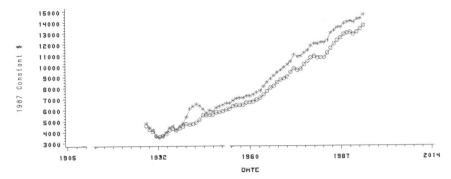

Personal Disposable Income=star
Consumption Expenditures=circle

Figure 9.14 Personal consumption expenditures driven by per capita personal disposable income in 1987 constant dollars. Source: U.S. Bureau of Economic Analysis.

functions respectively specified in the plot statement. Two series are overlaid and plotted against DATE in order to see which series may be driving the other from this similarity of pattern. Therefore, the relationship between them is explored further. Then six title statements replace the previous ones and two FOOTNOTEs are added to the bottom of the graph. The footnotes that are left blank below delete the two footnotes for subsequent procedures.

```
options ls=80;
title 'Per capita PDI => ce in 1987 dollars';
title2 'Source: U.S. Bureau of Economic Analysis';
title3 'National Income and Product Accounts of the United States';
title4 'Survey of Current Business July, 1994 and March 1995';
LIBNAME inp 'c:\stats\sas';
data new ;
   set inp.PDI_ce;

/* pre-processing the series */

if _n_ > 66 then delete;
date = INTNX('year','01jan1929'd,_n_-1);
drop year_ date_ tab1987;
format date year4.;
/* Examination of the data to be sure it is read correctly */
proc print;
run;

/* Preliminary Plotting of the Series */
axis1 label=(a=90 '1987 Constant $');
symbol1 i=join c=green v=star;
symbol2 i=join c=blue v=circle;
```

```
proc gplot;
   plot (PDI ce) * date/overlay vaxis=axis1;
title ' Personal Consumption Expenditures';
title2 'Driven by Per Capital Personal Disposable Income ';
title3 'in 1987 Constant $';
title4 'Source: US Bureau of Economic Analysis';
title5 'National Income and Product Accounts of the US';
title6 'Survey of Current Business, July 1994 & March 1995';
footnote1 justify=L 'Personal Disposable Income=star';
footnote2 justify=L 'Consumption Expenditures=circle';
run;
footnote1 justify=L '        ';
footnote2 justify=L '        ';
```

The next step in the Box–Jenkins approach is the preliminary testing of ARMA noise models for the input series. The strategy is to identify the best noise model for the input series. By trying several alternative models, the best fitting and most parsimonious model is selected. For each model tested specific PROC ARIMA syntax is employed.

This procedure begins with the PROC ARIMA command. The identification subcommand of the ARIMA procedure begins with I VAR=PDI(1) CENTER NLAG=25; . I abbreviates IDENTIFY. VAR=PDI indicates that the personal disposable income variable, PDI, is to be analyzed. PDI has to be first differenced in order to be rendered stationary, so a (1) is placed immediately after the variable name. The series is subsequently centered to simplify the modeling. The CENTER subcommand subtracts the mean from each observation of the input series PDI. The NLAG=25 option sets the number of lags to be reviewed at 25 and prints the ACF, IACF, and PACF of the first difference of PDI for evaluation.

Underneath the IDENTIFY statement is the ESTIMATE statement. Several alternative ARMA noise models for the input series are estimated. The ESTIMATE statements begin with an abbreviation E. Each of the estimations was performed with maximum likelihood with a maximum of 40 iterations. The NOINT options specified that no intercept was to be used because the series were already centered. The PRINTALL option specifies that the preliminary estimation, iteration history, and optimization summary be printed in addition to the final estimation results. The PLOT option requests the residual ACF and PACF plots. Each model estimation converged. For model 1, an AR(1) MA(10) model is estimated. For model 2, an AR(1) MA(4 10) model is estimated. Model 3 is an AR(1) MA(2 4 10) parameterization. All the models were evaluated by the SBC, the modified portmanteau test, and the ACF and PACF graphs. Title statements give numbers to the models and specify their SBC and residual results. Although model 1 has the lowest SBC, model 2 had ACF and PACF residuals that appeared to be more white noise. For this reason, model 2 was selected as the ARIMA noise model for the input series.

```
/* ********************************************************* */
/* Preliminary Identification of Input Series Noise Model   */
/* This is done to set up prewhitening inverse filter       */
/* ********************************************************* */

proc arima;
    identify var=ce;
    identify var=pdi;
title7 'Preliminary Noise Model Identification';
run;

proc arima;
    i var=pdi(1) center;
    e p=1 q=(10) noint printall plot method=ml maxit=40;
title7 'PDI estimation p=1 q=(10) SBC=890.3 - residual spike at lag 4';
title8 'Model 1';
run;

proc arima;
    i var=pdi(1) center;
    e p=1 q=(4 10) noint printall plot method=ml maxit=40;
title7 'PDI estimation p=1 q=(4 10) SBC= 891.5 - good residuals';
title8 'Model 2';
run;

proc arima;
    i var=pdi(1) center;
    e p=1 q=(2 4 10) noint printall plot method=ml maxit=40;
title7 'PDI estimation p=1 q=(2 4 10) good residuals';
title8 'Model 3 p=1 q=(2 4 10) SBC=893.1 vy good residuals';
run;

/* ********************************************************* */
/* Model 2 selected as input Noise model                    */
/* ********************************************************* */
```

Prewhitening is applied next. Prewhitening is invoked by the syntax of the full transfer function model. To form the prewhitening filter, the input (personal disposable income) series identification and estimation are performed first. Once formed, the prewhitening filter is then applied to both series. By stacking the identification and estimation commands of the input series on those of the output series, invoking the cross-correlation function option on the identification command for the output series, and by specifying the input series with the INPUT option in the estimation subcommand of the output series, the researcher applies the prewhitening filter to both series.

In the computer syntax for the full transfer model, the first two lines identify and estimate the input series noise model.

```
proc arima;
  i var=pdi(1) center esacf nlag=25;
  e p=1 q=(4 10) noint printall plot;
```

This noise model is used to prewhiten both series. The second identification and estimation subcommands

```
  i var=ce(1) center esacf nlaq=25 crosscorr=(pdi(1));
  e q=1 printall plot noint input=((10)pdi);
```

run the ACF and the PACF up to 25 lags of the first differenced and centered output series. They run the ACF and PACF of the residuals of the estimated model. They invoke the cross-correlation function. The CCF is invoked with the CROSSCORR=(PDI(1)) option. It should be noted that the PDI differencing is indicated by the (1) suboption. Pankratz argues that if the noise model requires differencing, then algebraically both the response (CE) and the input series (PDI) should be identically differenced (Pankratz, 1991).

```
title7 'Full Transfer Function Model ';
title8 'Per capita pdi = Consumption expenditures';

proc arima;
  i var=pdi(1) center esacf nlag=25;
  e p=1 q=(4 10) noint printall plot;
  i var=ce(1) center esacf nlag=25 crosscorr=(pdi(1));
  e q=1 printall plot noint input=((10)pdi);
  f lead=24 id=date interval=year out=fore;
run;
```

The cross-correlations between the differenced input and output series are indicated as prewhitened by the inverse filter in Fig. 9.15.

A direct estimation of the transfer function characteristics is undertaken from a review of these cross-correlations. In Fig. 9.16, it can be seen that there are no statistically significant negative spikes, which implies no feedback. In other words, the per capita personal disposable income does seem to be exogenous in this relationship. There are statistically significant spikes at lags 0 and 10, however. This would imply that there is an immediate effect of disposable income on the expenditures within that annual period. Also, that there is an effect that lags by about 10 years as well. The number of lags shown depends on the number attached to the NLAG= option in the identify subcommand. The cross-correlations for 10 lags are shown here.

At this point, the program syntax is modified, so that NLAG=24 in the ESTIMATE subcommand of the line with CROSSCORR in it. The cross-

ARIMA Procedure

Both variables have been prewhitened by the following filter:

Prewhitening Filter

Autoregressive Factors
Factor 1: 1 - 0.41038 B**(1)

Moving Average Factors
Factor 1: 1 - 0.22279 B**(4) - 0.46116 B**(10)

Figure 9.15

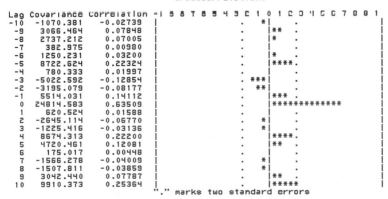

```
                                      Crosscorrelations            34
                                        11:30 Thursday, November 26, 1998

                      ARIMA Procedure

            Correlation of CE and PDI
            Variable PDI has been differenced.
            Period(s) of Differencing = 1.
            Both series have been prewhitened.
            Variance of transformed series = 36561.81 and 41755.98
            Number of observations =        65
            NOTE: The first observation was eliminated by
                  differencing.

                      Crosscorrelations

Lag Covariance Correlation -1 9 8 7 6 5 4 3 2 1 0 1 2 3 4 5 6 7 8 9 1
-10  -1070.381    -0.02739  |               .    *|  .          |
 -9   3066.464     0.07848  |               .    |** .          |
 -8   2737.212     0.07005  |               .    |*  .          |
 -7    382.975     0.00980  |               .    |   .          |
 -6   1250.231     0.03200  |               .    |*  .          |
 -5   8722.624     0.22324  |               .    |****.         |
 -4    780.333     0.01997  |               .    |   .          |
 -3  -5022.592    -0.12854  |               . ***|   .          |
 -2  -3195.079    -0.08177  |               .  **|   .          |
 -1   5514.031     0.14112  |               .    |***.          |
  0  24814.583     0.63509  |               .    |*************  |
  1    620.524     0.01588  |               .    |   .          |
  2  -2645.114    -0.06770  |               .   *|   .          |
  3  -1225.416    -0.03136  |               .   *|   .          |
  4   8674.313     0.22200  |               .    |****.         |
  5   4720.461     0.12081  |               .    |** .          |
  6    175.017     0.00448  |               .    |   .          |
  7  -1566.278    -0.04009  |               .   *|   .          |
  8  -1507.811    -0.03859  |               .   *|   .          |
  9   3042.440     0.07787  |               .    |** .          |
 10   9910.373     0.25364  |               .    |****          |
                            "." marks two standard errors
```

Figure 9.16

Crosscorrelation Check Between Series

To Lag	Chi Square	DF	Prob	Crosscorrelations					
5	33.24	6	0.000	0.651	0.004	0.125	-0.043	0.232	0.130
11	40.64	12	0.000	0.048	0.012	0.010	0.038	0.314	0.108
17	42.45	18	0.001	0.045	-0.008	0.113	0.004	0.020	0.112
23	44.86	24	0.006	-0.064	-0.072	0.160	0.018	0.040	0.011

Figure 9.17

correlation function and the Q tests for significance from lag 0 through lag 23 are printed (Fig. 9.17).

Because the cross-correlations reflect pulse response weights, ω_s regression terms are modeled at lags $s = 0$ and 10 (where the cross-correlation function reveals significant, sharply defined, and pronounced spikes) in the input statement of the ESTIMATE subcommand for the personal consumption expenditures series. INPUT = ((10) PDI) requests the following transfer function model: $(\omega_0 - \omega_1 L^{10})(1 - L)$ PDI$_t$. The $(1 - L)$ comes from the earlier differencing of the PDI series, assuring the analyst that the consumer expenditure pulse responses are found at lags 0 and 10. There are unpatterned spikes that suggest numerator regression coefficients. These lags suggest the lags of coefficients. This completes the tentative identification of transfer function parameters.

If the cross-correlations had more complicated shapes, the syntax for modeling these forms of input functions are given in Table 9.5. The syntax for modeling these transfer functions is contained in the corresponding input option of the identify command of the response variable of the table. At this point, it is reasonable to inquire how one programs the six transfer function models specified earlier. Table 9.5 illustrates how to formulate those transfer functions. Bresler *et al.* (1991) and Ege *et al.* (1993), as well as Bowerman and O'Connell (1993), give detailed explanations for other transfer function model parameterizations in SAS.

Table 9.5

SAS Transfer Function Model Syntax for $b = 3$

Model 1	$Y_t = \omega_0 X_{t-b}$	INPUT = (3\$X$_T$)
Model 2	$Y_t = (\omega_0 - \omega_1 L)X_{t-b}$	INPUT = (3\$ (1)X$_T$)
Model 3	$Y_t = (\omega_0 - \omega_1 L - \omega_2 L^2)X_{t-b}$	INPUT = (3\$(1,2)X$_T$)
Model 4	$Y_t = \dfrac{\omega_0}{1 - \delta_1 L} X_{t-b}$	INPUT = (3\$/(1)X$_T$)
Model 5	$Y_t = \dfrac{(\omega_0 - \omega_1 L)}{(1 - \delta_1 L)} X_{t-b}$	INPUT =((3\$(1)/(1))X$_T$)
Model 6	$Y_t = \dfrac{(\omega_0 - \omega_1 L - \omega_2 L^2)}{(1 - \delta_1 L - \delta_2 L^2)} X_{t-b}$	INPUT =(3\$ (1, 2)/(1 2)X$_T$)

The model is estimated by maximum likelihood estimation. Diagnosis of the transfer function model parameters is the next step. The significance tests of these transfer function model parameters are estimated, and if they are nonsignificant, they are trimmed from the model. This output is shown in Fig. 9.18.

ARIMA Procedure

Maximum Likelihood Estimation

Parameter	Estimate	Approx. Std Error	T Ratio	Lag	Variable	Shift
MA1,1	-0.39475	0.13090	-3.02	1	CE	0
NUM1	0.41493	0.08370	4.96	0	PDI	0
NUM1,1	-0.16703	0.07491	-2.23	10	PDI	0

```
Variance  Estimate  = 14853.2471
Std Error Estimate  = 121.873898
AIC                 = 687.496275
SBC                 = 693.518275
Number of Residuals=     55
```

Figure 9.18

All of the transfer function parameters are significant. The two transfer function model parameters called NUM1 and NUM1,1 at lags 0 and 10, respectively, have very significant t statistics. Inclusion of these parameters in the model reduces the magnitude of the AIC and SBC to 687.50 and 693.52, respectively.

In the meantime, it is noted that the MA parameters are no longer statistically significant. These were originally necessary to control for moving average variation in the input series. It can be argued because they are not part of input autocorrelation, they could be trimmed from the model prior to forecasting. Because they were part of the original prewhitening filter, they are left in the model in accordance with convention for simplified presentation of the modeling process. Examination of the correlations among the parameters and of the Q tests for the residuals of the transfer function model in Fig. 9.19 permits further diagnosis of the transfer function. The transfer function parameters are significant and account for all of the spikes in the cross-correlation function, from which we conclude that the transfer function has been successfully identified.

Diagnosis of the transfer function includes review of the correlation matrix among the parameters. Ideally, the transfer function and the noise model are independent of one another. Under such circumstances, the correlations between the transfer function and noise model parameters should be small. Given the prewhitening by the inverse filter on the consumer expenditures series, there are autoregressive and moving average noise model parameters along with the transfer function parameters in the correlation matrix. Owing to relatively low correlations between these components shown in Fig. 9.19, multicollinearity does not appear to pose a problem. Diagnosis, however, involves more than a multicollinearity review.

ARIMA Procedure

Correlations of the Estimates

Variable	Parameter	CE MA1,1	CE MA1,2	CE AR1,1	PDI NUM1	PDI NUM1,1
CE	MA1,1	1.000	0.072	-0.154	-0.027	0.009
CE	MA1,2	0.072	1.000	0.188	-0.123	0.054
CE	AR1,1	-0.154	0.188	1.000	-0.147	0.092
PDI	NUM1	-0.027	-0.123	-0.147	1.000	-0.443
PDI	NUM1,1	0.009	0.054	0.092	-0.443	1.000

Autocorrelation Check of Residuals

To Lag	Chi Square	DF	Prob		Autocorrelations				
6	3.95	3	0.267	0.050	-0.128	-0.154	-0.006	-0.135	-0.063
12	6.82	9	0.656	0.019	-0.017	0.171	0.019	-0.035	-0.100
18	9.66	15	0.840	0.078	-0.064	-0.097	0.039	0.121	-0.009
24	10.97	21	0.963	0.042	0.016	0.038	0.098	0.023	-0.016

Figure 9.19

A diagnostic review of the residuals is also in order. When this noise model is coupled with the transfer function model, nonsignificant residuals emerge. The nonsignificant residuals in Fig. 9.19 indicate the parameterization of the models accounts for all of the systematic variance. Hence, there seems to be no need to resort to the corner table for estimation of the structural parameters of the transfer function or the ESACF for estimation of the order of the ARMA noise model.

If the transfer function model together with the prewhitened noise model did not already account for all of the systematic variance, the residuals would not be nonsignificant. In that hypothetical case, then the next step in the modeling strategy is to remodel the noise model until the residuals are white noise. Figures 9.20 and 9.21, respectively, provide the ACF and PACF of the residuals from this model.

An additional transfer function diagnosis is necessary. In the transfer function model, there is an assumption that the noise model and the transfer function are independent of one another. We must validate this assumption before placing any faith in the model. A cross-correlation of noise model residuals and the transfer function is output for diagnostic examination (Fig. 9.22). If the assumption holds, there should be no statistically significant cross-correlation between them. If a statistically significant cross-correlation remains, then autocorrelation remains between the transfer function and the noise model residuals, potentially corrupting estimation, and either the transfer function or the noise model or both need to be reidentified. Once the residuals appear to be white noise, we can proceed to metadiagnosis and the final step of forecasting. To test this assumption, SAS performs Q tests of the cross-correlations of the residuals of this input

ARIMA Procedure

Autocorrelation Plot of Residuals

```
Lag Covariance Correlation -1 9 8 7 6 5 4 3 2 1 0 1 2 3 4 5 6 7 8 9 1
  0  14861.389   1.00000   |                    |********************|
  1    187.055   0.01259   |                .   |  .                 |
  2   -293.617  -0.01976   |                .   |  .                 |
  3  -2511.466  -0.16899   |             .  ***|  .                 |
  4  -2146.098  -0.14441   |             .  ***|  .                 |
  5  -2375.640  -0.15985   |             .  ***|  .                 |
  6  -1218.898  -0.08202   |             .   **|  .                 |
  7   1412.346   0.09503   |                .  |**  .               |
  8  -1018.295  -0.06852   |                . *|  .                 |
  9   3416.279   0.22988   |                .  |*****.              |
 10   1178.776   0.07932   |                .  |**  .               |
 11    408.669   0.02750   |                .  |*   .               |
 12  -1664.301  -0.11199   |                **|  .                 |
 13    156.989   0.01056   |                .  |    .               |
 14   -856.335  -0.05762   |                . *|  .                 |
 15  -1676.031  -0.11278   |                **|  .                 |
 16    850.711   0.05724   |                .  |*   .               |
 17   1348.864   0.09076   |                .  |**  .               |
 18   -312.487  -0.02103   |                .  |    .               |
 19    887.742   0.05973   |                .  |*   .               |
 20    533.071   0.03587   |                .  |*   .               |
 21    679.407   0.04572   |                .  |*   .               |
 22   1139.206   0.07666   |                .  |**  .               |
 23   -149.950  -0.01009   |                .  |    .               |
 24 -95.563864  -0.00643   |                .  |    .               |
 25  -1429.049  -0.09616   |                **|  .                 |
                          ".". marks two standard errors
```

Figure 9.20

ARIMA Procedure

Partial Autocorrelations

```
Lag Correlation -1 9 8 7 6 5 4 3 2 1 0 1 2 3 4 5 6 7 8 9 1
  1   0.01259   |                .   |  .                 |
  2  -0.01992   |                .   |  .                 |
  3  -0.16858   |             .  ***|  .                 |
  4  -0.14524   |             .  ***|  .                 |
  5  -0.17561   |             .****|  .                 |
  6  -0.13595   |             .  ***|  .                 |
  7   0.00130   |                .   |  .                 |
  8  -0.16974   |             .  ***|  .                 |
  9   0.15445   |                .  |***  .              |
 10   0.04826   |                .  |*   .               |
 11  -0.00092   |                .   |  .                 |
 12  -0.05746   |                . *|  .                 |
 13   0.07117   |                .  |*   .               |
 14  -0.00395   |                .   |  .                 |
 15  -0.07249   |                . *|  .                 |
 16   0.02413   |                .   |  .                 |
 17   0.11135   |                .  |**  .               |
 18  -0.09240   |                **|  .                 |
 19   0.05385   |                .  |*   .               |
 20   0.01397   |                .   |  .                 |
 21   0.11633   |                .  |**  .               |
 22   0.14146   |                .  |***  .              |
 23   0.01536   |                .   |  .                 |
 24   0.09847   |                .  |**  .               |
 25   0.02264   |                .   |  .                 |
```

Figure 9.21

Crosscorrelation Check of Residuals with Input PDI

To	Chi			Crosscorrelations					
Lag	Square	DF	Prob						
5	4.15	4	0.386	0.137	-0.046	-0.057	0.002	0.206	0.094
11	6.12	10	0.805	0.102	0.002	-0.097	0.065	-0.108	0.012
17	7.48	16	0.963	0.023	0.052	0.030	-0.068	0.047	0.122
23	10.41	22	0.982	-0.160	-0.077	0.133	-0.035	-0.011	0.064

Figure 9.22

noise model of personal disposable income and the transfer function. In the event that there is no additional cross-correlation, the Q tests will be nonsignificant. When residuals are diagnosed as white noise, as they are shown to be in Fig. 9.22, the transfer function plus noise model are inferred to be independent, to provide an acceptable fit, and to explain the process.

Once the model has been diagnosed and the residuals are found to be white noise, the next step entails generation and evaluation of the forecast from the model. It is at this point that any nonsignificant transfer function parameters may be trimmed from the model. The graph of the forecast profile generated is shown in Fig. 9.23.

The forecast data set, designated FORE in the forecast subcommand, is generated from the unprewhitened series by the forecast subcommand in the ARIMA procedure. That data set includes the variables: the observation number, OBS; the date variable, DATE; the response variable, CE; the forecast, FORECAST; the standard error, STD; the lower 95% confidence

Figure 9.23 Personal consumption expenditures driven by per capita personal disposable income in 1987 constant dollars. Source: U.S. Bureau of Economic Analysis.

limit, L95; the upper 95% confidence limit, U95; and the residual, RESIDUAL.

```
f lead=12 id=date interval=year out=fore;
```

The data set is then merged with the input data set so both PDI and CE are contained in the data set for graphing.

```
/* merging the Forecast series set with the estimation data set */

data new2;
  merge new fore; by date;
format date year.;
run;
```

In this case, the forecast is extended 12 years into the future. The identi-fying variable is DATE, which designates the year of the observation. The interval of year is specified. The output data set is called FORE. It is merged with the previous data set by date (in years). The SAS syntax in Fig. 9.24 graphs the two series and their forecast shown at the beginning of this modeling.

```
/* Figure 9.24   Graphing the Forecast Profile */
data new3;
  set new2;
/* eliminating series overlaps for graphing */
if _n_ < 66 then l95 = . ;
if _n_ < 66 then u95 =. ;
if _n_ < 66 then forecast = .;
proc print data=new3;
title 'new3';
run;
/* generates the x-axis value above the reference line */
data anno; informat date date9.;
  input date date9. text $ 12-35;
function='label'; angle = 90 ; xsys='2'; ysys='1';
x=date; y=40; position='B';
cards;
01jan1994  1994 Forecast begins
;
/* ****************************************************************/
/* Graphing the two series and the forecast from the transfer function */
/* + noise model                                                  */
/* ****************************************************************/
axis1 label=(a=90 '1987 Constant $');
symbol1 i=join c=green v=star;
symbol2 i=join c=blue v=plus;
symbol3 i=join c=red;
symbol4 i=join c=red;
symbol5 i=join c=black v=diamond;
proc gplot data=new3;
  plot (ce forecast l95 u95 pdi) * date/overlay vaxis=axis1 href='01jan1994'd
  annotate=anno;
title justify=L 'Figure 9.23          Personal Consumption Expenditures ';
title2 'Driven by Per Capita Personal Disposable Income ';
title3  'in Constant 1987 $';
title4 'Source: US Bureau of Economic Analysis';
title5 'National Income and Product Accounts of the US';
title6 'Survey of Current Business, July 1994 & March 1995';
footnote1 justify=L 'Personal Disposable income = Diamond';
footnote2 justify=L 'Consumption Expenditures = star';
footnote3 justify=L  'Forecast=Plus      Confidence limits=lines';
run;
```

Figure 9.24

   The question remains how one interprets the SAS output and translates it into a formula. Immediately after the cross-correlations between the transfer function model and the noise model, the estimated parameters are output. The noise model for the input series, PDI, is given in the first four of the following lines.

```
Model for variable PDI

Data have been centered by subtracting the value 152.13846154.
No mean term in this model.
Period(s) of Differencing = 1.

Autoregressive Factors
Factor 1: 1 - 0.41038 B**(1)

Moving Average Factors
Factor 1: 1 - 0.22279 B**(4) - 0.46116 B**(10)
```

From this input series, the prewhitening filter is developed and applied.

```
Both variables have been prewhitened by the following filter:

Prewhitening Filter

Autoregressive Factors
Factor 1: 1 - 0.41038 B**(1)

Moving Average Factors
Factor 1: 1 - 0.22279 B**(4) - 0.46116 B**(10)
```

But the final full (untrimmed) model for the transfer function and ARIMA noise model is given in the last part of the output, where it provides the model for personal consumption expenditures, CE.

```
Model for variable CE

Data have been centered by subtracting the value 141.01538462.
No mean term in this model.
Period(s) of Differencing = 1.

Autoregressive Factors
Factor 1: 1 - 0.27365 B**(1)

Moving Average Factors
Factor 1: 1 - 0.21738 B**(4) + 0.1275 B**(10)

Input Number 1 is PDI.
Period(s) of Differencing = 1.

The Numerator Factors are
Factor 1: 0.4721 + 0.17895 B**(10)
```

From this output, the following model is estimated.

$$(CE_t - 141.02)(1 - L) = (0.472 + 0.179L^{10})(PDI_t - 152.14)(1 - L)$$

$$+ \frac{(1 - 0.217L^4 + 0.128L^{10})}{1 - 0.274L} e_t,$$

(9.22)

where e_t is the innovation or random shock,
CE_t is the consumption expenditures, and
PDI_t is the personal disposable income.

If the series were not centered, there would be a constant noted in the output that would be entered as the first term on the right-hand side of Eq. (9.22). If there were denominator factors in the transfer function, these would be placed underneath the factor preceding the mean-centered PDI_t on the right-hand side of this equation. Additional trimming of this model before forecasting is left as an exercise. When the researcher trims nonsignificant moving average terms from the full model, he should observe carefully what happens to the model fit, the model parsimony, the mean square forecast error, and the forecast interval. When there is moderate or substantial correlation between transfer function and noise model parameters, spurious spikes can appear in the residuals of the noise or transfer function model parameters. The benefits and costs from trimming have to be weighed in determining how many erstwhile significant moving average terms can be prudently pruned. Parsimony is an important principle in this process, yet the final decision on what terms can be shaved from the model may depend on artistic taste as well as sound scientific judgement. In this way, SAS can be used to apply the Box–Jenkins strategy to transfer function modeling and forecasting.

9.4.2. The Linear Transfer Function Modeling Strategy

9.4.2.1. Dynamic Regression

This section focuses exclusively on the dynamic regression (DR) or linear transfer function (LTF) modeling strategy for transfer functions. Because the strategy does not involve prewhitening the series, it is easy to apply. Because it claims that there is no need for prewhitening, this strategy claims to be able to include model multiple inputs without unnecessary complication. Insofar as the several input series are not highly correlated, this modeling strategy is in fact very useful. Therefore, this approach is argued to be more robust than the classical Box–Jenkins one. This newer approach has been developed by Lui and Hanssens (1982) and Tsay (1985). If the researcher is using SPSS he should use this approach because SPSS cannot automatically prewhiten series selected for analysis. To overcome possible problems of omitted time-lagged input terms, autocorrelation in the disturbance series, and common correlation patterns among the input and output series that yield spurious correlations, the dynamic regression or linear transfer function modeling approach can be very helpful (Pankratz, 1991). Using this approach, the researcher can develop an ARIMA noise model and then include one or more unprewhitened input series in either SAS or SPSS. First, we entertain the underlying theoretical considerations and then a programming application.

First, preliminary background preparation is necessary. Before the modeling begins, the researcher should familiarize himself with the history, the theory, the empirical reality, and the relevant logic of processes and/or persons involved. From a literature review of the theory and/or history of the subject, the researcher/analyst may be able to tell which predictor series are needed to model his response series. From this knowledge, he may be able to tell which predictor series are not needed for this project. He may be able to tell from the history of the series which events are significantly or not significantly related to turning points in the response series. From this information, he can determine which events need to be included in the model and which events need not be included. He may also be able to tell how many predictor series he needs to model the process. He can interview key persons to check on his literature. He can graph the series against the time line and compare changes in the response to historical related events. He needs to be sure that they correspond. There is no substitute for expert knowledge in the beginning of the analysis.

Second, the researcher should graph the data and review the series for outliers. These may be holiday effects or singular events that affect the series. If he finds outlying observations, he should test them to confirm that they are outliers. If they are confirmed outliers, he should model, smooth, or remove them in accordance with his understanding of the historical data.

Third, preliminary identification of both series with a review of the ACF and PACF is helpful at this pont. Although Tsay (1995) contends that the LTF method is more robust to nonstationarity than the conventional Box–Jenkins approach, Pankratz (1991) suggests that nonstationarity would require the appropriate detrending or differencing between the two series. Preliminary transformations to obtain covariance stationarity may be necessary, lest such nonstationarity confound the forthcoming feedback test.

Fourth, the researcher needs to perform the Granger causality test for feedback, while making some allowances for sampling variation. If feedback is not logical or theoretically permissible, inputs that suggest apparent feedback may need to be considered for evaluation of sampling variation, spurious regression, and subsequent removal. It may be helpful to remove or model seasonality at this point.

Fifth, specification of a linear transfer function model involves formulation of an equation in the form

$$Y_t = \alpha + v(L) + e_t$$
$$= \alpha + v_0 X_t + v_1 X_{t-1} + v_2 X_{t-2} + \cdots + v_j X_{t-j} + e_t. \tag{9.23}$$

We choose the order j to be large enough to accommodate the significant lags of the exogenous series, X_t. By comparison of the actual pattern with the theoretical patterns of the transfer functions previously mentioned, we can discover the nature of the transfer function and its structural parameters. Small sample sizes may limit the number of lags that are modeled, so researchers prefer larger sample sizes for this kind of modeling. We can prune the free-form distributed lag to exclude nonsignificant lagged terms.

Sixth, we model a disturbance term, e_t, as a low-order autoregressive–seasonal autoregressive process.

$$(1 - \varphi_1 L)(1 - \Phi_s L^s) n_t - e_t$$
$$n_t = \frac{e_t}{(1 - \varphi_1 L)(1 - \Phi_s L^s)}. \tag{9.24}$$

Alternatively, a first-order nonseasonal AR term can be modeled, and if the ACF or PACF of the residuals reveal seasonality, a seasonal AR factor can be included. This identification and estimation will absorb the autocorrelation in the noise model so that it will not contaminate (render inefficient the significance tests of the impulse response weights) the cross-correlations between the input and output series. In other words, the new model is parameterized as a combination of Eqs. (9.23) and (9.24) and estimated as

$$Y_t - \alpha + v_0 X_t + v_1 X_{t-1} + \cdots v_f X_{t-j} + \frac{e_t}{(1 - \varphi_1 L)(1 - \Phi_s L^s)}. \tag{9.25}$$

Seventh, we can diagnose the ACF and PACF of the noise model to determine whether the model is nonstationary and requires further differencing. If we find the noise model AR term to be close to or equal to unity, differencing may be in order. The effect of this is to difference both the input and output series identically. If differencing is required, it may be first and/or seasonal differencing. If only first differencing is required, the new model would be

$$\nabla Y_t = a + v_0 \nabla X_t + v_1 \nabla X_{t-1} + \cdots + v_j \nabla X_{t-j}$$
$$+ \frac{e_t}{(1 - \varphi_1 L)(1 - \Phi_s L^s)}. \tag{9.26}$$

where $a = \nabla \alpha$.

Eighth, the ACF and PACF for the noise model residuals should reveal whether this model has accounted for all of the residual autocorrelation. If the residuals of the noise model are white noise after differencing, then the model has properly accounted for the residual autocorrelation. If, however, seasonal nonstationarity remains, and the residuals exhibit a cyclical pattern that tails off slowly, then the input and output series can be seasonally differenced. The first and seasonal differencing of the input and output series should then render the model stationary. The ACF and PACF of the residuals should taper off quickly or, if the noise model is properly identified and already estimated, then they should reveal only residual white noise. The researcher can overfit and underfit the model to attain the best noise model at this stage.

Ninth, after identifying and estimating the noise model, the researcher can graph the impulse response function. The model is first specified as a free-form distributed lag model with enough v_j weights to model the transfer function effect (Eq. 9.25). He compares this actual v_j weight pattern with theoretical transfer function patterns plotted earlier and formulated in Table 9.5, and then he attempts by direct estimation to identify the structural parameters of the transfer function. In this way, he can identify the delay, then the decay, and finally, the number of numerator regression parameters.

Tenth, he specifically examines the delay in impact. If the v_j weights are nonsignificant for five lags, then 5 is the delay parameter. b is then said to equal 5. If there is no delay before the v_j weights become statistically significant, then $b = 0$. In this manner, he identifies the b parameter.

Eleventh, he next examines the decay rate pattern of the impulse response function. If there is an instantaneous pulse or a sudden albeit permanent level shift, there is no decay at all and the r decay parameter equals 0. If there is simple exponential dampening of the magnitude of the v_j weights, this is first-order decay, in which case, the r decay parameter might equal 1. If there is compound dampening in the pattern of the response function, then this is evidence of higher-order decay. If, for example, he discerns cyclical variation in the pattern of impulse response weights, then the decay parameter might equal 2 or more. Most transfer function models have first- or second-order decay. Having identified the decay parameter, which models patterned spikes, he proceeds to examine the numerator coefficients, which represent unpatterned spikes.

Twelfth, he counts the number of unpatterned startup values. If there are two unpatterned spikes in the pattern of v_j weights, then $s + 1$ equals 2 so s equals unity. The number of numerator regression coefficients that is significant should be $s + 1$. Frequently, it is helpful to test one or two more to empirically observe the point at which these tail off in significance. Thus, he can identify the structural parameters of the impulse response function. As a check of his transfer function modeling, he can construct a corner table for assistance in confirming the order of the structural parameters of the impulse response function.

Thirteenth, after the noise model and the transfer function model have been estimated, he reviews the residuals. At this stage of the analysis, he checks the correlations between the parameters of the noise and the transfer function model. If these correlations are minimal, this evidence supports the assumption of independence between these two models. If the noise model is independent of the transfer function model, then the CCF between their residuals should show white noise. If there is some dependence, there might appear to be a relationship between e_t and e_{t-j} that could be interpreted as a θ_{t-j} in the noise model. A failure to model an ω_{t-j} regression parameter may result in a spike in the CCF at lag j. Sometimes theory or history is helpful in determining which of the parameters is the appropriate one to model this relationship. Lacking such hints, the researcher can revert to remodeling the linear transfer function in hope of removing residual spikes in the CCF. The CCF can reveal misspecification of the dynamic regression model and suggest the need for revision (Pankratz, 1991).

Fourteenth, he reviews the ARMA model for the disturbance term. He rechecks the residual ACF, PACF, and EACF. He looks for outliers that can confound the sample correlogram output and models or smooths them. If, for example, there is one statistically significant spike in the sample ACF and one statistically significant spike in the sample PACF, this pattern would suggest an MA(1) model for the stationary noise model. As a double-check, the sample EACF could be used. The upper left vortex of the triangle of zeros (at the 0.05 level of significance) would indicate at what order of AR rows and MA columns identifies the ARMA noise model. In this way, the sample EACF could be used to confirm identification of the order of the ARMA noise model. If the low-order AR and/or SAR terms already account for the autocorrelation, the EACF will confirm proper MA specification of the noise model. The model is then fine-tuned to produce white noise residuals.

Fifteenth, if diagnosis of the sample ACF, PACF, and EACF residuals discloses noise or transfer function misspecification, reformulation of the model is in order. The analyst can hypothesize and test noise model reformulation. An incorrect transfer function specification can induce autocorrelation in the noise model, although the remainder of the ARIMA specification is in fact correct. Trimming insignificant parameters may eliminate peculiar residual spikes in the correlograms (ACF, PACF, EACF, and CCF) and should be used to improve fit, parsimony, and prediction.

The model building process is iterative. The researcher may need to recycle through these steps until the model fulfills the assumptions, makes sense, and fits well. The model must be adequate and should be simple. After reiterating through this process, the model should be the best plausible model. In this way, the researcher builds a model to represent the dynamic relationship under consideration.

9.4.2.2. Advantages of the Dynamic Regression Modeling Strategy

The advantages of the LTF strategy are several according to the work of Liu and Hanssens (1982), Tsay (1985), Lui et al. (1992), and Pankratz (1991). The LTF (DR) method is easy to apply. It does not entail complicated prewhitening. It is more independent of sequence than the conventional Box–Jenkins method. The conventional method requires a rigid sequence for (1) proper specification of the ARIMA noise model, (2) proper prewhitening, and (3) the use of the CCF for transfer function model identification. If there is misspecification at any of these stages or if the order of the sequence is altered, the modeling process may become seriously undermined. These authors emphasize that the strategy can handle nonstationary models, multiple inputs without complicating prewhitening, and autocorrelation in the noise model without rendering the significance tests inefficient and biased. Like the Box–Jenkins method, it requires diagnosis and model reformulation before the fine-tuning of the mode is complete.

A principal advantage is that dynamic regression models can be applied to the combination of multiple forecasts. With each forecast constituting a driving or forcing series, the combination point forecast can be constructed. If the dynamic regression model possesses ARMA(p,q) residuals, then the residuals can be modeled with an ARIMA procedure according to the linear transfer function method described. If the dynamic regression model possesses only moving average or seasonal moving average errors, this is a regression with a form of time series errors that requires no prewhitening (Bresler et al., 1993). It is common, however, for models to have inputs or noise that are not free of autocorrelation. Because of its theoretical and applied advantages, the linear transfer function approach is recommended as a strategy to modeling multiple input series.

There are two basic approaches to modeling transfer functions with multiple input series. The researcher can either sequentially or simultaneously input the explanatory series. To sequentially model, one models the output series as a transfer function of the input series, in the manner just explained. In the next stage of input, the researcher may use the residuals from the previous model as output to the new input series. In this way, he can sequentially chain input series to the output series. In the Box–Jenkins approach, multiple inputs may be sequentially chained taken two at a time. An example of this type of model was set forth in McCleary et al. (1980) between the size of the Swedish harvest and the Swedish fertility rate, first, and between the fertility rate and the Swedish population next. For such analysis, the Box–Jenkins method, with its prewhitening, serves well. For modeling several series simultaneously, the linear transfer function approach, which does not require prewhitening, is recommended.

The advantages of LTF modeling are that they permit the development of multiple input models. They permit modeling multiple input models that can be sequential or simultaneous. If the models are sequential, then they are chains of bivariate sequences of variables strung together. They permit modeling of multicausal theoretical relationships among series. If there is not much multicollinearity, simultaneous multiple input models can be identified, estimated, diagnosed, metadiagnosed, and forecast with some precision. The newer modeling strategy can handle autocorrelation within the disturbance term and nonstationary series and instances of spurious correlation produced by common patterns in the within-series correlation. In essence, these developments have raised time series analysis to the level where different hypotheses can be tested as to the transfers from one series to another. The advantages show how one or several series can drive an endogenous series; they can form the basis of a path analysis of time series. They provide an opportunity to model, test, and explore the understanding of these relationships. They can be used to test predictive or concurrent validity, or they can be combined to fashion designed responses or to model complex situations. They are a powerful tool of analysis that forms the basis for much quality and engineering process control analysis.

9.4.2.3. Problems with Multiple-Input Models

If we use the regression approach for multiple inputs, some problems occasionally occur that merit consideration. When we are modeling two simultaneous inputs, these inputs cannot be highly correlated with one another without multicollinearity creating problems for the estimation process. Too much multicollinearity inflates standard errors and may preclude convergence of the estimation process. If simultaneous inputs are substantially correlated and the transfer functions are estimated, the question of how to handle this problem arises (Reilly, 1999).

A number of suggestions have been tendered. Centering the series reduces the number of terms in the model and therefore some of the possibility of multicollinearity. If the Box–Jenkins approach to modeling is used, Liu and Hanssens (1982) suggest using the same prewhitening filter. Priestley (1971) has recommended prewhitening both the input and output series before obtaining transfer function weights. If there is one endogenous series and two exogenous series, each of which has different levels of autocorrelation within it, then it could be very difficult to develop a common prewhitening filter that would be suitable for all three. The prewhitening filter developed from one might be inappropriate for the other. Liu and Hanssens (1982) have suggested that in some instances the use of double-precision computing may avoid these problems. When it does not matter whether the original variables are radically transformed prior to analysis, Tsiao and Tsay (1985) propose a canonical analysis to produce scalar components models, after which they subject these components to a vector ARMA approach to solving the problem, but this last suggestion is beyond the current scope of this text. Reiterative trimming reduces multicollinearity in the model (Liu et al., 1992). If the LTF strategy is applied, this approach need not involve problematic prewhitening with multiple input models.

9.4.2.4. Dynamic Regression Modeling with SAS and SPSS

It seems logical that construction contracts lead to housing starts. Our housing start and construction contract data extend from January 1983 through October 1989 and are available from Rob Hyndman?s Time Series Data Library, the World Wide Web address of which is http://www.maths.-monash.edu.au/~hyndman/TSDL/index.htm. A graph displaying these data may also be found in Makridakis et al. (1998). The presumption is that construction contracts naturally lead to housing starts.

We can use either SAS or SPSS to program the linear transfer function method. Let us consider the SAS syntax first. SAS `PROC ARIMA` can be used to build a transfer function model between the number of construction contracts and housing starts during this time period. The complete SAS programming syntax, with its major steps, is found in `C9PGM2.SAS`.

First, we conduct our background research and check the integrity of the data with a `PROC PRINT`.

Second, we graph the series, examine the plotted series, and find no obvious outliers (Figure 9.25).

Third, we perform preliminary identification of the series with the ACF and PACF to check for nonstationarity and seasonality. With the observation of seasonal spiking at lag 12, we see that seasonal differencing of order 12 is required to bring about stationarity (Figs. 9.26 and 9.27).

After 12th differencing of both the contracts input and housing starts output series, we can see from their ACFs that both series are stationary (Figs. 9.28 and 9.29).

These differenced autocorrelations tail off nicely. Both the input and output series are differenced at lag 12. No further transformation of either series appears to be needed.

Fourth, we perform a sample preliminary feedback check by regression of 12th differenced contracts on five distributed lags of 12th differenced

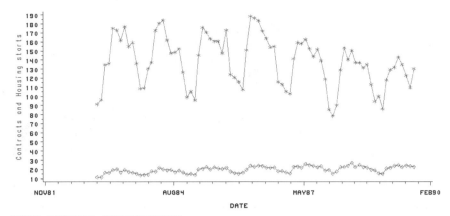

Figure 9.25 (*) Housing starts, (◇) construction contracts, January 1983 through October 1989.

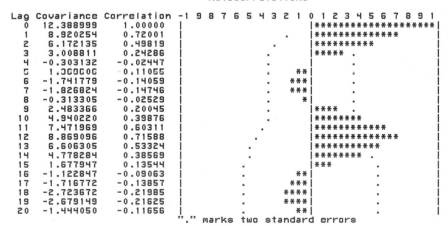

Figure 9.26 Step 3: Preliminary stationarity check—construction contracts.

```
ARIMA Procedure

Name of variable = HSTARTS.

Mean of working series = 137.9768
Standard deviation     = 27.50451
Number of observations =       82

                        Autocorrelations

Lag Covariance Correlation -1 9 8 7 6 5 4 3 2 1 0 1 2 3 4 5 6 7 8 9 1
  0   756.498    1.00000   |                        |********************|
  1   553.498    0.73166   |                   .    |***************     |
  2   290.885    0.38452   |                .       |********            |
  3  26.247882   0.03470   |                .       |*                   |
  4  -192.267   -0.25415   |               . *****  |                    |
  5  -244.180   -0.32278   |              .*****|    .                   |
  6  -249.838   -0.33026   |              ******|    .                   |
  7  -242.842   -0.32101   |             . ******|   .                   |
  8  -167.404   -0.22129   |              . ****|    .                   |
  9  17.264693   0.02282   |              .     |    .                   |
 10   229.849    0.30383   |              .     |****** .                |
 11   453.026    0.59885   |              .     |***********             |
 12   537.970    0.71113   |              .  .  |**************          |
 13   399.665    0.52831   |              .     |***********             |
 14   180.475    0.23857   |              .     |*****  .                |
 15  -58.945411 -0.07792   |              .    **|   .                   |
 16  -229.405   -0.30325   |              . ******|  .                   |
 17  -276.498   -0.36550   |              . *******| .                   |
 18  -271.107   -0.35837   |              . *******| .                   |
 19  -261.299   -0.34541   |              . *******| .                   |
 20  -192.157   -0.25401   |              . *****|    .                   |
                    "." marks two standard errors
```

Figure 9.27   Step 3: Preliminary stationarity check—housing starts.

housing starts with the program command statements

```
/* Step 4 Granger Causality Test */

proc reg;
  model difctrct = d12hst1 d12hst2 d12hst3 d12hst4 d12hst5;
title 'Step4 Granger Causality test';
title2 'Insignificant Regression Coefficients reveal no problem';
run;
```

We find no significant regression coefficients for this sample output (Fig. 9.30), though more lags are generally tested. In fact, there is evidence of minor irregularity, sampling variation, or feedback when more lags are tested. For our purposes, these aberrations are not considered serious and we proceed with the analysis.

If we have reason to believe that there might be feedback, we can examine as many lags might be relevant.

In step 5, we can model the regression of differenced and centered new housing starts on a distributed lag of differenced and centered construction contracts. Lagged contracts variables are constructed and equally differenced before serving as predictors of the similarly differenced Housing Starts series. For these monthly data, all series are differenced at lag 12 in the CROSSCORR statement in accordance with Eq. (9.26).

```
                           ARIMA Procedure


                  Name of variable = CONTRCTS.

                  Period(s) of Differencing = 12.
                  Mean of working series = 0.860786
                  Standard deviation     = 1.440926
                  Number of observations =       70
                  NOTE: The first 12 observations were eliminated by
                        differencing.

                          Autocorrelations

Lag Covariance Correlation -1 9 8 7 6 5 4 3 2 1 0 1 2 3 4 5 6 7 8 9 1
  0   2.076267   1.00000    |                    |********************|
  1   0.209748   0.10102    |                  . |**  .               |
  2   0.575546   0.27720    |                  . |******              |
  3   0.636676   0.30664    |                  . |******              |
  4   0.357923   0.17239    |                  . |***  .              |
  5   0.288106   0.13876    |                  . |***  .              |
  6   0.626925   0.30195    |                  . |******              |
  7   0.141293   0.06805    |                  . |*    .              |
  8   0.328636   0.15828    |                  . |***  .              |
  9   0.523094   0.25194    |                  . |*****.              |
 10  -0.063286  -0.03048    |                  . *|     .             |
 11   0.568243   0.27369    |                  . |*****.              |
 12  -0.093099  -0.04484    |                  . *|     .             |
 13  -0.015554  -0.00749    |                  . |     .             |
 14   0.469440   0.22610    |                  . |*****.              |
 15   0.276848   0.13334    |                  . |***  .              |
 16  -0.089570  -0.04314    |                  . *|     .             |
 17   0.398560   0.19196    |                  . |****  .             |
                    "." marks two standard errors
```

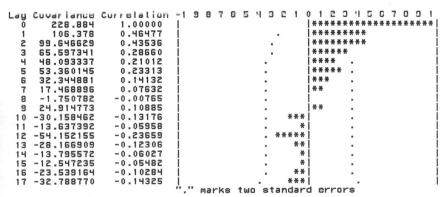

Figure 9.28   Step 3: Stationarity check—construction contracts.


```
                           ARIMA Procedure


                  Name of variable = HSTARTS.

                  Period(s) of Differencing = 12.
                  Mean of working series = -4.15857
                  Standard deviation     = 15.12893
                  Number of observations =       70
                  NOTE: The first 12 observations were eliminated by
                        differencing.

                          Autocorrelations

Lag Covariance Correlation -1 9 8 7 6 5 4 3 2 1 0 1 2 3 4 5 6 7 8 9 1
  0  228.884    1.00000     |                    |********************|
  1  106.378    0.46477     |                  . |*********           |
  2   99.646629 0.43536     |                  . |*********           |
  3   65.597341 0.28660     |                  . |******              |
  4   48.093337 0.21012     |                  . |****  .             |
  5   53.360145 0.23313     |                  . |*****.              |
  6   32.344881 0.14132     |                  . |***  .              |
  7   17.468896 0.07632     |                  . |**   .             |
  8   -1.750782 -0.00765    |                  . |     .             |
  9   24.914773 0.10885     |                  . |**   .             |
 10  -30.158462 -0.13176    |                . ***|     .             |
 11  -13.637392 -0.05958    |                  . *|     .             |
 12  -54.152155 -0.23659    |                . *****|     .            |
 13  -28.166909 -0.12306    |                  . **|     .            |
 14  -13.795572 -0.06027    |                  . *|     .             |
 15  -12.547235 -0.05482    |                  . *|     .             |
 16  -23.539164 -0.10284    |                  . **|     .            |
 17  -32.788770 -0.14325    |                . ***|     .             |
                    "." marks two standard errors
```

Figure 9.29   Step 3: Stationarity check—housing starts.

```
Model: MODEL1
Dependent Variable: D12CTRCT
```

Analysis of Variance

Source	DF	Sum of Squares	Mean Square	F Value	Prob>F
Model	5	9.95574	1.99115	0.967	0.4454
Error	59	121.48706	2.05910		
C Total	64	131.44279			

Root MSE	1.43496	R-square	0.0757
Dep Mean	0.74711	Adj R-sq	-0.0026
C.V.	192.06835		

Parameter Estimates

Variable	DF	Parameter Estimate	Standard Error	T for H0: Parameter=0	Prob > |T|
INTERCEP	1	0.836636	0.19709574	4.245	0.0001
D12HST1	1	0.025370	0.01512738	1.677	0.0988
D12HST2	1	-0.019282	0.01544643	-1.248	0.2168
D12HST3	1	0.008407	0.01556179	0.540	0.5911
D12HST4	1	0.012301	0.01462073	0.841	0.4035
D12HST5	1	-0.014178	0.01388162	-1.021	0.3113

Figure 9.30   Step 4: Preliminary Granger causality test.

```
proc arima;
  i var=hstarts(12) center crosscorr=(contrcts(12) contrL1(12) contrL2(12) contrL3(12)
    contrl4(12) contrl5(12) contrl6(12) contrl7(12) contrl8(12)
    contrl9(12) contrl10(12)) nlag=25;
  e input=(contrcts contrL1 contrl2 contrl3 contrl4 contrl5 contrl6 contrl7 contrl8
    contrl9 contrl10 ) printall plot noint method=ML maxit=40;
title 'Step 5 Free form distributed lagged model of exogenous terms';
title2 ' with review of the residuals';
Title3 'Step 5 LTF model approach ';
run;
```

The output suggests that only the first numerator at lag 0 is significant.

```
        Step 5 Free form distributed lagged model of exogenous terms
with review of the residuals
Step 5 LTF approach
23:22 Sunday, August 15, 1999

ARIMA Procedure

Maximum Likelihood Estimation
```

		Approx.				
Parameter	Estimate	Std Error	T Ratio	Lag	Variable	Shift
NUM1	3.23582	1.36103	2.38	0	CONTRCTS	0
NUM2	1.65967	1.36239	1.22	0	CONTRL1	0
NUM3	0.70663	1.34629	0.52	0	CONTRL2	0
NUM4	0.58266	1.32519	0.44	0	CONTRL3	0
NUM5	-0.65923	1.33267	-0.49	0	CONTRL4	0
NUM6	0.64851	1.33627	0.49	0	CONTRL5	0
NUM7	-0.15888	1.35718	-0.12	0	CONTRL6	0
NUM8	0.36643	1.36415	0.27	0	CONTRL7	0
NUM9	1.93708	1.37504	1.41	0	CONTRL8	0
NUM10	-0.23511	1.42390	-0.17	0	CONTRL9	0
NUM11	-0.57502	1.40027	-0.41	0	CONTRL10	0

An examination of the estimates of these impulse response weights reveals no dead time. The only significant response is at the beginning. There is an immediate peak and no significant gradual decay, indicative of a pulse input. The *SBC* is 510.000. From Figure 9.31, we observe that the shape of the response weights is identical to that of a single significant pulse at lag 0 and a possible additional pulse at lag 8. It can be argued that the pulse at lag 8 is not significant at first glance and should be dropped, but we first try to test a model that includes this second pulse.

There were some preliminary matters demanding attention. Although we tried a first-difference, we found it unnecessary to assure stationarity. We could either leave the significant lagged exogenous terms in or we could reexpress them as a function. A first-order decay parameter was found to be nonsignificant. To conserve degrees of freedom, we now trim the nonsignificant extraneous distributed lag terms with a zero-order pulse in the input statement INPUT = ((8)CONTRCTS) to match the pattern of the transfer function on the ESTIMATE subcommand, begun with the abbreviation, E. Differencing usually centers a series. The differenced exogenous inputs are lagged at periods 0 and 8 to represent a $(\omega_0 - \omega_8 L^8)(1 - L^{12})$ CONTRCTS$_t$ formulation. This formulation, given the relative magnitude of the parameters, effectively models the spikes and reduces the *SBC* to 492.69. If we do not recognize the empirical form to be one of the conventional theoretical patterns of the transfer function, we can always use the corner table for transfer function identification. If we find that there are several significant transfer functions separated, we can model a more complex combination to fit the pattern.

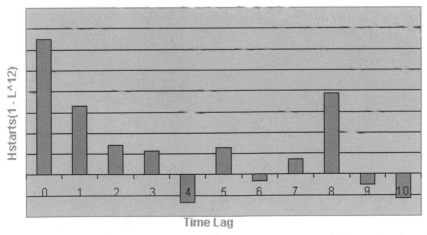

Figure 9.31 Step 5: First-order transfer function for housing starts (12) as a function of construction contracts (12).

In step 6, we note the residuals from this transfer function and the ACF spikes at lags 1 and 2. Therefore, we try a low-order ARSAR noise model, with the P=(1 2)(12) in the beginning of the ESTIMATE subcommand.

```
proc arima;
 i var=hstarts(12) center crosscorr=(contrcts(12)) nlag=25;
 e p=(1 2 )(12) input=((8)contrcts) printall plot noint method=ML maxit=40;
title 'Step 6 Test Lower Order AR terms to model input AR';
title7 'Using trimmed significant lags Regression approach ';
title8 'Test Low Order AR terms to get rid of corrupting influence';
run;
```

From this ARIMA procedure, we obtain an estimation that shows that all the transfer function terms and AR noise model terms are significant.

ARIMA Procedure

Maximum Likelihood Estimation
Approx.

Parameter	Estimate	Std Error	T Ratio	Lag	Variable	Shift
AR1,1	0.25280	0.12730	1.99	1	HSTARTS	0
AR1,2	0.35405	0.13152	2.69	2	HSTARTS	0
AR2,1	-0.30258	0.15073	-2.01	12	HSTARTS	0
NUM1	2.77137	0.94842	2.92	0	CONTRCTS	0
NUM1,1	-2.03126	0.93072	-2.18	8	CONTRCTS	0

```
Variance  Estimate = 105.691652
Std Error Estimate = 10.2806445
AIC                = 471.259174
SBC                = 481.894846
Number of Residuals= 62
```

In step 7, we trim the first-order autoregressive parameters from the model in the interest of parsimony in the next procedure by merely specifying a $p = (2)$, which means that only the second-order autoregressive term is estimated.

```
proc arima;
 i var=hstarts(12) center crosscorr=(contrcts(12)) nlag=25;
 e p=(2) input=((8)contrcts)  printall plot noint method=ML Maxit=40;
title  'Step 7 Trimming the Model';
title2 'LTF modeling Strategy ';
run;
```

The coefficients are reestimated with the following results.

ARIMA Procedure
Maximum Likelihood Estimation

Approx.

Parameter	Estimate	Std Error	T Ratio	Lag	Variable	Shift
AR1,1	0.27966	0.12649	2.21	1	HSTARTS	0
AR1.2	0.33000	0.12866	2.56	2	HSTARTS	0
NUM1	2.96772	0.95632	3.10	0	CONTRCTS	0
NUM1,1	-2.00643	0.93008	-2.16	8	CONTRCTS	0

```
Variance  Estimate = 112.621286
Std Error Estimate = 10.6123177
AIC                = 473.125455
SBC                = 481.633993
Number of Residuals=         62
```

The AIC and SBC are not much lower, but the model is more parsimonious and all of the parameters are significant. The parameters are not highly intercorrelated and a review of the residuals shows that this model fits.

```
                    ARIMA Procedure

              Autocorrelation Check of Residuals
  To   Chi                Autocorrelations
 Lag Square DF  Prob
  6   6.30    4 0.178   0.006 -0.048 -0.116  0.134  0.237 -0.040
 12  11.61   10 0.312   0.066 -0.101  0.101 -0.028 -0.083 -0.190
 18  17.39   16 0.361  -0.046  0.027 -0.032  0.089 -0.201 -0.117
 24  25.83   22 0.259  -0.169 -0.018 -0.033 -0.066  0.123 -0.188
```

If we overfit and underfit to be sure that we have the best model, we can arrive at an alternative model. We can review the residuals and discover a not quite significant moving average spike at lag 5. To model this spike, we can add a moving average parameter at lag 5 and trim the nonsignificant autoregressive terms at lags 1 and 12. The program for the fine-tuned of the model is

```
proc arima;
  i var=hstarts(12) center esact crosscorr=(contrcts(12)) nlag=25;
  e p=(1 2) q=(5) input=((8)contrcts) printall plot noint
  method=ML Maxit=40;
Title 'Step 8 Fine-tuning the LTF Model ';
run;
```

The coefficients are reestimated and the output from the model reveals a more parsimonious model with an SBC of 480.76, slightly better than that of the earlier model and substantially better than that of the model with only lagged exogenous predictors.

```
              Maximum Likelihood Estimation
```

		Approx.				
Parameter	Estimate	Std Error	T Ratio	Lag	Variable	Shift
MA1.1	-0.33328	0.12918	-2.58	5	HSTARTS	0
AR1.1	0.39355	0.12314	3.20	2	HSTARTS	0
NUM1	3.53925	0.91761	3.86	0	CONTRCTS	0
NUM1,1	-2.03453	0.89755	-2.27	8	CONTRCTS	0

```
Variance  Estimate  =  110.161259
Std Error Estimate  =  10.4957734
AIC                 =  472.257264
SBC                 =  480.765801
Number of Residuals=         62
```

The correlation among these parameters are small and not problematic.

Correlations of the Estimates

Variable	Parameter	HSTARTS MA1,1	HSTARTS AR1,1	CONTRCTS NUM1	CONTRCTS NUM1,1
HSTARTS	MA1.1	1.000	0.150	-0.038	-0.046
HSTARTS	AR1.1	0.150	1.000	0.098	-0.026
CONTRCTS	NUM1	-0.038	0.098	1.000	0.046
CONTRCTS	NUM1,1	-0.046	-0.026	0.046	1.000

The autocorrelation check of the residuals of this model reveals that they are clearly white noise.

Autocorrelation Check of Residuals

To Lag	Chi Square	DF	Prob	Autocorrelations					
6	4.38	4	0.357	0.201	-0.052	0.046	0.140	0.007	0.032
12	10.92	10	0.364	0.111	-0.066	0.015	-0.002	-0.155	-0.208
18	19.43	16	0.247	-0.029	0.011	-0.038	0.068	-0.157	-0.252
24	26.87	22	0.216	-0.184	-0.057	-0.098	-0.037	0.110	-0.128

A visual inspection in step 8 of the ACF and PACF confirms white noise (Figs. 9.32 and 9.33).

In step 9 of the programming, the researcher needs to be sure that there is no significant correlation between the transfer function parameters and the noise model; a final cross-correlation check between them is run. Special program syntax is required to invoke this check. In the same ARIMA procedure, the input series needs to be identified and estimated prior to the identification and estimation of the output series.

```
proc arima:
   i var=contrcts(12) center nlag=25;
   e printall plot noint:
   i var-hstarts(12) center crosscorr=(contrcts(12)) nlag=25;
   e p=(2) q=(5)input=((8) contrcts ) printall plot noint method=ML maxit=40;
   f lead=12 interval=month id=date out=fore;
title 'Step 9 Cross-Corr Check between Noise and TF Parms';
run;
```

From these results, it is clear that there is no significant cross-correlation between the separate components of the model.

Step 8   Fine-Tuning the LTF model

ARIMA Procedure

Autocorrelation Plot of Residuals

```
Lag Covariance Correlation -1 9 8 7 6 5 4 3 2 1 0 1 2 3 4 5 6 7 8 9 1
  0  110.161     1.00000   |                    |********************|
  1   22.184433   0.20138   |                .   |****.               |
  2   -5.775516  -0.05243   |                .  *|    .               |
  3    5.030154   0.04566   |                .   |*   .               |
  4   15.422232   0.14000   |                .   |*** .               |
  5    0.721568   0.00655   |                .   |    .               |
  6    3.559035   0.03231   |                .   |*   .               |
  7   12.275207   0.11143   |                .   |**  .               |
  8   -7.318295  -0.06643   |                . *|    .               |
  9    1.601489   0.01454   |                .   |    .               |
 10   -0.260872  -0.00237   |                .   |    .               |
 11  -17.079485  -0.15504   |              . ***|    .               |
 12  -22.905967  -0.20793   |              . ****|    .               |
 13   -3.240429  -0.02942   |                . *|    .               |
 14    1.194912   0.01085   |                .   |    .               |
 15   -4.145639  -0.03763   |                . *|    .               |
 16    7.522263   0.06828   |                .   |*   .               |
 17  -17.281068  -0.15687   |              . ***|    .               |
 18  -27.792028  -0.25228   |              .****|    .               |
 19  -20.284163  -0.18413   |              . ****|    .               |
 20   -6.282714  -0.05703   |                . *|    .               |
 21  -10.830444  -0.09831   |                . **|    .               |
 22   -4.022918  -0.03652   |                . *|    .               |
 23   12.104297   0.10988   |                .   |**  .               |
 24  -14.092575  -0.12793   |              . ***|    .               |
 25    1.624592   0.01475   |                .   |    .               |
```
"." marks two standard errors

Figure 9.32

Step 8   Fine-Tuning the LTF model

ARIMA Procedure

Partial Autocorrelations

```
Lag Correlation -1 9 8 7 6 5 4 3 2 1 0 1 2 3 4 5 6 7 8 9 1
  1   0.20138   |              .   |****.               |
  2  -0.09691   |              . **|    .               |
  3   0.08076   |              .   |**  .               |
  4   0.11361   |              .   |**  .               |
  5  -0.04243   |              . *|    .               |
  6   0.06174   |              .   |*   .               |
  7   0.08296   |              .   |**  .               |
  8  -0.12809   |            . ***|    .               |
  9   0.08297   |              .   |**  .               |
 10  -0.06017   |              . *|    .               |
 11  -0.17616   |            .****|    .               |
 12  -0.12056   |              . **|    .               |
 13  -0.00508   |              .   |    .               |
 14  -0.00563   |              .   |    .               |
 15   0.03696   |              .   |*   .               |
 16   0.11287   |              .   |**  .               |
 17  -0.20821   |            .****|    .               |
 18  -0.15051   |            . ***|    .               |
 19  -0.14015   |            . ***|    .               |
 20  -0.09168   |              . **|    .               |
 21  -0.04012   |              . *|    .               |
 22   0.02803   |              .   |*   .               |
 23   0.09459   |              .   |**  .               |
 24  -0.15002   |            . ***|    .               |
 25   0.14256   |              .   |*** .               |
```

Figure 9.33

Crosscorrelation Check of Residuals with Input CONTRCTS

To	Chi			Crosscorrelations					
Lag	Square	DF	Prob						
5	5.35	4	0.253	0.069	0.204	0.134	0.067	-0.097	0.091
11	8.46	10	0.584	-0.073	0.060	0.009	0.062	-0.025	-0.191
17	9.85	16	0.875	-0.094	-0.015	-0.059	-0.087	0.042	0.018
23	19.57	22	0.610	-0.265	-0.062	-0.222	-0.178	-0.046	-0.000

The output of the program yields the basis for the final formula.

Model for variable HSTARTS

Data have been centered by subtracting the value -4.158571429.
No mean term in this model.
Period(s) of Differencing =12.

Autoregressive Factors
Factor 1: 1 - 0.39355 B**(2)

Moving Average Factors
Factor 1: 1 + 0.33328 B**(5)

Input Number 1 is CONTRCTS.
Period(s) of Differencing = 12.

The Numerator Factors are
Factor 1: 3.539 + 2.0345 B**(8)

In other words, the formula for the model is

$$(1 - L^{12})(\text{Housing starts}_t + 4.159)$$

$$= (3.539 + 2.035L^8)(1 - L^{12})\text{Contracts}_t + \left(\frac{1 + 0.333L^5}{1 - 0.394L^2}\right) e_t. \quad (9.27)$$

In this manner, we apply the linear transfer function modeling strategy without prewhitening and recommend this method for modeling for multiple input models. Having fit the model, we can generate the forecast profile (Fig. 9.34).

The SPSS ARIMA commands use the WITH option to include predictor variables. The first line of SPSS ARIMA command syntax is almost identical to that shown in Chapter 8. Instead of an ARIMA procedure with a discrete variable, the ARIMA procedure with one or more continuous variables is specified. Suppose that there are two continuous predictor variables—

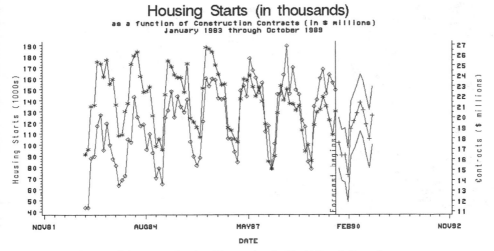

Figure 9.34

namely, X1 and X2 . They should be centered. The input variable is successively lagged. If necessary, both series are identically differenced to attain stationarity. After the ARIMA noise model is designed with low order autocorrelation terms to control the within-series autocorrelation, the first line of ARIMA command syntax, starting in column one of the syntax window, is simply modified to read

```
ARIMA Y WITH X1 X2 .
```

Multiple lags of X1, X2, and other predictor series can be constructed and sequentially added until they fail to retain statistical significance. In a later step, the nonsignificant predictors are pruned from the model and the ARIMA modeling discussed earlier is applied to the residuals. Whereas the SAS model can use a transfer function formulation, the final SPSS model substitutes lagged exogenous predictors. As a result, the SPSS ARIMA procedure in the TRENDS model performs a regression of lagged exogenous variables with time series errors. The SPSS syntax for the housing start analysis follows.

```
*Program C9PGM3.SPS .
*Step 1 Data Check.
*Step 2 Sequence Charts .
LOT VARIABLES= chstarts contrcts
   /ID= year_
   /NOLOG.
```

```
*Step 3 Preliminary Check for Stationarity .
ACF
  VARIABLES= chstarts ccntrcts
  /NOLOG
  /MXAUTO 20
  /SERROR=IND
  /PACF.
* Seasonal Differencing needed at order 12.
* Recheck for stationarity with Seasonal Differencing at d=12.
ACF
  VARIABLES= chstarts ccntrcts
  /NOLOG
  /SDIFF=1
  /MXAUTO 20
  /SERROR=IND
  /PACF.
*Step 4 Preliminary Granger Causality Check .
REGRESSION
  /MISSING LISTWISE
  /STATISTICS COEFF OUTS R ANOVA
  /CRITERIA=PIN(.05) POUT(.10)
  /NOORIGIN
  /DEPENDENT ccntrcts
  /METHOD=ENTER l1chstar l2chstar l3chstar l4chstar l5chstar .
*Step 5 Free Form Lags of Exogenous Variables.
* ARIMA.
TSET PRINT=DEFAULT CIN=95 NEWVAR=ALL .
PREDICT THRU END.
ARIMA chstarts WITH ccntrc1 ccntrc2 ccntrc_3 ccntrc_4 ccntrc_5 ccntrc_6
  ccntrc7 ccntrc8 ccntrc9 ccntrc10
  /MODEL=( 0 0 0 )( 0 1 0 ) NOCONSTANT
  /MXITER 10
  /PAREPS .001
  /SSQPCT .001
  /FORECAST EXACT .
*Residual Diagnosis of Free Form Distributed Lag Model.
ACF
  VARIABLES= err_11
  /NOLOG
  /MXAUTO 20
  /SERROR=IND
  /PACF.
* Step 6 ARIMA Testing Lower Order AR term to model disturbance.
TSET PRINT=DEFAULT CIN=95 NEWVAR=ALL .
PREDICT THRU END.
ARIMA chstarts WITH ccntrc1 ccntrc2 ccntrc_3 ccntrc_4 ccntrc_5 ccntrc_6
  ccntrc7 ccntrc8 ccntrc9 ccntrc10
  /MODEL=( 2 0 0 )( 0 1 0 ) NOCONSTANT
  /P=(2)
```

```
  /MXITER 10
  /PAREPS .001
  /SSQPCT .001
  /FORECAST EXACT .
*Residual Diagnosis for Model Fine-Tuning.
* Notice spikes at 7 and 12.
ACF
  VARIABLES= err_12
  /NOLOG
  /MXAUTO 20
  /SERROR=IND
  /PACF.
*Step 7 Fine-Tuning ARIMA noise model.
TSET PRINT=DEFAULT CIN=95 NEWVAR=ALL .
PREDICT THRU END.
ARIMA chstarts WITH ccntrc_1 ccntrc_2 ccntrc_3 ccntrc_4 ccntrc_5 ccntrc_6
  ccntrc_7 ccntrc_8 ccntrc_9 ccntrc10
  /MODEL=( 2 0 1 )( 0 1 0 ) NOCONSTANT
  /P=(2)
  /Q=(12)
  /MXITER 10
  /PAREPS .001
  /SSQPCT .001
  /FORECAST EXACT .
*Residual Analysis of model.
ACF
  VARIABLES= err_13
  /NOLOG
  /MXAUTO 20
  /SERROR=IND
  /PACF.
* Step 8 Several Overfitting and Trimming Steps are here.
* Step 9 Final model.
TSET PRINT=DEFAULT CIN=95 NEWVAR=ALL .
PREDICT THRU END.
ARIMA chstarts WITH ccntrc_1 ccntrc_4 ccntrc_5 ccntrc_7
  ccntrc8
  /MODEL=( 1 0 1 )( 0 1 0 ) NOCONSTANT
  /P=(2)
  /Q=(12)
  /MXITER 10
  /PAREPS .001
  /SSQPCT .001
  /FORECAST EXACT.
ACF
  VARIABLES= err_18
  /NOLOG
  /MXAUTO 20
  /SERROR=IND
  /PACF.
```

9.5. COINTEGRATION

In econometric modeling or dynamic regression modeling, there is a caveat. Independent series often seem to be related to one another. Granger and Newbold (1986) have noted that in a regression model of one series upon another, when the R^2 is higher than the Durbin–Watson d, there is frequently a chance of a spurious correlation or spurious regression. Specious relationships are more the rule than the exception when random walks or integrated moving average processes are regressed upon one another, and especially where there are a lot of independent variables in the model. Indeed, when autocorrelated errors can lead to artificially low standard errors and inflated R^2 and F tests among unrelated series, spurious trends may emerge. The parameter estimates of those values can be large (Granger and Newbold, 1986). The researcher should be careful to properly specify his model to avoid regressing integrated series upon one another (Maddala, 1992). How these goodness of fit and significance tests become distorted will be discussed in more detail in the next chapter.

From the examination of the relationship between personal disposable income (PDI) and personal consumption expenditures (CE), we can see that these two variables seem to cling to one another over time. They appear to have a common trend and to be interrelated by a long-run dynamic equilibrium. There are sometimes pairs or larger sets of variables that appear to be in equilibrium with one another (Davidson and MacKinnon, 1993). Prices of the same commodity in different countries, money supply and prices, or wages and prices might be other examples of paired series whose values follow one another. When series share a common trend, these series may be integrated at order one I(1), or at I(2) if the trend is quadratic. As such, they are nonstationary and require transformation before becoming amenable to econometric modeling. Occasionally, particular combinations of these variables exist that render the combination stationary or I(0). Series that can be combined in this way are said to be cointegrated.

Sometimes a regression model can combine these trending series in such a way as to produce a combination that in and of itself is I(0). A regression model that combines nonstationary series and yields stationary residuals is called a cointegrating regression. For example, earlier we examined personal disposable income (PDI) and consumer expenditures (CE). To demonstrate that PDI and CE are nonstationary when taken by themselves, we examine the results of the ADF unit root tests (Chapter 3, Sections 3.6.1 through 3.6.4). Both PDI and CE have t probabilities in the ADF tests that are nonsignificant indicating that series are individually found to be nonstationary and I(1).

When centered consumer expenditures, CCE_t, is regressed on centered personal disposable income, $CPDI_t$, using the data from C9PGM1.SAS, the following cointegrated regression model is estimated:

$$CCE_t = 0.91CPDI_t + e_t, \qquad (9.28)$$

where

CCE_t = centered personal consumer expenditures
$CPDI_t$ = centered personal disposable income.

If these series share a common trend, this cointegrating regression represents the long-run equilibrium around that trend between the two series. The residuals, representing long-run disequilibrium error, of this cointegrating regression should be found to be I(0). We perform a cointegration test. These residuals, e_t, are analyzed with an ACF and tested by an augmented Dickey–Fuller test. They appear to attenuate rapidly and to have a significant t probability. Therefore, they now appear to be stationary. The cointegrating parameter is -0.91 and because $e_t = CCE_t - 0.91CPDI_t$, the cointegrating vector of (CCE, CPDI)' is therefore $(1, -0.91)$. Using this technique, linear combinations of sets of series can be found to render those combinations stationary and amenable to conventional time series analysis.

9.6. LONG-RUN AND SHORT-RUN EFFECTS IN DYNAMIC REGRESSION

If two series, say CCE and CPDI, are I(1), the relationship between them found in the cointegrating regression in Eq. (9.28) defines the long-run dynamics of the relationship. When series are differenced, they lose their long-run interpretation. The differences of these series represent short-run marginal changes. Nonetheless, the first differences need to be employed to render the series stationary. When specifying regressions in time series, all the series in the equation have to be integrated by the same order (Maddala, 1992). Engel and Granger (1987) suggest a two-step estimation procedure. First, with ordinary least squares, we can estimate the cointegrating parameter or vector from the long-run equation. Second, we can include the error correction mechanism (in the previous time period) in the short-run differenced equation, permitting us to capture both long-run and short-run changes in the same regression model as Eq. (9.29).

$$\nabla CCE_t = \beta \nabla CPDI_i + \gamma(CCE - 0.91CPDI)_{t-1} + \nu_t, \qquad (9.29)$$

where

  ∇CCE_t = differenced centered personal consumption expenditures
  ∇CPDI_t = differenced centered personal disposable income
  CCE_t = centered personal consumption expenditures
  CPDI_t = centered personal disposable income
  $(\text{CCE} - 0.91\text{CPDI})_{t-1}$ = error correction mechanism.

This error correction model relates the change in consumption expenditures to the change in the last period of personal disposable income and the long-run adjustment from disequilibrium during the past period. By rendering these effects stationary, cointegration permits the modeling of both long-run equilibrium and short-run disequilibrium, which has utility in many fields, some examples of which are the study of rational expectations, differential market efficiency, and purchasing power parity (Maddala, 1992).

9.7. BASIC CHARACTERISTICS OF A GOOD TIME SERIES MODEL

Regardless of the modeling strategy chosen, the resulting model should have certain essential characteristics. A good time series model would have some basic theoretical and statistical qualities. A good theoretical model would have acceptable theoretical scope, power, reliability, parsimony, and appeal. Good theoretical scope can be measured by specification error tests or F tests for parameter or variance encompassing. The theoretical power of the model deals with the completeness of explanation of phenomena being analyzed. Power may be measured by the R^2, minimum residuals sums of squares, minimum information criteria, Hausman, or White's general test for specification error. For reliability, the theoretically important parameters should be stable, regardless of changes of and in auxiliary theoretical variables (Hansen, 1992; Leamer, 1983). The theoretical appeal derives from the parsimony, simplicity, and depth of the explanation.

A good time series model should also have certain fundamental statistical qualities. It should be sufficiently statistically powerful with sample size characteristics, a subject to be discussed in detail in Chapter 12. Outliers should be trimmed, replaced, or modeled. A respectable model should have good explanatory power, fit, parsimony, stability, and forecasting capability. ARMA models need to have stationary and invertible parameters. For a model to have good explanatory power, it would have to encompass the essential theoretical components and they would have to explain most of the variation of the response series. For the model to fit well, it should have statistically independent residuals. For it to fit better than other models, it should have maximum adjusted R^2 and minimum information criteria. For the model to be parsimonious, it would have to be fit with the minimum number of statistically significant parameters. For the model to have good stability, the model should be tested with a split-sample Chow test. For transfer function models to be stable, the decay parameters need to be stable. If there are level shifts, then models need to account for such shifts by modeling the splines over time. If the process exhibits seasonal pulses or local trends, these should be modeled as well. For the model to have good forecasting capability, it would have to have minimum forecast error variance in the near term and have acceptable minimum forecast error variance over as long a forecast horizon as possible (Granato, 1991; Hansen, 1992; Tovar, 1998).

REFERENCES

Bowerman, B. L., and O'Connell, R.T. (1993). *Forecasting and Time Series: An Applied Approach,* 3rd ed., Belmont, CA: Duxbury Press, pp. 657–705.

Box, G. E. P., and Jenkins, G. M. (1976). *Time Series Analysis: Forecasting and Control,* 2nd ed., Oakland, CA, pp. 350–351, 370–381.

Box, G. E. P., Jenkins, G. M., and Reinsel, G. C. (1994). *Time Series Analysis: Forecasting and Control,* 3rd ed., Englewood Cliffs, NJ, pp. 416, 483–532.

Bresler, L. E., Cohen, B. L., Ginn, J. M., Lopes, J., Meek, G. R., and Weeks, H. (1991). *SAS/ETS Software: Application Guide 1: Time Series Modeling and Forecasting, Financial Reporting and Loan Analysis.* Version 6, 1st ed., Cary, NC: SAS Institute, Inc., pp. 175–189.

Brocklebank, J., and Dickey, D. (1994). *Forecasting Techniques using SAS/ETS Software: Course Notes.* Cary, NC: SAS Institute Inc., pp. 394–412.

Davidson, R., and MacKinnon, J. G. (1993). *Estimation and Inference in Econometrics.* New York: Oxford University Press, pp. 715–730.

Ege, G., Erdman, D. J., Killam, B., Kim, M., Lin., C. C., Little, M., Narter, M. A., and Park, H. J. (1993). *SAS/ETS User's Guide. Version 6,* 2nd ed. Cary, NC: SAS Institute Inc., pp. 100–182.

Engel, R. W., and Granger, C. W. J. (1987). "Cointegration and Error Correction: Representation, Estimation and Testing," *Econometrica* **55,** 251–256.

Fildes, R., Hibon, M., Makridakis, S., and Meade, N. (1998). "Generalizing about Univariate Forecasting Methods: Further Empirical Evidence," *The International Journal of Forecasting* **14**(3), Science., p. 342.

Granato, J. (1991). "An Agenda for Econometric Model Building," *Political Analysis* **3,** 123–154.

Granger, C. W. J., and Newbold, P. (1986). *Forecasting Economic Time Series* 2nd ed. San Diego, CA: Academic Press, pp. 205–215.

Granger, C. W. J. (1999, January). "Comments on the Evaluation of Econometric Models and of Forecasts," Paper presented at Symposium on Forecasting, New York University, New York, NY.

Hansen, B. (1992). "Testing for Parameter Instability in Linear Models," *Journal of Policy Modeling* **14,**(4), 517–533.

Hyndman, R. Time Series Data Library, World Wide Web URL: http://www.maths. monash.edu.au/~hyndman/TSDL/index.htm. Data sets are used with permission of author and John Wiley & Sons, Inc.

Leamer, E. E. (1983). "Let's Take the Con Out of Econometrics," *American Economic Review,* **73**(1), 31–43.

Liu, L. M., and Hanssens, D. H. (1982). "Identification of Multiple-Input Transfer Function Models," *Communications in Statistics—Theory and Methods* **11**(3), 297–314.

Lui, L. M., Hudak, G. B., Box, G. E. P., Muller, M., and Tiao, G. C. (1992). *Forecasting and Time Series Analysis using the S. C. A. System.* (1) Oak Brook, IL: Scientific Computing Associates Corp., 8.29.

Maddala, G. S. (1992). *Introduction to Econometrics.* New York: Macmillan, pp. 260–262, 601.

Makridakis, S., Wheelwright, S. C., and McGee, V. (1983). *Forecasting: Methods and Applications,* 2nd ed. New York: Wiley, pp. 485, 501.

Makridakis, S., Wheelwright, S., and Hyndman, R. (1998). *Forecasting: Methods and Applications,* 3rd ed. New York: John Wiley and Sons, pp. 415–416. The construction contract and housing start data are used with the permission of Professor Hyndman and John Wiley and Sons, Inc.

McCleary, R., and Hay, R., Jr. with Meidinger, E. E., and McDowell, D. (1980). *Applied Time Series Analysis for the Social Sciences.* Beverly Hills: Sage, pp. 18–83, 315–316. Data used with permission of author.

Mills, T. C. (1990). *Time Series Techniques for Economists.* New York: Cambridge University Press, pp. 249, 257.

Pankratz, A. (1991). *Forecasting with Dynamic Regression Models.* New York: John Wiley and Sons, Chapter 4, pp. 177–179, 184–189, 202–215.

Priestley, M. B. (1971). "Fitting Relationships between Time Series Models using Box–Jenkins Philosophy." Automatic Forecasting Systems, Hatboro, PA. Cited in Lui and Hanssens (1982), "Identification of Multiple-Input Transfer Function Models." *Communications in Statistics—Theory and Methods* **11**(3), 298.

Reilly, D. (1999) Automatic Forecasting Systems, Hatboro, PA. Personal communication (June 26–30, 1999).

Tovar, G. (1998). Giovanni pointed out the need for mentioning these characteristics (Personal communication), Feb. 10, 1998.

Tsay, R. S. (1985). "Model Identification in Dynamic Regression (Distributed Lag) Models," *Journal of Business and Economic Statistics* **3**(3), pp. 228–237.

Tsay, R. S., and Tiao, G. C. (1985). "Use of Canonical Analysis in Time Series Model Identification," *Biometrika* **72,** 299–315.

Vandaele, W. (1983). *Applied Time Series and Box–Jenkins Models.* Orlando, FL: Academic Press, pp. 259–265.

Wei, W. S. (1993). *Time Series Analysis: Univariate and Multivariate Methods.* New York: Addison-Wesley, p. 292.

Woodward, D. (1997). SAS Institute, Inc., Cary, N.C., personal communication (November 18, 1997).

Chapter 10

Autoregressive Error Models

10.1. THE NATURE OF SERIAL CORRELATION OF ERROR

In time series analysis, researchers often prefer to use multiple-input dynamic regression models to explain processes of interest. The dependent series y_t is the subject of interest and the input series x_{1t}, x_{2t}, ... serve as indicators of plausible alternative explanations of that subject. The error term e_t represents whatever has not been explained by the model of the predictor input series. While bivariate time series models are relatively easy to explain, with their separate endogenous and exogenous series, the causal system is in reality rarely so closed that it is monocausalistic. Other series commonly affect the response series. Dynamic regression models with multiple inputs have an added advantage of easily permitting the hypothesis testing of the plausible alternative causes. By permitting the simultaneous testing of significance and magnitude of the hypothesized input series, the dynamic regression analysis allows more sophisticated model building of dynamic causal systems.

$$Y_t = \alpha + b_1 x_{1t} + b_2 x_{2t} + \cdots + e_t,$$ (10.1)

where

$$x_i = \text{particular input series.}$$

Furthermore, multiple input dynamic regression models can also provide more stable long run forecasts than many other methods. A problem that commonly plagues dynamic regression models, however, is that of autocorrelation (serial correlation) of the error.

This chapter examines the implications and corrections for serial correlation of error. First, it reviews basic linear regression analysis and its conventional assumptions (Hanushek and Jackson, 1977; Goldberger, 1991; Gujarati, 1995; Theil, 1971). When autocorrelation violates those assumptions, the efficiency of estimation is impaired. In particular, it corrupts the computation of the error variance, significance testing, confidence interval estimation, forecast interval estimation, and the R^2 calculation. It elaborates on how serial correlation corrupts this estimation, and it examines sources, tests, and corrections for serial correlation. Under conditions of autocorrelated error, we can use the structure of that correlated error to improve prediction. Finally, the chapter presents programming options and examples of autoregression procedures designed to deal with autocorrelated errors.

10.1.1. REGRESSION ANALYSIS AND THE CONSEQUENCES OF AUTOCORRELATED ERROR

How do autocorrelated residuals violate the basic assumptions of linear regression analysis? Among the basic assumptions of ordinary least squares estimation in regression analysis are four that relate to autocorrelation of the errors. These four assumptions are homogeneity of variance of errors, independent observations, zero sum of the errors, and nonstochastic independent variables. Two of these assumptions, homogeneity of variance of the errors and independent errors, specify the structure of the ordinary least squares error variance–covariance matrix. Homogeneity of variance of the residuals indicates that the error variance in the principal diagonal of the matrix is constant and is equal to σ^2. Independent observations indicates noncorrelation of the errors, and this in turn indicates that the off-diagonal elements of the matrix are all equal to zero.

Therefore, the structure of the error variance–covariance matrix appears as follows:

$$E(ee') = \sigma_e^2 \begin{pmatrix} 1 & 0 & \cdots & 0 \\ 0 & 1 & \cdots & 0 \\ \cdot & \cdot & \cdots & \cdot \\ 0 & 0 & \cdots & 1 \end{pmatrix}. \tag{10.2}$$

Significance testing of the parameters is also dependent on an error structure that is $\sigma^2 I$, shown in Eq. (10.2). It is imperative that the errors be uncorrelated to preclude irregular fluctuation of the magnitude of the standard errors and consequent inefficient estimation.

The intercept and regression coefficients in a linear regression equation $y_t = a + bx_t + e_t$ can be shown to be related to their variance. The variance of these parameters is shown by Makridakis *et al.* (1983), Johnston (1984), and Kamenta (1986), among others, to be dependent on the magnitude of the error variance. From the formula for the simple bivariate regression equation, we can rearrange terms and solve for the error term e_t. We can then square e_t, sum over the cases, take the partial differential of the sum of squared errors with respect to the regression coefficient, b, and by solving the first-order condition, obtain the formula for the regression coefficient.

Let $Y_t = y_t - \bar{y}$ and $X_t = x_t - \bar{x}$.

For $Y_t = bX_t + e_t$,

$$e_t = Y_t - bX_t,$$

and

$$e_t^2 = (Y_t - bX_t)^2.$$

Therefore, $\Sigma e_t^2 = \Sigma(Y_t - bX_t)^2 = \Sigma(Y_t^2 - 2bX_tY_t + b^2X_t^2)$.

Taking the partial derivative with respect to b,

$$\frac{\partial \Sigma e_t^2}{\partial b} = -2\Sigma X_t Y_t + 2b \Sigma X_t^2.$$

Setting $\dfrac{\partial \Sigma e_t^2}{\partial b} = 0$ to obtain a minimum,

$$-2 \Sigma X_t Y_t + 2b \Sigma X_t^2 = 0, \tag{10.3}$$

and

$$b = \frac{\Sigma(X_t)(Y_t)}{\Sigma(X_t)^2}$$

$$= \frac{\Sigma(x_t - \bar{x})(y_t - \bar{y})}{\Sigma(x_t - \bar{x})^2} = r_{yx}\frac{s_y}{s_x}.$$

The variance of the regression coefficient can be expressed as

Because $y_t = a + bx_t + e_t$

and the expectation $E(b) = \beta$ and $E(a) = \alpha$,

$$b = \frac{\Sigma X_t(\alpha + \beta x_t + e_t)}{(x_t - \bar{x})^2}$$

$$= \beta + \frac{\Sigma(x_t - \bar{x})e_t}{(x_t - \bar{x})^2}. \tag{10.4}$$

Because $\text{Var}(b) = E(b - \beta)^2$

$$= E\left(\Sigma(x_t - \bar{x})e_t\right)^2 = \frac{\sigma_e^2 \Sigma(x_t - \bar{x})^2}{\Sigma(x_t - \bar{x})^4}$$

$$= \frac{\sigma_e^2}{\Sigma(x_t - \bar{x})^2}.$$

The standard error of the regression parameter is a function of the error of the model.

$$SE_b = \sigma_b = \frac{\sigma_e}{\sqrt{\Sigma X_t^2}}. \tag{10.5}$$

The significance test for the intercept is a t test that is also dependent on the standard error of the parameter estimate.

$$t = \frac{b - \beta}{SE_b} = \frac{b - \beta}{s/\sqrt{(x_t - \bar{x})^2}}$$

where

$$s = \frac{\Sigma e_t^2}{(T - k)} \tag{10.6}$$

T = sample size
k = number of parameters tested
$df = T - k$.

The variance of the intercept of the regression model can also be shown to be a function of that error variance. First, the formula for the intercept is obtained from

$$y_t = a + bx_t + e_t$$

$$e_t = y_t - a - bx_t$$

$$\Sigma e_t = \Sigma(y_t - a - bx_t)$$

$$\Sigma e_t = \Sigma y_t - \Sigma a - b\,\Sigma x_t$$

Because $\Sigma e_t = 0$ and $\Sigma a = Ta$, $\qquad\qquad$ (10.7)

$$\Sigma y_t - Ta - b\,\Sigma x_t = 0,$$

and

$$Ta = \Sigma y_t - b\,\Sigma x_t,$$

$$a = \bar{y} - b\bar{x}.$$

Johnston (1984) shows that the expected variance of the intercept is

Because $E(b) = \beta$, $E(e_t) = 0$, $E(a) = \alpha$
and $a = \bar{Y} - b(x_t - \bar{x})$

$$= \alpha + \beta(x_t - \bar{x}) + \bar{e} - b(x_t - \bar{x}), \qquad\qquad (10.8)$$

$$\mathrm{Var}(a) = E(a - \alpha)^2$$

$$= E(b - \beta)^2 (x_t - \bar{x})^2 + E(\bar{e})^2 - 2E[(b - \beta)(x_t - \bar{x})e_t].$$

Because $E(\bar{e}^2) = \dfrac{\sigma_e^2}{T}$, these terms may be reexpressed as

$$\mathrm{Var}(a) = \frac{\sigma_e^2 (x_t - x)^2}{\Sigma(x_i - \bar{x})^2} + \frac{\sigma_e^2}{T}$$

$$\qquad\qquad (10.9)$$

$$= \sigma_e^2 \left(\frac{(x_t - \bar{x})^2}{\Sigma(x_i - \bar{x})^2} + \frac{1}{T} \right).$$

The square root of this estimate yields the standard error of the regression parameter, which is clearly a function of the equation error:

$$SE_a = \sigma_e \sqrt{\left(\frac{(x_t - \bar{x})^2}{\Sigma(x_i - \bar{x})^2} + \frac{1}{T} \right)}. \qquad\qquad (10.10)$$

The t test for the significance of the regression parameter estimate is a function of this standard error, and that in turn is a function of the equation error:

$$t = \frac{a - \alpha}{s \sqrt{\dfrac{(x_t - \bar{x})^2}{\Sigma(x - \bar{x})^2} + \dfrac{1}{T}}} \qquad\qquad (10.11)$$

$$df = T - 2.$$

Hence, the significance tests of the parameters depend on the accurate estimate of the variance of the parameter and that of the error.

To be sure, the R^2 and F test are also functions of the error variance, σ_e^2. The F test is the ratio of the variance explained by the model to the error variance. The smaller the error variance, all other things remaining equal, the larger the F value. The larger the error variance, all other things remaining equal, the smaller the F value. The error variance is equal to $1 - R^2/(T - k - 1)$, where T equals the number of observations in a regression model and k equals the number of regressors in the model. Therefore, the larger the R^2, all other things being equal, the smaller the error variance and vice versa. Therefore, both the R^2 and the F value of the model are functions of the error variance.

The confidence intervals around the parameters and the confidence intervals around the predicted value of Y are functions of the error variance as well. Makridakis *et al.* (1983) show how the variance of the mean forecast is a function of the model error variance:

$$\sigma_{\hat{Y}_i}^2 = E(\hat{Y}_i - E(\hat{Y}_i))^2$$

$$= E[a - bX_i - E(a) - E(b)X_i]^2$$

Letting $\alpha = E(a)$ and $\beta = E(b)$

$$= E[(a - \alpha) - X_i(b - \beta)]^2$$

$$= \sigma_a^2 + X_i^2 \sigma_b^2 - 2X_i \text{Cov}(a,b) \tag{10.12}$$

and obtaining σ_a^2 and σ_b^2 from (10.4) and (10.9)

$$= \left(\frac{1}{T} + \frac{\overline{X}_i^2}{\Sigma(x_t - \overline{x})^2} \right) \sigma_e^2 + \frac{X_i^2}{\Sigma(x_t - \overline{x})^2} \sigma_e^2 - \frac{2\overline{X}X_i}{\Sigma(x_t - \overline{x})^2} \sigma_e^2$$

$$= \left(\frac{1}{T} + \frac{(X_i - \overline{X})^2}{\Sigma(x_t - x_i)^2} \right) \sigma_e^2.$$

If the error variance were not properly estimated, very important aspects of the regression analysis would be in error. Although the estimates of the parameters would not be biased, assessments regarding their variances would be incorrect. The goodness of fit tests, the significance tests, and confidence intervals of the parameters and the forecasts would be in error. We can now examine how autocorrelation corrupts model estimation.

In time series regression models these assumptions are relaxed so as to commonly exhibit autocorrelation in the disturbance term. When one observation is correlated with the previous observation in that series and measurement of the observation is less than perfect, errors associated with the observation at one time period are a function of the errors of the observation at a previous time period. The error (disturbance or shock) of the system does not evaporate at the time period of its impact, though trend and seasonality may have been removed. With first-order autocorrelation, the effect of the error does not dissipate until after the subsequent time period has elapsed. From the dynamic linear regression, the shock or error has an inertial memory of one period.

$$y_t = a + b_1 x_{1t} + b_2 x_{2t} + e_t$$
$$e_t = \rho e_{t-1} + v_t,$$
$$(10.13)$$

where

$$|\rho| < 1$$
$$\rho = \text{autocorrelation of error}$$

For stationary processes, it should be remembered that $\rho < 1$. If the autocorrelation is positive, it may have a smoothing effect on the error, as can be seen in Fig.10.1. A second-order autoregressive error process is a function of the errors of the previous two time lags in the series; the inertial memory of error is a function of its order. This second-order autocorrelation of error is

$$e_t = \rho e_{t-1} + \rho^2 e_{t-2} + v_t$$
$$(10.14)$$

where

$$|\rho| < 1.$$

The larger the order, the longer the memory of the autocorrelated error process. Under conditions of autocorrelation of disturbances, the uncorrected errors, e_t, are not serially independent. Kamenta (1986) notes that if this autocorrelation of the error is positive in direction, it will exhibit a form of inertial reinforcement of previous error. The existence of negative autocorrelation tends to produce regular alternations in the direction of error.

It is helpful to examine the effect of autocorrelation on the error variance and the error covariance to see how this renders estimation inefficient. The apparent variance of autocorrelated error, $E(e_t^2)$ can be represented as

$$E(e_t, e_t) = E(\rho e_{t-1} + v_t)(\rho e_{t-1} + v_{t-1})$$
$$\sigma_e^2 = \rho^2 \sigma_e^2 + \sigma_v^2$$
$$\sigma_v^2 = (1 - \rho^2)\sigma_e^2,$$
$$(10.15)$$

where
$\sigma_e^2 = $ apparent (uncorrected autocorrelated) error variance
$\sigma_v^2 = $ actual identically, independently distributed error variance.

Clearly, with no first-order autocorrelation, the real errors are independent of one another and their variance is constant over time. Such constant error variance is easily and efficiently estimated with ordinary least squares. The larger the magnitude of first-order positive autocorrelation, the more the error variance aggregates unto itself portions of variance carried over from the earlier time periods. Each time period that passes, the portion of the variance carried over from the first time period declines by a power of the number of time periods that have elapsed since the first period. The successive aggregation prevents the error variance from remaining constant and augments the error variance over what would have been estimated at period one.

$$\frac{\sigma_v^2}{(1 - \rho^2)} = \sigma_v^2(1 + \rho^2 + \rho^4 + \cdots). \tag{10.16}$$

This effect of the positive autocorrelation on the error variance decreases the estimated standard errors and biases significance tests toward false positive significance of the parameter estimates. The F tests and R^2 become inflated. The forecast interval becomes artificially inflated. Without correction for autocorrelation, the model and forecast error variances are larger than those estimated by least squares. Estimation with other than minimal error variances is inefficient and usually leads to erroneous inference.

How model efficiency is impaired requires elaboration. It is useful to review the covariance of the errors to gain a better understanding of the process. To gain a sense of the error covariance structure and its effect on the variance of the parameters, one can expand $E(e_t e_{t-1})$, into factors and then multiply.

If $y_t = \beta x_t + e_t$ and errors are AR(1),
error variances are not efficient because

$$E(e_t, e_{t-1}) = E(e_t + v_t)(e_{t-1} + v_{t-1})$$
$$= E(e_t + v_t)(\rho e_t + v_{t-1})$$
$$= \rho \sigma_e^2.$$

If $y_t = \beta x_t + e_t$ and errors are AR(s),
autocorrelation shrinks apparent error because

$$E(e_t, e_{t-2}) = \rho \rho \sigma_2 = \rho^2 \sigma_e^2$$

$$E(e_t, e_{t-3}) = \rho^3 \sigma_e^2$$

$$\quad . \qquad\qquad .$$

$$\quad . \qquad\qquad .$$

$$E(e_t, e_{t-s}) = \rho^s \sigma_e^2,$$

(10.17)

where s is the order of autocorrelation.

In Eq. 10.18, Maddala (1992) also derives the warping factor by which autocorrelation contributes alters the parameter variance.

$$\text{Because } \beta = \frac{\Sigma(x_t - \bar{x})(y_t - \bar{y})}{\Sigma(x_t - \bar{x})^2} \text{ and } E(\hat{\beta} - \beta) = 0, (\hat{\beta} - \beta) = \frac{\Sigma(X_t)(e_t)}{\Sigma(X_t)^2}$$

$$E(\hat{\beta} - \beta)^2 = Var(\hat{\beta} - \beta)^2 = \frac{Var \, \Sigma(X_t)(e_t)}{\Sigma(X_t^2)^2}$$

$$= \frac{\sigma_e^2}{\Sigma(X_t^2)^2} \Sigma(X_t c_t)$$

$$= \frac{\sigma_e^2}{\Sigma(X_t^2)^2} (\Sigma X_t^2 + 2\rho \Sigma X_t X_{t-1} + 2\rho^2 \Sigma X_t X_{t-2} + \cdots)$$

$$= \frac{\sigma_e^2}{\Sigma(X_t^2)} \left(1 + 2\rho \frac{\Sigma X_t X_{t-1}}{\Sigma(X_t^2)} + 2\rho^2 \frac{\Sigma X_t X_{t-2}}{\Sigma(X_t^2)} + \cdots \right) \qquad (10.18)$$

$$= \frac{\sigma_e^2}{\Sigma(X_t^2)} (1 + 2\rho r + 2\rho^2 r^2 + \cdots)$$

$$= \frac{\sigma_e^2}{\Sigma(X_t^2)} \left(1 + \frac{2\rho r}{(1 - \rho r)}\right)$$

$$= \frac{\sigma_e^2}{\Sigma(X_t^2)} \left(\frac{(1 + \rho r)}{(1 - \rho r)}\right),$$

where the right hand factor represents bias due to parameter variance when errors exhibit autocorrelation.

The higher the order of autocorrelation in general, the higher the order by which the apparent error variance shrinks. Johnston (1984) and Ostrom (1990) have explained how to compute the bias in the parameter variance from the factor $(1 + \rho r)/(1 - \rho r)$ induced by the autocorrelation in an AR(*1*) model. The amount of change in error variance is a function of its magnitude as well as its sign, and the deviation of error from white noise shown in (Figs. 10.1 and 10.2) has implications on the fit and significance tests.

As a matter of fact, it is reasonable in dynamic time series regression models to expect that the error terms will be correlated. The errors e_t are generated by a process described by a first or higher-order autoregressive process. Instead of having a homogeneous error variance–covariance matrix with a minimal σ^2 in the principal diagonal, the error variance–covariance matrix for a regression model with autocorrelated errors is

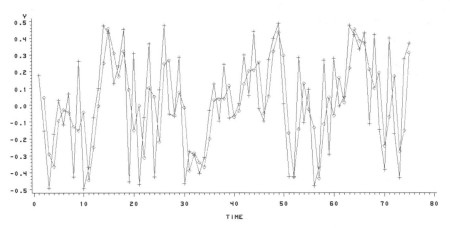

Figure 10.1 White noise and AR(1) simulation: ($+$) white noise, ($\diamond$) positively autocorrelated error.

given by

$$
E(ee') = E
\begin{pmatrix}
e_1^2 & e_1 e_2 & \cdots & e_1 e_s \\
e_2 e_1 & e_2^2 & \cdots & e_2 e_s \\
\cdot & \cdot & \cdots & \cdot \\
\cdot & \cdot & \cdots & \cdot \\
\cdot & \cdot & \cdots & \cdot \\
e_s e_1 & e_s e_2 & \cdots & e_s^2
\end{pmatrix}
$$

(10.19)

$$
= \frac{\sigma_v^2}{(1 - \rho^2)}
\begin{pmatrix}
1 & \rho & \rho^2 & \cdots & \rho^{s-1} \\
\rho & 1 & \rho & \cdots & \rho^{s-2} \\
\cdot & \cdot & \cdots & \cdot \\
\cdot & \cdot & \cdots & \cdot \\
\cdot & \cdot & \cdots & \cdot \\
\rho^{s-1} & \rho^{s-2} & \rho^{s-3} & \cdots & 1
\end{pmatrix}.
$$

In this matrix, the error variance is periodically deflated by the power of the autocorrelation. To illustrate, we focus attention on a first-order positive autocorrelated error process, where each error is expressed in terms of its temporal predecessor and a random error term.

The artificial compression of the estimated least squares regression error

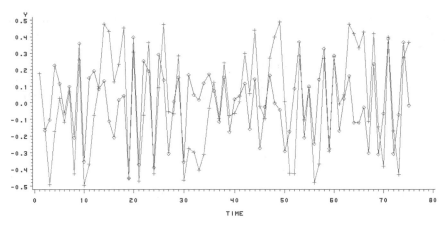

Figure 10.2 White noise and AR(1) simulation: (+) white noise, (◊) negatively autocorrelated error.

variance, unless corrected, will produce inefficient and erroneous estimates of standard errors. The larger the positive autocorrelation, the more serious the relative compression of the standard errors, the more likely the false significance tests, and the more inflated R^2 of the model. If the error variances are artificially compressed, then the forecast error will be deflated and the forecast intervals erroneously constricted. Residual variances from earlier periods when unmodeled gives rise to aggregation of forecast error variance. Correction for autocorrelation reduces the forecast error. Otherwise, inaccurate forecasts can follow. The parameter estimates are not as efficient under these circumstances as they would be if the autocorrelation were controlled for in the model. Finally, the consequences impede accurate prediction (Ege *et al.,* 1993). Therefore, the violation of the regression analysis assumption that the errors are independent of one another can have serious consequences.

Even when there is autocorrelation of the errors, as long as there is no lagged endogenous variable, the parameter estimates remain unbiased and consistent. For proofs, the reader is referred to Kamenta (1986) or Johnston (1984).

10.2. SOURCES OF AUTOREGRESSIVE ERROR

Gujarati (1995) gives several reasons for autocorrelation of the errors. Some time series possess inertia or momentum built into their processes. When measurement of a process is imperfect and when what happens at one time period depends on what took place at the previous time period, this error in measurement manifests itself as serial correlation in the error.

Misspecification may derive from whole variables being excluded from the model. When variables are omitted from the model, they become part of the error term. When the dependent variable is explained by unspecified series that are autocorrelated, the error term becomes autocorrelated (Griffiths, Hill, and Judge, 1993).

Misspecification of functional form may produce autocorrelated error. If the data-generating process follows a quadratic functional form, while the analyst models a linear relationship between the dependent and independent variables, serial correlation may follow from a lack of squared or other polynomial terms in the model. The excluded squared component is correlated with the included linear component. The omission of that series may result in a positive correlation of the included variable with the error term. The excluded error may expand quadratically over time, giving rise to autocorrelated, heteroskedastic error. Such models may be deemed to produce a form of specification bias as well.

Gujarati (1995) writes that latency effects can produce autocorrelation of the errors. A cyclical relationship may lead to such serial correlation. A gestation period may be required before a reaction in another variable may develop. For example, the amount of crops farmers plant may depend on the price of the crop during the previous year. If the crop price was high the previous year, farmers may plant more and harvest more the next year. Hence, crop prices may influence crop planting and crop harvest one year later. This is an example of a lagged effect on the part of other variables. If any of these series are inaccurately measured or erroneously omitted in the specification of the model, they become part of the error term and bestow autocorrelation upon the error as well.

Much the same can be said for counter-cyclical effects discovered in the data. Overproduction during the previous year may result in reduction of planting this year. Underproduction during the previous year may cause the farmers to plant more seed this year. Counter-cyclical effects may produce a cobweb phenomenon, a U-shaped appearance on a graph of high one year, low the next, and high the following year. But there are other types of delayed effects. Estimation is made inefficient by such sources of serial correlation of the error.

If the series is a function of time (trend) or seasonality, estimation of these parameters may depend on correct standard errors. With the autocorrelation inherent in the series, the standard error bias in the first stage of analysis may corrupt specification of seasonality or trend. This functional trend may be linear, quadratic, cubic, or of a higher power. As pointed out earlier, such trends are forms of nonstationarity that should be controlled for before subsequent Box–Jenkins analysis. With the standard errors inflated, estimation of trend and seasonality parameters may be incorrect as well (Wonnacott and Wonnacott, 1979).

Sometimes an analyst wishes to study two series, one of which is monthly and the other of which is quarterly. Usually, he will combine the three months of the monthly series so that he can analyze two quarterly series. In the process of aggregating the monthly series, he is smoothing the data. While this smoothing eliminates some variation, it may introduce artificial autocorrelation. If this happens, the smoothed series may now have an AR(p) aggregation bias (Gujarati, 1995). One, some, or any of these phenomena may force the researcher to consider the consequences of autocorrelation for his analysis.

10.3. AUTOREGRESSIVE MODELS WITH SERIALLY CORRELATED ERRORS

Autoregressive models with lagged endogenous variables are sometimes used to handle autocorrelation of the process and of the error. Many natural and social phenomena contain inertia. Technological modernization is a form of emulation, when measured, that contains inertia as well. Cultural fashions, styles, fads, movements, and trends are inertial phenomena (Maddala, 1992). Inertial effects have built-in lag, and these lagged phenomena may be analyzed with autoregressive models. If these phenomena are imprecisely measured or even omitted, the errors possess autocorrelation as well.

When models possess a lagged endogenous variable as well as serially correlated errors, there is a complicated warping of the error. The autoregression in the structural portion of the model generates a geometric lag of the exogenous variables along with a change in the error structure that renders the estimation biased, inconsistent, and inefficient. For a more detailed treatment of such models, the reader is referred to Greene (1997).

10.4. TESTS FOR SERIAL CORRELATION OF ERROR

There are several tests by which we can detect autocorrelation of the residuals. From a regression on a time trend, with possible inclusion of seasonal dummy variables, we can compute residuals for graphical and statistical analysis. An examination of time plots, ACF, PACF, with their modified portmanteau tests, should reveal the order and type of autocorrelation (Greene, 1997).

A test of first-order autocorrelation is the Durbin–Watson d test. The formula is similar to that of a χ^2 test on the difference between the current and first lagged residual. This test is not applicable if there are lagged dependent variables. The Durbin–Watson d is applicable only to first-order, and not higher order, autocorrelation, though it is somewhat robust to violations of homoskedasticity or normality (Kamenta, 1986):

$$\text{Durbin–Watson } d = \frac{\sum\limits_{i=2}^{T} (e_t - e_{t-1})^2}{\sum\limits_{t=1}^{T} e_t^2}. \tag{10.20}$$

The range of the d is from 0 to 4. The d will tend to be smaller ($d < 2$) for positively autocorrelated residuals and larger ($d > 2$) for negatively autocorrelated ones. If the d approximates zero, there will be no first-order autocorrelation among the residuals.

The Durbin–Watson tables include upper (d_U) and lower (d_L) bounds. The Durbin–Watson scale ranges from 0 to 4. Within this range there are 5 segments. They consecutively extend from (1) 0 to d_L, (2) from d_L to d_U, (3) from d_U to $4 - d_U$, (4) from $4 - d_U$ to $4 - d_L$, and (5) from $4 - d_L$ to 4. In general if $d < d_L$, there is positive first-order autocorrelation. The test is inconclusive if either $d_L < d < d_U$ or if $4 - d_U < d < 4 - d_L$. If d is 2, there is no first-order autocorrelation of the errors. For $d > 2.0$, the residuals are negatively autocorrelated ones. Another way of conceptualizing the Durbin–Watson d is

$$d \approx 2(1 - r). \tag{10.21}$$

The significance levels vary for the number of regressors in the equation and according to the upper and lower bounds of significance for the Durbin–Watson d. Again, this test loses power if there are lagged dependent variables and the d becomes inappropriate (Johnston, 1984; Gujarati, 1995).

Other tests can be applied for first or higher order autocorrelation, among which is the Breusch-Godfrey test or Durbin M test. This test procedure is to regress the dependent series on the exogenous variables with OLS and obtain the current residual e_t plus the lagged residuals from $t - 1$ to $t - p$. The null hypothesis is that $\rho_{t-i} = 0$, where $i = 1$ to p. If the disturbance term e_t is a significant function of a higher order autocorrelation, then at least one ρ will be significant in

$$e_t = \rho_1 e_{t-1} + \rho_2 e_{t-2} + \cdots + \rho_p e_{t-p} + v_t. \tag{10.22}$$

The e_t is a random error term with a mean of 0 and a constant variance. If the sample size is large, then this test is equivalent to a LaGrange multiplier test with

$$TR^2 \sim \chi^2$$
with $df = p$ (number of parameters in model)   (10.23)
and T = sample size.

If $p = 1$, this test is called the Durbin M test (Greene, 1997).

10.5. CORRECTIVE ALGORITHMS FOR REGRESSION MODELS WITH AUTOCORRELATED ERROR

Transformations of the regression equation with autocorrelated errors may render those errors independent of one another and may permit best linear unbiased parameter estimation. Among these corrective algorithms are the Cochrane–Orcutt, Hildreth–Lu, Prais–Winsten, and maximum likelihood methods. There are two-step and iterative versions of these methods. In the two-step Cochrane–Orcutt algorithm, the OLS regression is run and the residuals are saved. From the residuals, the first-order autocorrelation, $\hat{\rho}$, among the residuals is estimated with

$$e_t = \rho e_{t-1} + v_t$$

$$\hat{\rho} = \frac{\Sigma e_t e_{t-1}}{\Sigma e_t^2}.$$ (10.24)

Since there is no predecessor to the first observation, this process cannot use the first observation for the computation of the first-order autocorrelation. All other observations are utilized to estimate $\hat{\rho}_1$. Then this estimate is applied to the model in the next equation to obtain the parameter estimates for α and β by least squares estimation:

$$(Y_t - \hat{\rho}Y_{t-1}) = \alpha(1 - \hat{\rho}) + \beta(X_t - \hat{\rho}X_{t-1}) + v_t.$$ (10.25)

Alternatively, a solution can be obtained by iterative minimization of the squared residuals, v_t^2. At that point convergence is reached and the parameter estimates are output.

The Hildreth–Lu algorithm, sometimes referred to as unweighted least squares or nonlinear least squares, performs a grid search along a parameter space to try different $\hat{\rho}_1$ values—say, from $\hat{\rho}_1 = .1$ to 1.0 by .2—to obtain a sum of squared residuals for each tested parameter estimate. When the value of the sum of the squared residuals converges upon a minimum, the parameter at this point in the parameter space becomes the final estimate (Pindyck, R. S. and Rubinfeld, D. L, 1993).

The Prais–Winsten or estimated generalized least squares algorithm has both a two-step and an iterative form. Whereas the Cochrane–Orcutt estimator of $\hat{\rho}$ can be obtained by minimizing the sum of the squared residuals,

$$SS_{\text{error}} = \sum_{t=2}^{T} (e_t - \rho e_{t-1})^2, \tag{10.26}$$

it loses the first observation. In the Prais–Winsten algorithm, the objective criterion of the adjusted sum of squared residuals, S_{pw}, is minimized, with the following adjusted utilization of the first observation:

$$SS_{pw\,\text{adjusted error}} = (1 - \rho^2)e_1^2 + \sum_{t=2}^{T} (e_t - \rho e_{t-1})^2. \tag{10.27}$$

Once this criterion is minimized, the ρ_{pw} associated with that minimum can be found according to

$$\hat{\rho}_{pw} = \frac{\sum_{t=2}^{T} e_t e_{t-1}}{\sum_{t=3}^{T} e_{t-1}^2}. \tag{10.28}$$

Recall that in Eq. (10.13), the formula is given for first-order autocorrelated error where

$$e_t^2 = \frac{v_t^2}{(1 - \rho^2)}, \text{ therefore } e_t = \frac{v_t}{\sqrt{(1 - \rho^2)}}.$$

It follows that

$$Y_t = a + \beta X_t + \frac{v_t}{\sqrt{(1 - \rho^2)}}; \tag{10.29}$$

and multiplication by the common factor yields

$$\sqrt{(1 - \rho^2 L)}Y_t = \sqrt{(1 - \rho^2 L)}a + \sqrt{(1 - \rho^2 L)}\beta X_t + v_t.$$

When the Prais–Winsten transformation is applied, the transformed variables, designated by asterisks, in the transformed model

$$Y_t^* = a^* + \beta_t X_t^* + v_t,$$

where $\tag{10.30}$

$$Y_t^*, a^*, X_t^*, \text{ and } v_t \text{ are the transformed variables,}$$

become the best linear unbiased estimators. In the two-step procedure, the estimated autocorrelation is computed and plugged into the transformation.

In the iterative versions, various ρ values are searched, then squared so identically, independently distributed residuals, ν_t^2, may be found, and the equation is solved by minimization of the sum of squared residuals. This iterative Prais–Winsten algorithm generally yields excellent results.

Another algorithm that yields good results is that of maximum likelihood. The likelihood function is premultiplied by the Prais–Winsten transformation. The natural log is taken,

$$\ln(\text{Likelihood}) = -\frac{N}{2}\ln(2\pi) - \frac{N}{2}\ln(\sigma_\nu^2) - \frac{1}{2}\ln(|V|) - \frac{S}{2\sigma_\nu^2},$$

so minimization of $\dfrac{S}{2\sigma_\nu^2}$ is performed,

where $S = (y - X\beta)V^{-1}(y - X\beta)$, (10.31)

y is the y matrix,

X is the x matrix,

β is the β matrix,

V^{-1} is estimated from $\dfrac{\sigma_\nu^2}{1 - \rho^2}$,

and S is the sum of squares of transformed residuals.

with minimization performed by a Marquardt algorithm (Ege *et al.*, 1993).

Kamenta (1986) notes that in general, algorithms that do not lose the first observation perform better than those that drop that observation. He writes that this algorithm, which does not lose the first observation, in Monte Carlo studies has produced results with relatively small samples as good as those yielded by maximum likelihood (Ege *et al.*, 1993). Moreover, he writes that these results are usually better than those of ordinary least squares, and he suggests that the iterative procedures successively improve on their estimates and in general are to be preferred to the two-step procedures.

10.6. FORECASTING WITH AUTOCORRELATED ERROR MODELS

If the model were one with independently, identically, and normally distributed disturbances, OLS would be the most efficient procedure by which to estimate the model. When the model possesses autocorrelated errors, then a form of generalized least squares is a more efficient estimation procedure. In this form of generalized least squares, the model is first estimated by OLS, and then the variables and OLS residuals are transformed to correct for the autocorrelation of the residuals. The autocorrelation in the error may be estimated preliminarily or iteratively as the estimation proceeds to convergence of minimization of squared error. Corrective weights are constructed as functions of the autocorrelated error and its variance. These weights are formed from the V^{-1} matrix displayed in Eq. (10.31). When these weights are used, the effect is that of transforming the original variables by the Prais–Winsten transformation in Eq. (10.29). When least squares estimation is performed on these transformed variables and identically independently distributed (*i.i.d.*) error, it has been called estimated generalized least squares or feasible generalized least squares. At this juncture, it is helpful to review the nature of that bias and how the first-order correction rectifies the error variance inflation in the forecasting process.

The regression model at time, t, is

$$y_t = a_t + \beta_t x_t + e_t$$
$$e_t = \rho e_{t-1} + \nu_t$$

so

$$
\begin{aligned}
y_t &= a_t + \beta x_t + \rho e_{t-1} + \nu_t \\
y_{t+1} &= a_t + \beta x_{t+1} + \rho e_t + \nu_{t+1} \\
y_{t+2} &= a_t + \beta x_{t+2} + \rho^2 e_t + \rho e_{t+1} + \nu_{t+2} \\
y_{t+h} &= a_t + \beta x_{t+h} + \rho^h e_t + \rho^{h-1} e_{t+1} + \cdots + \nu_{t+h}.
\end{aligned}
\tag{10.32}
$$

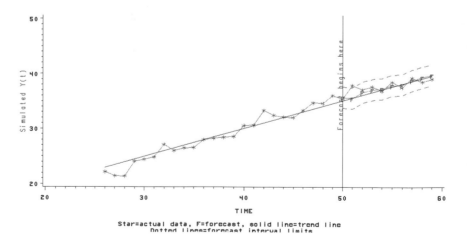

Star=actual data, F=forecast, solid line=trend line
Dotted lines=forecast interval limits

Figure 10.3 Forecast profile of regression with AR(1) errors: (*) actual data, (F) forecast, (solid line) trend line, (dotted line) forecast interval limits.

To predict h periods ahead, the best predictor in terms of the current error at the time t is $\rho^h e_t$. As h increases, the amount of error incrementally added to the forecast exponentially attenuates until an asymptote is approximated. The forecast interval attenuation characteristic of the AR(1) forecast profile is very rapid and cannot always be observed (Fig. 10.3). The programming of the model and the interpretation of the parameterization is essential to the proper application of this technique.

10.7. PROGRAMMING REGRESSION WITH AUTOCORRELATED ERRORS

10.7.1. SAS PROC AUTOREG

A regression model with first-order autocorrelated errors can be analyzed with either SAS or SPSS. The procedures utilized for this analysis are SAS PROC AUTOREG and SPSS AREG. Although the same analysis can be performed with SAS PROC ARIMA or the SPSS ARIMA program, attention is directed to the procedures dedicated to handling regression models with autocorrelated errors because the ARIMA procedures have already been discussed. The data are generated by a simulation of first-order autocorrelated disturbances along a time trend, displayed in Fig. 10.1. If the analyst has reason to suspect that the errors are autocorrelated, he may save and test the residuals from his model with ARIMA, AUTOREG, or AREG procedures. Although it is easy to apply the autoregression procedures in either statistical package, SPSS users should be cautioned at the writing of this text that AREG handles only first-order autocorrelation problems. For higher order autocorrelation or ARMA error problems, they would have to employ SPSS ARIMA or SAS PROC AUTOREG.

We turn to a presentation of the SAS PROC AUTOREG program syntax. In program C10PGM2.SAS, the data are obtained from the data set designated genauto1.sd2 and in line 31 and specifies the model statement as a regression on a time trend.

$$Y_t = a + b_1 \text{time} + e_t,$$

where   (10.33)

$$e_t = \rho e_{t-1} + v_t.$$

The program strategy entails first detrending the series to be able to analyze the residuals. The next step is to diagnose autocorrelation of the residuals and then to correct for them. In this way, the estimation will become efficient, the standard errors will be corrected, the goodness of fit tests will be more precise, and prediction will be rendered more accurate. Of course, other exogenous variables could also be included. But in this case, only the time trend is included on the right-hand side of the equation.

We examine the programming syntax. The log file of program C10PGM.SAS gives the commands and their associated line numbers. The first part of this program includes a subroutine (in lines 4 through 12) to generate autocorrelated error for later analysis. Lines 18 through 22 graph the output of this subroutine. In lines 24 through 28, we check the autocorrelation of the residuals with the ACF and PACF of an ARIMA to be sure that the residuals are correctly generated. In the later lines, the series is detrended by a regression against time, and the autocorrelated errors are modeled. Although there might be other exogenous variables modeled in other cases, the problem of autocorrelation of the disturbances should be handled with either a SAS AUTOREG or ARIMA procedure. After presentation of the SAS program syntax, we elaborate on the AUTOREG procedure.

```
1    options ls=80;
2    title 'Chapter 10 Simulation of AR(1) error';
3    title2 'Generation of the First estimation sample';
4    data genauto1;
5      u1=0;
6      do time=-5 to 100;
7      u=.5*u1+rannor(123456);
8      y=10+.5*time+u;
9      if time>0 then output;
10     u1=u;
11    end;
12   run;
```

NOTE: The data set WORK.GENAUTO1 has 100 observations and 4 variables.
NOTE: The DATA statement used 0.23 seconds.

```
16   proc print data=genauto1;
17   run;
```

NOTE: The PROCEDURE PRINT used 0.02 seconds.

```
18   symbol1 i=join c=green v=star;
19   symbol2 i=r c=blue v=none;
20   proc gplot;
21     plot Y*time=1 Y*time=2/overlay;
22   run;
```

NOTE: Regression equation: Y=9.837328+0.505249*TIME.

```
NOTE: The PROCEDURE GPLOT used 0.38 seconds.

24  proc arima data=genauto1;
25    i var=u center;
26    e p=1 printall plot ;
27  title2 'Check of Autocorrelated error in Original data';
28  run;
29


NOTE: The PROCEDURE ARIMA used 0.17 seconds.

30  proc autoreg data=genauto1;
31    model y = time/nlag=6 method=ml dwprob dw=6 backstep;
32    output out=resdat2 r=resid2 ucl=ucl lcl=lcl p=forecast pm=ytrend;
33  title2 'Autoregression Model of the data';
34  run;

NOTE: The data set WORK.RESDAT2 has 100 observations and 9 variables.
NOTE: The PROCEDURE AUTOREG used 0.2 seconds.

36  data resck;
37    set resdat2;
38  title2 'The Residual Data Set';

NOTE: The data set WORK.RESCK has 100 observations and 9 variables.
NOTE: The DATA statement used 0.17 seconds.

39  proc print;
40  run;

NOTE: The PROCEDURE PRINT used 0.02 seconds.

42  proc arima;
43    i var=resid2;
44  title2 'Check of autocorrelation of residuals of Autoregression';
45  run;

NOTE: The PROCEDURE ARIMA used 0.01 seconds.

46  data together;
47    set resdat2;
48    if time < 50 then forecast=.;
49    if time < 50 then ucl=.;
50    if time < 50 then lcl=.;
51  run;

NOTE: The data set WORK.TOGETHER has 100 observations and 9 variables.
NOTE: The DATA statement used 0.19 seconds.
```

```
58  /* generates the x-axis value above the reference line */
59  data anno;
60    input time 1-4 text $ 5-58;
61    function='label'; angle = 90 ; xsys='2'; ysys='1';
62    x=time; y=60; position='B';
63  cards;
```

NOTE: The data set WORK.ANNO has 1 observations and 9 variables.
NOTE: The DATA statement used 0.14 seconds.

```
66  axis1 label=(a=90 'Simulated Y(t)');
67  symbol1 i=join c=blue v=star;
68  symbol2 i=join c=green v=F;
69  symbol3 i=join c=black line=1;
70  symbol4 i=join c=red line=20;
71  symbol5 i=join c=red line=20;
72  proc gplot data=together;
73    plot (y forecast ytrend lcl ucl) * time/overlay vaxis=axis1 href=50
74    annotate=anno;
75    where time > 25 & time < 60;
76  title 'Figure 10.3 Forecast Profile of Regression with AR(1) Errors';
77  footnote1 'Star=actual data, F=forecast, solid line=trend line';
78  footnote2 'Dotted lines=forecast interval limits';
79  run;
```

Let us focus on the principal portions of the program. In lines 4 through 12 of the program, in the data step called GENAUTO1, a simulation generates first-order autocorrelated residuals. First, the data set is printed out beginning on the first page of output, C10PGM2.LST, to permit inspection for obvious problems. A partial listing of the data is presented to facilitate interpretation.

Chapter 10 Simulation of AR(1) error
Generation of the First estimation sample

OBS	U1	TIME	U	Y
1	0.82577	1	−0.08182	10.4182
2	−0.08182	2	−1.10301	9.8970
99	2.12288	99	−0.38309	59.1169
100	−0.38309	100	0.20595	60.2059

Even if the researcher has reason to suspect autocorrelation of the residuals, he may wish to empirically test and confirm it. This test is performed with the ARIMA procedure following the data generation on the error series, U. An ACF and PACF confirm a first- and sixth-order autocorrelation.

At this point, the AUTOREG procedure is run on the data series, Y. Lines 30 through 34 provide the SAS program syntax for setting up this model. The autoregression procedure may be invoked (lines 30 through 34 in the log file) to perform the analysis

```
30 proc autoreg data=genauto1;
31    model y = time/nlag=6 method=ml dwprob dw=6 backstep;
32    output out=resdat2 r=resid2 ucl=ucl lcl=lcl p=forecast pm=ytrend;
33 title2 'Autoregression Model of the data';
34 run; .
```

In line 30, the data to be analyzed with the autoregression procedure are drawn from data set GENAUTO1. In line 31 the model statement specifies that a dependent series Y is to be autoregressed on a linear trend called TIME. TIME is a counter that increases by one unit with each period of time and is created by the loop statement in line 6. The number of lags for which autocorrelated error are diagnosed is six, specified with NLAG=6. The method of estimation is that of maximum likelihood, selected with METHOD=ML; otherwise, the Yule–Walker estimation is used by default. To request Durbin–Watson significance levels for each of the requested six Durbin–Watson tests, the DW=6 and the DWPROB options are specified. In this case, the backward elimination procedure is requested with the BACKSTEP option. This process begins at the lag specified with the NLAG option and successively eliminates non-significant (with a default significance level of 0.05) autocorrelated error terms. Yule-Walker estimation is used during the backward elimination to obtain the preliminary model order and then maximum likelihood estimation is used for the rest of the parameters.

An output data set is constructed in line 32. The name of the output data set is RESDAT2. In addition to the regular variables within that data set, new names are given to the five auxiliary output variables. The residual is called RESDAT2, the predicted scores are called FORECAST, the trend line is called YTREND, and the upper and lower 95% confidence limits are called UCL and LCL respectively. With the syntax in lines 66 through 79, the output variables are plotted to produce Fig. 10.3.

The first part of the AUTOREG output presents the OLS estimates, shown in Fig. 10.4. The dependent variable is identified as Y. The regression R^2 (REG RSQUARE) is the R^2 of the structural part of the model, after transforming for autocorrelation correction. In short, the regression R^2 is a measure of the transformed regression model. The total R^2 (TOTAL RSQUARE) is the R^2 of the transformed intercept, transformed variables,

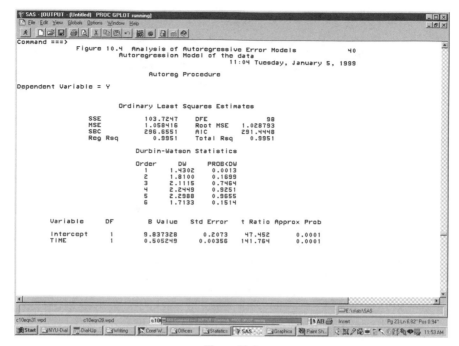

448   10/Autoregressive Error Models

Figure 10.4

and the autocorrelation correction. Therefore, the total R^2 is a measure of how well the next value can be predicted from the complete model. (Ege *et al.*, 1993). When there is no correction for autocorrelation, as is the case in the OLS estimation, these R^2 remain the same (Fig. 10.4). The Durbin–Watson tests suggest that a first-order autoregressive error model may be in order. The model suggested by the OLS estimation is $Y_t = 9.844 + 0.505$Time $+ e_t$ with both R^2 equal to 0.9951.

One good way to determine the order of autoregressive error is to employ the backward elimination procedure, the output for which is shown in Fig. 10.5. This output reveals the autocorrelation, the standard error, and the T ratio for each of the parameters tested. Significant autocorrelation is found at lags 1, 5, and 6.

The maximum likelihood estimation output is contained in Figs. 10.6 and 10.7. In Fig. 10.6, the error variance (MSE) is shown to be 0.92 and the regression R^2 is 0.9917 while the total R^2 is now 0.9959. The regression R^2 is less than the total here by a small amount which indicates that there is some difference owing to the autocorrelation correction. The Durbin–Watson statistics in Fig. 10.6 are given for the corrected model, and their

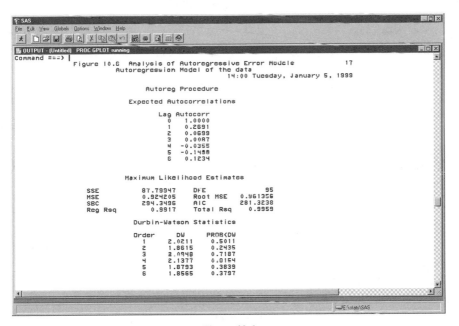

```
 SAS                                                                          _□×
File Edit View Globals Options Window Help
 ★ D|⊜|▥|⊜|◨|Ж|▨|▨|▨|▨|◈|⊞|◈
 OUTPUT - [Untitled]  PROC GPLOT running                                      _□×
Command ===> |
          Figure 10.5  Analysis of Autoregressive Error Models        16
                      Autoregression Model of the data
                                    12:58 Tuesday, January 5, 1999

                          Autoreg Procedure

                     Estimates of Autocorrelations

   Lag  Covariance  Correlation -1 9 8 7 6 5 4 3 2 1 0 1 2 3 4 5 6 7 8 9 1

    0   1.037247    1.000000  |                    |********************|
    1   0.29537     0.284763  |                    |******              |
    2   0.091108    0.087837  |                    |**                  |
    3  -0.08286    -0.079886  |                  **|                    |
    4  -0.15235    -0.146882  |                 ***|                    |
    5  -0.1857     -0.179036  |                ****|                    |
    6   0.115721    0.111566  |                    |**                  |


              Backward Elimination of Autoregressive Terms

                   Lag   Estimate   t-Ratio    Prob
                    2   -0.051298   -0.4909   0.6247
                    3    0.051221    0.5088   0.6121
                    4    0.093911    0.9472   0.3459


                   Preliminary MSE = 0.887616

           Estimates of the Autoregressive Parameters

           Lag   Coefficient    Std Error     t Ratio
            1    -0.29515236     0.096971      -3.044
            5     0.19860599     0.099525       1.996
            6    -0.22096452     0.100062      -2.208
◄|                                                                           ►|
 ◄|                                                               ►
                                                          □E:\stats\SAS
```

Figure 10.5

```
 SAS                                                                          _□×
File Edit View Globals Options Window Help
 ★ D|⊜|▥|⊜|◨|Ж|▨|▨|▨|▨|◈|⊞|◈
 OUTPUT - [Untitled]  PROC GPLOT running                                      _□×
Command ===> |
          Figure 10.6  Analysis of Autoregressive Error Models        17
                      Autoregression Model of the data
                                    14:00 Tuesday, January 5, 1999

                          Autoreg Procedure

                      Expected Autocorrelations

                        Lag  Autocorr
                         0    1.0000
                         1    0.2691
                         2    0.0699
                         3    0.0087
                         4   -0.0355
                         5   -0.1458
                         6    0.1234


                 Maximum Likelihood Estimates

        SSE        87.79947    DFE             95
        MSE        0.924205    Root MSE   0.961356
        SBC        294.3496    AIC         281.3238
        Reg Rsq      0.9917    Total Rsq     0.9959

                 Durbin-Watson Statistics

              Order    DW      PROB<DW
                1    2.0211    0.5011
                2    1.8615    0.2435
                3    2.0948    0.7187
                4    2.1377    0.0154
                5    1.8793    0.3839
                6    1.8565    0.3797
◄|                                                                           ►|
 ◄|                                                               ►
                                                          □E:\stats\SAS
```

Figure 10.6

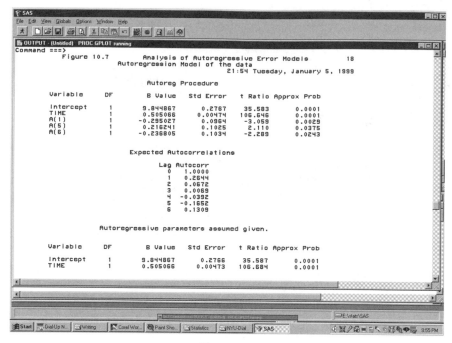

Figure 10.7

probabilities reveal no significant residual autocorrelation, suggesting that residual autocorrelation has been corrected.

In Fig.10.7, the maximum likelihood model estimates are given for the structural model as well as for the error components under the rubric "Autoreg Procedure." The variable, its degrees of freedom, the coefficient, the standard error, the t ratio, and the approximate probability of the parameter are given first for the structural model and then for the errors. The errors are named $A(t)$ where t is the order of the error term. For the model estimated in Fig. 10.7, the equation is

$$Y_t = 9.844 + 0.505 time + e_t \tag{10.34}$$
$$e_t = 0.295 e_{t-1} - 0.216 e_{t-5} + 0.237 e_{t-6} + v_t.$$

At this point, a caveat is noteworthy. The signs in the lower error equation of Eq. 10.34 are the reverse of those shown in the output (Bresler *et al.*, 1991; Ege *et al.*, 1993).

The method requested is that of maximum likelihood. If this algorithm is not requested, the Yule–Walker method, a form of estimated generalized least squares, is used by default. The estimated generalized least squares for AR(1) uses the Prais–Winsten two-step technique. For the iterative version, the ITYW option must be employed. The ULS option is a more advanced version of the Hildreth–Lu estimation technique (Ege *et al.*, 1993). The BACKSTEP option invokes backward elimination to eliminate all nonsignificant autoregressive parameters. This procedure trims the model of potentially intercorrelated insignificant predictor variables and autocorrelated error terms.

Three things remain to be done. First, an output data set is constructed, with the residuals, forecast, trend line, and the forecast confidence limits saved. From the program log, the command in line 32 performs these tasks. The output data set is called RESDAT2 and the residuals from this analysis are called RESID2. Key variables are constructed and added to the series in the data set. Among these newly constructed variables are the forecast, given the same name; the trend, called YTREND; and respectively the upper and lower confidence limits of the forecast. Line 36 creates a data set called RESCK in which RESDAT2 is subset on the next line. Second, the autoregression model residuals are double-checked with an ARIMA procedure in lines 42 and 43 of the log file. The ACF and PACF for the residuals of the estimated AR model are generated along with Q statistics confirming white noise. In this way, the model is shown to fit. Third, the forecast profile is plotted in Fig. 10.3.

The programming of the forecast profile plot is done in lines 46 through 79 in the program log file. The forecasting profile begins at time $t = 50$ here, so in lines 48 through 50 in the log, the data for the forecast and its confidence limits are set to missing. Lines 58 through 60 set up the annotation of the reference line. Lines 66 through 79 set up the forecast plot. Line 66 defines the axis label for the vertical axis. Lines 67 through 71 define the different symbols for the forecast graph. The GPLOT command then plots a forecast profile for the components of the autoregression output. Lines 73 through 75 instruct the GPLOT to overlay the actual data, the forecast, the trend line, the lower, and the upper confidence limits on the graph. A vertical reference line is positioned on the horizontal time axis at period 50. That line is then annotated according to the data found in the data set called ANNO. To prevent the forecast from becoming too small for close inspection, a window of resolution is defined between times 25 and 60 to be displayed. These graphs greatly facilitate interpretation.

The SAS autoregression procedure is very flexible and powerful. Not only can AUTOREG model ordinary exogenous series with simple AR(p) error structures, it can also model seasonal dummies, lagged dependent variables, and generalized autoregressive heteroskedasticity conditional on time. Built into it are a variety of tests for the different assumptions of autoregressive models. For ordinary autoregressive models, the most recent version of AUTOREG contains tests for higher order serial correlation of errors, for normality of the residuals, for stability of the model, for unit roots, and for different orders of heteroskedasticity. In the event lagged dependent variables are used, it contains the Durbin h test for first-order autocorrelation of the lagged dependent variable, and if there are different orders of heteroskedasticity, it contains a La Grange multiplier test for determining the order of the heteroskedasticty. This SAS procedure for autoregressive models is powerful and flexible.

10.7.2. SPSS ARIMA Procedures for Autoregressive Error Models

At the time of this writing, the SPSS AREG procedure can model time series regressions with only first-order autocorrelation of the residuals. It performs Cochrane–Orcutt, Prais–Winsten, and maximum likelihood estimation of these model parameters. Because AREG cannot handle higher order autocorrelated error structures, SPSS ARIMA is invoked. After a preliminary invocation of AREG, in paragraph 1 of the SPSS command syntax, the syntax for the ARIMA models is contained in paragraphs following paragraph 2 in the SPSS command syntax below.

In the following SPSS program (c10pgm3sps), the SPSS AREG programming commands are given in the first paragraph. They model an autoregression of Y on time, with the assumption of first-order error autocorrelation. The output is shown in Fig. 10.8. The first-order correction is invoked with AREG, the parameters are estimated with maximum likelihood, but higher order serial correlation remains. Despite this, the SPSS AREG algorithm is not yet capable of correcting for it. To test for such residual autocorrelation, the residuals, ERR_1, from this model are reviewed in paragraph 2 of the command syntax. From sequential diagnosis of the residuals, we see that there are significant autocorrelations at lags 5 and 6, as can be seen in Fig. 10.9. To permit modeling of these higher order autocorrelations, SPSS ARIMA is invoked. The command syntax shown next models an autoregression on time with first-, fifth-, and sixth-order autoregressive errors in the third from last paragraph and the diagnosis in the final paragraph that such a models leaves white noise residuals. The output for the parameter estimates and their residuals are contained in Figs. 10.10 and 10.11.

```
Autoregression with AR(1) error against Time .
TSET PRINT=DEFAULT CNVERGE=.001 CIN=95 NEWVAR=ALL .
PREDICT THRU END.
```

```
AREG y WITH time
  /METHOD=ML
  /CONSTANT
  /RHO=0
  /MXITER=10.
*ACF and PACF reveal spikes at Lag =5.
ACF
  VARIABLES= err_1
  /NOLOG
  /MXAUTO 16
  /SERROR=IND
  /PACF.
*ARIMA against Time P=(1,5) model.
TSET PRINT=DEFAULT CIN=95 NEWVAR=ALL .
PREDICT THRU END.
ARIMA y WITH time
  /MODEL=( 1 0 0 )CONSTANT
  /P=(1,5)
  /MXITER 10
  /PAREPS .001
  /SSQPCT .001
  /FORECAST EXACT .
*Diagnosis of Residuals shows spike at lag=6.
ACF
  VARIABLES= err_2
  /NOLOG
  /MXAUTO 16
  /SERROR=IND
  /PACF.
*ARIMA against time P=(1,5,6) model.
TSET PRINT=DEFAULT CIN=95 NEWVAR=ALL .
PREDICT THRU END.
ARIMA y WITH time
  /MODEL=( 1 0 0 )CONSTANT
  /P=(1,5,6)
  /MXITER 10
  /PAREPS .001
  /SSQPCT .001
  /FORECAST EXACT .
*Final Diagnosis indicates white noise.
ACF
  VARIABLES= err_3
```

```
/NOLOG
/MXAUTO 16
/SERROR=IND
/PACF.
*ARIMA Final Model Forecast Generation.
TSET PRINT=DEFAULT CIN=95 NEWVAR=ALL .
PREDICT THRU 100 .
ARIMA y WITH time
/MODEL=( 1  0  0 )CONSTANT
/P=(1,5,6)
/MXITER 10
/PAREPS .001
/SSQPCT .001
/FORECAST EXACT .
*Forecast Plot based on Final Model.
TSPLOT VARIABLES= y lcl_5 ucl_5 fit_5
/ID= time
/NOLOG.
*ARIMA.
TSET PRINT=DEFAULT CIN=95 NEWVAR=ALL .
PREDICT THRU 125 .
ARIMA y WITH time
/MODEL=( 1  0  0 )CONSTANT
/P=(1,5,6)
/MXITER 10
/PAREPS .001
/SSQPCT .001
/FORECAST EXACT .
```

The SPSS autoregression analysis output appears in Fig. 10.8. The method selected is that of maximum likelihood. The initial ρ value is set at 0 and found through iteration. When the residuals, ERR_1, exhibit fifth order autocorrelation, then a more sophisticated and more flexible SPSS ARIMA procedure has to be invoked (Fig. 10.9).

We attempt an ARIMA procedure modeling those autocorrelations. The residuals, ERR_2, upon review show that there is also an autocorrelation at lag 6. After repeated diagnosis, it is revealed that first-, fifth- and sixth-order autoregressive errors are significant and they are modeled in the third from last paragraph of the SPSS command syntax.

```
ARIMA y WITH time
/MODEL=( 1  0  0 )CONSTANT
/P=(1,5,6)
```

```
/MXITER 10
/PAREPS .001
/SSQPCT .001
/FORECAST EXACT .
```

When the SPSS ARIMA procedure is finally invoked, the Y process is identified, estimated, and diagnosed as an AR model with spikes at lags 1, 5 and 6. Hence, we attempt a final ARIMA model regressed on time, with errors autocorrelated at lags 1, 5, and 6. This model fits. With the /FORECAST EXACT subcommand, we generate the forecast for later graphing.

In sum, for complex AR(p) error structure analysis, the researcher can utilize the SPSS ARIMA procedure. An ARIMA model with AR parameters is generated by the subcommand P=(1,5). The output of this analysis is shown, but the ACF and PACF of ERR_2 reveals significant spikes at lag 6. Therefore, a third ARIMA model is run with AR parameters set by P=(1,5,6). Now the ARIMA modeled with time as an independent variable has AR(1), AR(5), and AR(6) parameters in the model. They are

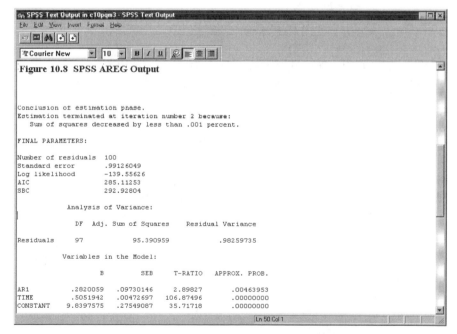

Figure 10.8 SPSS AREG output.

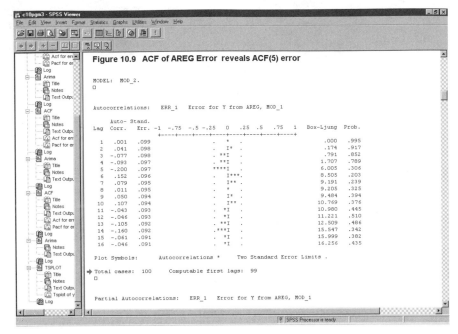

Figure 10.9   ACF of AREG error reveals ACF(5) error.

all significant at the $p < 0.05$ level. The error now is diagnosed as white noise as can be seen from the residuals in Fig. 10.10.

The original SPSS AREG procedure output (Fig. 10.8) specifies the model to be

$$Y_t = 9.844 + 0.505time + e_t$$
$$e_t = 0.282e_{t-1} + v_t$$

(10.35)

(SPSS, 1994; 1996). The residuals are not white noise and are modeled here without a sign reversal in the output. After switching to the ARIMA procedure regressing Y_t on time, with maximum likelihood estimation, we obtain the following significant parameters: AR1, AR5, AR6, Time, and a Constant. The last two parameters pertain to the principal equation, whereas the first three autoregressive parameters define the error structure of the model. The output of this model is shown in Fig. 10.11. The equation obtained is essentially identical to that obtained by SAS on the same data as shown in Eq. (10.34). These parameters are not highly correlated with one another. They appear to be stable. When we diagnose the residuals (ERR_3) of this model with an ACF and PACF, we find them to be without

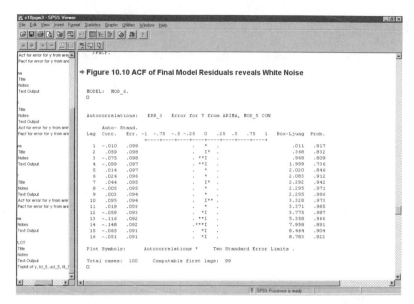

Figure 10.10 ACF of final model residuals reveals white noise.

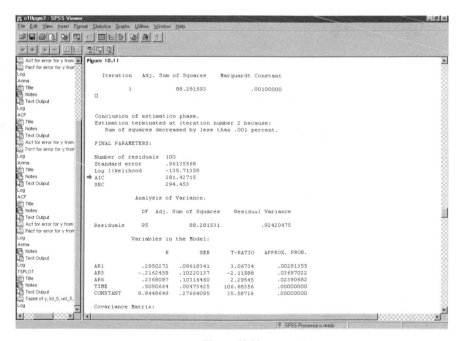

Figure 10.11

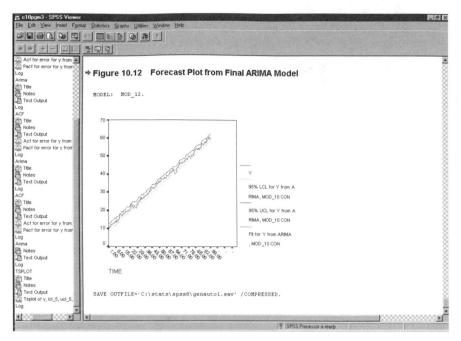

Figure 10.12   Forecast plot from final ARIMA model.

any statistically significant spikes. In other words, they appear to be white noise, indicating that the model has been fully explained by these parameters. At this juncture, we re-estimate the model. We extend the forecast along with its confidence limits to the end of the data set and save them. We plot these data in Figure 10.12.

10.8. AUTOREGRESSION IN COMBINING FORECASTS

Granger and Ramanathan (1984) have suggested the use of regression and regression controlling for autocorrelated errors as models to combine forecasts. Others, such as Diebold (1996, 1998) and Clements and Hendry (1998), have followed suit. In the early years of the U.S. economy, farms and plantations predominated. Eventually, during the later nineteenth and early twentieth centuries, industry developed and factory workers predominated. Since the Second World War, the U.S. economy has become for the most part a service economy. Thus, the average hourly wage of the service worker is of interest.

The data are divided into an historical and an evaluation period. The historical period extends from January 1964 through December 1991; the evaluation period extends from January 1992 through February 1999, shown in Fig. 10.13. In program C10PGM4.SAS, two different models, each formed on the historical data, are used to generate forecasts. Although this example combines two forecasts, at least five forecasts can be combined if they are actually available (Armstrong, in press). The two forecasts generated by these models span the time horizon of the evaluation data set.

The first model is that of an exponential smoothing with a linear trend. The equation for this model is (Smoothed Mean Hourly Wage)$_t$ = 0.10557 Current value of Service Worker Mean Hourly Wage$_t$ + (1 − 0.10557) (Smoothed Service Worker Average Hourly Wage)$_{t-1}$. This model fits the data nicely with an R^2 of 0.998 and produces an excellent forecast and a very small forecast interval, as shown in Fig. 10.14.

The second model, graphed in Fig. 10.15, is a polynomial autoregression model, with time and time-squared used as predictors. SAS PROC AUTOREG is employed with backward stepwise elimination of the non-significant autocorrelations, revealing multiple significant remaining autoregressive errors at lags 1, 2, 23, 24, 26, and 27. The maximum likelihood estimation in SAS corrects the standard errors for bias of the significance tests that would otherwise contaminate the model. The model that emerges from this analysis is (Mean Hourly Wage of Service worker)$_t$ = 1.725 + 0.012Time + 0.0004Time^2 + e_t, with each of these parameters significant at $p < 0.001$. The autoregressive error structure is represented by e_t = 0.760e_{t-2} + 0.264e_{t-2} + 0.208e_{t-23} − 0.372e_{t-24} + 0.306e_{t-26} − 0.185e_{t-27} + v_t, where v_t is the uncorrelated error. It is important to remember that these signs of the autoregressive parameters in the maximum likelihood

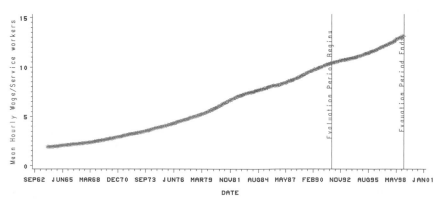

Figure 10.13 Average hourly earnings of service workers in the United States. Seasonally adjusted (Bureau of Labor Statistics data: censinfo@bls.gov).

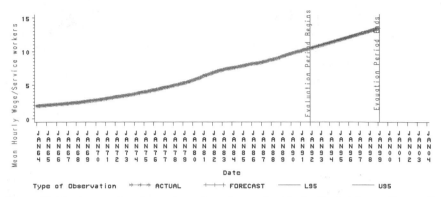

Figure 10.14 Average hourly earnings of service workers in the United States. Seasonally adjusted (Bureau of Labor Statistics data: censinfo@bls.gov). Model 1 exponential smoothing with linear trend forecast.

output change when these terms appear in this equation because of a rearrangement of terms in the error structure. Also, the model fits very well with a high $R^2 = 0.906$ after correction. With this second model a forecast is generated that extends till February 1992, and this forecast profile is also displayed in Figure 10.15.

These two forecasts, which extend over the evaluation sample, are then combined by autoregression, to form a more accurate forecast profile (Fig. 10.16). Autoregression is used to adjust for the autocorrelation inherent in the actual and forecast series. The first forecast, called F_1, is produced by an exponential smoothing model. The second forecast, called F_2, comes

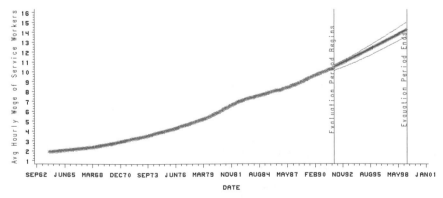

Figure 10.15 Average hourly earnings of service workers in the United States. Seasonally adjusted (Bureau of Labor Statistics data: censinfo@bls.gov). Model 2 series EES80000006 autoregression forecast.

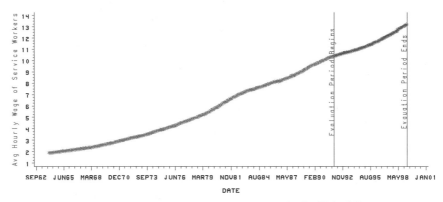

Figure 10.16 Average hourly earnings of service workers in the United States. Seasonally adjusted (Bureau of Labor Statistics data: censinfo@bls.gov). Graph of combined autoregression forecast.

from the polynomial autoregression analysis. Within the evaluation period, the actual data are regressed on the two forecasts and the autoregression adjusts the model for serial correlation in the error structure. The fundamental formula for the combining autoregression is

$$CF_t = a + b_1 F_{1t} + b_2 F_{2t} + e_t$$

$$e_t = v_t - \varphi_1 v_{t-1} - \cdots - \varphi_p v_{t-p},$$

where   (10.36)

CF_t is the Combined forecast
F_{1t} is the Model 1 forecast
F_{2t} is the Model 2 forecast.

In this example, the actual U.S. service worker mean hourly wage is used as the dependent variable in the autoregression on the two forecasts. The model estimated has a high R^2 of 0.996 with each of the forecast parameters having a significance level of $p < 0.001$. The model estimated is

```
Average Hourly Wage (of US service worker)_{t+h} = 14.986 - 5.959
F_{1t+h} + 5.475F_{2t+h} + e_t,
    where e_t=.824e_{t-1} + v_t.
```

The combined forecast consists of a set of predicted scores generated from this autoregression model. The combined forecast profile corrects for the autocorrelation in the series to render a less biased estimate than would emerge from an OLS regression combination. This forecast fits the data well, and is evaluated by comparing the actual data within the evaluation window to the combined forecast (Meyer, 1998).

Table 10.1

Forecast Evaluation

Type of model	Mean square forecast error	Mean absolute percentage error
Model 1 exponential smoothing forecast	0.0217	3.121
Model 2 polynomial autoregression forecast	0.094	6.730
Autoregression combined forecast	0.00009	0.087

These methods for combining forecasts are optimized when there is no autocorrelation. Granger (1989) recommended that serial correlation be taken into consideration when combining forecasts. In 1998, Diebold recommended not only contemplation of serial correlation of the errors, but also of lagged endogenous variables to capture all of the dynamics in the forecast by the combining method. To do this, he recommends using the regression method just described, with an important modification. He suggests saving the residuals from the regression combination and modeling those residuals as an ARMA(p,q) process. He maintains that this process need not be linear. For nonlinear models, there can be interaction terms, polynomial terms, or even polynomial interactions on the right-hand side of the model.

In the comparative evaluation of the separate forecasts and the combined forecast, these forecasts are compared with the actual data within the evaluation period. The MSFE or the MAPE are general criteria that can be used to make this comparison. The MSFE and the MAPE for each of the two forecasts and the combining forecast are presented in Table 10.1, from which we see that according to both criteria the combining autoregression greatly improves the forecast accuracy. Accordingly, in the graph of the forecast generated from this autoregression combination of forecasts, the forecast interval around the prediction scores is so small that it is difficult to see.

An ARIMA procedure models MA errors in the residuals if any exist and then produces the forecast interval data. The ARIMA also supplies the upper and lower 95% confidence limits for the forecast profile. Although not demonstrated here, the AUTOREG procedure can be used to model changes in variance as well.

10.9. MODELS WITH STOCHASTIC VARIANCE

An assumption of a valid regression model is that it possesses constant error variance. To be sure that the model is valid, we must test the assumptions. There is really no reason to believe that the errors are white noise without testing (Granger and Ramanathan, 1984). Engle (1982) has written that under some circumstances, the "error variance may change over time and be predicted by past forecast errors." Processes with such an autoregressive heteroskedasticity have found particular application in matters of financial econometrics and the analysis of inflation (Engle, 1982; Diebold and Lopez, 1996; Figlewski, 1999b). With inflation, volatility in the value of a stock option may increase. Where error variance of a stock option profit model increases over time, the risk of the investment increases. In cases of regression models, where the value of error variance is a function of the time lag, an autoregressive model with conditional heteroskedastic (ARCH) error variance may be in the appropriate model to model that risk or volatility (Bollerslev, 1984). Engle, Granger, and Kraft (1984) suggest that combining forecasts can be accomplished with ARCH models (Peel *et al.*, 1990). For these reasons, the subject of ARCH models is briefly introduced.

10.9.1. ARCH AND GARCH MODELS

ARCH process have error variances that can be expressed in a simple functional form. If Y_t is a model that has a variance, h_t, that is conditional on the error variance at a previous time periods, that model, with its conditional variance, can be expressed as

If $Y_t = \beta_1 x_t + e_t$,

and $e_t \sim N(0, h_t)$,

$$h_t = \text{Var}(e_t) = \alpha_0 + \alpha_1 e_{t-1}^2 \qquad (10.37)$$

or when the model is of order q—that is, $ARCH(q)$:

$$h_t = \text{VAR}(e_t) = \alpha_0 + \sum_{j=1}^{q} \alpha_{t-q}^2.$$

Bollerslev (1984) extended the ARCH model to generalized version called a GARCH model. The GARCH model is one where the variance is a function of previous conditional variances as well as previous innovations. The fundamental formulation of a GARCH (q, p) model is

$$h_t = \text{Var}(e_t) = \alpha_0 + \sum_{j=1}^{q} \alpha_j e_{t-q}^2 + \sum_{i=1}^{p} \beta_i h_{t-i}^2. \qquad (10.38)$$

A basic test for ARCH errors is a test for the significance of α_1. After estimating the model, save the residuals and regress the squared residuals on past lags of the squared residuals. If the hypothesis that $|\alpha_1| > 0$ is confirmed, then there are ARCH errors. The tests for the order of ARCH or GARCH are performed with a LaGrange multiplier test. Estimation of these models is performed with maximum likelihood; a BHH algorithm is preferred for estimation of ARCH or GARCH models.

10.9.2. ARCH MODELS FOR COMBINING FORECASTS

Engle, Granger and Kraft (1984) have suggested that ARCH models be used for combining forecasts. They use a relatively complicated ARCH model to generate time-varying combining weights.

They introduce a bivariate ARCH model, based on Eq. (10.37). The forecasts are autocorrelated, so autoregression is preferable to OLS regression. The conditional heteroskedasticity is modeled as well. To allow for the covariance of the errors, the matrix equation for the variance of errors is specified in quadratic form:

$$H(e_{t-1}) = H_t = [H_{ijt}]$$
$$H_{ijt} = a_{i,j0} + e'_{t-1} C_{ij} e_{t-1}. \tag{10.39}$$

For all possible combinations of e_1 and e_2 for an ARCH(1) model to be specified, Engle *et al.* (1984) express the process as

$$h_t = \begin{bmatrix} H_{11t} \\ H_{21t} \\ H_{22t} \end{bmatrix} = \begin{bmatrix} a_{01} \\ a_{02} \\ a_{03} \end{bmatrix} + \begin{bmatrix} a_{11} & a_{12} & a_{13} \\ a_{21} & a_{22} & a_{23} \\ a_{31} & a_{32} & a_{33} \end{bmatrix} \begin{bmatrix} e^2_{1t-1} \\ e_{1t-1} e_{2t-1} \\ e^2_{1t-1} \end{bmatrix} \tag{10.40}$$

$$\equiv a_0 + A h_{t-1}$$

Alternatively, the improved combined forecast, f_{ct}, is obtained from a combination of the forecast from one model, f_{1t}, and the forecast from another model, f_{2t}, with combining weights β_0, β_1 and β_2:

$$F_{ct} = \beta_0 + \beta_1 F_{1t} + \beta_2 F_{2t} + e_{ct}. \tag{10.41}$$

Because $e_{it} = Y_t - f_{iv}$, where Y_{it} is the actual data from the evaluation sample. This combination implies a forecast error:

$$e_{ct} = Y_{it} - \beta_0 - \beta_1 F_{1t} - \beta_2 F_{2t}. \tag{10.42}$$

The error variance forms the basis of the forecast error and confidence intervals. If in testing for ARCH(q) in the error variance, the researcher finds it and can estimate

$$e_{ct}^2 = \alpha_0 + \sum_{j=1}^{q} \alpha_{t-q}^2, \tag{10.43}$$

then this ARCH(q) model can explain the risk structure in the combined forecast.

A caveat is in order here. ARCH and GARCH models, which involve more than one equation, are relatively complex and difficult to fit. They require large data sets. Only models with a small number of parameters appear to be well behaved, and these models have more parameters than others. The parameters need to be stable, lest they fall apart in out-of-sample tests. They may be good for one-step-ahead forecasts and not for multistep forecasts. The incremental utility of the improvement in fit that they obtain is not always worth the extra investment of time and energy (Figlewski, 1999a). For these reasons, simpler algorithms, such as a combining regression with ARMA errors, may well be preferred. Nonetheless, as the value of modeling time-dependent risk grows, the more the advanced theory and programming of GARCH and other models becomes an objective worthy of serious study.

REFERENCES

Armstrong, J. S. (1999). "Combining Forecasts" in Armstrong, J. S. (Ed.) Principles of Forecasting: A Handbook for Researchers and Practitioners. Norwell, MA. Academic Publishers, in press.

Bollerslev, T. (1984). "Generalized Autoregressive Conditional Heteroskedasticity." In *ARCH Selected Readings.* Engle, R. F. (Ed.) New York: Oxford University Press. 1995, pp. 42 60.

Bresler, L., Cohen, B. L., Ginn, J. M., Lopes, J., Meek, G. R., and Weeks, H. (1991). *SAS/ETS®* Software: Applications Guide 1: Time Series Modeling and Forecasting, Financial Reporting, and Loan Analysis Version 6, 1st ed. Cary, NC: SAS Institute, Inc. Chapter 3, pp. 35–65.

Clements, M. P., and Hendry, D. F. (1998). *Forecasting Economic Time Series.* Cambridge, UK: Cambridge University Press, pp. 231–232.Diebold, F. (1998). *Elements of Forecasting.* Cincinnati: Southwestern College Publishing. Chapter 12, pp. 339–374.

Diebold, F. X. (1998) *Elements of Forecasting.* Cincinnati: Southwestern College Publishing, Chapter 12, pp. 339–374.

Diebold, F. X., and Lopez, J. A. (1996). "Forecast Evaluation and Combination." In *Handbook of Statistics Statistical Methods in Finance, Vol. 14.* Maddala, G. S., and Rao, C. R. (Eds.). Amsterdam: Elsevier Science. pp. 241–269.

Ege, G., Erdman, D. J., Killam, B., Kim, M., Lin, C., Little, M. R., Narter, M. A., and Park, H. J. (1993). *SAS ETS User's Guide,* 2nd ed. Cary, NC: SAS Institute, Inc. pp. 201–213, 214–217, 218–222, 223–253.

Engle, R. F. (1982). "Autoregressive Conditional Heteroskedasticity with Estimates of the

Variances of United Kingdom Inflation." In *ARCH Selected Readings* Engle, R. F. (Ed.) New York: Oxford University Press, 1995, pp. 1–24.

Engle, R. F., Granger, C. W. J., and Kraft, D. (1984). "Combining Competing Forecasts of Inflation using a Bivariate ARCH Model," *Journal of Economic Dynamics and Control,* **8.** pp. 151–165.

Figlewski, Stephen (1999a). "Forecasting Volatility," a presentation at Sixth International Conference on Computational Finance, Stern School of Business, New York University, January 6, 1999.

Figlewski, Stephen (1999b). "Forecasting Volatility," *Financial Markets, Institutions, and Instruments.* **6**(1), pp. 1–88.

Goldberger, A. S. (1991). *A Course in Econometrics.* Cambridge, MA: Harvard University Press. pp. 300–307.

Granger, C. W. J., and Newbold, P. (1986). *Forecasting Economic Time Series,* 2nd ed. San Diego: Academic Press, pp. 13–25, 187–196.

Granger, C. W. J. (1989) "Combining Forecasts—Twenty Years Later," *Journal of Forecasting,* **8,** pp. 167–173.

Granger, C. W. J., and Ramanathan (1984). "Improved Methods of Combining Forecasts," *Journal of Forecasting,* **3,** pp. 197–204.

Greene, W. H. (1997). *Econometric Analysis.* Englewood Cliffs, NJ: Prentice Hall, pp. 594–595.

Griffiths, W. E., Hill, R., Carter, and Judge, G. G. (1993). *Learning and Practicing Econometrics.* New York: John Wiley and Sons, p. 517.

Gujarati, D. (1995). *Basic Econometrics.* New York: McGraw-Hill, pp. 401–404, 422–423.

Hanushek, E. A., and Jackson, J. E. (1977). *Statistical Methods for Social Scientists.* New York: Academic Press, pp. 155–156.

Holden, K., Peel, D. A., and Thompson, J. L. (1990) *Economic Forecasting: An Introduction.* New York: Cambridge University Press, pp. 85–106.

Johnston, J. J. (1984). *Econometric Methods..* 3rd ed. New York: McGraw-Hill, pp. 27–45, 309–310.

Kamenta, J. (1986). *Elements of Econometrics,* 2nd ed. New York: Macmillan, pp. 212–220, 298–304, 331.

Maddala, G. S. (1992). *Introduction to Econometrics,* 2nd ed. New York: Macmillan Publishing, pp. 230, 241–243, 528.

Makridakis, S., Wheelwright, S. C., and McGee, V. E. (1983). *Forecasting: Methods and Applications,* 2nd ed. New York: John Wiley and Sons, pp. 232–238.

Newbold, P., and Granger, C. W. J. (1974). "Experience with Forecasting Univariate Time Series and the Combination of Forecasts." *Journal of the Royal Statistical Society,* A. 137, Part 2, pp. 131–146.

Meyer, Kevin (1998). Cary, NC: SAS Institute, Inc. Personal communication (Nov–Dec, 1998).

Ostrom, C. (1990). *Time Series Regression Techniques,* 2nd ed. Newberry Park, CA: Sage Publications, pp. 21–26, 32–35.

Pindyck, R. S., and Rubinfeld, D. L. (1991). *Economic Models and Economic Forecasts,* 3rd ed. New York: McGraw-Hill, p. 138.

SPSS, Inc. (1994). *SPSS Trends 6.1.* Chicago: SPSS, Inc., pp. 111–136, 260–271.

SPSS, Inc. (1996). *SPSS 7.0 Statistical Algorithms Chicago:* SPSS, Inc. pp. 37–50.

Theil, H. (1971). *Principals of Econometrics.* New York: John Wiley and Sons, p. 251.

Wonnacott, R. J., and Wonnacott, T. H. (1979). *Econometrics.* New York: John Wiley, p. 212.

Chapter 11

A Review of Model and Forecast Evaluation

11.1. MODEL AND FORECAST EVALUATION

Two principal purposes of time series analysis are explanation and forecasting. Throughout this book, the models discussed range from the simple to the more complex. As we examine the different approaches, the time series models become more sophisticated. Not only can they handle more inputs, they can also handle more complicated inputs. The more complicated models become vehicles for theoretical explanation and theory testing. Larger models have the potential to be more theoretically encompassing (Harvey *et al.*, 1998). The analyst must develop competing models and comparatively evaluate them.

This chapter addresses evaluation with respect to explanation as well as prediction. We evaluate the explanatory model, refine it, compare it with alternative models, and select the best specified model. We comparatively evaluate the explanatory and forecasting capabilities of different models, and then compare and contrast combined models with respect to forecasting accuracy. From the assessments of the models addressed in this book, we find that different models have specific advantages and drawbacks. Focusing on these relative advantages of the moving average, exponential smoothing, X11-X12, ARIMA, seasonal ARIMA, intervention, transfer function, dynamic regression, autoregression, and combined models provides a guide for the analyst. Where combinations of models outperform individual models, it behooves the analyst to know which combinations provide maximal advantage. Consequently, a comparative analysis of relative advantages and disadvantages of specific approaches and their combinations provides a guide for their proper application.

11.2. MODEL EVALUATION

Whether the model is an ARIMA model or a dynamic regression model, there are general criteria by which it can be evaluated. The model must be consistent with theory. The model should explain the process as simply as possible, but not more simply than that, as Albert Einstein was reported to have remarked (Parzen, 1982). The better model will be theoretically more encompassing in scope. The model should have some goodness of fit. It should be well specified. It should be parsimonious. Its parameters should be stationary, stable, and invertible; they should, however, not be collinear. The model should be stable over time and robust to changes in its auxiliary parameters. It should have good predictive power over a variety of forecast horizons. If an ARIMA model shares these characteristics, it has utility.

If the model is a dynamic regression model, most of the criteria are the same. The good model is derived from a good data set, which consists of sufficient sample size that has been properly measured, equally spaced, consistently collected, double-checked, and cleaned of typographical errors. Outliers have been identified and corrected, replaced, or modeled. The good model has the proper dynamic specification. It has the right number of AR terms for each of the exogenous variables. The exogenous series included have been tested for exogeneity. The parameters should be constant over the time period. The parameters should not be substantially collinear. Residual autocorrelation should be properly modeled (Pankratz, 1991). Such dynamic regression models should have been built with Hendry and Richard's general-to-specific approach to avoid the pernicious effects of specification error. The constructed model should be stable and reliable. It should be robust to regime shift. The parameters should be constant over such shifts. If auxiliary variables are interchanged, the key theoretical parameters should exhibit robustness, stability, and constancy (Leamer, 1985). The parameters by themselves and the model as a whole should be consistent with theory. The model should explain most of the variance of the dependent variable. Misspecification should be minimized. There should be goodness of fit. The model should have maximum encompassing of both variance and theory (Granato, 1991). It should be parsimonious. The model should not only have the explanatory power that comes from fulfilling such requirements, it should have sufficient predictive power to be able to forecast to a validation sample with minimal error of prediction and minimal mean absolute forecast error over a sufficient time horizon at a minimal cost. The model should be subject to crossvalidation (Maddala, 1992). It should possess these qualities both in the short and the long run (Gujarati, 1995).

11.3. COMPARATIVE FORECAST EVALUATION

In addition to explanation, one of the fundamental objectives of social science is prediction. Forecasting is one means of predictively validating theory. Theoretically elaborate models may not always predict as well as simpler ones. In this chapter, we compare the forecasting capability of the time series models. In general, forecast evaluation is performed by subsetting the series into an estimation subsample and a validation subsample. The model is developed using the estimation or historical subsample, whereas its forecast, extended into the validation subsample, is evaluated on the basis of the latter subsample (Granger and Newbold, 1986). It could also be validated on data collected later. Evaluation of the forecast is essential to this validation process.

In the process, the forecasting capability can be evaluated with reference to various standards. The standards by which forecasts are evaluated include statistical measures of accuracy and assessments of cost in terms of time, money, and effort involved in preparation of the data and fine-tuning a model. The statistical standards include the required size of the information set, the definition and specification of the variables in the forecast model, bias, mean square forecast error (which may not reflect parameter constancy), mean absolute percentage error, ability to detect turning points, accuracy over different forecast horizons, stability of the model, and encompassing scope of the model. The availability, quantity (sample size), and quality of the data needed should be examined. The cost in time and money of data set preparation must be considered. More specifically, the cost of sampling, data collection, managerial oversight, verification, and cleaning necessary for data set preparation, the number and kind of transformations of the variables involved, and the number of runs needed to prepare a functional model are practical considerations that should not be overlooked (Makridakis *et al.*, 1983; Montgomery *et al.*, 1990; Sullivan and Claycombe,

1977; Clements and Hendry, 1994). The practical standards are useful in planning, while the statistical standards are useful in evaluating the forecasting model.

From a number of forecasting competitions, researchers have come to basic conclusions about which models are more accurate. Makridakis has held several forecasting model competitions since 1982, called the M competitions. He has found that no one model, regardless of criteria of evaluation and the circumstances, outperforms all others. Some models perform well when evaluated by one criterion while other models perform better when evaluated by other criteria. Sometimes simpler models outperform the more sophisticated ones. Moreover, different models forecast more effectively with different kinds of data (Gilchrist, 1976; Makridakis, 1984).

Several factors were found to influence forecast accuracy. When the sampling variability of the estimation data set differs, the forecasts will differ. The size and type of the data set required is another criterion. Some data sets are nonseasonal, others are both nonseasonal and seasonal, while others are seasonally adjusted with nonseasonal models. Outliers or seasonal pulses can make a difference. Regime or level shifts can also make a difference. Time trends can also make a difference in the data. Some series will have local or piecewise trends and others will have global trends. Some time trends are deterministic, whereas others are stochastic. Not only the type of data, but also the forecast horizons over which these approaches may be evaluated differ. Various combinations of data type and forecast horizons are more amenable to some forecast models than others.

In general, the further ahead into the time horizon the forecast is made, the less accurate it is (Granger, 1989). It behooves the researcher to examine his series to see which aspects dominate in the short-, middle-, and long-term forecast horizons. In the short run, which usually extends to approximately the first six temporal periods of the forecast horizon, the random error and the seasonality may predominate. Extrapolative methods, the more sophisticated of which take local time trend and seasonality into account, can be useful in providing reasonably accurate forecasts in the near term with relative ease of computation (Makridakis et al., 1997). In the middle range, from about 7 to 18 periods into the forecast horizon, while random error still is important, cycle and seasonality become salient and trend becomes increasingly important. Cycles, often difficult to precisely forecast, gain prominence in this time range and render forecasting even more hazardous. In the longer term, the cycle may decline while the global time trend may grow in prominence (Makridakis et al., 1983), even though systemic regime shifts may render these forecasted trends useless. For this reason, long-run forecasting often becomes more difficult, doubtful, and dangerous than midterm prediction (Makridakis et al., 1997). The amount and proportion of randomness in the data may be responsible for differences in model performance (Makridakis and Hibon, 1984). In other words, performance of different models may depend on components of trend, cycles, seasonality, and random error exhibited by the data.

11.3.1. CAPABILITIES OF FORECAST METHODS

Scholars have commented on the relative advantages and disadvantages of different models with respect to their forecast capability. Some scholars describe these attributes of the different models with respect to forecasting over various horizons. They refer to forecast accuracy as well as ability to detect turning points. Sometimes they refer to the data requirements of the models. They refer to the cost of the method as well as the ease of computation. They refer to the time it takes to develop the model and the applications to which such models are put. Some methods have better capability in the short run. Others have better capability in the middle range, while still others have better long-term capability.

In this section, general and tentative descriptions of the different forecasting capabilities of the methods emphasized in this work are presented. Sullivan and Claycombe (1977) write that moving averages have varying accuracy. They claim that the accuracy of moving averages is poor to good in the short run and worse in the medium and long run. To be sure, they require stationary data. A minimum of 2 years of data for seasonal analysis is recommended. They also note that calculation of moving averages requires little sophistication and expense. While the computations may take less than a day to estimate, turning point detection is poor. Nonetheless, this method frequently finds application in areas of inventory control (Sullivan and Claycombe, 1977).

Exponential smoothing exhibits better accuracy than moving averages in the short run. The accuracy of the simpler exponential smoothing procedure typically goes from good to poor in the medium range, and gets worse in the long run. The data required by the simpler smoothing methods needs to be stationary as well. These simpler methods do a poor job in the identification of turning points and require a minimum of 2 years of data for seasonal material and less for nonseasonal data. Single exponential smoothing may do better than most methods with small data sets. Although exponential smoothing requires a little more sophistication than does the moving average method, it is still simple and easy to apply in its simpler forms without computers. The more sophisticated types of exponential smoothing that account for local time trend and seasonality are more easily calculated with computers than others. This procedure can be automated and performed routinely to generate many forecasts with relatively little cost in terms of data, computer storage, computer time, or labor. It may take a day or less to estimate, depending on the complexity of the data, the length of the forecast horizon, and the method. Still, the relative ease and amenability to automation are reasons that exponential smoothing methods are commonly used for inventory and production control, and simple kinds of financial data analysis. With very small data sets of 30 observations or less, the Holt–Winters method is considered by some to be about the only one acceptable (Granger and Newbold, 1986). The Holt–Winters exponential smoothing method is said to perform well with 40 to 50 observations (Newbold and Granger, 1974). As the forecast horizon was extended, the Holt–Winters method outperformed the stepwise autoregression more often (Newbold and Granger, 1974; Makridakis et al., 1983). Not only have Holt exponential smoothing procedures done well in the M competitions, another simple method, called the Theta method, developed by V. Assimokopoulos, that combines linear trend and moving average estimates, has also performed well in the M3 competition (Fildes et al., 1998; Hibon, 1999, June). Other forecasting packages that earned honorable mention in some of the M3 competition were Forecast Pro (Goodrich, 1999) and Autobox (Reilly, 1999).

If the forecasting method used is that of classical decomposition or Census X-11, the method breaks down the series into component parts of trend, cycle, seasonality, and random error. Because there is no guarantee that the series components will in reality remain the same, it is necessary to gather enough data to test parameter constancy. The problem is that there is no guarantee how much data is needed for this purpose, although 5 to 6 years of observations is generally considered advisable. Decomposition methods are generally effective in extracting the trend, cycle, and seasonality from the irregular component of a series, although they have more difficulty in isolating trend, cycle, and seasonal subpatterns (Makridakis et al., 1983). Census X-11 has been widely used by governments around the world since the 1950s to deseasonalize data prior to forecasting. This method is useful in making medium-range predictions, where other factors remain relatively stable (Makridakis et al., 1997). This method is being replaced by Census X-12.

Census X-12, not yet part of SAS or SPSS, contains a number of innovations over earlier X-11 and the 1988 update, X-11-ARIMA, developed by E. Dagum et al. at Statistics Canada. With X-11-ARIMA, Dagum introduced the use of backcasting and forecasting to reduce bias at the ends of the series. The new X-12 program contains more "systematic and focused diagnostics for assessing the quality of seasonal adjustments." X-12 has a wide variety of filters from which to choose in order to extract trend and seasonal patterns, plus a set of asymmetric filters to be used for the ends of the series. Some of the diagnostics assess the stability of the extracted components of the series. Optional power transformations permit optimal modeling of the series. X-12 contains a linear regression with ARIMA errors (REGARIMA) that forecasts, backcasts, and preadjusts for sundry (moving holiday and Leap Year) effects. The corrected *AIC* (see AICC in glossary) is used to detect the existence of trading day effects. This REGARIMA can partial out the effects of regime shifts, explanatory variables prior to decomposition, as well as better test for seasonal patterns and sundry calendar effects—including trading day, moving holiday, and Leap Year effects. In this way, it can partial out user-defined effects and thereby eliminate corruption from such sources of bias (Findley *et al.,* 1998; Makridakis *et al.,* 1997). REGARIMA provides for enhanced outlier detection of and protection from additive outliers and level shifts (including transient ramps). Moreover, the X-12 program incorporates an option for automatic model selection based on the best *AICC* (Findley *et al.,* 1998; Soukamp, 1999). X-12 may soon become the global standard for deseasonalization of series data.

The Box–Jenkins method combines comprehensive moving average and autoregressive capability. If there is a univariate or a unidirectional bivariate model to define and forecast, the Box–Jenkins model often provides a good forecast, especially in the short run. If there are just a few uncorrelated inputs, then the Box–Jenkins model may serve nicely. Box–Jenkins modeling requires a sound mathematical background, some experience at ARIMA modeling, and access to good computer software and hardware (Sullivan and Claycombe, 1977).

Box–Jenkins models exhibit forecasts that decline in accuracy over the forecast horizon. In the short run, their accuracy is reportedly good to excellent (Anderson and Weiss, 1984). In medium term, their accuracy is reportedly good to poor, and in the long term, their accuracy tends to be poor. The more data they have, the better their models. Scholars disagree over how many observations are necessary for ARIMA models. ARIMA models require more data than some prominent scholars have claimed. The data have to have been already detrended or detrendable by differencing. Although some scholars maintain that ARIMA models can be based on as few as 30 observations (Makridakis *et al.,* 1983), others claim that they require 50 to 100 equally spaced observations (Box and Jenkins, 1976; Box *et al.,* 1994; Granger, 1989). Seasonal models require more data than nonseasonal ones, and, with that data, may extend the accuracy of forecast further into the forecast horizon (Newbold and Granger, 1974). Box-Jenkins-Tiao intervention models require more data than nonintervention models, but can significantly improve the models when the data are plagued by singular or unusual events. To clarify the confusion and help resolve the controversy over this matter, Monnie McGee analyzes the sample size requirements of common time series models in Chapter 12.

Although intervention models perform well in the short and midrun and may improve upon ARIMA models over those horizons, they may fall behind simpler models for long-run forecasting under some circumstances. They have the capability to identify impacts as well as trend, seasonal, and cyclical patterns. They may require a few days to model—especially to diagnose and metadiagnose (Granger and Newbold, 1986; Makridakis et al., 1997). In general, it appears that the Box–Jenkins methods outperform both the stepwise autoregressive models in the short and early part of the medium range (Granger and Newbold, 1986; Makridakis et al., 1983, 1997). For longer forecast horizons, especially with trends in the data, the Holt models may provide more accurate forecasts than the ARIMA models (Fildes et al., 1998).

Transfer function models can have reasonably good predictive accuracy as far as predicting the continuation of the data-generating process. They may be used to test theoretical hypotheses, especially when extended to include transfer functions of multiple inputs. When these models employ leading indicators to forecast the turnaround of the economy, they may have less accuracy than others. In the near and medium terms, their accuracy is reportedly good to poor, but when they are used to forecast turning points their reliability becomes even more suspect. Whether the causes of business cycles stem from environmental problems or problematic economic conditions, the lengths of and variations in business cycles, estimated by some scholars to worsen the human condition with troughs of depression and misery (or peaks of inflation) lasting from 2 to 10 or more years, often renders such prediction a real challenge. That is to say, leading indicator models have had less than complete success in predicting turning points in the economy. Data requirements for these models include several years of data and in some cases a 5+ to 10-year historical data set (Sulliivan and Claycombe, 1977). With the linear transfer function modeling strategy, dynamic regression models with multiple response functions may be developed for multiple input series as long as the inputs remain relatively uncorrelated. Such models have substantial theoretical explanatory power.

Regression models forecast better than the other techniques over medium- and long-run forecasting horizons. Over the short run, they often have limited forecasting ability. They can have good to very good forecast accuracy over the longer range. They can be used to extract an average trend. As long as the trend is global rather than local, the regression on trend can prove useful. These trends may be linear or polynomial. Spline regressions can be used to deal with multiple local trends. With dummy variables the models can capture seasonality and regime shifts, and with trigonometric functions they can capture deterministic cyclicity.

Regression models may be used to forecast long-term trends alone or those coupled with cycles. The quality of prediction depends on the starting position of the data and whether there are enough data in the series from which to form a pattern from which a long-term prediction can be made. In general, long-term predictions are founded on a *ceteris paribus* assumption that usually does not hold over the very long run. Over the very long range, systemic regime or level shifts may come about that render the informational set from which the predictions are made inappropriate as a basis for such forecasts. Even in the better long-run regression forecasts, the farther into the future one predicts, the less certain one is of the outcome. A nonparametric method, called robust trend, which is based on median change, has also performed well in the M3-competition (Hibon, 1999). The growth of risk with expanding forecast error often makes such soothsaying questionable. Very long-term forecasts are extremely precarious at best.

Autoregressive models attempt to compensate for autocorrelation bias in the least squares trend extrapolation. They model trends while compensating for serial correlation bias in the error. They can handle multiple inputs easily and therefore possess an advantage in theory testing where parameter encompassing is important. More advanced programs such as SAS can even handle lagged endogenous variables in such models as well. These models have the added advantage of lending themselves to fully automatic variable selection, model construction, and model refinement. The process of determining the number of lags of the exogenous variable to be included in the model building may be performed with a minimum information criterion or a LaGrange multiplier test. In a stepwise autoregression model, successive inclusion of lagged terms can proceed until the fit no longer significantly improves (Payne, 1973), and these models may function well with data sets of at least 30 observations (Newbold and Granger, 1974; Granger and Newbold, 1986). A better automatic approach would be initial overparameterization of lagged terms followed by backward elimination of nonsignificant lags.

Other forms of autoregression include regression models with stochastic error volatility. There are many kinds of autoregressive conditionally heteroskedastic (ARCH) models, whose error variance may be modeled by an autoregressive function. A more general kind of ARCH model is the general autoregressive conditionally heteroskedastic (GARCH) model, whose error variance may be modeled by both autoregressive and moving average components. Both of these kinds of models are frequently used and show great promise for modeling the structure of risk in the field of computational finance.

11.4. COMPARISON OF INDIVIDUAL FORECAST METHODS

To compare the average forecast accuracy of different models, Makridakis in his M competition uses the mean absolute percentage error (MAPE), which is less sensitive to outlier distortion than the mean square forecast error (MSFE). He compares the moving average, single exponential smoothing, Holt's, Winter's, and Box–Jenkins methods for three different forecast horizons. Very rarely was the moving average method superior to any of the exponential smoothing methods (Hibon, 1984). Makridakis *et al.* (1983) speak of the average forecasting accuracy of the particular models in terms of four different forecast horizons. These horizons span less than a month, 1 to 3 months, 3 months to 2 years, and longer than 2 years, respectively. In the M-competitions, they arrived at some basic conclusions about comparative forecast capabilities of these approaches. Although the Box–Jenkins method is found to generally outperform the others in MAPE accuracy, Winter's exponential smoothing method is the second best in all three lengths of forecast horizons in the M competition. Single exponential smoothing appears to be next best in terms of MAPE accuracy, while Holt's method, the next best, generally outperforms the moving average method. If the data set is very small, however, single exponential smoothing may outperform either Box–Jenkins or stepwise autoregression models (Granger, 1989).

There are, however, qualifications to these conclusions. Sometimes the amount of random error in the series may determine which model forecasts more accurately. Sometimes the forecast horizon may determine which model is better. Makridakis and Hibon (1984) find that simpler (single or Holt–Winters) exponential smoothing models may occasionally outperform the Box–Jenkins models when there is more random error in the series. The single exponential smoothing does better with monthly and microlevel data, whereas the Holt and Winters methods do better with yearly data. The Holt method performed better on data that were already deseasonalized than the Winters method did on seasonal data, but the difference between these two methods under such circumstances is small (Hibon, 1984).

Another qualification is that some of the exponential smoothing models have ARIMA functional equivalents. Simple exponential smoothing forecasts are functionally equivalent to ARIMA(0,1,1) models. Holt's linear method is the same as an ARIMA(0,2,2) model. Moreover, the Holt–Winters additive model is the cognate to an ARIMA$(0,1,s+1)(0,1,0)_s$ model. The more randomness in the data, the more the Box–Jenkins models may overfit the data and the better single exponential smoothing performs compared to more complex methods. With such series, the farther the forecast horizon, the more possible it is that exponential smoothing models may outperform Box–Jenkins models (Hibon, 1984; Makridakis *et al.*, 1997). Although the Box–Jenkins method provides for a more comprehensive model, the simpler exponential smoothing methods may allow for simpler, easier, cheaper, and more automatic forecasts.

Granger (1989) writes that Box–Jenkins methods do better when focusing on one-step-ahead or short-range forecasts. Granger and Newbold (1986) have found that the Box–Jenkins models generally yield superior forecasts to the others. They find, more specifically, that Box–Jenkins methods generally outperform exponential smoothing and stepwise autoregression models in terms of forecast accuracy. The percentage of time that the Box–Jenkins forecasts remain superior to exponential smoothing forecasts declines as the forecast horizon is extended. The same can be said for the comparison of Box–Jenkins forecasts to those of stepwise autoregression. Exponential smoothing may outperform stepwise autoregression in dealing with local trend in the short run, but stepwise autoregression may be more accurate than exponential smoothing in the one-step-ahead forecast (Granger, 1989).

Granger infers some general rules concerning forecasting. The farther into the future the forecast is made, the less accurate it is. He also maintains that the larger the information set, the better the forecast as long as the extra information is relevant. The more data analysis that the forecaster does, the better in general is the forecast, he continues. He qualifies this conclusion by saying that sometimes smaller models do indeed produce better forecasts. Finally, he holds that when separate forecasts are combined, the combined forecast usually yields more accurate results (Granger, 1989).

11.5. COMPARISON OF COMBINED FORECAST MODELS

In general, methods of combining forecasts do better than suboptimal individual methods, as long as the one individual method does not encompass the others. Three common methods of combining forecasts were explained in Chapter 7: the simple average, the variance–covariance method of Bates and Granger (1969) and the regression (with intercept) method of combination of Granger and Ramanathan (1984). The question arises as to which combinations of these methods outperformed others in accuracy. Granger, seeking an answer to this question in the near term, compares combinations of models against individual models and shows that combinations of models usually outperform any individual model. Occasionally, the Box–Jenkins model will outperform combinations pitted against it.

Generally, the combination is more accurate when it contains the Box–Jenkins method (which are called Box–Jenkins combinations). The Box–Jenkins combinations are found to outperform all single methods more than 50% of the time. Stepwise autoregression combinations generally outperform exponential smoothing methods. Stepwise autoregression and exponential smoothing combinations do not outperform Box–Jenkins methods, however.

Among new regression methods for optimally combining forecasts are those that involve the use of regression with ARIMA errors, autoregression, or ARCH or GARCH models with time-varying combining weights. Diebold (1988) suggested that autoregression be employed to combine forecasts whose errors are serially correlated, to achieve more efficient estimation than mere OLS as a combining tool. Diebold (1998) recommended that regression combinations allow for lagged dependent variables and serially correlated errors to capture any of the dynamics not yet modeled in the combining formula. To do so, he suggested that the regression combination be modeled with $ARMA(p,q)$ errors. For a more efficient combination of such forecasts, time-varying combining weights have been advocated using ARCH models (Engle $et\ al.$, 1984) or switching or smoothing regression models (Deutsch $et\ al.$, 1984) to improve the combination and to reduce the error.

Figlewski (1998) has cautioned about the utility of the GARCH modeling. Although these models may be necessary for forecasting volatility, these models are more complex than others. They require large data sets. They depend on parameter constancy. They are better for one-step ahead forecasts. There is too much parameter uncertainty with multistep forecasts. They are difficult to fit, but their power and flexibility in modeling the variance has great utility in risk assessment and financial analysis, fascinating subjects to be examined in another text (Figlewski, 1997). Unless the forecasting of volatility is necessary, Figlewski suggests using simpler solutions.

REFERENCES

Anderson, O. D. (Ed.) *Forecasting: Proceedings of the Institute of Statisticians,* Annual Conference, Cambridge, 1976.

Anderson, A., and Weiss, A. (1984). "Forecasting: The Box–Jenkins Approach." In *The Forecasting Accuracy of Major Time Series Methods.* Makridakis, S., *et al.* (Eds.) Chichester: John Wiley and Sons, Ltd., p. 200.

Bates, J. M., and Granger, C. W. J. (1969). "The Combination of Forecasts," *Operation Research Quarterly,* **20,** 451–486.

Box, G. E. P., and Jenkins, G. M. (1976). *Time Series Analysis: Forecasting and Control.* 2nd ed. San Francisco: Holden-Day, p. 18.

Box, G. E. P., Jenkins, G. M., and Reinsel, G. C. (1994). *Time Series Analysis Forecasting and Control,* 3rd. Ed. Englewood Cliffs, NJ: Prentice Hall, p. 17.

Clements, M. P., and Hendry, D. F. (1994). "Toward a Theory of Economic Forecasting," In *Nonstationary Time Series Analysis and Cointegration* (Colin P. Hargreaves, Ed.). New York: Oxford University Press, pp. 9–51.

Deutsch, M., Granger, C. W. J., Teräsvirta, T. (1984). "The Combination of Forecasts Using Changing Weights," *International Journal of Forecasting,* **10,** 47–57.

Diebold, F. X. (1988). "Serial Correlation and the Combination of Forecasts," *Journal of Business and Economic Statistics* **8** (1), 105–111.

Diebold, F. X. (1998) *Elements of Forecasting.* Cincinnati, Ohio: Southwestern College Publishing, Chapter 12, pp. 339–372.

Engle, R., Granger, C. W. J., and Kraft, D. (1984). "Combining Competing Forecasts of Inflation using a Bivariate Arch Model," *Journal of Economic Dynamics and Control* **8,** 151–165.

Figlewski, S. (1997). Forecasting Volatility. *Financial Markets, Institutions and Instruments.* **6**(1), 2–88.

Figlewski, S. (1999, January). "Forecasting Volatility," A Paper presented at Sixth International Conference on Computational Finance. Stern School of Business, New York University, New York.

Fildes, R., Hibon, M., Makridakis, S., and Meade, N. (1998). "Generalizing about Univerariate Forecasting Methods: Further Empirical Evidence," *International Journal of Forecasting* **14,** 339–358.

Findley, D. F., *et al.* (1998). "New Capabilities and Methods of the X-12-ARIMA Seasonal Adjustment Program," *Journal of Business and Economic Statistics* **16** (2), 127–152.

Gilchrist, W. G. (1976). *Statistical Forecasting.* New York: John Wiley and Sons, p. 36.

Goodrich, R. (1999). Robert Goodrich is the author of FORECAST PRO. Belmont, MA, personal communication (June 26–29, 1999).

Granato, J. (1991). "An Agenda for Model Building," *Political Analysis,* **3,** 123–154.

Granger, C. W. J. (1989). *Forecasting in Business and Economics,* 2nd ed. San Diego: Academic Press, pp. 104, 192–193.

Granger, C. W. J. (1993). "Forecasting in Economics." In *Time Series Prediction: Forecasting the Future and Understanding the Past* (Weigand, A. S., and Gershenfeld, N., Eds.). Proceedings of the NATO Advanced Workshop on Comparative Time Series Analysis Vol. 15, Boston: Addison-Wesley, pp. 529–536.

Granger, C. W. J., and Newbold, P. (1974). "Spurious Regressions in Econometrics," *Journal of Econometrics* **2,** 111–120.

Granger, C. W. J., and Newbold, P. (1986). *Forecasting Economic Time Series,* 2nd ed. San Diego: Academic Press, pp. 179–186, 265–296.

Granger, C. W. J., and Ramanathan, R. (1984). "Improved Methods of Combining Forecasts." *Journal of Forecasting.* Vol 3., pp. 197–204.

Gujarati, D. N. (1995). *Basic Econometrics,* 3rd ed. New York: McGraw-Hill, pp. 454–494.

Harvey, D. I., Leybourne, S. J., and Newbold, P. (1998). "Tests for Forecast Encompassing," *Journal of Business and Economic Statistics* **16**(2), 254.

Hibon, M. (1984). "Naive, Moving Average, Exponential Smoothing, and Regression Methods" In *The Forecasting Accuracy of Major Time Series Methods* Makridakis, S., *et al.* Eds.) Chichester: John Wiley and Sons, Ltd., pp. 240–243.

Hibon, M., and Makridakis, S. (1999, June). M3-Competion. Paper presented at the 19th International Symposium on Forecasting, Washington, D.C.

Leamer, E. E. (1985). "Sensitivity Analysis Would Help." In *Modeling Economic Series* (Granger, C. W. J., Ed.). New York: Oxford University Press, pp. 88–96.

Maddala, G. S. (1992). *Introduction to Econometrics,* 2nd ed. New York: Macmillan, p. 582.

Makridakis, S., Wheelwright, S., and McGee, V. E. (1983). *Forecasting: Methods and Applications,* 2 ed. New York: John Wiley and Sons., pp. 47–567, 716–731, 772–778.

Makridakis, S., Wheelwright, S., and Hyndman, R. J. (1997). *Forecasting: Methods and Applications,* 3rd ed. New York: John Wiley and Sons, pp. 373, 520–542, 553–574.

Makridakis, S. (1984). "Forecasting: State of the Art." In Makridakis, S. (1984). *The Forecasting Accuracy of Major Time Series Methods.* Chichester: John Wiley, p. 5.

Makridakis, S., and Hibon, M. (1984). "The Accuracy of Forecast: An Empirical Investigation (with discussion)." In Makridakis, S. *et al.* (1984). *The Forecasting Accuracy of Major Time Series Methods.* Chichester: John Wiley, p. 43.

Montgomery, D. C., Johnson, L. A., and Gardiner, J. S. (1990). *Forecasting and Time Series Analysis,* 2nd ed. New York: McGraw-Hill, pp. 16, 288–290.

Newbold, P., and Granger, C. W. J. (1974). Experience with Forecasting Univariate Time Series and the Combination of Forecasts. *Journal of the Royal Statistical Society,* A 137, 131–146.

Pankratz, A. (1991). *Forecasting with Dynamic Regression Models.* New York: John Wiley, p. 174.

Parzen, E. (1982). "ARARMA Models for Time Series Analysis and Forecasting," *Journal of Forecasting* **1,** p. 68.

Payne, D. J. (1973) "The Determination of Regression Relationships Using Stepwise Regression Techniques." Ph.D. Thesis, Dept. of Mathematics, Univ. of Nottingham, cited in Granger, C. W. J., and Newbold, P. (1986). *Forecasting Economic Time Series.* San Diego: Academic Press, p. 183.

Reilly, D. P. (1999). David P. Reilly, author of AUTOBOX, has pointed out modeling problems stemming from local time trends, seasonal pulses, and level shifts in personal communication at the 19th International Symposium on Forecasting. Washington, D.C. June 26–30, 1999.

Soukamp, R. (1999). Presentation on new Regarima features of Census X-12 program at the 19th International Symposium on Forecasting. Washington, D.C. June 26–30, 1999.

Sullivan, W. G., and Claycombe, W. W. (1977). *Fundamentals of Forecasting.* Reston, VA: Reston Publishing Co., pp. 36, 37, 39, 222–223.

Chapter 12

Power Analysis and Sample Size Determination for Well-Known Time Series Models

Monnie McGee

Statistical consultants are often asked how many subjects are needed in order to have reliable and valid results. If the sample size for an experiment is too large, the results may show a statistically significant difference, even when no real difference exists in the population. Likewise, an investigator would not be able to detect a real difference with a sample size that is too small. Real differences are those that are truly present in the population from which the sample for the analysis is drawn. These problems pervade time series analysis, regardless of the statistical program being used to do the analysis.

In the Box–Jenkins time series context, a common null hypothesis is that the residuals of the process are *white noise,* although this is not always the case. Suppose the analyst incorrectly models a realization of a time series data-generating process. For example, suppose he models a series as an AR(1) process when it is really an AR(2) process. In that case, the

481

residuals of the series probably would not be white noise. A hypothesis test on the residuals would, one hopes, indicate that the model is incorrect. However, if the test is not statistically powerful enough, then it might falsely indicate that the residuals *are* white noise, when in fact they *are not*. Making such an error would affect any predictions calculated with the incorrect model.

In determining a sample size for an experiment, the statistician must consider several things: Type I error, Type II error, effect size, and power. Type I error (also known as α-level or level of significance) is the probability of incorrectly rejecting a correct null hypothesis. Type II error (often denoted by β) is the probability of incorrectly accepting a false null hypothesis. Statistical power is the converse of Type II error: the probability that a real difference will be detected in a sample given that such a difference really exists in the population. The ability of a test to detect an alternative also depends on the effect size, which is a measure of how grossly the alternative departs from the null. Large differences are easier to detect (and thus require smaller sample sizes) than small differences.

In this section, the sample size and power properties for Census X-11, Box–Jenkins (ARIMA) models, unit-root tests, intervention analysis, transfer functions, and regression with autocorrelated errors are discussed. Instead of focusing on the power of certain tests, we choose to focus on minimum sample sizes required to detect specific effect sizes while maintaining good power (>80%) at a given level of significance. The results in this chapter are meant to be applied to estimation of historical time series data used to build a model.

12.1. CENSUS X-11

The X-11 variant of the Census Method II of Seasonal Adjustment, commonly known as "Census X-11," is a computationally intensive procedure for removing seasonal components from time series data (see Chapter 2). X-11 works by passing each realization of a time series through many filters in an iterative fashion, in order to separate the series into seasonal, trend, and irregular components. The filtered data are then tested after each pass through a filter to ascertain whether or not the resulting irregular component is white noise. Sometimes, tests are also done to make sure that the data have not been overdifferenced. Testing the irregular component is most often done via a portmanteau test, a modified portmanteau, or a unit-root test (Scott, 1992, and Dagum, 1981). These tests can have low power for certain alternatives. More specific results are given in later sections.

Not much has been said about the optimal length of a series for X-11. In the journal articles surveyed, the series were anywhere from 5 to 20 years of monthly data in length, with a majority of the series having at least 12 years of data. Most of the data used to evaluate the performance of X-11 versus other methods of analyzing seasonal components were chosen for their "empirical and practical" interest (Kenny and Durbin, 1972; Abraham and Chatterjee, 1983; McKenzie and Stith, 1981). Wallis (1982) discusses types of data that are well modeled by X-11, but he does not discuss sample size.

12.2. BOX–JENKINS MODELS

The Box–Jenkins method of modeling and forecasting, discussed in Chapters 3 through 7, has four main steps. These are model identification, parameter estimation, diagnostic checking, and comparison of alternative models. It is the third stage where most of the hypothesis testing takes place. Once a candidate model has been identified and its parameters have been estimated, it remains to be determined whether or not that model fits the data (*i.e.,* whether or not the model residuals form a white noise process).

In order to examine the hypothesis that the residuals are white noise rather than an unspecified alternative, Box and Pierce (1970) developed the *portmanteau* test, also called the Box–Pierce test. This test involves a sum of a certain number of autocorrelations from the series, and this number is denoted by M. Several papers have explored the statistical properties of the portmanteau test, and the consensus is that the test has poor statistical properties. For example, Davies *et al.* (1977) found that the size (of the Type I error) of the Box–Pierce test was much less than it should be for AR(1) series with 50, 100, and 200 observations. In a series with 50 observations, the empirical Type I error rate was 0.013 when the theoretical rate was 0.05. This means that the test rejects the null hypothesis of white noise much more often than it should, which means that the analyst is told to search for another model when the one under consideration truly does have white noise residuals. Surprisingly, only for AR(1) data with 500 observations did the probability of type I error approach 0.05. These simulations were done for $M = 20$.

In order to address the issue of the poor properties of the Box–Pierce test, the *modified portmanteau* (Ljung–Box) test was developed (Ljung and Box, 1978). Davies and Newbold (1979) found that it also has poor properties. They ran simulations with 24 ARMA models with AR orders of 1 or 4 and MA orders of 2. The simulations calculated power and one-step-ahead forecast misspecification error for sample sizes $T = 50, 100,$ and

Table 12.1

Sample Sizes Needed to Obtain Power of 80% at 5% Significance Level

Alternative	$M = 5$		$M = 10$		$M = 15$		$M = 20$		$M = 30$	
	P	MP	P	MP	P	MP	P	MP	P	MP
Undermodeling	200	50	150	100	200	100	250	200	250	250
Overmodeling	>250	>250	>250	>250	>250	>250	>250	>250	>250	>250
Underdifferencing	50	150	150	200	150	250	150	250	250	250
Overdifferencing	250	250	250	250	250	250	250	250	250	250
Overestimation of seasonality	100	50	100	50	250	150	250	150	250	200

M = number of autocorrelations tested
P = Portmanteau test; MP = modified Portmanteau test

200, using significance levels of 0.05 and 0.10. In all cases, M was set to 20. For $T = 50$, most of the powers were below 50% (and often below 25%), even in cases when the forecast misspecification error was large. Low power implies that the test would fail to reject even when the residuals are clearly not white noise, thus leading the analyst to use an incorrect model for forecasting. For $T = 200$, the simulated powers were greater than 80% in about half of the 24 series. The series for which the test had the most power were those with strong correlation in the AR part of the model.

These results, although informative, are also incomplete. Table 12.1 gives various results of sample sizes needed to achieve at least 80% power at a significance level of 0.05 for the portmanteau (P) and modified portmanteau (MP) tests. Simulations were done for various values of M, and for several alternative hypotheses. For M equal to 5, 10, 15, 20, 30, 50, and 70, series of various lengths were generated for each alternative hypothesis. Their residuals were tested using both P and MP tests, and the p-value for each test was recorded. This procedure was repeated 1000 times for each of the scenarios. The resulting power was determined by dividing the number of times that the test correctly rejected a false null hypothesis by 1000. All simulations were performed on a Gateway E-3000 computer using S-plus for Windows 95. Sample size requirements for Box–Jenkins models will be examined under the following alternatives; incorrect order, underdifferencing, overdifferencing, presence of seasonality, and nonstationarity. Although the portmanteau test is designed to be a test against an unspecified alternative, it does better against some alternatives than others. Each of the results is discussed in turn.

Undermodeling: This represents a special case of incorrect order when a series is modeled by another series with a smaller order than is correct.

Specifically, an AR(2) series with coefficients 0.85 and -0.5 was simulated, and then modeled by an AR(1) series with coefficient 0.85. By the time the sample size reached 100, both tests were rejecting approximately 98% of the time; however, the type I error for the Portmanteau test was too small. At $M = 30$ and $T = 250$, the probability of type I error approached 0.10 for both tests. The MP test generally outperformed the P test.

Overmodeling: This is the reverse case of under-modeling. An AR(1) series with coefficient 0.85 was simulated and modeled by an AR(2) series with coefficients 0.85 and -0.1. The performance of both P and MP tests was abysmal for this scenario. Even with 250 data points, the power did not reach 1%. With 750 observations, powers were typically from 10 to 15%.

Overdifferencing: This represents the scenario when a series with a linear trend is modeled by one with a quadratic trend. In this case, an ARIMA(1,1,0) series was modeled by an ARIMA(1,2,0) series. For both series, the AR coefficient was 0.85. Even with 500 observations, the power of either test to detect overdifferencing did not exceed 62%. In addition, for the modified portmanteau test, the α-level of the test consistently exceeded 0.06.

Underdifferencing: This represents the opposite case of overdifferencing. An ARIMA(1,2,0) series was modeled by an ARIMA(1,1,0) series. Both tests are able to detect this type of misspecification very well, even with as few as 50 observations. Unfortunately, S-plus had a great deal of trouble with this simulation because of the nonlinearities present in underdifferenced data. The simulations were repeated only 100 times in this case because the software often balked after the 99th iteration. Note that P outperforms MP.

Overestimation of seasonality: This scenario is the case when too many seasonal components are extracted from a series. Both tests seem to have trouble detecting a series that has been modeled with an incorrect seasonal component, although the problem is not as severe as it is in the case of overdifferencing. For these situations, an ARIMA$((0,1,1)(0,1,1)_{12})$ was modeled with an ARIMA $((0,1,1)(0,1,1)_4)$ model. It was not possible to run simulations for the case of underestimation because of the nonlinearities present in models with unextracted seasonal components.

Recall from the previous section that Census X-11 uses both the P and MP tests to examine the irregular components of models produced through its series of filters on seasonal data. Because of the poor performance of these tests for small sample sizes, one must be sure to have adequate amounts of data when using X-11. In addition, X-11 uses tests for nonstationary, which we will now discuss.

12.3. TESTS FOR NONSTATIONARITY

One important problem in econometrics is the question of the nonstationarity of certain important series. This question underlies controversies over perfect market theory, permanent income theory of consumption, and real business cycles, for example (De Jong *et al.*, 1992). Recall that in order for a series to be considered stationary, and therefore be a candidate for Box–Jenkins modeling, it must not have any parameter estimates inside the unit circle (parameter estimates with complex roots), and it must oscillate around a constant mean with a constant variance. Several tests have been developed to test for changes in the mean (structural shifts) or zeros inside the unit circle (unit roots). The Chow *F*-test, the Dickey–Fuller (DF) test, and the augmented DF test are tests for nonstationarity of various types. These tests were introduced in Chapter 3. In this section, we will discuss their statistical properties.

Chow (1960) developed a test for structural shifts under the assumption that the series has the same variance both before and after the shift. This test is called the Chow *F*-test, and it was discussed earlier in chapter three. However, structural shifts often are accompanied by changes in variance as well. If the variance changes after the shift, the Chow *F*-test will not give reliable results (Toyoda, 1974; Schmidt and Sickles, 1977). Various attempts have been made to develop a test in which one can have different variances. Gupta (1982) and Zellner (1962) introduced likelihood ratio tests to deal with the problem of heteroskedasticity, while Jayatissa (1977) introduced a large sample test for this phenomenon. Simulation studies for the performance of the Chow F-test, along with various modifications of it, are given in Ali and Silver (1985). The authors find that the actual size of the Chow *F*-test is often very much above or below the theoretical size; therefore, they do not perform a power analysis on it. However, they present a modification to the Chow *F*-test that has a maximum of 70% power at a 5% significance level when the sample size is 100. This power holds in the heteroskedastic scenario when the variance after the shift represents 20% of the sum of the variances before and after the shift.

There is an ongoing study of the performance of the unit root tests in the econometric literature. De Jong *et al.* (1992) examine the performance of these tests under the assumption that the time series data have autocorrelated errors. These errors can have either an AR(1) or MA(1) structure (the authors do not examine the possibility of an ARMA structure). For the DF tests, in the presence of AR and MA errors with moderate correlation, the power against trend stationary alternatives is close to 0. In addition, the size of the test is distorted by as much as 30% compared to the size of the test under white noise errors. All simulations were done for a series with 100 observations at a significance level of 0.05. The authors note that increasing sample size by increasing sampling frequency (*e.g.,* using monthly instead of quarterly data for the same series) does not increase power. No mention is made of the effect on power if one increases sample size while maintaining the same sampling frequency.

Although its performance is anything but perfect, the augmented Dickey-Fuller (ADF) test has the best power to detect a unit root in the presence of positively autocorrelated errors. For roots close to the unit circle and positive AR (or MA) coefficients in the error term, the minimum power is 0.06 (0.19 for MA errors). When the coefficient in the error term is negative, the ADF performs worse than the Phillips-Perron (PP) test, especially as the roots approach the unit circle. De Jong *et al.* (1992) recommend the ADF test as the best overall test for a unit root in the presence of autocorrelated errors, mainly because it does not suffer size distortions under overparameterization, extreme autocorrelation, and increased sampling frequency. Similar findings are reported in Ghysels and Perron (1993) and Nabeya and Tanaka (1990).

12.4. INTERVENTION ANALYSIS AND TRANSFER FUNCTIONS

Box–Tiao intervention analysis (Box and Tiao, 1975) is explained in Chapter 8. Outliers (or interventions) come in two flavors: additive and innovative. Additive interventions are those that affect only one observation. Innovative outliers represent shocks at a certain time that have lingering effects on the data at subsequent time points. Chang *et al.* (1988) give results of a simulation of the power of the intervention analysis model to detect both additive and innovative (and mixed) disturbances in an AR(1) model with coefficient 0.6 and variance 1, and in an MA(1) model with coefficient 0.6 and variance 1. Two different sizes of outliers, 3.5 and 5 standard deviations from the mean (called moderate and large, respectively), are used in the simulation with sample sizes of 50, 100, and 150. It is assumed that the disturbance occurs in the middle of each series. Another simulation is performed to detect two disturbances with much the same specifications, assuming that the shocks occur at the one-third and two-thirds points from the beginning of the series. Simulations for the MA(1) model are not performed in the two-outlier case.

In general, the power of the procedure in the one-intervention case is quite good. One can detect a moderate shock ($>\bar{x} \pm 3.5$ SD) correctly at least 80% of the time when there are 100 observations. A large shock can be detected with only 50 observations. For two moderate outliers, the procedure is very likely to miss at least one of them even if the sample size is 150. The power is much better for large interventions, requiring only 50 observations when the outliers are either both innovative or mixed; however, 100 observations are required when both outliers are additive.

Transfer function models are an extension of intervention models. Transfer functions, as discussed in Chapter 9, are meant to describe the relationship between two series when one series drives the other. In other words, a data value of a driving series (called the input series) 1 month ago influences a value of the second series (the output series) today. One models a transfer function pair by modeling the input series and using an inverse of that model as a filter for the series (a process called prewhitening). Then, one obtains the cross-correlation function (CCF) for the prewhitened series. Significant spikes (spikes that are greater than 1.96 times the standard deviation of the series) in the cross-correlation function appear at certain lags, indicating how much the output series is lagged behind the input series. For example, if a significant spike appears at a lag of 2, this indicates that the output series is driven by the value of the input series two lags ago. Once the series is modeled, one can test the residuals for white noise with one of the tests mentioned above However, those tests appear to have rather poor properties for even moderate sample sizes.

Since the cross-correlation function is such an integral part of modeling a transfer function pair, one might ask how well it is able to detect the correct number of lags for the output series. To test its detection ability, we simulated several transfer function pairs separated by a known number of lags and ran each simulation 1000 times in order to count the number of times a significant spike occurred at the correct lag. Several transfer function models with various inputs were used, and their equations are given next. These models were taken from a table on page 349 of Box and Jenkins (1976). In this notation $\nabla Y_t = Y_t - Y_{t-1}$.

$$Y_t = aX_{t-3} \tag{12.1}$$
$$(1 + a\nabla)Y_t = X_{t-3} \tag{12.2}$$
$$Y_t = (1 - a\nabla)X_{t-3} \tag{12.3}$$
$$(1 - 0.25\nabla + 0.5\nabla^2)Y_t = X_{t-3} \tag{12.4}$$
$$Y_t = (1 - 0.25\nabla + 0.5\nabla^2)X_{t-3} \tag{12.5}$$
$$(1 - a\nabla)Y_t = (1 - 0.5\nabla)X_{t-3} \tag{12.6}$$
$$(1 - 0.5\nabla)Y_t = (1 - a\nabla)X_{t-3} \tag{12.7}$$

These equations represent, respectively, the following types of transfer functions: pulse, AR(1), MA(1), AR(2), MA(2), ARMA(1,1) with the variable AR coefficient, and ARMA(1,1) with a changing MA coefficient. Input series are denoted by X_t, with $t = 1, \ldots, T,$ and output series are given by

Y_t, with $t = 1, \ldots, T$. Different coefficients for each AR and MA model were used in order to test the influence of the strength of the correlation on the ability of the cross-correlation function to detect it. These coefficients are denoted by a in the above equations, where $a = 0.2, 0.4, 0.6, 0.8$, and 1.0. The input series used ranged from simple pulse inputs to white noise to AR(1) models, with coefficients 0.1, 0.5, and 0.9.

The results are fairly simple to interpret. The more complicated the model and the smaller the AR (or MA, or both) coefficients, the more difficult it is to obtain a significant spike at the appropriate lag (lag 3, in this case). When the input series was a pulse or white noise, the CCF had a significant spike at lag 3, no matter what the transfer function. For the pulse transfer function model, the program detected a spike at the third lag 100% of the time, regardless of the structure of the input series, the strength of the correlation within the input series, or the value of a. This occurred at sample sizes as low as 30.

For AR(1) input series with AR(1) and MA(1) transfer functions, the CCF was significant at the correct lag 100% of the time with a sample size of 50. Interestingly, at this sample size the CCF also had significant spikes at lag 4 more than 88% of the time when a was 0.8 or 1.0 and the AR(1) input series coefficient was 0.1, and more than 98% of the time for all AR(1) input series coefficients of 0.5 and 0.9 at all values of a. With an MA(1) transfer function, sample sizes of 50 were sufficient to give the correct results 100% of the time, except for the case when the AR(1) input coefficient was 0.1 and $a = 0.6$ or 0.8. (When the AR(1) input coefficient was 0.5 and a was 0.8, a series with 50 observations had a significant CCF at the third lag 86% of the time.) With 100 observations, the model with $a = 0.6$ was correct 100% of the time, and the model with $a = 0.8$ was correct 89% of the time. Although there were some instances in which the CCF was not significant at lag 3 (the correct lag) with $T = 50$, in all cases the CCF was significant at lag 4 more than 98% of the time for the MA transfer function.

For the second-order AR transfer function model, the CCF was significant no more than 65% of the time at the third lag for any combination of coefficients. Sample sizes tested were 30, 50, and 100. Significant spikes at lags 4 and 5 of the CCF occurred approximately the same percentage of the time as did the spikes at the correct lag. For the MA(2) transfer function model, the third lag was significant more than 98% of the time even at a sample size of 30. Significant spike in the CCF at lags 4 and 5 were also detected over 90% of the time when $T = 100$. Performance of the ARMA(1,1) models with a varying AR(1) coefficients are reliable for all combinations of coefficients, even at a sample size of 30. When we vary the MA(1) coefficient, the CCF has significant values more than 80% of the time at lag 3 for $T = 30$ only for $a = 0.2$, and 0.4. The sample size must be larger than 50 for there to be more than 80% accuracy for $a = 0.6$. Finally, the sample size should be 100 for the same amount of accuracy to be achieved when $a = 0.8$ or 1.0. These numbers are true regardless of the value of the AR(1) input series coefficient. Spurious significant spikes appear at lag 4 100% of the time for both models when there are as few as 30 observations.

From these results, it seems that the complexity of the model plays a large role in the accuracy of the CCF at a given lag. The coefficients of the transfer function also determine how often the CCF has a significant spike at the correct lag. The coefficients of the input function, at least in the AR(1) case, do not seem to matter as much. A more interesting result is that significant spikes appear at incorrect lags nearly as often as significant values occur at the lag of interest.

12.5. REGRESSION WITH AUTOREGRESSIVE ERRORS

Sometimes there are certain variables that are believed to influence the path of a time series. The time series data can then be modeled in a regression context with several independent variables representing exogenous variables in a hypothesized causal model. This method of modeling was discussed in Chapter 10. The residuals of this type of regression are usually correlated with time in some manner, and this autocorrelation can affect the precision of any least squares estimates of the parameters of the regression model. There are several widely used procedures for time series regression modeling. These are Cochrane–Orcutt, Hildreth–Lu, and Prais–Winsten. All of these involve a transformation of the variables and residuals estimated from the data in order to obtain noncorrelated residuals. Specific forms of the transformation for each procedure have been discussed in Chapter 10.

Taylor (1981) discusses the efficiency of the Cochrane–Orcutt (CO) estimator relative to ordinary least squares (OLS) in time series regression models. One estimator is more efficient than another one if fewer observations are needed to achieve the same accuracy for the estimation of the parameters. His conclusion is that CO is more efficient than OLS except when the realization of the exogenous variable process is strongly trended. He also concludes that a generalized least squares (GLS) estimator beats both CO and OLS when the coefficients of the exogenous variable process are known. It is generally accepted that all iterative methods outperform one-stage or two-stage estimators.

Rao and Griliches (1969) discuss the properties of several estimators for small samples. These authors calculate small sample efficiency for GLS, OLS, CO, Durbin (D), Prais-Winsten (PW), and a nonlinear estimator (NL). Their model is a regression with a single exogenous variable and an AR(1) error process. All calculations are done for data sets with 100 observations. The authors first begin with the estimation of the coefficient of the AR(1) error process, since this is calculated in different ways for each of the estimators. They find that no estimator has uniformly smaller bias than the others, as far as the estimation of this coefficient is concerned. All results are given relative to the GLS estimator. Note that GLS is unattainable if the coefficients of the error process are unknown. However, when the coefficients of the error processes are known, it is the most efficient estimator. The simulation results show that no estimator attains the efficiency of the GLS estimator, and that the OLS estimator performs the worst, with only 15% efficiency for strongly correlated errors. As for the other estimators, none consistently outperforms the others, although they all are better than NL.

Magee (1985) examines the problem analytically. He compares the iterated PW, the two-stage PW, and maximum likelihood estimates of a regression model with AR(1) disturbances. The three estimators are found to be equivalent in terms of the MSE of the estimates.

12.6. CONCLUSION

In their 1976 book, Box and Jenkins stated that 50 to 100 observations were necessary to ensure adequate power for model testing (Box and Jenkins, 1976). This viewpoint has been supported in other time series texts (Cook and Campbell, 1979; McCain and McCleary, 1979; and McCleary *et al.*, 1980). However, several papers and our simulations have shown that the minimum number of observations is more likely to be between 100 and 250. This observation is certainly true for tests of model fit, such as the portmanteau tests and unit root tests. The discrepancy between the minimum number of observations for valid hypothesis tests in this section and that of the common belief indicates that much of the analysis based on such short series may be in need of reestimation. Clearly, more research in this field is necessary, especially since previous authors tended to fix the sample size and examine the power, rather than fix the power and examine the sample size. With contemporary computing speed, various sample sizes can be used to find how many observations are necessary to achieve acceptable power at an acceptable significance level.

REFERENCES

Abraham, B., and Chatterjee, C. (1983). "Seasonal Adjustment with X-11-ARIMA and Forecast Efficiency." In *Time Series Analysis: Theory and Practice 4*, (O. D. Anderson, Ed.). Elsevier Science, pp. 13–22.

Ali, M. M., and Silver, J. L. (1985). "Tests for Equality Between Sets of Coefficients in two Linear Regressions Under Heteroskedasticity," *Journal of the American Statistical Association* **80**, 730–735.

Box, G. E. P., and Jenkins, G. M. (1976). *Time Series Analysis Forecasting and Control*. Oakland, CA: Holden Day, p. 18.

Box, G. E. P., and Pierce, D. A. (1970). "Distribution of Residual Autocorrelations in Autoregressive-Integrated Moving Average Time Series Models," *Journal of the American Statistical Association* **65**, pp. 332, 1509–1526.

Box, G. E. P., and Tiao, G. (1975). "Intervention Analysis with Applications to Economic and Environmental Problems," *Journal of the American Statistical Association* **70**, 70–79.

Chang, I., Tiao, G., and Chen, C. (1988). "Estimation of Time Series Parameters in the Presence of Outliers," *Technometrics* **30**, 193–204.

Chow, G. C. (1960). "Tests of Equality between Sets of Coefficients in Two Linear Regressions," *Econometrica* **28**, 591–605.

Cook, T. D., and Campbell, D. T. (1979). *Quasi-Experimentation: Design and Analysis for Field Settings*. Boston: Houghton Mifflin & Co., p. 228.

Dagum, E. B. (1981). "Diagnostic Checks for the ARIMA Models of the X-11-ARIMA Seasonal Adjustment Method." In *Time Series Analysis*, (O. D. Anderson and M. R. Perryman, Eds.). North-Holland, pp. 133–145.

Davies, N., and Newbold, P. (1979). "Some Power Studies of a Portmanteau Test of Time Series Model Specification," *Biometrika* **66**(1), 153–155.

Davies, N., Triggs, C. M., and Newbold, P. (1977). "Significance Levels of the Box–Pierce Portmanteau Statistics in Finite Samples," *Biometrika* **64**(3), 517–522.

De Jong, D. N., Nankervis, J. C., Savin, N. E., and Whiteman, C. H. (1992). "The Power Problems of Unit Root Tests in Time Series with Autoregressive Errors," *Journal of Econometrics* **53**, 323–343.

Gupta, S. A. (1982). "Structural Shift: A Comparative Study of Alternative Tests." In *Proceedings of the Business and Economic Statistics Section, American Statistical Association*, pp. 328–331.

Jayatissa, W. A. (1977). "Tests of Equality Between Sets of Coefficients in Two Linear Regressions when Disturbances are Unequal," *Econometrica* **45**, 1291–1292.

Kenny, P. B., and Durbin, J. (1982). "Local Trend Estimation and Seasonal Adjustment of Economic and Social Time Series with Discussion," *Journal of the Royal Statistical Society, Series A* **145**, 1–41.

Ljung, G. M., and Box, G. E. P. (1978). "On a Measure of Lack of fit in Time Series Models," *Biometrika* **65**(2), 297–303.

Magee, L. (1985). "Efficiency of Iterative Estimators in the Regression Model with AR(1) Disturbances," *Journal of Econometrics* **29**, 275–287.

McCain, L. J., and McCleary, R. (1979). "The Statistical Analysis of the Simple Interrupted Time Series Quasi-Experiment." In *Quasi-Experimentation* by Cook and Campbell, p. 235n.

McCleary, R., and Hay, R. with Merdinger, E. E. and McDowell, D. (1980). *Applied Time Series Analysis for the Social Sciences*. Sage Publications.

McKenzie, S., and Stith, J. (1981). "A Preliminary Comparison of Several Seasonal Adjustment

Techniques." In *Time Series Analysis* (O. D. Anderson and M. R. Perryman, Eds.). North-Holland, pp. 327–343.

Nabeya, S., and Tanaka, K. (1990). "Limiting Power of Unit-Root Tests in Time-Series Regression." *Journal of Econometrics* **46,** 247–271.

Rao, P., and Griliches, Z. (1969). "Small-Sample Properties of Several Two-Stage Regression Methods in the Context of Auto-Correlated Errors," *Journal of the American Statistical Association* **64,** 253–272.

Saikkonen, P. (1989). "Asymptotic Relative Efficiency of the Classical Test Statistics Under Misspecification," *Journal of Econometrics* **42,** 351–369.

Schmidt, P., and Sickles, R. (1977). "Some Further Evidence on the Use of the Chow Test under Heteroscedasicity," *Econometrica* **45,** 1293–1298.

Scott, S. (1992). "An Extended Review of the X11-ARIMA Seasonal Adjustment Package," *International Journal of Forecasting* **8,** 627–633.

Taylor, W. (1981). "On the Efficiency of the Cochrane–Orcutt Estimator," *Journal of Econometrics* **17,** 67–82.

Toyoda, T. (1974). "Use of the Chow Test under Heteroskedasticity," *Econometrica* **42,** 601–608.

Wallis, K. F. (1982). "Seasonal Adjustment and Revision in Current Data: Linear Filters for the X-11 Method," *Journal of the Royal Statistical Society A* **145,** 74–85.

Whittle, P. (1952). "Tests of Fit in Time Series," *Biometrika* **39,** 309–318.

Zellner, A. (1962). "An Efficient Method of Estimating Seemingly Unrelated Regressions and Tests for Aggregation Bias," *Journal of the American Statistical Association* **57,** 348–368.

Appendix A

Empirical Cumulative Distribution of $\hat{\tau}$ for $\rho = 1$

Sample Size n	Probability of Smaller Value								
	0.01	0.03	0.05	0.10	0.50	0.90	0.95	0.98	0.99
				$\hat{\tau}$ with no constant					
25	−2.65	−2.26	−1.95	−1.60	−0.47	0.92	1.33	1.70	2.15
50	−2.62	−2.25	−1.95	−1.61	−0.49	0.91	1.31	1.66	2.08
100	−2.60	−2.24	−1.95	−1.61	−0.50	0.90	1.29	1.64	2.04
250	−2.58	−2.24	−1.95	−1.62	−0.50	0.89	1.28	1.63	2.02
500	−2.58	−2.23	−1.95	−1.62	−0.50	0.89	1.28	1.62	2.01
∞	−2.58	−2.23	−1.95	−1.62	−0.51	0.89	1.28	1.62	2.01
				$\hat{\tau}$ with constant					
25	−3.75	−3.33	−2.99	−2.64	−1.53	−0.37	0.00	0.34	0.71
50	−3.59	−3.23	−2.93	−2.60	−1.55	−0.41	−0.04	0.28	0.66
100	−3.50	−3.17	−2.90	−2.59	−1.56	−0.42	−0.06	0.26	0.63
250	−3.45	−3.14	−2.88	−2.58	−1.56	−0.42	−0.07	0.24	0.62
500	−3.44	−3.13	−2.87	−2.57	−1.57	−0.44	−0.07	0.24	0.61
∞	−3.42	−3.12	−2.86	−2.57	−1.57	−0.44	−0.08	0.23	0.60
				$\hat{\tau}$ with constant and trend					
25	−4.38	−3.95	−3.60	−3.24	−2.14	−1.14	−0.81	−0.50	−0.15
50	−4.16	−3.80	−3.50	−3.18	−2.16	−1.19	−0.87	−0.58	−0.24
100	−4.05	−3.73	−3.45	−3.15	−2.17	−1.22	−0.90	−0.62	−0.28
250	−3.98	−3.69	−3.42	−3.13	−2.18	−1.23	−0.92	−0.64	−0.31
500	−3.97	−3.67	−3.42	−3.13	−2.18	−1.24	−0.93	−0.65	−0.32
∞	−3.96	−3.67	−3.41	−3.13	−2.18	−1.25	−0.94	−0.66	−0.32

The table was constructed by David A. Dickey with Monte Carlo Methods. It is reprinted with permission of Wayne Fuller and John Wiley and Sons, Inc.
Source: Wayne A. Fuller, *Introduction to Statistical Time Series,* 2nd ed. New York: John Wiley & Sons, Inc., 642.

Glossary

ACF (Autocorrelation function) The autocorrelation structure of a series over time, where time is measured in lags 0 through p, where p is the highest time lag.

Additive model A model whose terms are added together; these models have neither multiplicative factors nor interaction product terms.

Additive outlier See outlier.

ADF See augmented Dickey-Fuller test.

Adjusted R square Shrunken R^2. When variables are added to a model, they tend to inflate the R^2. The adjusted R^2 is shrunken by a degree of freedom correction that compensates for the variable inflation.

ADL model See autoregressive distributed lag model.

AIC (Akaike information criterion) Akaike's goodness of fit measure. This criterion is equal to minus 2 times the log likelihood function plus 2 times the number of free parameters in the model.

AICC (Akaike information criterion corrected) Used in Census X-12, the formula for this informa-

tion criterion is $= -2\ Ln(L) + 2m(T/(T - m - 1))$, where $L =$ estimated likelihood function, $m =$ number of model parameters, and $T =$ sample size.

A posteriori **analysis** An analysis conducted after the forecast has been analyzed.

A priori **analysis** An analysis conducted prior to the forecasting and its assessment.

ARCH model Introduced by Engle in 1982, these models are models where the error variance is conditional on past squared disturbances. There are many variations of ARCH models. See also GARCH.

AREG An SPSS procedure that performs first-order autoregression error correction analysis.

ARIMA An autoregressive integrated moving average analysis developed by George Box and Gwilym Jenkins. In this analysis, a series is transformed to a condition of covariance stationary, and then it is identified, estimated, diagnosed, possibly metadiagnosed, and forecast. The ARIMA model is represented by

ARIMA(p, d, q). In this notation, the parameters inside the parentheses represent the order of (p) autoregression, (d) for differencing, and (q) for moving average in the model.

ARIMA Procedure SAS and SPSS programs to perform ARIMA analysis.

ARMA model A model of a series that contains both autoregressive and moving average components of a stationary series that needs no differencing.

Asymptotically unbiased estimation Parameter estimation whose bias approaches zero as the sample size approaches infinity.

Augmented Dickey–Fuller (ADF) test A Dickey–Fuller test for unit roots that removed serial correlation in the series by adding autoregressive parameters to control for it.

Autocorrelation serial correlation The correlation of observations at particular temporal distances from one another within the same series.

Autocorrelated errors Autocorrelation of the errors, innovations, or shocks of a model.

Autocorrelation function (ACF) See ACF.

AUTOREG An SAS autoregression procedure that estimates autoregressive, ARCH, and GARCH models.

Autoregression A regression of a series on past lags of itself.

Autoregressive conditional heteroskedasticity See ARCH model.

Autoregressive distributed lag (ADL) model A transfer function model. A type of dynamic regression model that includes a ratio of two polynomials multiplied by distributed lags of one or more exogenous variable(s).

Autoregressive error model A regression model with serial correlation of error.

Backcasting Forecasting the initial values or preliminary values of a series from the remainder of the series. Sometimes referred to as backforecasting.

Backshift operator The lag operator, L, which invokes a backstep in time on a series. $(L)Y_t = Y_{t-1}$; $(L^2)Y_t = Y_{t-2}$. Sometimes a B is used instead of L.

Bias Difference between the expectation of a statistic and the true population parameter.

BIC See *SBC*.

Bounds of invertibility The limits within which the value of a moving average parameter may vary if an autoregressive representation of it is to be able to converge.

Bounds of stability The limits within which the value of the decay parameter in a transfer function may vary so that the transfer function process will not become chaotic.

Bounds of stationarity The limits within which an autoregressive parameter may vary for the model to be able to converge.

Box–Jenkins analysis See ARIMA. A methodology developed in the 1930s of combining autoregression and moving average models after they have been transformed and differenced to attain a condition of covariance stationarity, which permits the estimations to converge. In the 1970s, G.E.P. Box and Gwilym Jenkins applied this methodology to time series analysis, forecasting, and control of processes.

Box-Jenkins-Tiao methodology A response function analysis of the impact on a series of an external event or intervention, expounded by Box and Tiao in 1975.

Box-Ljung Q statistic A test of significance of autocorrelated errors found in an ACF or PACF. A modification of the Box–Pierce Q statistic with a degree of freedom correction to enhance accuracy in correlogram analysis in smaller samples.

Box-Pierce Q statistic A test of significance of autocorrelated errors found in an ACF or PACF. The Q statistic is the sample size times the sum of the autocorrelations. It is distributed as a chi-square with m degrees of freedom. Sometimes called a portmanteau test.

Breusch–Godfrey test A large sample test for higher order autocorrelation of error, involving a regression of the error term from a regression on the regressors plus lags of the error term from the first regression. A chi-square test of significance indicates presence or absence of higher order autocorrelation.

Business cycle Periodic variation in a series indicating some aspect of business activity. Business cycles traditionally refer to periods of downswings, depressions, upswings, and prosperity. Business cycle theorists seek to predict turning points and to determine the troughs and peaks of the cycle.

Causal modeling With a presumption of closure of a system, the events of series X_{it} are unidirectionally associated with and followed by those of series Y_t, where i = the number of the exogenous variable and t = time period. Intervention and transfer function models are examples of such causal modeling.

Census I See classical decomposition. This decomposition is performed by the SPSS Season procedure.

Census II An upgrading of classical decomposition that was incorporated into Census X-11.

Census X-11 See X-11.

Census X-12 See X-12.

Chow test A test for the constant variance in a series.

Classical decomposition A procedure that extracts from a series trend, cycle, seasonal, and irregular components. The procedure can be additive or multiplicative, so that the components are added together or multiplied together to reconstitute the original series. This decomposition is performed by the SPSS SEASON procedure.

Cochrane–Orcutt algorithm An algorithm for first-order correction of inefficiency caused by serially correlated error in a regression model. The algorithm involves transforming the variables by multiplying them by the factor $(1 - \rho_1)$, where ρ_1 = first-order autocorrelation of errors. This algorithm is performed by SAS AUTOREG and SPSS AREG.

Combining forecasts See forecast combination methods.

Compound transfer function A combination of two or more impulse response functions.

Concurrent validity   Validation against a known, tested, and accepted criterion at the same time.

Conditional forecast   An *ex post* forecast. This forecast is conditional on the series model, its explanatory variables, and its assumptions about how it extends over the forecast horizon.

Conditional least squares (CLS)   An algorithm for estimation of ARIMA models that backcasts the starting values and proceeds to estimate the parameters by minimization of the sum of squared errors. This algorithm is an option in SAS ARIMA modeling and SPSS ARIMA forecasting.

Confidence interval   An interval formed by the probability distribution around a parameter, that extends over a distance of two standard errors on either side of a particular parameter, and should bracket the true value of the parameter 95% of the time. When applied to a forecast, this interval is called the forecast interval.

Cointegrating parameter (or vector)   A regression parameter that permits cointegration of nonstationary series. The cointegration regression yields a stationary series that may be used in the analysis. See error correction mechanism.

Cointegrating regression   The regression, the residuals of which constitute the stationary series computed from the two nonstationary series. See error correction mechanism.

Cointegration   A combination of two or more nonstationary series into a stationary one that can be modeled with dynamic regression or ARIMA methodology.

Consistency   A property of an estimate

that the bias tends toward zero as the sample size becomes very large. The estimate is said to converge in probability to the true parameter.

Constant   The intercept in an equation. The value of the response variable where the value of the exogenous variables equal zero. See Chapter 4 for how this constant can differ from the mean of the model.

Corner method   A method, sometimes referred to as a C-array, proposed by Lui and Hanssens (1982) to identify the structure of a transfer function. See Chapter 9 for details.

Correlogram   A plot of a correlation against time, called a correlation function. Examples are the ACF and PACF.

Covariance stationarity   Threefold property of a series that includes equilibrium about the mean, variance, and autocovariance. Constant mean, variance, and covariance is also known as weak stationarity or second-order stationarity.

Cross-correlation function (CCF)   A functional correlation between two series over time, used in identifying transfer function structure and unidirectionality of association (See Chapter 9 for details).

Cross-validation   Repeated forecasting of sequentially deleted observations in a model to determine the mean square of the model. Comparison of models is conducted according to this criterion.

Cyclicity   Periodic temporal variation within a series, especially that which spans more than a year. This pattern of temporal variation has a downswing, a trough, an upswing, and a peak. Analysts usually try to predict the imminence or incidence of its turning points.

Data-generating process (DGP) The underlying process that yields the realization from which the model is built.

Dead time (delay time) The time between onset of an input and reaction in the response variable.

Decay rate parameter A rate parameter in a transfer or response function.

Decision Time An SPSS module for time series analysis that performs automatic modeling of and forecasting from exponential smoothing, ARIMA, intervention, and LTF models.

Decomposition methods The extraction of and extrapolation from cycle, trend, seasonality, and irregular components of a series.

Degrees of freedom The number of elements that are free to vary in the computation of a statistic or estimation of a statistical model.

Delphi method A qualitative forecast extracted from the collective judgment of a panel of experts.

Deterministic trend A trend that is a function of another variable.

DF test See Dickey-Fuller test.

Diagnosis A stage of ARIMA modeling in which the model is assessed for goodness of fit and its estimated parameters are assessed for retention or deletion. In this stage, the model is examined for fulfillment of assumptions concerning residuals and the model parameters are examined for their estimated values, permissible range, statistical significance, direc-

tion, multicollinearity, and theoretical meaning.

Dickey–Fuller test A test for unit roots (nonstationarity), developed by Wayne Fuller and David Dickey, whose test distribution depends on whether the model under examination has no constant, has a constant, or has a constant with a deterministic trend. An augmented version (ADF) controls for serial correlation within the process. (A Dickey-Fuller critical value table can be found in Appendix A).

Differencing A transformation of a variable from levels to changes. This transformation is accomplished by first or generalized differencing. First differencing subtracts the lagged observation from its successor. Second differencing involves taking a difference of a difference. Generalized differencing involves taking a second or higher order difference.

Difference stationary A property of being transformable to stationarity through differencing.

Disturbance An innovation, shock, or error. "Well-behaved" disturbances are identically, independently distributed with mean of zero and normal variance. See error.

Double moving average A moving average of a moving average. See moving average.

Drift Random variation about a non-zero level.

Dummy variable A variable with two values, coded as 0 and 1, to indicate the presence or absence of an event, intervention, or observational outlier in a model.

Durbin *h* test   A test for higher order autocorrelation of error designed for use in autoregressive models with lagged dependent variables. The $h = \hat{\rho}(T/(1 - T[\text{var}(\beta)]))$, where $d =$ Durbin–Watson d; $\hat{\rho} \approx 1 - 2/d$; $T =$ sample size, and $\text{var}(\beta) =$ variance of coefficient of lagged dependent variable.

Durbin *M* test   A test for higher order autocorrelation of error.

Durbin-Watson *d* test   A test for first-order autocorrelation of error. The statistic d is equal to the sum of squared differences between successive errors, divided by the sum of the squared errors. The Durbin–Watson d has an approximation of $d \approx 2(1 - \hat{\rho})$.

Dynamic regression model   A dynamic regression is a regression of one response time series on other input series. These models assume the data to be observed at equally spaced intervals and that there is no feedback exist between the response series and the input series.

EACF   Extended autocorrelation function. Used for identifying orders of ARMA models.

Efficiency   Minimum variance of estimation.

Encompassing   There are several types of encompassing. There is theory, variance, and forecast encompassing. If a first model theoretically encompasses a second model, the first model explains whatever the second model explains. Among nested models, this may be measured by explained variance. The encompassing model may contain all the explanatory variables that the encompassed model contains. The encompassing model explains all of the variance of the response variable that the encompassed model explains. Among nonnested models, the preferred model may be one with the better fit. The more preferred model may explain more of the theoretically important response variance. If the forecast of the first model contains all of the pertinent information of the second model, the first model forecast encompasses the second model forecast, and nothing would be gained by attempting to combine the two forecasts.

Endogenous variable   A response variable determined within the system. Common name for a variable influenced by other variables within the simultaneous dynamic equation model or path analytic model.

Equilibrium   Baseline value around which a series may vary or cycle.

Ergodicity   A condition of a time series realization whose sample moments approach the population parameters as the length of the realization approaches infinity.

Error   A difference between predicted and observed value. Another meaning of error is a disturbance or innovation. See disturbance.

Error correction mechanism   A mechanism in a model that corrects for long-run equilibrium error. In a cointegrated regression, the model is that of the difference of y_t regressed on the difference of x_t plus an error correction mechanism. That mechanism is a factor that captures the long-run equilibrium correction. In the equation, $\nabla y_t = \nabla x_t + b(y - \lambda x)_{t-1} + e_t$, the error correction mechanism is

$b(y - \lambda x)_{t-1}$ and the cointegrating parameter (or vector) is λ.

Error correction model (ECM) A model where there is a correction for past, current, or expected disequilibrium. Adaptive expectations, partial adjustment, and cointegrated regression models are sometimes referred to as ECMs.

Error cost function The relationship of cost of a forecast to the size of forecast error.

Estimation Computational determination of the values of the parameters by an algorithm that minimizes a criterion of error. Common algorithms used include conditional least squares, unconditional least squares, or maximum likelihood estimation.

Estimation Sample See historical sample.

Event analysis See intervention analysis.

Exogeneity The independence of a variable in a system or model. Lack of feedback from other variables in the model of a causal system. Strict exogeneity holds when the values of an exogenous variable for each time period are independent of the random errors of all other variables at all time periods. Weak exogeneity obtains when inference on a set of parameters can be made conditionally from a particular variable without loss of information.

Exogenous variable A variable whose values are determined outside the model or system. Common description of an independent variable in the context of a simultaneous dynamic equation model or path model.

EXPAND procedure SAS interpolation procedure.

Expected value The mean of a distribution of a random variable or series.

Exponential smoothing Models of weighted averages that give varying weights to the most recent or the set of past observations. These weights usually decline exponentially in magnitude with the passage of time. Holt's smoothing can handle trends, whereas Winter's smoothing can handle both trend and seasonality.

Extended sample autocorrelation function (ESACF or EACF) A table of correlations used to identify the order of an ARMA model.

Extrapolative forecasting Forecasting, broadly construed, as using exponential smoothing, decomposition, or trend models.

***Ex ante* forecast** A prediction using data available at the time of prediction. An unconditional forecast.

***Ex post* forecast** A prediction using data collected after the forecast horizon begins. A conditional forecast.

Feedback Simultaneity. The influence of the endogenous variables on the exogenous variables in a model.

Filter A function or algorithm that screens out particular components of a series and lets other components of a series pass through a system.

Forecast Prediction over a forecast horizon. The forecast can be a point forecast, an interval forecast, or a probability density forecast.

Forecast combination methods Averaging (simple and weighted), variance–covariance (conventional or adaptive), and regression (OLS,

WLS, autoregression, ARCH, and regARIMA) methods for combining forecasts to improve accuracy.

Forecast horizon The prediction window. The time over which a forecast is made.

Forecast interval The confidence interval around the point forecast. See confidence interval.

Forecast model A model that is used for forecasting.

Forecast profile The pattern of forecast defined by the point and interval forecast; the profile at times can include the probability density forecast.

Fourier analysis Time series analysis in the frequency domain. Spectral analysis. The analysis of the time series by decomposition into sines, cosines, amplitudes, wavelengths, and phase angles.

Frequency The number of cycles within a period of time, usually a year.

GARCH model A generalized autoregressive conditional heteroskedastic model introduced by Bollerslev in 1986 that has heteroskedasticity conditional on past error variances and past variances.

Goodness of fit A measure of the proportion of variance explained, the proportion of variance unexplained, the amount of error, or the parsimony of the model. Typical indicators of goodness of fit of models are R^2, *AIC*, *SBC*, *RMSE*. Typical indicators of goodness of fit of the forecast in the holdout sample are *MSFE* or *MAPE*.

Granger causality A test for exogeneity of variables, where the each variable is regressed on the current and lagged values of the other variables. Given two time-dependent variables, Y_t and X_t, nonsignificant regression coefficients of the regression of X_t on the current and lagged values of Y_t suggest lack of feedback from Y_t to X_t and Granger causality from X_t to Y_t. This noncausal condition is called Granger noncausality of Y_t to X_t.

Heteroskedasticity Unequal variance of the errors of the model.

Hildreth–Lu algorithm An algorithm for performing a correction for serial correlation of error that finds the optimal first-order autocorrelation by iterating through the range of autocorrelation correction factors and selects the coefficient that minimizes the sum of squared residuals.

Historical sample The segment of the sample reserved for estimation and model building.

Holdout sample The portion of a sample that is reserved for evaluation, validation, or testing of the model.

Holt's method An exponential smoothing method that models trend as part of the analysis of the series.

Homogeneity of variance of residuals An assumption of ordinary least squares regression analysis that there is equality of variance of the residuals across the predicted scores.

Homogeneous stationarity Equality of variance in a time series.

Homoskedasticity See homogeneity of variance of residuals.

Identification In Box–Jenkins time series analysis, a process of examining the ACF and PACF to determine the nature of the process under consideration.

Impulse response function A function displaying the structure of the response to a pulse, step, or continuous input in a dynamic regression model. There are several kinds of these response functions, including simple, higher-order, compound, and multiple.

Impulse response weights Coefficients of lagged exogenous variables in a dynamic regression model.

Independence of observations An assumption of ordinary least squares regression analysis.

Innovation An error or disturbance or random shock.

Innovational outlier See outlier.

Integrated A summative process that must be removed by differencing or regression before the series can be modeled with Box–Jenkins analysis. If $w_t = \Sigma y_t$, so that $\nabla w_t = y_t$, then w_t is integrated before differencing.

Intercept See constant.

Intervention analysis Impact analysis with deterministic step or pulse input, expounded by G. E. P. Box and G. C. Tiao in 1975. A dynamic analysis of the impact of the occurrence of an event.

Inverse autocorrelation function An autocorrelation function developed by Cleveland to help identify need for differencing and identify overdifferencing in models.

Invertibility The ability to invert an MA series to obtain a convergent autoregressive series and vice versa.

Lag operator The lag operator, L, which invokes a backstep in time on a series. $(L)Y_t = Y_{t-1}$. $(L^2)Y_t = Y_{t-2}$.

Same as backshift or backstep operator, B.

Lagging indicator An economic indicator that lags behind the part of the business cycle that is under examination—usually the downswing, trough, or upswing.

La Grange multiplier test An asymptotic test for higher order autocorrelation of errors that is distributed as a χ^2 test with the degrees of freedom equal to the number of parameters, used in identifying the order of (G)ARCH errors.

Lead A projection forward in time. See lead operator.

Leading indicator An indicator whose value is supposed to indicate the onset of a part of a business cycle under examination before that onset takes place.

Lead operator The lead operator is sometimes indicated as F. $(F)Y_t = Y_{t+1}$.

Linear transfer function (LIF) method A method of modeling transfer functions that uses low order autoregressive terms and distributed lags of the exogenous variables, along with the corner method to identify the structure of the transfer function in a dynamic regression model. This method handles multiple transfer functions better than the classical Box–Jenkins approach to such modeling.

Ljung–Box Test A modified portmanteau Q test for significance of serial correlation.

LTF See linear transfer function method.

MAE Mean absolute error.

MAPE Mean absolute percentage error.

Maximum likelihood estimation methods Asymptotic iterative estimation of parameters by maximizing the likelihood that the model fits the data. This procedure usually involves modeling the likelihood, taking its natural log, finding its minimum, and estimating the parameter values that minimize that lack of fit (maximize the likelihood that the model fits the data).

Mean The average.

Mean absolute percentage error (MAPE) The average of the sum of the absolute values of the percentage errors. A measure of forecast accuracy used in the M competitions, not as susceptible to outlier distortion as mean square forecast error.

Mean square error (MSE) Error variance.

Mean square forecast error (MSFE) Forecast error variance.

Mean stationarity The property of a constant level.

Metadiagnosis The fourth stage of ARIMA modeling where alternative models are compared to determine which of the adequate models is optimal. Some analysts would include comparative forecast evaluation as part of this stage.

M competitions A series of forecasting competitions run by Makridakis (1982, 1993) and the International Institute of Forecasters (1997), in which different methods are compared for their forecast accuracy over different forecasting horizons on 111, 29, and 3003 time series, respectively.

Misspecification Proper specification of the variables, polynomial terms, and cross-products in the model as well as specification of the autocorrelation and heteroskedasticity in the model.

Mixed model A time series model includes both autoregressive and moving average processes or both seasonal and nonseasonal parameters.

Model A symbolic representation of an empirical process. The representation commonly specifies the component variables and the nature of their interrelationships.

Moving average A series comprising an average of x time periods that slides its span as time progresses. A process that entails a linear combination of current and previous random shocks or errors.

Multicollinearity Correlation among explanatory variables in a model. If this condition is sufficiently severe, it can bias downward statistical tests of significance.

Multiplicative model An ARIMA model with nonseasonal and seasonal factors multiplied by one another.

Multivariate time series models Time series models with multiple endogenous series, including structural equation models, vector autoregression and state space models, etc. Multivariate models with multiple response series are distinguished from univariate models, with one endogenous series.

Naive forecast (NF) A forecast based on the last actual observation. NF 1 uses the last real observation, whereas NF 2 seasonally adjusts the data and uses the last observation of the seasonally adjusted data.

Noise Random variation or error in the values of a series.

Nonseasonal model A model without seasonal parameters or factors.

Nonstationarity Drift, random walk, or trend in a series. These series are said to have unit roots, where the parameters have reached the limit of invertibility or stationarity.

Normality of residuals An assumption of ordinary least squares regression analysis that the errors of a model arc normally distributed.

Ordinary least squares (OLS) A method of estimation of parameters in a model that involves minimizing the sum of the squared errors. This estimation method is commonly used in analysis of variance and regression analysis.

Outlier A data value that is more than 3 standard errors away from the expected value of the parameter being estimated is an outlier.

Overdifferencing The differencing of a series at a higher order than is necessary to render the series stationary.

Parameter Target population characteristics that are estimated with sample statistics.

Parameter constancy Parameter stability when a model is subjected to predictive validation.

Parsimony "As simple a model as is possible, but no simpler" (Albert Einstein).

Partial autocorrelation function (PACF) An autocorrelation function that identifies the k lag magnitude of the autocorrelation between the t and $t - k$, controlling for all intervening autocorrelations.

Peak Maximum value of the observations in a cycle of observations.

Periodicity Fixed length of cycle in Fourier or spectral analysis.

Phillips–Perron (PD) test A nonparametric test for unit roots.

Prais–Winsten algorithm A method of transforming variables in a regression with autocorrelated errors to correct for the serial correlation of error. In a model with first-order autocorrelation, this transformation involves multiplying each of the variables in the regression by $\sqrt{1 - \rho^2}$, where ρ^2 is the square of the first-order autocorrelation coefficient.

Prediction A point, interval, or probability density forecast.

Predictive validation Reliability testing of the model estimated on a historical data set by assessment of its goodness of fit with a validation data set.

Prewhitening Application of an inverse filter, designed to neutralize the contaminating serial correlation in the input series, to both input and output series prior to examining the relationship between the series with a cross-correlation function.

Pulse input An instantaneous change in value of an input.

Pulse response function A response over time to a pulse input.

Purely seasonal model A model with only between-period effects.

Q statistic test See Box–Pierce or Box–Ljung test of significance.

R square Coefficient of determination. The proportion of variance of a dependent variable explained by the model, used as a measure of fit. See adjusted R square.

Randomness Unsystematic variation.

Random sampling Sampling the elements of a population so that every element selected for the sample has a known or equal probability of selection. Random samples of sufficiently large size should possess statistics that more or less reflect the characteristics of the population parameters. The sampling error reflects the extent that the sample is not representative of the population.

Random walk Random movement from one point to another in time.

REGARIMA OLS regression with time series (ARIMA) modeling of residuals.

Regression An explanatory model with a dependent (criterion) variable being explained by explanatory variables, called independent, regressors, or predictor variables. Regression can be bivariate or multiple. It can be univariate or multivariate. It can be linear or nonlinear. The regression can be cross-sectional or dynamic.

Regression parameters Variables in the population that are estimated in a regression model.

Regression with ARIMA errors See RegARIMA.

Residual An error. The difference between the observed value and that predicted by the model.

Sample Subsetting Segmenting the sample into two subsets. The first subset is an estimation (historical) subsample. The model is built on this subset of data. The second subset is the test, evaluation, or predictive validation subsample. This post-sample evaluation compares the forecast to the actual data in this subsample.

SAS® Statistical Analysis System from SAS Institute, Inc., Cary, NC.

SBC (Schwartz Bayesian criterion) A measure of goodness of fit. This criterion is equal to the number of free parameters times the natural log of the number of residuals minus 2 times the natural log of the likelihood function. This measure is often used for order selection of models.

Seasonal adjustment Removal of seasonality from a time series. Governments around the world have customarily used Census X-11 or X11-ARIMA procedures to perform the seasonal adjustment. Census X-12 is now coming into use as the program of choice for deseasonalization of series.

Seasonal differencing Differencing at seasonal lags.

Seasonal integration Integration at seasonal lags.

Seasonal moving average Periodic moving average model.

Seasonal model An ARIMA model with seasonal parameters. A purely seasonal model contains only seasonal parameters; a mixed multiplicative model contains nonseasonal as well as seasonal parameters.

Seasonal pulse A pulse that occurs during particular seasons in an input series.

Seasonal unit root Seasonal integration of a series.

Seasonal autoregression Periodic autoregression model.

Seasonality Annual variation within a series.

Second-order stationarity See covariance stationarity.

Serial correlation of error Autocorrelated error.

Shock An innovation, random fluctuation, or error.

Slope The regression coefficient of a function: the amount of increase in the response variable per unit increase in the explanatory variable.

Specification error Failure to properly specify a model by not including of all essential significant explanatory variables and/or improperly defining the functional form of the relationship between the dependent variable(s) and explanatory variables.

Spectral analysis See Fourier analysis.

SPSS® Statistical Package for the Social Sciences. A popular statistical package for general statistical analysis developed by SPSS, Inc. Chicago, IL.

State space models Models of jointly stationary multivariate time series processes that have dynamic interactions and that are formed from two basic equations. The state transition equation consists of a state vector of auxiliary variables as a function of a transition matrix and an input matrix, whereas the measurement equation consists of a state vector canonically extracted from observable variables. These vector models are estimated with a recursive protocol and can be used for multivariate forecasting.

Stationarity A condition of mean, variance, and covariance equilibrium. These properties of a series render it amenable to ARIMA analysis. There are two basic types: strict and weak. Weak (covariance or second-order) stationarity includes constant mean, variance, and autocovariance. Strict stationarity adds another requirement to the series, and that is normality.

Statistics Sample characteristics that estimate population parameters.

Step input A sudden and permanent change in the input.

Step response function A sudden and permanent response over time to a step input.

Stochastic trend A systematic change of level in a series that also has random variation within it.

Strict stationarity A weak stationarity conjoined with normal distribution of its observations.

Super-consistency Rapid convergence to a limit as a sample grows in size at a higher than normal rate. Least squares estimation of parameters with unit roots exhibit super-consistency and downward bias. Super-consistency also characterizes convergence of estimation of parameters in cointegrating regressions.

Theil's U statistic The ratio of a one-step-ahead sum of squared forecast errors to the sum of squared errors for a random walk.

Time-varying combining weights Weights that change over time as weighted forecasts are combined to improve the predictive reliability.

TIMEPLOT An SAS procedure for a plot of the series over time.

Time series A realization of a data-generating process, where observations are equally spaced across time.

Trading day Census X-11 and X-12 correct for the number of workdays in the month and year in performing their computations.

Transfer function A model of a functional relationship between an input and output time series. These models can have pulse, step, or continuous inputs. They have decay rates that are ordered according to the number of decay rate parameters in the model: Zero-order transfer functions are have no decay parameter. Their responses are level shifts or pulses. First-order transfer functions have an impulse response function with one rate parameter and they exhibit exponential attenuation of growth or diminution. Second-order impulse response function with two rate parameters and exhibit undulation of decay. Higher-order response functions have more than two decay rate parameters and exhibit complex attenuation.

Trend A systematic change in level over time. Trends are classified according to their type and length. There are deterministic and stochastic types of trends. There are local (short run) and global (basic, long run) trends. Holt-Winter exponential smoothing models estimate local or recent trend better than regression models. Regression models estimate basic or long run trends better than exponential smoothing models.

Trend analysis The extraction and extrapolation of trend from a series, often performed with ordinary least squares linear regression analysis, to make a forecast. This analysis focuses on absolute or relative direction, magnitude, significance, linearity, and stability of the trends. See Trend.

Trend stationary A series that needs to be detrended by regressing the series over time and saving the residuals.

Trough Minimum value of a cycle.

Turning point A change in direction of seasonal, trend, or cyclical value of a series.

Unbiasedness The equality of a mean of a statistic and the true population parameter, when the mean of the statistic is computed from a large number of samples.

Unconditional forecast See *ex ante* forecast.

Uncorrelated errors An assumption of ordinary least squares regression estimation.

Unit root A model parameter whose value reaches or exceeds the bounds of stationarity or invertibility. A parameter value that renders its model nonstationary.

Unweighted least squares (ULS) An algorithm for estimation of models that is based on minimization of the sum of squared errors. Starting values are set to zero.

Validation Concurrent validation can be measured by the fit of the model on the historical sample. Predictive validation is measured by the fit of the forecast to the real data in the holdout sample.

Variance stationarity Constant variance in a series.

Vector autoregression A multivariate autoregression analysis that regresses an N-dimensional vector of variables on p lags of itself and past lags of the other variables as well. This procedure allows errors to be correlated and for multivariate interactions.

Wavelength The shortest temporal distance between two identical parts of a cycle.

Weak stationarity See Covariance stationarity.

What If An SPSS module that performs contingency analysis with exponential smoothing, ARIMA, intervention, and dynamic regression time series models. Used in conjunction with the SPSS Decision Time module.

White noise A process with only random or unsystematic variation residing within it.

White's general test A test developed by Halbert White for heteroskedasticity and specification error that is used for diagnosis of regression models.

Winter's method Exponential smoothing that models both trend and seasonality in additive and multiplicative models.

Wold decomposition theorem A theorem that every covariance stationary, nondeterministic, stochastic process can be written as a sum of a sequence of uncorrelated variables.

X11 SAS procedure for performing Census X-11 series decomposition and seasonal adjustment.

X12 SAS procedure in version 8 for performing Census X-12 series decomposition and seasonal adjustment. See X-12 and X-12-ARIMA.

X-11 Census X-11. A method of decomposing a time series into trend, cycle, seasonality, and irregular components, by which governments have seasonally adjusted data. Developed by Shiskin *et al.* at the U.S. Bureau of the Census, Department of Commerce.

X-12 Census X-12. An enhanced version of Census X-11 that incorporates new filters, regression with time series errors, and new diagnostics. See Chapter 2.

X-11-ARIMA An update of Census X-11 developed by E. B. Dagum at Statistics Canada in 1988. Later adopted by the U.S. Census Bureau.

X-12-ARIMA Latest improvement of X-11-ARIMA developed in 1998 by the U. S. Bureau of the Census, now replacing X-11-ARIMA. See X-12.

Index